W9-BDD-807

Numerical Recipes
in Fortran 90
Second Edition

Volume 2 of
Fortran Numerical Recipes

Numerical Recipes
in Fortran 90

The Art of *Parallel* Scientific Computing

Second Edition

Volume 2 of
Fortran Numerical Recipes

William H. Press

Harvard-Smithsonian Center for Astrophysics

Saul A. Teukolsky

Department of Physics, Cornell University

William T. Vetterling

Polaroid Corporation

Brian P. Flannery

EXXON Research and Engineering Company

Foreword by

Michael Metcalf

CERN, Geneva, Switzerland

Published by the Press Syndicate of the University of Cambridge
The Pitt Building, Trumpington Street, Cambridge CB2 1RP
40 West 20th Street, New York, NY 10011-4211, USA
10 Stamford Road, Oakleigh, Melbourne 3166, Australia

Numerical Recipes in Fortran 90: The Art of Parallel Scientific Computing,
Volume 2 of Fortran Numerical Recipes, Second Edition, first published 1996.
Reprinted with corrections 1999.
The code in this volume is corrected to software version 2.08

Printed in the United States of America
Typeset in TEX

Library of Congress Cataloging-in-Publication Data

Numerical recipes in Fortran 90 : the art of parallel scientific computing / William H. Press ... [et al.]. – 2nd ed.
p. cm.
Includes bibliographical references and index.
ISBN 0-521-57439-0 (hardcover)
1. FORTRAN 90 (Computer program language) 2. Parallel programming (Computer science) 3. Numerical analysis–Data processing.
I. Press, William H.
QA76.73.F25N85 1996
519.4′0285′52–dc20 96-5567
CIP

A catalog record for this book is available from the British Library.

ISBN 0 521 57439 0 Volume 2 (this book)
ISBN 0 521 43064 X Volume 1
ISBN 0 521 43721 0 Example book in FORTRAN
ISBN 0 521 57440 4 FORTRAN diskette (IBM 3.5″)
ISBN 0 521 57608 3 CDROM (IBM PC/Macintosh)
ISBN 0 521 57607 5 CDROM (UNIX)

Contents

Contents of Volume 1: Numerical Recipes in Fortran 77

Preface to Volume 2

Fortran 90 is not just the long-awaited updating of the Fortran language to modern computing practices. It is also the vanguard of a much larger revolution in computing, that of multiprocessor computers and widespread parallel programming. Parallel computing has been a feature of the largest supercomputers for quite some time. Now, however, it is rapidly moving towards the desktop.

As we watched the gestation and birth of Fortran 90 by its governing "X3J3 Committee" (a process interestingly described by a leading committee member, Michael Metcalf, in the Foreword that follows), it became clear to us that the right moment for moving Numerical Recipes from Fortran 77 to Fortran 90 was sooner, rather than later.

Fortran 90 compilers are now widely available. Microsoft's Fortran PowerStation for Windows 95 brings that firm's undeniable marketing force to PC desktop; we have tested this compiler thoroughly on our code and found it excellent in compatibility and performance. In the UNIX world, we have similarly tested, and had generally fine experiences with, DEC's Fortran 90 for Alpha AXP and IBM's xlf for RS/6000 and similar machines. NAG's Fortran 90 compiler also brings excellent Fortran 90 compatibility to a variety of UNIX platforms. There are no doubt other excellent compilers, both available and on the way. Fortran 90 is completely backwards compatible with Fortran 77, by the way, so you don't have to throw away your legacy code, or keep an old compiler around.

There have been previous special versions of Fortran for parallel supercomputers, but always specific to a particular hardware. Fortran 90, by contrast, is designed to provide a general, architecture-independent framework for parallel computation. Equally importantly, it is an international standard, agreed upon by a large group of computer hardware and software manufacturers and international standards bodies.

With the Fortran 90 language as a tool, we want this volume to be your complete guide for learning how to "think parallel." The language itself is very general in this regard, and applicable to many present and future computers, or even to other parallel computing languages as they come along. Our treatment emphasizes general principles, but we are also not shy about pointing out parallelization "tricks" that have frequent applicability. These are not only discussed in this volume's principal text chapters (Chapters 21–23), but are also sprinkled throughout the chapters of Fortran 90 code, called out by a special "parallel hint" logo (left, above). Also scattered throughout the code chapters are specific "Fortran 90 tips," with their own distinct graphic call-out (left). After you read the text chapters, you might want simply to browse among these hints and tips.

A special note to C programmers: Right now, there is no effort at producing a parallel version of C that is comparable to Fortran 90 in maturity, acceptance, and stability. We think, therefore, that C programmers will be well served by using this volume for an educational excursion into Fortran 90, its parallel programming constructions, and the numerical algorithms that capitalize on them. C and C++ programming have not been far from our minds as we have written this volume, and we think that you will find that time spent in absorbing its principal lessons (in Chapters 21–23) will be amply repaid in the future, as C and C++ eventually develop standard parallel extensions.

A final word of truth in packaging: **Don't buy this volume unless you also buy (or already have) Volume 1** (now retitled *Numerical Recipes in Fortran 77*). Volume 2 does not repeat any of the discussion of what individual programs actually do, or of the mathematical methods they utilize, or how to use them. While our Fortran 90 code is thoroughly commented, and includes a header comment for each routine that describes its input and output quantities, these comments are *not* supposed to be a complete description of the programs; the complete descriptions are in Volume 1, which we reference frequently. But here's a money-saving hint to our previous readers: If you already own a Second Edition version whose title is *Numerical Recipes in FORTRAN* (which doesn't indicate either "Volume 1" or "Volume 2" on its title page) then take a marking pen and write in the words "Volume 1." There! (Differences between the previous reprintings and the newest reprinting, the one labeled "Volume 1," are minor.)

Acknowledgments

We continue to be in the debt of many colleagues who give us the benefit of their numerical and computational experience. Many, though not all, of these are listed by name in the preface to the second edition, in Volume 1. To that list we must now certainly add George Marsaglia, whose ideas have greatly influenced our new discussion of random numbers in this volume (Chapter B7).

With this volume, we must acknowledge our additional gratitude and debt to a number of people who generously provided advice, expertise, and time (a great deal of time, in some cases) in the areas of parallel programming and Fortran 90. The original inspiration for this volume came from Mike Metcalf, whose clear lectures on Fortran 90 (in this case, overlooking the beautiful Adriatic at Trieste) convinced us that Fortran 90 could serve as the vehicle for a book with the larger scope of parallel programming generally, and whose continuing advice throughout the project has been indispensable. Gyan Bhanot also played a vital early role in the development of this book; his first translations of our Fortran 77 programs taught us a lot about parallel programming. We are also grateful to Greg Lindhorst, Charles Van Loan, Amos Yahil, Keith Kimball, Malcolm Cohen, Barry Caplin, Loren Meissner, Mitsu Sakamoto, and George Schnurer for helpful correspondence and/or discussion of Fortran 90's subtler aspects.

We once again express in the strongest terms our gratitude to programming consultant Seth Finkelstein, whose contribution to both the coding and the thorough testing of all the routines in this book (against multiple compilers and in sometimes-buggy, and always challenging, early versions) cannot be overstated.

WHP and SAT acknowledge the continued support of the U.S. National Science Foundation for their research on computational methods.

February 1996

William H. Press
Saul A. Teukolsky
William T. Vetterling
Brian P. Flannery

Foreword
by Michael Metcalf

Sipping coffee on a sunbaked terrace can be surprisingly productive. One of the *Numerical Recipes* authors and I were each lecturing at the International Center for Theoretical Physics in Trieste, Italy, he on numerical analysis and I on Fortran 90. The numerical analysis community had made important contributions to the development of the new Fortran standard, and so, unsurprisingly, it became quickly apparent that the algorithms for which *Numerical Recipes* had become renowned could, to great advantage, be recast in a new mold. These algorithms had, hitherto, been expressed in serial form, first in Fortran 77 and then in C, Pascal, and Basic. Now, nested iterations could be replaced by array operations and assignments, and the other features of a rich array language could be exploited. Thus was the idea of a "Numerical Recipes in Fortran 90" first conceived and, after three years' gestation, it is a delight to assist at the birth.

But what *is* Fortran 90? How did it begin, what shaped it, and how, after nearly foundering, did its driving forces finally steer it to a successful conclusion?

The Birth of a Standard

Back in 1966, the version of Fortran now known as Fortran 66 was the first language ever to be standardized, by the predecessor of the present American National Standards Institute (ANSI). It was an all-American affair. Fortran had first been developed by John Backus of IBM in New York, and it was the dominant scientific programming language in North America. Many Europeans preferred Algol (in which Backus had also had a hand). Eventually, however, the mathematicians who favored Algol for its precisely expressible syntax began to defer to the scientists and engineers who appreciated Fortran's pragmatic, even natural, style. In 1978, the upgraded Fortran 77 was standardized by the ANSI technical committee, X3J3, and subsequently endorsed by other national bodies and by ISO in Geneva, Switzerland. Its dominance in all fields of scientific and numerical computing grew as new, highly optimizing compilers came onto the market. Although newer languages, particularly Pascal, Basic, PL/1, and later Ada attracted their own adherents, scientific users throughout the 1980s remained true to Fortran. Only towards the end of that decade did C draw increasing support from scientific programmers who had discovered the power of structures and pointers.

During all this time, X3J3 kept functioning, developing the successor version to Fortran 77. It was to be a decade of strife and contention. The early plans, in the late 1970s, were mainly to add to Fortran 77 features that had had to be left out of that standard. Among these were dynamic storage and an array language, enabling it to map directly onto the architecture of supercomputers, then coming onto the market. The intention was to have this new version ready within five years, in 1982. But two new factors became significant at that time. The first was the decision that the next standard should not just codify existing practice, as had largely been the case in 1966 and 1978, but also extend the functionality of the language through innovative additions (even though, for the array language, there was significant borrowing from John Iverson's APL and from DAP Fortran). The second factor was that X3J3 no longer operated under only American auspices. In the course of the 1980s, the standardization of programming languages came increasingly under the authority of the international body, ISO. Initially this was in an advisory role, but now ISO is the body that, through its technical committee WG5 (in full, ISO/IEC JTC1/SC22/WG5), is responsible for determining the course of the language. WG5 also steers the work of the development body, then as now, the highly skilled and competent X3J3. As we shall see, this shift in authority was crucial at the most difficult moment of Fortran 90's development.

The internationalization of the standards effort was reflected in the welcome given by X3J3 to six or seven European members; they, and about one-third of X3J3's U.S. members, provided the overlapping core of membership of X3J3 and WG5 that was vital in the final years in bringing the work to a successful conclusion. X3J3 membership, which peaked at about 45, is restricted to one voting member per organization, and significant decisions require a majority of two-thirds of those voting. Nationality plays no role, except in determining the U.S. position on an international issue. Members, who are drawn mainly from the vendors, large research laboratories, and academia, must be present or represented at two-thirds of all meetings in order to retain voting rights.

In 1980, X3J3 reported on its plans to the forerunner of WG5 in Amsterdam, Holland. Fortran 8x, as it was dubbed, was to have a basic array language, new looping constructs, a bit data type, data structures, a free source form, a mechanism to "group" procedures, and another to manage the global name space. Old features, including `COMMON`, `EQUIVALENCE`, and the arithmetic-`IF`, were to be consigned to a so-called obsolete module, destined to disappear in a subsequent revision. This was part of the "core plus modules" architecture, for adding new features and retiring old ones, an aid to backwards compatibility. Even though Fortran 77 compilers were barely available, the work seemed well advanced and the mood was optimistic. Publication was intended to take place in 1985. It was not to be.

One problem was the sheer number of new features that were proposed as additions to the language, most of them worthwhile in themselves but with the totality being too large. This became a recurrent theme throughout the development of the standard. One example was the suggestion of Lawrie Schonfelder (Liverpool University), at a WG5 meeting in Vienna, Austria, in 1982, that certain features already proposed as additions could be combined to provide a full-blown derived data type facility, thus providing Fortran with abstract data types. This idea was taken up by X3J3 and has since come to be recognized, along with the array language, as one of the two main advances brought about by what became Fortran 90. However, the ramifications go very deep: all the technical details of how to handle arrays of objects of derived types that in turn have array components that have the pointer attribute, and so forth, have to be precisely defined and rigorously specified.

Conflict

The meetings of X3J3 were often full of drama. Most compiler vendors were represented as a matter of course but, for many, their main objective appeared to be to maintain the status quo and to ensure that Fortran 90 never saw the light of day. One vendor's extended (and much-copied) version of Fortran 77 had virtually become an industry standard, and it saw as its mission the maintenance of this lead. A new standard would cost it its perceived precious advantage. Other large vendors had similar points of view, although those marketing supercomputers were clearly keen on the array language. Most users, on the other hand, were hardly prepared to invest large amounts of their employers' and their own resources in simply settling for a trivial set of improvements to the existing standard. However, as long as X3J3 worked under a simple-majority voting rule, at least some apparent progress could be made, although the underlying differences often surfaced. These were even sometimes between users — those who wanted Fortran to become a truly modern language and those wanting to maintain indefinite backwards compatibility for their billions of lines of existing code.

At a watershed meeting, in Scranton, Pennsylvania, in 1986, held in an atmosphere that sometimes verged on despair, a fragile compromise was reached as a basis for further work. One breakthrough was to weaken the procedures for removing outdated features from the language, particularly by removing no features whatsoever from the next standard and by striking storage association (i.e., `COMMON` and `EQUIVALENCE`) from the list of features to be designated as obsolescent (as they are now known). A series of votes definitively removed from the language all plans to add: arrays of arrays, exception handling, nesting of internal procedures, the `FORALL` statement (now in Fortran 95), and a means to access skew array sections. There were other features on this list that, although removed, were reinstated at later meetings: user-defined operators, operator overloading, array and structure constructors, and vector-valued subscripts. After many more travails, the committee voted, a year later, by 26 votes to 9, to forward the document for what was to become the first of three periods of public comment.

While the document was going through the formal standards bureaucracy and being placed before the public, X3J3 polished it further. X3J3 also prepared procedures for processing the comments it anticipated receiving from the public, and to each of which, under the rules, it would have to reply individually. It was just as well. Roughly 400 replies flooded in, many of them very detailed and, disappointingly for those of us wanting a new standard quickly, unquestionably negative towards our work. For many it was too radical, but many others pleaded for yet more modern features, such as pointers.

Now the committee was deadlocked. Given that a document had already been published, any further change required not a simple but a two-thirds majority. The conservatives and the radicals could each block a move to modify the draft standard, or to accept a revised one for public review — and just that happened, in Champagne-Urbana, Illinois, in 1988. Any change, be it on the one hand to modify the list of obsolescent features, to add the pointers or bit data type wanted by the public, to add multi-byte characters to support Kanji and other non-European languages or, on the other hand, to emasculate the language by removing modules or operator overloading, and hence abstract data types, to name but some suggestions, none of these could be done individually or collectively in a way that would achieve consensus. I wrote:

"In my opinion, no standard can now emerge without either a huge concession by the users to the vendors (`MODULE` / `USE`) and/or a major change in the composition of the committee. I do not see how members who have worked for up to a decade or more, devoting time and intellectual energy far beyond the call of duty, can be expected to make yet more personal sacrifices if no end to the work is in sight, or if that end is nothing but a travesty of what had been designed and intended as a modern scientific programming language. ... I think the August meeting will be a watershed — if no progress is achieved there will be dramatic resignations, and ISO could even remove the work from ANSI, which is failing conspicuously in its task."

(However, the same notes began with a quotation from *The Taming of the Shrew*: "And do as adversaries do in law, / Strive mightily, but eat and drink / as friend." That we always did, copiously.)

Resolution

The "August meeting" was, unexpectedly, imbued with a spirit of compromise that had been so sadly lacking at the previous one. Nevertheless, after a week of discussing four separate plans to rescue the standard, no agreement was reached. Now the question seriously arose: Was X3J3 incapable of producing a new Fortran standard for the international community, doomed to eternal deadlock, a victim of ANSI procedures?

Breakthrough was achieved at a traumatic meeting of WG5 in Paris, France, a month later. The committee spent several extraordinary days drawing up a detailed list of what *it* wanted to be in Fortran 8x. Finally, it set X3J3 an ultimatum that was unprecedented in the standards world: The ANSI committee was to produce a new draft document, corresponding to WG5's wishes, within five months! Failing that, WG5 would assume responsibility and produce the new standard itself.

This decision was backed by the senior U.S. committee, X3, which effectively directed X3J3 to carry out WG5's wishes. And it did! The following November, it implemented most of the technical changes, adding pointers, bit manipulation intrinsic procedures, and vector-valued subscripts, and removing user-defined elemental functions (now in Fortran 95). The actual list of changes was much longer. X3J3 and WG5, now collaborating closely, often in gruelling six-day meetings, spent the next 18 months and two more periods of (positive) public comment putting the finishing touches to what was now called Fortran 90, and it was finally adopted, after some cliff-hanging votes, for forwarding as a U.S. and international standard on April 11, 1991, in Minneapolis, Minnesota.

Among the remaining issues that were decided along the way were whether pointers should be a data type or be defined in terms of an attribute of a variable, implying strong typing (the latter was chosen), whether the new standard should coexist alongside the old one rather than definitively replace it (it coexisted for a while in the U.S., but was a replacement elsewhere, under ISO rules), and whether, in the new free source form, blanks should be significant (fortunately, they are).

Fortran 90

The main new features of Fortran 90 are, first and foremost, the array language and abstract data types. The first is built on whole array operations and assignments, array sections, intrinsic procedures for arrays, and dynamic storage. It was designed with optimization in mind. The second is built on modules and module procedures, derived data types, operator overloading and generic interfaces, together with pointers. Also important are the new facilities for numerical computation including a set of numeric inquiry functions, the parametrization of the intrinsic types, new control constructs — `SELECT CASE` and new forms of `DO`, internal and recursive procedures and optional and keyword arguments, improved I/O facilities, and many new intrinsic procedures. Last but not least are the new free source form, an improved style of attribute-oriented specifications, the `IMPLICIT NONE` statement, and a mechanism for identifying redundant features for subsequent removal from the language. The requirement on compilers to be able to identify, for example, syntax extensions, and to report why a program has been rejected, are also significant. The resulting language is not only a far more powerful tool than its successor, but a safer and more reliable one too. Storage association, with its attendant dangers, is not abolished, but rendered unnecessary. Indeed, experience shows that compilers detect errors far more frequently than before, resulting in a faster development cycle. The array syntax and recursion also allow quite compact code to be written, a further aid to safe programming.

No programming language can succeed if it consists simply of a definition (witness Algol 68). Also required are robust compilers from a wide variety of vendors, documentation at various levels, and a body of experience. The first Fortran 90 compiler appeared surprisingly quickly, in 1991, especially in view of the widely touted opinion that it would be very difficult to write one. Even more remarkable was that it was written by one person, Malcolm Cohen of NAG, in Oxford, U.K. There was a gap before other compilers appeared, but now they exist as native implementations for almost all leading computers, from the largest to PCs. For the most part, they produce very efficient object code; where, for certain new features, this is not the case, work is in progress to improve them.

The first book, *Fortran 90 Explained*, was published by John Reid and me shortly before the standard itself was published. Others followed in quick succession, including excellent texts aimed at the college market. At the time of writing there are at least 19 books in English and 22 in various other languages: Chinese, Dutch, French, Japanese, Russian, and Swedish. Thus, the documentation condition is fulfilled.

The body of experience, on the other hand, has yet to be built up to a critical size. Teaching of the language at college level has only just begun. However, I am certain that this present volume will contribute decisively to a significant breakthrough, as it provides models not only of the numerical algorithms for which previous editions are already famed, but also of an excellent Fortran 90 style, something that can develop only with time. Redundant features are abjured. It shows that, if we abandon these features and use new ones in their place, the appearance of code can initially seem unfamiliar, but, in fact, the advantages become rapidly apparent. This new edition of *Numerical Recipes* stands as a landmark in this regard.

Fortran Evolution

The formal procedures under which languages are standardized require them either to evolve or to die. A standard that has not been revised for some years must either be revised and approved anew, or be withdrawn. This matches the technical pressure on the language developers to accommodate the increasing complexity both of the problems to be tackled in scientific computation and of the underlying hardware on which programs run. Increasing problem complexity requires more powerful features and syntax; new hardware needs language features that map onto it well.

Thus it was that X3J3 and WG5, having finished Fortran 90, began a new round of improvement. They decided very quickly on new procedures that would avoid the disputes that bedevilled the previous work: WG5 would decide on a plan for future standards, and X3J3 would act as the so-called development body that would actually produce them. This would be done to a strict timetable, such that any feature that could not be completed on time would have to wait for the next round. It was further decided that the next major revision should appear a decade after Fortran 90 but, given the somewhat discomforting number of requests for interpretation that had arrived, about 200, that a minor revision should be prepared for mid-term, in 1995. This should contain only "corrections, clarifications and interpretations" and a very limited number (some thought none) of minor improvements.

At the same time, scientific programmers were becoming increasingly concerned at the variety of methods that were necessary to gain efficient performance from the ever-more widely used parallel architectures. Each vendor provided a different set of parallel extensions for Fortran, and some academic researchers had developed yet others. On the initiative of Ken Kennedy of Rice University, a High-Performance Fortran Forum was established. A coalition of vendors and users, its aim was to produce an ad hoc set of extensions to Fortran that would become an informal but widely accepted standard for portable code. It set itself the daunting task of achieving that in just one year, and succeeded. Melding existing dialects like Fortran D, CM Fortran, and Vienna Fortran, and adopting the new Fortran 90 as a base, because of its array syntax, High-Performance Fortran (HPF) was published in 1993 and has since become widely implemented. However, although HPF was designed for data parallel codes and mainly implemented in the form of directives that appear to non-HPF processors as comment lines, an adequate functionality could not be achieved without extending the Fortran syntax. This was done in the form of the `PURE` attribute for functions — an assertion that they contain no side effects — and the `FORALL` construct — a form of array assignment expressed with the help of indices.

The dangers of having diverging or competing forms of Fortran 90 were immediately apparent, and the standards committees wisely decided to incorporate these two syntactic changes also into Fortran 95. But they didn't stop there. Two further extensions, useful not only for their expressive power but also to access parallel hardware, were added: elemental functions, ones written in terms of scalars but that accept array arguments of any permitted shape or size, and an extension to allow nesting of `WHERE` constructs, Fortran's form of masked assignment. To readers of *Numerical Recipes*, perhaps the most relevant of the minor improvements that Fortran 95 brings are the ability to distinguish between a negative and a positive real zero, automatic deallocation of allocatable arrays, and a means to initialize the values of components of objects of derived data types and to initialize pointers to null.

The medium-term objective of a relatively minor upgrade has been achieved on schedule. But what does the future hold? Developments in the underlying principles of procedural programming languages have not ceased. Early Fortran introduced the concepts of expression abstraction (`X=Y+Z`) and later control expression (e.g., the `DO` loop). Fortran 77 continued this with the `if-then-else`, and Fortran 90 with the `DO` and `SELECT CASE` constructs. Fortran 90 has a still higher level of expression abstraction (array assignments and expressions) as well as data structures and even full-blown abstract data types. However, during the 1980s the concept of objects came to the fore, with methods bound to the objects on which they operate. Here, one particular language, C++, has come to dominate the field. Fortran 90 lacks a means to point to functions, but otherwise has most of the necessary features in place, and the standards committees are now faced with the dilemma of deciding whether to make the planned Fortran 2000 a fully object-oriented language. This could possibly jeopardize its powerful, and efficient, numerical capabilities by too great an increase in language complexity, so should they simply batten down the hatches and not defer to what might be only a passing storm? At the time of writing, this is an open issue. One issue that is not open is Fortran's lack of in-built exception handling. It is virtually certain that such a facility, much requested by the numerical community, and guided by John Reid, will be part of the next major revision. The list of other requirements is long but speculative, but some at the top of the list are conditional compilation, command line argument handling, I/O for objects of derived type, and asynchronous I/O (which is also planned for the next release of HPF). In the meantime, some particularly pressing needs have been identified, for the handling of floating-point exceptions, interoperability with C, and allowing allocatable arrays as structure components, dummy arguments, and function results. These have led WG5 to begin processing these three items using a special form of fast track, so that they might become optional but standard extensions well before Fortran 2000 itself is published in the year 2001.

Conclusion

Writing a book is always something of a gamble. Unlike a novel that stands or falls on its own, a book devoted to a programming language is dependent on the success of others, and so the risk is greater still. However, this new *Numerical Recipes in Fortran 90* volume is no ordinary book, since it comes as the continuation of a highly successful series, and so great is its significance that it can, in fact, influence the outcome in its own favor. I am entirely confident that its publication will be seen as an important event in the story of Fortran 90, and congratulate its authors on having performed a great service to the field of numerical computing.

Geneva, Switzerland Michael Metcalf
January 1996

License Information

Read this section if you want to use the programs in this book on a computer. You'll need to read the following Disclaimer of Warranty, get the programs onto your computer, and acquire a Numerical Recipes software license. (Without this license, which can be the free "immediate license" under terms described below, the book is intended as a text and reference book, for reading purposes only.)

Disclaimer of Warranty

We make no warranties, express or implied, that the programs contained in this volume are free of error, or are consistent with any particular standard of merchantability, or that they will meet your requirements for any particular application. They should not be relied on for solving a problem whose incorrect solution could result in injury to a person or loss of property. If you do use the programs in such a manner, it is at your own risk. The authors and publisher disclaim all liability for direct or consequential damages resulting from your use of the programs.

How to Get the Code onto Your Computer

Pick one of the following methods:

- You can type the programs from this book directly into your computer. In this case, the *only* kind of license available to you is the free "immediate license" (see below). You are not authorized to transfer or distribute a machine-readable copy to any other person, nor to have any other person type the programs into a computer on your behalf. We do not want to hear bug reports from you if you choose this option, because experience has shown that *virtually all* reported bugs in such cases are typing errors!
- You can download the Numerical Recipes programs electronically from the Numerical Recipes On-Line Software Store, located at our Web site (`http://www.nr.com`). They are packaged as a password-protected file, and you'll need to purchase a license to unpack them. You can get a single-screen license and password immediately, on-line, from the On-Line Store, with fees ranging from $50 (PC, Macintosh, educational institutions' UNIX) to $140 (general UNIX). Downloading the packaged software from the On-Line Store is also the way to start if you want to acquire a more general (multiscreen, site, or corporate) license.
- You can purchase media containing the programs from Cambridge University Press. Diskette versions are available in IBM-compatible format for machines running Windows 3.1, 95, or NT. CDROM versions in ISO-9660 format for PC, Macintosh, and UNIX systems are also available; these include both Fortran and C versions (as well as versions in Pascal

and BASIC from the first edition) on a single CDROM. Diskettes purchased from Cambridge University Press include a single-screen license for PC or Macintosh only. The CDROM is available with a single-screen license for PC or Macintosh (order ISBN 0 521 576083), or (at a slightly higher price) with a single-screen license for UNIX workstations (order ISBN 0 521 576075). Orders for media from Cambridge University Press can be placed at 800 872-7423 (North America only) or by email to orders@cup.org (North America) or trade@cup.cam.ac.uk (rest of world). Or, visit the Web sites `http://www.cup.org` (North America) or `http://www.cup.cam.ac.uk` (rest of world).

Types of License Offered

Here are the types of licenses that we offer. Note that some types are automatically acquired with the purchase of media from Cambridge University Press, or of an unlocking password from the Numerical Recipes On-Line Software Store, while other types of licenses require that you communicate specifically with Numerical Recipes Software (email: orders@nr.com or fax: 781 863-1739). Our Web site `http://www.nr.com` has additional information.

- ["Immediate License"] If you are the individual owner of a copy of this book and you type one or more of its routines into your computer, we authorize you to use them on that computer for your own personal and noncommercial purposes. You are not authorized to transfer or distribute machine-readable copies to any other person, or to use the routines on more than one machine, or to distribute executable programs containing our routines. This is the only free license.
- ["Single-Screen License"] This is the most common type of low-cost license, with terms governed by our Single Screen (Shrinkwrap) License document (complete terms available through our Web site). Basically, this license lets you use Numerical Recipes routines on any one screen (PC, workstation, X-terminal, etc.). You may also, under this license, transfer pre-compiled, executable programs incorporating our routines to other, unlicensed, screens or computers, providing that (i) your application is noncommercial (i.e., does not involve the selling of your program for a fee), (ii) the programs were first developed, compiled, and successfully run on a licensed screen, and (iii) our routines are bound into the programs in such a manner that they cannot be accessed as individual routines and cannot practicably be unbound and used in other programs. That is, under this license, your program user must not be able to use our programs as part of a program library or "mix-and-match" workbench. Conditions for other types of commercial or noncommercial distribution may be found on our Web site (`http://www.nr.com`).
- ["Multi-Screen, Server, Site, and Corporate Licenses"] The terms of the Single Screen License can be extended to designated groups of machines, defined by number of screens, number of machines, locations, or ownership. Significant discounts from the corresponding single-screen

prices are available when the estimated number of screens exceeds 40. Contact Numerical Recipes Software (email: orders@nr.com or fax: 781 863-1739) for details.

- ["Course Right-to-Copy License"] Instructors at accredited educational institutions who have adopted this book for a course, and who have already purchased a Single Screen License (either acquired with the purchase of media, or from the Numerical Recipes On-Line Software Store), may license the programs for use in that course as follows: Mail your name, title, and address; the course name, number, dates, and estimated enrollment; and advance payment of $5 per (estimated) student to Numerical Recipes Software, at this address: P.O. Box 243, Cambridge, MA 02238 (USA). You will receive by return mail a license authorizing you to make copies of the programs for use by your students, and/or to transfer the programs to a machine accessible to your students (but only for the duration of the course).

About Copyrights on Computer Programs

Like artistic or literary compositions, computer programs are protected by copyright. Generally it is an infringement for you to copy into your computer a program from a copyrighted source. (It is also not a friendly thing to do, since it deprives the program's author of compensation for his or her creative effort.) Under copyright law, all "derivative works" (modified versions, or translations into another computer language) also come under the same copyright as the original work.

Copyright does not protect ideas, but only the expression of those ideas in a particular form. In the case of a computer program, the ideas consist of the program's methodology and algorithm, including the necessary sequence of steps adopted by the programmer. The expression of those ideas is the program source code (particularly any arbitrary or stylistic choices embodied in it), its derived object code, and any other derivative works.

If you analyze the ideas contained in a program, and then express those ideas in your own completely different implementation, then that new program implementation belongs to you. That is what we have done for those programs in this book that are not entirely of our own devising. When programs in this book are said to be "based" on programs published in copyright sources, we mean that the ideas are the same. The expression of these ideas as source code is our own. We believe that no material in this book infringes on an existing copyright.

Trademarks

Several registered trademarks appear within the text of this book: Sun is a trademark of Sun Microsystems, Inc. SPARC and SPARCstation are trademarks of SPARC International, Inc. Microsoft, Windows 95, Windows NT, PowerStation, and MS are trademarks of Microsoft Corporation. DEC, VMS, Alpha AXP, and ULTRIX are trademarks of Digital Equipment Corporation. IBM is a trademark of International Business Machines Corporation. Apple and Macintosh are trademarks of Apple Computer, Inc. UNIX is a trademark licensed exclusively through X/Open

Attributions

The fact that ideas are legally "free as air" in no way supersedes the ethical requirement that ideas be credited to their known originators. When programs in this book are based on known sources, whether copyrighted or in the public domain, published or "handed-down," we have attempted to give proper attribution. Unfortunately, the lineage of many programs in common circulation is often unclear. We would be grateful to readers for new or corrected information regarding attributions, which we will attempt to incorporate in subsequent printings.

Chapter 21. Introduction to Fortran 90 Language Features

21.0 Introduction

Fortran 90 is in many respects a backwards-compatible modernization of the long-used (and much abused) Fortran 77 language, but it is also, in other respects, a new language for parallel programming on present and future multiprocessor machines. These twin design goals of the language sometimes add confusion to the process of becoming fluent in Fortran 90 programming.

In a certain trivial sense, Fortran 90 is strictly backwards-compatible with Fortran 77. That is, any Fortran 90 compiler is supposed to be able to compile any legacy Fortran 77 code without error. The reason for terming this compatibility trivial, however, is that you have to tell the compiler (usually via a source file name ending in "`.f`" or "`.for`") that it is dealing with a Fortran 77 file. If you instead try to pass off Fortran 77 code as native Fortran 90 (e.g., by naming the source file something ending in "`.f90`") it will not always work correctly!

It is best, therefore, to approach Fortran 90 as a new computer language, albeit one with a lot in common with Fortran 77. Indeed, in such terms, Fortran 90 is a fairly *big* language, with a large number of new constructions and intrinsic functions. Here, in one short chapter, we do not pretend to provide a complete description of the language. Luckily, there are good books that do exactly that. Our favorite one is by Metcalf and Reid [1], cited throughout this chapter as "M&R." Other good starting points include [2] and [3].

Our goal, in the remainder of this chapter, is to give a good, working description of those Fortran 90 language features that are not immediately self-explanatory to Fortran 77 programmers, with particular emphasis on those that occur most frequently in the Fortran 90 versions of the Numerical Recipes routines. This chapter, by itself, will not teach you to write Fortran 90 code. But it ought to help you acquire a reading knowledge of the language, and perhaps provide enough of a head start that you can rapidly pick up the rest of what you need to know from M&R or another Fortran 90 reference book.

CITED REFERENCES AND FURTHER READING:

Metcalf, M., and Reid, J. 1996, *Fortran 90/95 Explained* (Oxford: Oxford University Press). [1]

Kerrigan, J.F. 1993, *Migrating to Fortran 90* (Sebastopol, CA: O'Reilly). [2]

Brainerd, W.S., Goldberg, C.H., and Adams, J.C. 1996, *Programmer's Guide to Fortran 90*, 3rd ed. (New York: Springer-Verlag). [3]

21.1 Quick Start: Using the Fortran 90 Numerical Recipes Routines

This section is for people who want to jump right in. We'll compute a Bessel function $J_0(x)$, where x is equal to the fourth root of the Julian Day number of the 200th full moon since January 1900. (Now *there's* a useful quantity!)

First, locate the important files `nrtype.f90`, `nrutil.f90`, and `nr.f90`, as listed in Appendices C1, C1, and C2, respectively. These contain *modules* that either are (i) used by our routines, or else (ii) describe the calling conventions of our routines to (your) user programs. Compile each of these files, producing (with most compilers) a `.mod` file and a `.o` (or similarly named) file for each one.

Second, create this main program file:

```
PROGRAM hello_bessel
USE nrtype
USE nr, ONLY: flmoon, bessj0
IMPLICIT NONE
INTEGER(I4B) :: n=200,nph=2,jd
REAL(SP) :: x,frac,ans
call flmoon(n,nph,jd,frac)
x=jd**0.25_sp
ans=bessj0(x)
write (*,*) 'Hello, Bessel: ', ans
END PROGRAM
```

Here is a quick explanation of some elements of the above program:

The first USE statement includes a module of ours named `nrtype`, whose purpose is to give symbolic names to some kinds of data types, among them single-precision reals ("`sp`") and four-byte integers ("`i4b`"). The second USE statement includes a module of ours that defines the calling sequences, and variable types, expected by (in this case) the Numerical Recipes routines `flmoon` and `bessj0`.

The `IMPLICIT NONE` statement signals that we want the compiler to require us explicitly to declare all variable types. *We strongly urge that you always take this option.*

The next two lines declare integer and real variables of the desired kinds. The variable `n` is initialized to the value 200, `nph` to 2 (a value expected by `flmoon`).

We call `flmoon`, and take the fourth root of the answer it returns as `jd`. Note that the constant 0.25 is typed to be single-precision by the appended `_sp`.

We call the `bessj0` routine, and print the answer.

Third, compile the main program file, and also the files `flmoon.f90`, `bessj0.f90`. Then, link the resulting object files with also `nrutil.o` (or similar system-dependent name, as produced in step 1). Some compilers will also require you to link with `nr.o` and `nrtype.o`.

Fourth, run the resulting executable file. Typical output is:

```
Hello, Bessel:   7.3096365E-02
```

21.2 Fortran 90 Language Concepts

The Fortran 90 language standard defines and uses a number of standard terms for concepts that occur in the language. Here we summarize briefly some of the most important concepts. Standard Fortran 90 terms are shown in *italics*. While by no means complete, the information in this section should help you get a quick start with your favorite Fortran 90 reference book or language manual.

A note on capitalization: Outside a character context, Fortran 90 is not case-sensitive, so you can use upper and lower case any way you want, to improve readability. A variable like SP (see below) is the same variable as the variable sp. We like to capitalize keywords whose use is primarily at compile-time (statements that delimit program and subprogram boundaries, declaration statements of variables, fixed parameter values), and use lower case for the bulk of run-time code. You can adopt any convention that you find helpful to your own programming style; but we strongly urge you to adopt and follow *some* convention.

Data Types and Kinds

Data types (also called simply *types*) can be either *intrinsic data types* (the familiar INTEGER, REAL, LOGICAL, and so forth) or else *derived data types* that are built up in the manner of what are called "structures" or "records" in other computer languages. (We'll use derived data types very sparingly in this book.) Intrinsic data types are further specified by their *kind parameter* (or simply *kind*), which is simply an integer. Thus, on many machines, REAL(4) (with kind = 4) is a single-precision real, while REAL(8) (with kind = 8) is a double-precision real. *Literal constants* (or simply *literals*) are specified as to kind by appending an underscore, as 1.5_4 for single precision, or 1.5_8 for double precision. [M&R, §2.5–§2.6]

Unfortunately, the specific integer values that define the different kind types are not specified by the language, but can vary from machine to machine. For that reason, one almost never uses literal kind parameters like 4 or 8, but rather defines in some central file, and imports into all one's programs, symbolic names for the kinds. For this book, that central file is the *module* named nrtype, and the chosen symbolic names include SP, DP (for reals); I2B, I4B (for two- and four-byte integers); and LGT for the default logical type. You will therefore see us consistently writing REAL(SP), or 1.5_sp, and so forth.

Here is an example of declaring some variables, including a one-dimensional array of length 500, and a two-dimensional array with 100 rows and 200 columns:

```
USE nrtype
REAL(SP) :: x,y,z
INTEGER(I4B) :: i,j,k
REAL(SP), DIMENSION(500) :: arr
REAL(SP), DIMENSION(100,200) :: barr
REAL(SP) :: carr(500)
```

The last line shows an alternative form for array syntax. And yes, there *are* default kind parameters for each intrinsic type, but these vary from machine to machine and can get you into trouble when you try to move code. We therefore specify all kind parameters explicitly in almost all situations.

Array Shapes and Sizes

The *shape* of an *array* refers to both its dimensionality (called its *rank*), and also the lengths along each dimension (called the *extents*). The shape of an array is specified by a rank-one array whose elements are the extents along each dimension, and can be queried with the shape intrinsic (see p. 949). Thus, in the above example, shape(barr) returns an array of length 2 containing the values $(100, 200)$.

The *size* of an array is its total number of elements, so the intrinsic size(barr) would return 20000 in the above example. More often one wants to know the extents along each dimension, separately: size(barr,1) returns the value 100, while size(barr,2) returns the value 200. [M&R, §2.10]

Section §21.3, below, discusses additional aspects of arrays in Fortran 90.

Memory Management

Fortran 90 is greatly superior to Fortran 77 in its memory-management capabilities, seen by the user as the ability to create, expand, or contract workspace for programs. Within *subprograms* (that is, *subroutines* and *functions*), one can have *automatic arrays* (or other *automatic data objects*) that come into existence each time the subprogram is entered, and disappear (returning their memory to the pool) when the subprogram is exited. The size of automatic objects can be specified by arbitrary expressions involving values passed as *actual arguments* in the calling program, and thus received by the subprogram through its corresponding *dummy arguments*. [M&R, §6.4]

Here is an example that creates some automatic workspace named carr:

```
SUBROUTINE dosomething(j,k)
USE nrtype
REAL(SP), DIMENSION(2*j,k**2) :: carr
```

Finer control on when workspace is created or destroyed can be achieved by declaring *allocatable arrays*, which exist as names only, without associated memory, until they are *allocated* within the program or subprogram. When no longer needed, they can be *deallocated*. The *allocation status* of an allocatable array can be tested by the program via the allocated intrinsic function (p. 952). [M&R, §6.5]

Here is an example in outline:

```
REAL(SP), DIMENSION(:,:), ALLOCATABLE :: darr
...
allocate(darr(10,20))
...
deallocate(darr)
...
allocate(darr(100,200))
...
deallocate(darr)
```

Notice that darr is originally declared with only "slots" (colons) for its dimensions, and is then allocated/deallocated twice, with different sizes.

Yet finer control is achieved by the use of *pointers*. Like an allocatable array, a pointer can be allocated, at will, its own associated memory. However, it has the additional flexibility of alternatively being *pointer associated* with a *target* that already exists under another name. Thus, pointers can be used as redefinable aliases for other variables, arrays, or (see §21.3) *array sections*. [M&R, §6.12]

Here is an example that first associates the pointer parr with the array earr, then later cancels that association and allocates it its own storage of size 50:

```
REAL(SP), DIMENSION(:), POINTER :: parr
REAL(SP), DIMENSION(100), TARGET :: earr
...
parr => earr
...
nullify(parr)
allocate(parr(50))
...
deallocate(parr)
```

Procedure Interfaces

When a procedure is *referenced* (e.g., called) from within a program or subprogram (examples of *scoping units*), the scoping unit must be told, or must deduce, the procedure's *interface*, that is, its calling sequence, including the types and kinds of all dummy arguments, returned values, etc. The recommended procedure is to specify this interface via an *explicit interface*, usually an *interface block* (essentially a declaration statement for subprograms) in the calling subprogram or in some *module* that the calling program includes via a USE statement. In this book all interfaces are explicit, and the module named nr contains interface blocks for all of the Numerical Recipes routines. [M&R, §5.11]

Here is a typical example of an interface block:

```
INTERFACE
   SUBROUTINE caldat(julian,mm,id,iyyy)
   USE nrtype
   INTEGER(I4B), INTENT(IN) :: julian
   INTEGER(I4B), INTENT(OUT) :: mm,id,iyyy
   END SUBROUTINE caldat
END INTERFACE
```

Once this interface is made known to a program that you are writing (by either explicit inclusion or a USE statement), then the compiler is able to flag for you a variety of otherwise difficult-to-find bugs. Although interface blocks can sometimes seem overly wordy, they give a big payoff in ultimately minimizing programmer time and frustration.

For compatibility with Fortran 77, the language also allows for *implicit interfaces*, where the calling program tries to figure out the interface by the old rules of Fortran 77. These rules are quite limited, and prone to producing devilishly obscure program bugs. We strongly recommend that implicit interfaces never be used.

Elemental Procedures and Generic Interfaces

Many *intrinsic procedures* (those defined by the language standard and thus usable without any further definition or specification) are also *generic*. This means that a single procedure name, such as log(x), can be used with a variety of types and kind parameters for the argument x, and the result returned will have the same type and kind parameter as the argument. In this example, log(x) allows any real or complex argument type.

Better yet, most generic functions are also *elemental*. The argument of an elemental function can be an array of arbitrary shape! Then, the returned result is an array of the same shape, with each element containing the result of applying the function to the corresponding element of the argument. (Hence the name *elemental*, meaning "applied element by element.") [M&R, §8.1] For example:

```
REAL(SP), DIMENSION(100,100) :: a,b
b=sin(a)
```

Fortran 90 has no facility for creating new, user-defined elemental functions. It does have, however, the related facility of *overloading* by the use of *generic interfaces*. This is invoked by the use of an interface block that attaches a single *generic name* to a number of distinct subprograms whose dummy arguments have different types or kinds. Then, when the generic name is referenced (e.g., called), the compiler chooses the specific subprogram that matches the types and kinds of the actual arguments used. [M&R, §5.18] Here is an example of a generic interface block:

```
INTERFACE myfunc
   FUNCTION myfunc_single(x)
   USE nrtype
   REAL(SP) :: x,myfunc_single
   END FUNCTION myfunc_single

   FUNCTION myfunc_double(x)
   USE nrtype
   REAL(DP) :: x,myfunc_double
   END FUNCTION myfunc_double
END INTERFACE
```

A program with knowledge of this interface could then freely use the function reference `myfunc(x)` for `x`'s of both type `SP` and type `DP`.

We use overloading quite extensively in this book. A typical use is to provide, under the same name, both scalar and vector versions of a function such as a Bessel function, or to provide both single-precision and double-precision versions of procedures (as in the above example). Then, to the extent that we have provided all the versions that you need, you can pretend that our routine is elemental. In such a situation, if you ever call our function with a type or kind that we have *not* provided, the compiler will instantly flag the problem, because it is unable to resolve the generic interface.

Modules

Modules, already referred to several times above, are Fortran 90's generalization of Fortran 77's common blocks, `INCLUDE`d files of parameter statements, and (to some extent) statement functions. Modules are *program units*, like main programs or subprograms (subroutines and functions), that can be separately compiled. A module is a convenient place to stash global data, *named constants* (what in Fortran 77 are called "symbolic constants" or "`PARAMETER`s"), interface blocks to subprograms and/or actual subprograms themselves (*module subprograms*). The convenience is that a module's information can be incorporated into another program unit via a simple, one-line `USE` statement. [M&R, §5.5]

Here is an example of a simple module that defines a few parameters, creates some global storage for an array named `arr` (as might be done with a Fortran 77 common block), and defines the interface to a function `yourfunc`:

```
MODULE mymodule
   USE nrtype
   REAL(SP), PARAMETER :: con1=7.0_sp/3.0_sp,con2=10.0_sp
   INTEGER(I4B), PARAMETER :: ndim=10,mdim=9
   REAL(SP), DIMENSION(ndim,mdim) :: arr
   INTERFACE
      FUNCTION yourfunc(x)
      USE nrtype
      REAL(SP) :: x,yourfunc
      END FUNCTION yourfunc
   END INTERFACE
END MODULE mymodule
```

As mentioned earlier, the module `nr` contains `INTERFACE` declarations for all the Numerical Recipes. When we include a statement of the form

```
USE nr, ONLY: recipe1
```

it means that the program uses the additional routine `recipe1`. The compiler is able to use the explicit interface declaration in the module to check that `recipe1` is invoked with arguments of the correct type, shape, and number. However, a weakness of Fortran 90 is that there is no fail-safe way to be sure that the interface module (here `nr`) stays synchronized with the underlying routine (here `recipe1`). You might think that you could accomplish this by putting `USE nr, ONLY: recipe1` into the `recipe1` program itself. Unfortunately, the compiler interprets this as an erroneous double definition of `recipe1`'s interface, rather than (as would be desirable) as an opportunity for a consistency check. (To achieve this kind of consistency check, you can put the procedures themselves, not just their interfaces, into the module.)

CITED REFERENCES AND FURTHER READING:

Metcalf, M., and Reid, J. 1996, *Fortran 90/95 Explained* (Oxford: Oxford University Press).

21.3 More on Arrays and Array Sections

Arrays are the central conceptual core of Fortran 90 as a *parallel* programming language, and thus worthy of some further discussion. We have already seen that arrays can "come into existence" in Fortran 90 in several ways, either directly declared, as

```
REAL(SP), DIMENSION(100,200) :: arr
```

or else allocated by an *allocatable* variable or a *pointer* variable,

```
REAL(SP), DIMENSION(:,:), ALLOCATABLE :: arr
REAL(SP), DIMENSION(:,:), POINTER :: barr
...
allocate(arr(100,200),barr(100,200))
```

or else (not previously mentioned) passed into a subprogram through a dummy argument:

```
SUBROUTINE myroutine(carr)
USE nrtype
REAL(SP), DIMENSION(:,:) :: carr
...
i=size(carr,1)
```

```
j=size(carr,2)
```

In the above example we also show how the subprogram can find out the size of the actual array that is passed, using the `size` intrinsic. This routine is an example of the use of an *assumed-shape array*, new to Fortran 90. The actual extents along each dimension are inherited from the calling routine at run time. A subroutine with assumed-shape array arguments *must* have an explicit interface in the calling routine, otherwise the compiler doesn't know about the extra information that must be passed. A typical setup for calling `myroutine` would be:

```
PROGRAM use_myroutine
USE nrtype
REAL(SP), DIMENSION(10,10) :: arr
INTERFACE
   SUBROUTINE myroutine(carr)
   USE nrtype
   REAL(SP), DIMENSION(:,:) :: carr
   END SUBROUTINE myroutine
END INTERFACE
...
call myroutine(a)
```

Of course, for the recipes we have provided all the interface blocks in the file `nr.f90`, and you need only a `USE nr` statement in your calling program.

Conformable Arrays

Two arrays are said to be *conformable* if their shapes are the same. Fortran 90 allows practically all operations among conformable arrays and elemental functions that are allowed for scalar variables. Thus, if `arr`, `barr`, and `carr` are mutually conformable, we can write,

```
arr=barr+cos(carr)+2.0_sp
```

and have the indicated operations performed, element by corresponding element, on the entire arrays. The above line also illustrates that a scalar (here the constant `2.0_sp`, but a scalar variable would also be fine) is deemed conformable with *any* array — it gets "expanded" to the shape of the rest of the expression that it is in. [M&R, §3.11]

In Fortran 90, as in Fortran 77, the default lower bound for an array subscript is 1; however, it can be made some other value at the time that the array is declared:

```
REAL(SP), DIMENSION(100,200) :: farr
REAL(SP), DIMENSION(0:99,0:199) :: garr
...
farr = 3.0_sp*garr + 1.0_sp
```

Notice that `farr` and `garr` are conformable, since they have the same shape, in this case $(100, 200)$. Also note that when they are used in an array expression, the operations are taken between the corresponding elements *of their shapes*, not necessarily the corresponding elements *of their indices*. [M&R, §3.10] In other words, one of the components evaluated is,

```
farr(1,1) = 3.0_sp*garr(0,0) + 1.0_sp
```

This illustrates a fundamental aspect of array (or data) parallelism in Fortran 90. Array constructions should *not* be thought of as merely abbreviations for do-loops over indices, but rather as genuinely parallel operations on same-shaped objects, abstracted of their indices. This is why the standard makes no statement about the order in which the individual operations in an array expression are executed; they might in fact be carried out simultaneously, on parallel hardware.

By default, array expressions and assignments are performed for all elements of the same-shaped arrays referenced. This can be modified, however, by use of a `where` construction like this:

```
where (harr > 0.0_sp)
   farr = 3.0_sp*garr + 1.0_sp
end where
```

Here `harr` must also be conformable to `farr` and `garr`. Analogously with the Fortran `if`-statement, there is also a one-line form of the `where`-statement. There is also a `where ... elsewhere ... end where` form of the statement, analogous to `if ... else if ... end if`. A significant language limitation in Fortran 90 is that nested `where`-statements are not allowed. [M&R, §6.8]

Array Sections

Much of the versatility of Fortran 90's array facilities stems from the availability of *array sections*. An array section acts like an array, but its memory location, and thus the values of its elements, is actually a subset of the memory location of an already-declared array. Array sections are thus "windows into arrays," and they can appear on either the left side, or the right side, or both, of a replacement statement. Some examples will clarify these ideas.

Let us presume the declarations

```
REAL(SP), DIMENSION(100) :: arr
INTEGER(I4B), DIMENSION(6) :: iarr=(/11,22,33,44,55,66/)
```

Note that `iarr` is not only declared, it is also initialized by an *initialization expression* (a replacement for Fortran 77's `DATA` statement). [M&R, §7.5] Here are some array sections constructed from these arrays:

Array Section	*What It Means*
`arr(:)`	same as `arr`
`arr(1:100)`	same as `arr`
`arr(1:10)`	one-dimensional array containing first 10 elements of `arr`
`arr(51:100)`	one-dimensional array containing second half of `arr`
`arr(51:)`	same as `arr(51:100)`
`arr(10:1:-1)`	one-dimensional array containing first 10 elements of `arr`, but in *reverse order*
`arr( (/10,99,1,6/) )`	one-dimensional array containing elements 10, 99, 1, and 6 of `arr`, in that order
`arr(iarr)`	one-dimensional array containing elements 11, 22, 33, 44, 55, 66 of `arr`, in that order

Now let's try some array sections of the two-dimensional array

```
REAL(SP), DIMENSION(100,100) :: barr
```

Array Section	*What It Means*
`barr(:,:)`	same as `barr`
`barr(1:100,1:100)`	same as `barr`
`barr(7,:)`	one-dimensional array containing the 7th row of `barr`
`barr(7,1:100)`	same as `barr(7,:)`
`barr(:,7)`	one-dimensional array containing the 7th column of `barr`
`barr(21:30,71:90)`	two-dimensional array containing the sub-block of `barr` with the indicated ranges of indices; the shape of this array section is $(10, 20)$
`barr(100:1:-1,100:1:-1)`	two-dimensional array formed by flipping `barr` upside down and backwards
`barr(2:100:2,2:100:2)`	two-dimensional array of shape $(50, 50)$ containing the elements of `barr` whose row and column indices are both even

Some terminology: A construction like `2:100:2`, above, is called a *subscript triplet*. Its integer pieces (which may be integer constants, or more general integer expressions) are called *lower*, *upper*, and *stride*. Any of the three may be omitted. An omitted stride defaults to the value 1. Notice that, if (*upper* − *lower*) has a different sign from *stride*, then a subscript triplet defines an empty or *zero-length* array, e.g., `1:5:-1` or `10:1:1` (or its equivalent form, simply `10:1`). Zero-length arrays are not treated as errors in Fortran 90, but rather as "no-ops." That is, no operation is performed in an expression or replacement statement among zero-length arrays. (This is essentially the same convention as in Fortran 77 for do-loop indices, which array expressions often replace.) [M&R, §6.10]

It is important to understand that array sections, when used in array expressions, match elements with other parts of the expression *according to shape*, not according to indices. (This is exactly the same principle that we applied, above, to arrays with subscript lower bounds different from the default value of 1.) One frequently exploits this feature in using array sections to carry out operations on arrays that access neighboring elements. For example,

```
carr(1:n-1,1:n-1) = barr(1:n-1,1:n-1)+barr(2:n,2:n)
```

constructs in the $(n-1) \times (n-1)$ matrix `carr` the sum of each of the corresponding elements in $n \times n$ `barr` added to its diagonally lower-right neighbor.

Pointers are often used as aliases for array sections, especially if the same array sections are used repeatedly. [M&R, §6.12] For example, with the setup

```
REAL(SP), DIMENSION(:,:), POINTER :: leftb,rightb
```

```
leftb=>barr(1:n-1,1:n-1)
rightb=>barr(2:n,2:n)
```

the statement above can be coded as

```
carr(1:n-1,1:n-1)=leftb+rightb
```

We should also mention that array sections, while powerful and concise, are sometimes not quite powerful enough. While any row or column of a matrix is easily accessible as an array section, there is no good way, in Fortran 90, to access (e.g.) the diagonal of a matrix, even though its elements are related by a linear progression in the Fortran storage order (by columns). These so-called *skew-sections* were much discussed by the Fortran 90 standards committee, but they were not implemented. We will see examples later in this volume of work-around programming tricks (none totally satisfactory) for this omission. (Fortran 95 corrects the omission; see §21.6.)

CITED REFERENCES AND FURTHER READING:

Metcalf, M., and Reid, J. 1996, *Fortran 90/95 Explained* (Oxford: Oxford University Press).

21.4 Fortran 90 Intrinsic Procedures

Much of Fortran 90's power, both for parallel programming and for its concise expression of algorithmic ideas, comes from its rich set of intrinsic procedures. These have the effect of making the language "large," hence harder to learn. However, effort spent on learning to use the intrinsics — particularly some of their more obscure, and more powerful, optional arguments — is often handsomely repaid.

This section summarizes the intrinsics that we find useful in numerical work. We omit, here, discussion of intrinsics whose exclusive use is for character and string manipulation. We intend only a summary, not a complete specification, which can be found in M&R's Chapter 8, or other reference books.

If you find the sheer number of new intrinsic procedures daunting, you might want to start with our list of the "top 10" (with the number of different Numerical Recipes routines that use each shown in parentheses): `size` (254), `sum` (44), `dot_product` (31), `merge` (27), `all` (25), `maxval` (23), `matmul` (19), `pack` (18), `any` (17), and `spread` (15). (Later, in Chapter 23, you can compare these numbers with our frequency of using the short utility functions that we define in a module named `nrutil` — several of which we think ought to have been included as Fortran 90 intrinsic procedures.)

The type, kind, and shape of the value returned by intrinsic functions will usually be clear from the short description that we give. As an additional hint (though not necessarily a precise description), we adopt the following codes:

Hint	*What It Means*
[Int]	an INTEGER kind type
[Real]	a REAL kind type
[Cmplx]	a COMPLEX kind type
[Num]	a numerical type and kind
[Lgcl]	a LOGICAL kind type
[Iarr]	a one-dimensional INTEGER array
[argTS]	same type and shape as the first argument
[argT]	same type as the first argument, but not necessarily the same shape

Numerical Elemental Functions

Little needs to be said about the numerical functions with identical counterparts in Fortran 77: `abs`, `acos`, `aimag`, `asin`, `atan`, `atan2`, `conjg`, `cos`, `cosh`, `dim`, `exp`, `log`, `log10`, `max`, `min`, `mod`, `sign`, `sin`, `sinh`, `sqrt`, `tan`, and `tanh`. In Fortran 90 these are all *elemental* functions, so that any plausible type, kind, and shape of argument may be used. Except for `aimag`, which returns a real type from a complex argument, these all return [argTS] (see table above).

Although Fortran 90 recognizes, for compatibility, Fortran 77's so-called *specific names* for these functions (e.g., `iabs`, `dabs`, and `cabs` for the generic `abs`), these are entirely superfluous and should be avoided.

Fortran 90 corrects some ambiguity (or at least inconvenience) in Fortran 77's `mod(a,p)` function, by introducing a new function `modulo(a,p)`. The functions are essentially identical for positive arguments, but for negative `a` and positive `p`, `modulo` gives results more compatible with one's mathematical expectation that the answer should always be in the positive range 0 to `p`. E.g., `modulo(11,5)=1`, and `modulo(-11,5)=4`. [M&R, §8.3.2]

Conversion and Truncation Elemental Functions

Fortran 90's conversion (or, in the language of C, casting) and truncation functions are generally modeled on their Fortran 77 antecedents, but with the addition of an optional second integer argument, `kind`, that determines the kind of the result. Note that, if kind is omitted, you get a default kind — not necessarily related to the kind of your argument. The kind of the argument is of course known to the compiler by its previous declaration. Functions in this category (see below for explanation of arguments in slanted type) are:

[Real] `aint(a,`*`kind`*`)`
  Truncate to integer value, return as a real kind.

[Real] `anint(a,`*`kind`*`)`
  Nearest whole number, return as a real kind.

[Cmplx] `cmplx(x,`*`y,kind`*`)`

Convert to complex kind. If y is omitted, it is taken to be 0.

[Int] `int(a,kind)`
Convert to integer kind, truncating towards zero.

[Int] `nint(a,kind)`
Convert to integer kind, choosing the nearest whole number.

[Real] `real(a,kind)`
Convert to real kind.

[Lgcl] `logical(a,kind)`
Convert one logical kind to another.

We must digress here to explain the use of *optional arguments* and *keywords* as Fortran 90 language features. [M&R, §5.13] When a routine (either intrinsic or user-defined) has arguments that are declared to be optional, then the dummy names given to them also become keywords that distinguish — independent of their position in a calling list — which argument is intended to be passed. (There are some additional rules about this that we will not try to summarize here.) In this section's tabular listings, we indicate optional arguments in intrinsic routines by printing them in smaller slanted type. For example, the intrinsic function

```
eoshift(array,shift,boundary,dim)
```

has two required arguments, `array` and `shift`, and two optional arguments, `boundary` and `dim`. Suppose we want to call this routine with the actual arguments `myarray`, `myshift`, and `mydim`, but omitting the argument in the `boundary` slot. We do this by the expression

```
eoshift(myarray,myshift,dim=mydim)
```

Conversely, if we wanted a `boundary` argument, but no `dim`, we might write

```
eoshift(myarray,myshift,boundary=myboundary)
```

It is always a good idea to use this kind of keyword construction when invoking optional arguments, even though the rules allow keywords to be omitted in some unambiguous cases. Now back to the lists of intrinsic routines.

A peculiarity of the `real` function derives from its use both as a type conversion and for extracting the real part of complex numbers (related, but not identical, usages): If the argument of `real` is complex, and `kind` is omitted, then the result *isn't* a default real kind, but rather *is* (as one generally would want) the `real` kind type corresponding to the kind type of the complex argument, that is, single-precision real for single-precision complex, double-precision for double-precision, and so on. [M&R, §8.3.1] We recommend *never* using `kind` when you intend to extract the real part of a complex, and *always* using `kind` when you intend conversion of a real or integer value to a particular kind of `REAL`. (Use of the deprecated function `dble` is not recommended.)

The last two conversion functions are the exception in that they *don't* allow a `kind` argument, but rather return default integer kinds. (The X3J3 standards committee has fixed this in Fortran 95.)

[Int] `ceiling(a)`
Convert to integer, truncating towards more positive.

[Int] `floor(a)`
Convert to integer, truncating towards more negative.

Reduction and Inquiry Functions on Arrays

These are mostly the so-called *transformational functions* that accept array arguments and return either scalar values or else arrays of lesser rank. [M&R, §8.11] With no optional arguments, such functions act on all the elements of their single array argument, regardless of its shape, and produce a scalar result. When the optional argument `dim` is specified, they instead act on all one-dimensional sections that span the dimension `dim`, producing an answer one rank lower than the first argument (that is, omitting the `dim` dimension from its shape). When the optional argument `mask` is specified, only the elements with a corresponding true value in `mask` are scanned.

[Lgcl] `all(mask,`*`dim`*`)`
Returns true if all elements of `mask` are true, false otherwise.

[Lgcl] `any(mask,`*`dim`*`)`
Returns true if any of the elements of `mask` are true, false otherwise.

[Int] `count(mask,`*`dim`*`)`
Counts the true elements in `mask`.

[Num] `maxval(array,`*`dim,mask`*`)`
Maximum value of the array elements.

[Num] `minval(array,`*`dim,mask`*`)`
Minimum value of the array elements.

[Num] `product(array,`*`dim,mask`*`)`
Product of the array elements.

[Int] `size(array,`*`dim`*`)`
Size (total number of elements) of `array`, or its extent along dimension `dim`.

[Num] `sum(array,`*`dim,mask`*`)`
Sum of the array elements.

The use of the `dim` argument can be confusing, so an example may be helpful. Suppose we have

$$\texttt{myarray} = \begin{bmatrix} 1 & 2 & 3 & 4 \\ 5 & 6 & 7 & 8 \\ 9 & 10 & 11 & 12 \end{bmatrix}$$

where, as always, the `i` index in `array(i,j)` numbers the rows while the `j` index numbers the columns. Then

$$\texttt{sum(myarray,dim=1)} = (15, 18, 21, 24)$$

that is, the `i` indices are "summed away" leaving only a `j` index on the result; while

$$\texttt{sum(myarray,dim=2)} = (10, 26, 42)$$

that is, the `j` indices are "summed away" leaving only an `i` index on the result. Of course we also have

$$\texttt{sum(myarray)} = 78$$

Two related functions return the location of particular elements in an array. The returned value is a one-dimensional integer array containing the respective subscript of the element along each dimension. Note that when the argument object is a *one*-dimensional array, the returned object is an integer *array of length 1*, not simply an integer. (Fortran 90 distinguishes between these.)

[Iarr] `maxloc(array,`*`mask`*`)`
Location of the maximum value in an array.

[Iarr] `minloc(array,`*`mask`*`)`
Location of the minimum value in an array.

Similarly returning a one-dimensional integer array are

[Iarr] `shape(array)`
Returns the shape of `array` as a one-dimensional integer array.

[Iarr] `lbound(array,`*`dim`*`)`
When `dim` is absent, returns an array of lower bounds for each dimension of subscripts of `array`. When `dim` is present, returns the value only for dimension `dim`, as a scalar.

[Iarr] `ubound(array,`*`dim`*`)`
When `dim` is absent, returns an array of upper bounds for each dimension of subscripts of `array`. When `dim` is present, returns the value only for dimension `dim`, as a scalar.

Array Unary and Binary Functions

The most powerful array operations are simply built into the language as operators. All the usual arithmetic and logical operators (`+`, `-`, `*`, `/`, `**`, `.not.`, `.and.`, `.or.`, `.eqv.`, `.neqv.`) can be applied to arrays of arbitrary shape or (for the binary operators) between two arrays of the same shape, or between arrays and scalars. The types of the arrays must, of course, be appropriate to the operator used. The result in all cases is to perform the operation element by element on the arrays.

We also have the intrinsic functions,

[Num] `dot_product(veca,vecb)`
Scalar dot product of two one-dimensional vectors `veca` and `vecb`.

[Num] `matmul(mata,matb)`
Result of matrix-multiplying the two two-dimensional matrices `mata` and `matb`. The shapes have to be such as to allow matrix multiplication. Vectors (one-dimensional arrays) are additionally allowed as either the first or second argument, but not both; they are treated as row vectors in the first argument, and as column vectors in the second.

You might wonder how to form the *outer* product of two vectors, since `matmul` specifically excludes this case. (See §22.1 and §23.5 for answer.)

Array Manipulation Functions

These include many powerful features that a good Fortran 90 programmer should master.

[argTS] `cshift(array,shift,`*`dim`*`)`

If `dim` is omitted, it is taken to be 1. Returns the result of circularly left-shifting every one-dimensional section of `array` (in dimension `dim`) by `shift` (which may be negative). That is, for positive `shift`, values are moved to smaller subscript positions. Consult a Fortran 90 reference (e.g., [M&R, §8.13.5]) for the case where `shift` is an array.

[argTS] `merge(tsource,fsource,mask)`

Returns same shape object as `tsource` and `fsource` containing the former's components where `mask` is true, the latter's where it is false.

[argTS] `eoshift(array,shift,`*`boundary,dim`*`)`

If `dim` is omitted, it is taken to be 1. Returns the result of end-off left-shifting every one-dimensional section of `array` (in dimension `dim`) by `shift` (which may be negative). That is, for positive `shift`, values are moved to smaller subscript positions. If `boundary` is present as a scalar, it supplies elements to fill in the blanks; if it is not present, zero values are used. Consult a Fortran 90 reference (e.g., [M&R, §8.13.5]) for the case where `boundary` and/or `shift` is an array.

[argT] `pack(array,mask,`*`vector`*`)`

Returns a one-dimensional array containing the elements of `array` that pass the `mask`. Components of optional `vector` are used to pad out the result to the size of `vector` with specified values.

[argT] `reshape(source,shape,`*`pad,order`*`)`

Takes the elements of `source`, in normal Fortran order, and returns them (as many as will fit) as an array whose shape is specified by the one-dimensional integer array `shape`. If there is space remaining, then `pad` must be specified, and is used (as many sequential copies as necessary) to fill out the rest. For description of `order`, consult a Fortran 90 reference, e.g., [M&R, 8.13.3].

[argT] `spread(source,dim,ncopies)`

Returns an array whose rank is one greater than `source`, and whose `dim` dimension is of length `ncopies`. Each of the result's `ncopies` array sections having a fixed subscript in dimension `dim` is a copy of `source`. (That is, it spreads `source` into the `dim`th dimension.)

[argT] `transpose(matrix)`

Returns the transpose of `matrix`, which must be two-dimensional.

[argT] `unpack(vector,mask,field)`

Returns an array whose type is that of `vector`, but whose shape is that of `mask`. The components of `vector` are put, in order, into the positions where `mask` is true. Where `mask` is false, components of `field` (which may be a scalar or an array with the same shape as `mask`) are used instead.

Bitwise Functions

Most of the bitwise functions should be familiar to Fortran 77 programmers as longstanding standard extensions of that language. Note that the bit *positions* number from zero to one less than the value returned by the `bit_size` function. Also note that bit positions number *from right to left*. Except for `bit_size`, the following functions are all elemental.

[Int] `bit_size(i)`
Number of bits in the integer type of `i`.

[Lgcl] `btest(i,pos)`
True if bit position `pos` is 1, false otherwise.

[Int] `iand(i,j)`
Bitwise logical and.

[Int] `ibclr(i,pos)`
Returns `i` but with bit position `pos` set to zero.

[Int] `ibits(i,pos,len)`
Extracts `len` consecutive bits starting at position `pos` and puts them in the low bit positions of the returned value. (The high positions are zero.)

[Int] `ibset(i,pos)`
Returns `i` but with bit position `pos` set to 1.

[Int] `ieor(i,j)`
Bitwise exclusive or.

[Int] `ior(i,j)`
Bitwise logical or.

[Int] `ishft(i,shift)`
Bitwise left shift by `shift` (which may be negative) with zeros shifted in from the other end.

[Int] `ishftc(i,shift)`
Bitwise circularly left shift by `shift` (which may be negative).

[Int] `not(i)`
Bitwise logical complement.

Some Functions Relating to Numerical Representations

[Real] `epsilon(x)`
Smallest nonnegligible quantity relative to 1 in the numerical model of `x`.

[Num] `huge(x)`
Largest representable number in the numerical model of `x`.

[Int] `kind(x)`

Returns the kind value for the numerical model of `x`.

[Real] `nearest(x,s)`
Real number nearest to `x` in the direction specified by the sign of `s`.

[Real] `tiny(x)`
Smallest positive number in the numerical model of `x`.

Other Intrinsic Procedures

[Lgcl] `present(a)`
True, within a subprogram, if an optional argument is actually present, otherwise false.

[Lgcl] `associated(pointer,`*`target`*`)`
True if `pointer` is associated with `target` or (if `target` is absent) with any target, otherwise false.

[Lgcl] `allocated(array)`
True if the allocatable `array` is allocated, otherwise false.

There are some pitfalls in using `associated` and `allocated`, having to do with arrays and pointers that can find themselves in *undefined* status [see §21.5, and also M&R, §3.3 and §6.5.1]. For example, pointers are always "born" in an undefined status, where the `associated` function returns unpredictable values.

For completeness, here is a list of Fortran 90's intrinsic procedures not already mentioned:

Other Numerical Representation Functions: `digits`, `exponent`, `fraction`, `rrspacing`, `scale`, `set_exponent`, `spacing`, `maxexponent`, `minexponent`, `precision`, `radix`, `range`, `selected_int_kind`, `selected_real_kind`.

Lexical comparison: `lge`, `lgt`, `lle`, `llt`.

Character functions: `ichar`, `char`, `achar`, `iachar`, `index`, `adjustl`, `adjustr`, `len_trim`, `repeat`, `scan`, `trim`, `verify`.

Other: `mvbits`, `transfer`, `date_and_time`, `system_clock`, `random_seed`, `random_number`. (We will discuss random numbers in some detail in Chapter B7.)

CITED REFERENCES AND FURTHER READING:

Metcalf, M., and Reid, J. 1996, *Fortran 90/95 Explained* (Oxford: Oxford University Press).

21.5 Advanced Fortran 90 Topics

Pointers, Arrays, and Memory Management

One of the biggest improvements in Fortran 90 over Fortran 77 is in the handling of arrays, which are the cornerstone of many numerical algorithms. In this subsection we will take a closer look at how to use some of these new array features effectively. We will look at how to code certain commonly occurring elements of program design, and we will pay particular attention to avoiding "memory leaks," where — usually inadvertently — we keep cumulatively allocating new storage for an array, every time some piece of code is invoked.

Let's first review some of the rules for using allocatable arrays and pointers to arrays. Recall that a pointer is born with an undefined status. Its status changes to "associated" when you make it refer to a target, and to "disassociated" when you `nullify` the pointer. [M&R, §3.3] You can also use `nullify` on a newly born pointer to change its status from undefined to disassociated; this allows you to test the status with the `associated` inquiry function. [M&R, §6.5.4] (While many compilers will not produce a run-time error if you test an undefined pointer with `associated`, you can't rely on this *laissez-faire* in your programming.)

The initial status of an allocatable array is "not currently allocated." Its status changes to "allocated" when you give it storage with `allocate`, and back to "not currently allocated" when you use `deallocate`. [M&R, §6.5.1] You can test the status with the `allocated` inquiry function. Note that while you can also give a pointer fresh storage with `allocate`, you can't test this with `allocated` — only `associated` is allowed with pointers. Note also that nullifying an allocated pointer leaves its associated storage in limbo. You must instead `deallocate`, which gives the pointer a testable "disassociated" status.

While allocating an array that is already allocated gives an error, you are allowed to allocate a pointer that already has a target. This breaks the old association, and could leave the old target inaccessible if there is no other pointer associated with it. [M&R, §6.5.2] Deallocating an array or pointer that has not been allocated is always an error.

Allocated arrays that are local to a subprogram acquire the "undefined" status on exit from the subprogram unless they have the `SAVE` attribute. (Again, not all compilers enforce this, but be warned!) Such undefined arrays cannot be referenced in any way, so you should explicitly deallocate all allocated arrays that are not saved before returning from a subprogram. [M&R, §6.5.1] The same rule applies to arrays declared in modules that are currently accessed only by the subprogram. While you can reference undefined pointers (e.g., by first nullifying them), it is good programming practice to deallocate explicitly any allocated pointers declared locally before leaving a subprogram or module.

Now let's turn to using these features in programs. The simplest example is when we want to implement global storage of an array that needs to be accessed by two or more different routines, and we want the size of the array to be determined at run time. As mentioned earlier, we implement global storage with a `MODULE` rather than a `COMMON` block. (We ignore here the additional possibility of passing global variables by having one routine CONTAINed within the other.) There are two good ways of handling the dynamical allocation in a MODULE. Method 1 uses an allocatable array:

```
MODULE a
REAL(SP), DIMENSION(:), ALLOCATABLE :: x
END MODULE a

SUBROUTINE b(y)
USE a
REAL(SP), DIMENSION(:) :: y
...
allocate(x(size(y)))
... [other routines using x called here] ...
END SUBROUTINE b
```

Here the global variable x gets assigned storage in subroutine b (in this case, the same as the length of y). The length of y is of course defined in the procedure that calls b. The array x is made available to any other subroutine called by b by including a USE a statement. The status of x can be checked with an allocated inquiry function on entry into either b or the other subroutine if necessary. As discussed above, you must be sure to deallocate x before returning from subroutine b. If you want x to retain its values between calls to b, you add the SAVE attribute to its declaration in a, and *don't* deallocate it on returning from b. (Alternatively, you could put a USE a in your main program, but we consider that bug-prone, since forgetting to do so can create all manner of difficult-to-diagnose havoc.) To avoid allocating x more than once, you test it on entry into b:

```
if (.not. allocated(x)) allocate(x(size(y)))
```

The second way to implement this type of global storage (Method 2) uses a pointer:

```
MODULE a
REAL(SP), DIMENSION(:), POINTER :: x
END MODULE a

SUBROUTINE b(y)
USE a
REAL(SP), DIMENSION(:) :: y
REAL(SP), DIMENSION(size(y)), TARGET :: xx
...
x=>xx
... [other routines using x called here] ...
END SUBROUTINE b
```

Here the *automatic array* xx gets its temporary storage automatically on entry into b, and automatically gets deallocated on exit from b. [M&R, §6.4] The global pointer x can access this storage in any routine with a USE a that is called by b. You can check that things are in order in such a called routine by testing x with associated. If you are going to use x for some other purpose as well, you should nullify it on leaving b so that it doesn't have undefined status. Note that this implementation does not allow values to be saved between calls: You can't SAVE automatic arrays — that's not what they're for. You would have to SAVE x in the module, and allocate it in the subroutine instead of pointing it to a suitable automatic array. But this is essentially Method 1 with the added complication of using a pointer, so Method 1 is simpler when you want to save values. When you don't need to save values between calls, we lean towards Method 2 over Method 1 because we like the automatic allocation and deallocation, but either method works fine.

An example of Method 1 (allocatable array) is in routine `rkdumb` on page 1297. An example of Method 1 with `SAVE` is in routine `pwtset` on p. 1265. Method 2 (pointer) shows up in routines `newt` (p. 1196), `broydn` (p. 1199), and `fitexy` (p. 1286). A variation is shown in routines `linmin` (p. 1211) and `dlinmin` (p. 1212): When the array that needs to be shared is an argument of one of the routines, Method 2 is better.

An extension of these ideas occurs if we allocate some storage for an array initially, but then might need to increase the size of the array later without losing the already-stored values. The function `reallocate` in our utility module `nrutil` will handle this for you, but it expects a pointer argument as in Method 2. Since no automatic arrays are used, you are free to `SAVE` the pointer if necessary. Here is a simple example of how to use `reallocate` to create a workspace array that is local to a subroutine:

```
SUBROUTINE a
USE nrutil, ONLY : reallocate
REAL(SP), DIMENSION(:), POINTER, SAVE :: wksp
LOGICAL(LGT), SAVE :: init=.true.
if (init) then
   init=.false.
   nullify(wksp)
   wksp=>reallocate(wksp,100)
end if
...
if (nterm > size(wksp)) wksp=>reallocate(wksp,2*size(wksp))
...
END SUBROUTINE a
```

Here the workspace is initially allocated a size of 100. If the number of elements used (`nterm`) ever exceeds the size of the workspace, the workspace is doubled. (In a realistic example, one would of course check that the doubled size is in fact big enough.) Fortran 90 experts can note that the `SAVE` on `init` is not strictly necessary: Any local variable that is initialized is automatically saved. [M&R, §7.5]

You can find similar examples of `reallocate` (with some further discussion) in `eulsum` (p. 1070), `hufenc` (p. 1348), and `arcode` (p. 1350). Examples of reallocate used with global variables in modules are in `odeint` (p. 1300) and `ran_state` (p. 1144).

Another situation where we have to use pointers and not allocatable arrays is when the storage is required for components of a derived type, which are not allowed to have the allocatable attribute. Examples are in `hufmak` (p. 1346) and `arcmak` (p. 1349).

Turning away from issues relating to global variables, we now consider several other important programming situations that are nicely handled with pointers. The first case is when we want a subroutine to return an array whose size is not known in advance. Since dummy arguments are not allocatable, we must use a pointer. Here is the basic construction:

```
SUBROUTINE a(x,nx)
REAL(SP), DIMENSION(:), POINTER :: x
INTEGER(I4B), INTENT(OUT) :: nx
LOGICAL(LGT), SAVE :: init=.true.
if (init) then
```

```
      init=.false.
      nullify(x)
   else
      if (associated(x)) deallocate(x)
   end if
   ...
   nx=...
   allocate(x(nx))
   x(1:nx)=...
   END SUBROUTINE a
```

Since the length of x can be found from size(x), it is not absolutely necessary to pass nx as an argument. Note the use of the initial logic to avoid memory leaks. If a higher-level subroutine wants to recover the memory associated with x from the last call to SUBROUTINE a, it can do so by first deallocating it, and then nullifying the pointer. Examples of this structure are in zbrak (p. 1184), period (p. 1258), and fasper (p. 1259). A related situation is where we want a function to return an array whose size is not predetermined, such as in voltra on (p. 1326). The discussion of voltra also explains the potential pitfalls of functions returning pointers to dynamically allocated arrays.

A final useful pointer construction enables us to set up a data structure that is essentially an array of arrays, independently allocatable on each part. We are not allowed to declare an array of pointers in Fortran 90, but we can do this indirectly by defining a derived type that consists of a pointer to the appropriate kind of array. [M&R, §6.11] We can then define a variable that is an allocatable array of the new type. For example,

```
TYPE ptr_to_arr
   REAL(SP), DIMENSION(:), POINTER :: arr
END TYPE
TYPE(ptr_to_arr), DIMENSION(:), ALLOCATABLE :: x
...
allocate(x(n))
...
do i=1,n
   allocate(x(i)%arr(m))
end do
```

sets up a set x of n arrays of length m. See also the example in mglin (p. 1334).

There is a potential problem with dynamical memory allocation that we should mention. The Fortran 90 standard does not require that the compiler perform "garbage collection," that is, it is not required to recover deallocated memory into nice contiguous pieces for reuse. If you enter and exit a subroutine many times, and each time a large chunk of memory gets allocated and deallocated, you could run out of memory with a "dumb" compiler. You can often alleviate the problem by deallocating variables in the reverse order that you allocated them. This tends to keep a large contiguous piece of memory free at the top of the heap.

Scope, Visibility, and Data Hiding

An important principle of good programming practice is *modularization*, the idea that different parts of a program should be insulated from each other as much as possible. An important subcase of modularization is *data hiding*, the principle that actions carried out on variables in one part of the code should not be able to affect the values of variables in other parts of the code. When it is necessary for one "island" of code to communicate with another, the communication should be through a well-defined interface that makes it obvious exactly what communication is taking place, and prevents any other interchange from occurring. Otherwise, different sections of code should not have access to variables that they don't need.

The concept of data hiding extends not only to variables, but also to the names of procedures that manipulate the variables: A program for screen graphics might give the user access to a routine for drawing a circle, but it might "hide" the names (and methods of operation) of the primitive routines used for calculating the coordinates of the points on the circumference. Besides producing code that is easier to understand and to modify, data hiding prevents unintended side effects from producing hard-to-find errors.

In Fortran, the principal language construction that effects data hiding is the use of subroutines. If all subprograms were restricted to have no more than ten executable statements per routine, and to communicate between routines only by an explicit list of arguments, the number of programming errors might be greatly reduced! Unfortunately few tasks can be easily coded in this style. For this and other reasons, we think that too much procedurization is a bad thing; one wants to find the *right* amount. Fortunately Fortran 90 provides several additional tools to help with data hiding.

Global variables and routine names are important, but potentially dangerous, things. In Fortran 90, global variables are typically encapsulated in modules. Access is granted only to routines with an appropriate `USE` statement, and can be restricted to specific identifiers by the `ONLY` option. [M&R, §7.10] In addition, variable and routine names within the module can be designated as `PUBLIC` or `PRIVATE` (see, e.g., `quad3d` on p. 1065). [M&R, §7.6]

The other way global variables get communicated is by having one routine `CONTAINed` within another. [M&R, §5.6] This usage is potentially lethal, however, because *all* the outer routine's variables are visible to the inner routine. You can try to control the problem somewhat by passing some variables back and forth as arguments of the inner routine, but that still doesn't prevent inadvertent side effects. (The most common, and most stupid, is inadvertent reuse of variables named `i` or `j` in the `CONTAINed` routine.) Also, a long list of arguments reduces the convenience of using an internal routine in the first place. We advise that internal subprograms be used with caution, and only to carry out simple tasks.

There are some good ways to use `CONTAINS`, however. Several of our recipes have the following structure: A principal routine is invoked with several arguments. It calls a subsidiary routine, which needs to know some of the principal routine's arguments, some global variables, and some values communicated directly as arguments to the subsidiary routine. In Fortran 77, we have usually coded this by passing the global variables in a `COMMON` block and all other variables as arguments to the subsidiary routine. If necessary, we copied the arguments of the primary routine before passing them to the subsidiary routine. In Fortran 90, there is a more elegant way of accomplishing this, as follows:

```
SUBROUTINE recipe(arg)
REAL(SP) :: arg
REAL(SP) :: global_var
call recipe_private
CONTAINS
```

```
SUBROUTINE recipe_private
...
call subsidiary(local_arg)
...
END SUBROUTINE recipe_private
SUBROUTINE subsidiary(local_arg)
...
END SUBROUTINE subsidiary
END SUBROUTINE recipe
```

Notice that the principal routine (recipe) has practically nothing in it — only declarations of variables intended to be visible to the subsidiary routine (subsidiary). All the real work of recipe is done in recipe_private. This latter routine has visibility on all of recipe's variables, while any additional variables that recipe_private defines are *not* visible to subsidiary — which is the whole purpose of this way of organizing things. Obviously arg and global_var can be much more general data types than the example shown here, including function names. For examples of this construction, see amoeba (p. 1208), amebsa (p. 1222), mrqmin (p. 1292), and medfit (p. 1294).

Recursion

A subprogram is recursive if it calls itself. While forbidden in Fortran 77, recursion is allowed in Fortran 90. [M&R, §5.16–§5.17] You must supply the keyword RECURSIVE in front of the FUNCTION or SUBROUTINE keyword. In addition, if a FUNCTION calls itself directly, as opposed to calling another subprogram that in turn calls it, you must supply a variable to hold the result with the RESULT keyword. Typical syntax for this case is:

```
RECURSIVE FUNCTION f(x) RESULT(g)
REAL(SP) :: x,g
if ...
   g=...
else
   g=f(...)
end if
END FUNCTION f
```

When a function calls itself directly, as in this example, there always has to be a "base case" that does not call the function; otherwise the recursion never terminates. We have indicated this schematically with the if...else...end if structure.

On serial machines we tend to avoid recursive implementations because of the additional overhead they incur at execution time. Occasionally there are algorithms for which the recursion overhead is relatively small, and the recursive implementation is simpler than an iterative version. Examples in this book are quad_3d (p. 1065), miser (p. 1164), and mglin (p. 1334). Recursion is much more important when parallelization is the goal. We will encounter in Chapter 22 numerous examples of algorithms that can be parallelized with recursion.

SAVE Usage Style

A quirk of Fortran 90 is that any variable with initial values acquires the SAVE attribute automatically. [M&R, §7.5 and §7.9] As a help to understanding an algorithm, we have elected to put an explicit SAVE on all variables that really do need to retain their values between calls to a routine. We do this even if it is redundant because the variables are initialized. Note that we generally prefer to assign initial values with initialization expressions rather than with DATA statements. We reserve DATA statements for cases where it is convenient to use the repeat count feature to set multiple occurrences of a value, or when binary, octal, or hexadecimal constants are used. [M&R, §2.6.1]

Named Control Structures

Fortran 90 allows control structures such as do loops and if blocks to be named. [M&R, §4.3–§4.5] Typical syntax is

```
name:do i=1,n
   ...
end do name
```

One use of naming control structures is to improve readability of the code, especially when there are many levels of nested loops and if blocks. A more important use is to allow exit and cycle statements, which normally refer to the innermost do loop in which they are contained, to transfer execution to the end of some outer loop. This is effected by adding the name of the outer loop to the statement: exit name or cycle name.

There is great potential for misuse with named control structures, since they share some features of the much-maligned goto. We recommend that you use them sparingly. For a good example of their use, contrast the Fortran 77 version of simplx with the Fortran 90 version on p. 1216.

CITED REFERENCES AND FURTHER READING:

Metcalf, M., and Reid, J. 1996, *Fortran 90/95 Explained* (Oxford: Oxford University Press).

21.6 And Coming Soon: Fortran 95

One of the more positive effects of Fortran 90's long gestation period has been the general recognition, both by the X3J3 committee and by the community at large, that Fortran needs to evolve over time. Indeed, as we write, the process of bringing forth a minor, but by no means insignificant, updating of Fortran 90 — named Fortran 95 — is well under way.

Fortran 95 will differ from Fortran 90 in about a dozen features, only a handful of which are of any importance to this book. Generally these are extensions that will make programming, especially parallel programming, easier. In this section we give a summary of the anticipated language changes. In §22.1 and §22.5 we will comment further on the implications of Fortran 95 to some parallel programming tasks; in §23.7 we comment on what differences Fortran 95 will make to our nrutil utility functions.

No programs in Chapters B1 through B20 of this book edition use any Fortran 95 extensions.

FORALL Statements and Blocks

Fortran 95 introduces a new forall control structure, somewhat akin to the where construct, but allowing for greater flexibility. It is something like a do-loop, but with the proviso that the indices looped over are allowed to be done in any order (ideally, in parallel). The forall construction comes in both single-statement and block variants. Instead of using the do-loop's comma-separated triplets of lower-value, upper-value, and increment, it borrows its syntax from the colon-separated form of array sections. Some examples will give you the idea.

Here is a simple example that could alternatively be done with Fortran 90's array sections and transpose intrinsic:

```
forall (i=1:20, j=1:10:2) x(i,j)=y(j,i)
```

The block form allows more than one executable statement:

```
forall (i=1:20, j=1:10:2)
   x(i,j)=y(j,i)
   z(i,j)=y(i,j)**2
end forall
```

Here is an example that cannot be done with Fortran 90 array sections:

```
forall (i=1:20, j=1:20) a(i,j)=3*i+j**2
```

forall statements can also take optional masks that restrict their action to a subset of the loop index combinations:

```
forall (i=1:100, j=1:100, (i>=j .and. x(i,j)/=0.0) ) x(i,j)=1.0/x(i,j)
```

forall constructions can be nested, or nested inside where blocks, or have where constructions inside them. An additional new feature in Fortran 95 is that where blocks can themselves be nested.

PURE Procedures

Because the inside iteration of a forall block can be done in any order, or in parallel, there is a logical difficulty in allowing functions or subroutines inside such blocks: If the function or subroutine has *side effects* (that is, if it changes any data elsewhere in the machine, or in its own saved variables) then the result of a forall calculation could depend on the order in which the iterations happen to be done. This can't be tolerated, of course; hence a new PURE attribute for subprograms.

While the exact stipulations are somewhat technical, the basic idea is that if you declare a function or subroutine as PURE, with a syntax like,

```
PURE FUNCTION myfunc(x,y,z)
```

or

```
PURE SUBROUTINE mysub(x,y,z)
```

then you are guaranteeing to the compiler (and it will enforce) that the only values changed by mysub or myfunc are returned function values, subroutine arguments with the INTENT(OUT) attribute, and automatic (scratch) variables within the procedure.

You can then use your pure procedures within forall constructions. Pure functions are also allowed in some specification statements.

ELEMENTAL Procedures

Fortran 95 removes Fortran 90's nagging restriction that only intrinsic functions are elemental. The way this works is that you write a pure procedure that operates on scalar values, but include the attribute ELEMENTAL (which automatically implies PURE). Then, as long as the function has an explicit interface in the referencing program, you can call it with any shape of argument, and it will act elementally. Here's an example:

```
ELEMENTAL FUNCTION myfunc(x,y,z)
REAL :: x,y,z,myfunc
...
myfunc = ...
END
```

In a program with an explicit interface for myfunc you could now have

```
REAL, DIMENSION(10,20) :: x,y,z,w
...
w=myfunc(x,y,z)
```

Pointer and Allocatable Improvements

Fortran 95, unlike Fortran 90, requires that any allocatable variables (except those with SAVE attributes) that are allocated within a subprogram be automatically deallocated by the compiler when the subprogram is exited. This will remove Fortran 90's "undefined allocation status" bugaboo.

Fortran 95 also provides a method for pointer variables to be born with disassociated association status, instead of the default (and often inconvenient) "undefined" status. The syntax is to add an initializing => NULL() to the declaration, as:

```
REAL, DIMENSION(:,:), POINTER :: mypoint => NULL()
```

This does not, however, eliminate the possibility of undefined association status, because you have to remember to use the null initializer if want your pointer to be disassociated.

Some Other Fortran 95 Features

In Fortran 95, maxloc and minloc have the additional optional argument DIM, which causes them to act on all one-dimensional sections that span through the named dimension. This provides a means for getting the locations of the values returned by the corresponding functions maxval and minval in the case that their DIM argument is present.

The sign intrinsic can now distinguish a negative from a positive real zero value: sign(2.0,-0.0) is -2.0.

There is a new intrinsic subroutine cpu_time(time) that returns as a real value time a process's elapsed CPU time.

There are some minor changes in the namelist facility, in defining minimum field widths for the I, B, O, Z, and F edit descriptors, and in resolving minor conflicts with some other standards.

Chapter 22. Introduction to Parallel Programming

22.0 Why Think Parallel?

In recent years we Numerical Recipes authors have increasingly become convinced that a certain revolution, cryptically denoted by the words "parallel programming," is about to burst forth from its gestation and adolescence in the community of supercomputer users, and become the mainstream methodology for all computing.

Let's review the past: Take a screwdriver and open up the computer (workstation or PC) that sits on your desk. (Don't blame us if this voids your warranty; and be sure to unplug it first!) Count the integrated circuits — just the bigger ones, with more than a million gates (transistors). As we write, in 1995, even lowly memory chips have one or four million gates, and this number will increase rapidly in coming years. You'll probably count at least dozens, and often hundreds, of such chips in your computer.

Next ask, how many of these chips are CPUs? That is, how many implement von Neumann processors capable of executing arbitrary, stored program code? For most computers, in 1995, the answer is: about one. A significant number of computers do have secondary processors that offload input-output and/or video functions. So, two or three is often a more accurate answer, but only one is usually under the user's direct control.

Why do our desktop computers have dozens or hundreds of memory chips, but most often only one (user-accessible) CPU? Do CPU chips intrinsically cost more to manufacture? No. Are CPU chips more expensive than memory chips? Yes, primarily because fixed development and design costs must be distributed over a smaller number of units sold. We have been in a kind of economic equilibrium: CPU's are relatively expensive because there is only one per computer; and there is only one per computer, because they are relatively expensive.

Stabilizing this equilibrium has been the fact that there has been no standard, or widely taught, methodology for parallel programming. Except for the special case of scientific computing on supercomputers (where large problems often have a regular or geometric character), it is not too much of an exaggeration to say that nobody *really knows how* to program multiprocessor machines. Symmetric multiprocessor operating systems, for example, have been very slow in developing; and efficient, parallel methodologies for query-serving on large databases are even now a subject of continuing research.

However, things are now changing. We consider it an easy prognostication that, by the first years of the new century, the typical desktop computer will have 4 to 8 user-accessible CPUs; ten years after that, the typical number will be between 16 and 512. It is not coincidence that these numbers are characteristic of supercomputers (including some quite different architectures) in 1995. The rough rule of ten years' lag from supercomputer to desktop has held firm for quite some time now.

Scientists and engineers have the advantage that techniques for parallel computation in their disciplines *have* already been developed. With multiprocessor workstations right around the corner, we think that now is the right time for scientists and engineers who use computers to start *thinking parallel*. We don't mean that you should put an axe through the screen of your fast serial (single-CPU) workstation. We do mean, however, that you should start programming somewhat differently on that workstation, indeed, start thinking a bit differently about the way that you approach numerical problems in general.

In this volume of *Numerical Recipes in Fortran*, our pedagogical goal is to show you that there are conceptual and practical benefits in parallel thinking, even if you are using a serial machine today. These benefits include conciseness and clarity of code, reusability of code in wider contexts, and (not insignificantly) increased portability of code to today's parallel supercomputers. Of course, on parallel machines, either supercomputers today or desktop machines tomorrow, the benefits of thinking parallel are much more tangible: They translate into significant improvements in efficiency and computational capability.

Thinking Parallel with Fortran 90

Until very recently, a strong inhibition to thinking parallel was the lack of any standard, architecture-independent, computer language in which to think. That has changed with the finalization of the Fortran 90 language standard, and with the availability of good, optimizing Fortran 90 compilers on a variety of platforms.

There is a significant body of opinion (with which we, however, disagree) that there is no such thing as architecture-independent parallel programming. Proponents of this view, who are generally committed wizards at programming on one or another particular architecture, point to the fact that algorithms that are optimized to one architecture can run hundreds of times more slowly on other architectures. And, they are correct!

Our opposing point of view is one of pragmatism. We think that it is not hard to learn, in a general way, what kinds of architectures are in general use, and what kinds of parallel constructions work well (or poorly) on each kind. With this knowledge (much of which we hope to develop in this book) the user can, we think, write good, general-purpose parallel code that works on a variety of architectures — including, importantly, on purely serial machines. Equally important, the user will be aware of when certain parts of a code can be significantly improved on some, but not other, architectures.

Fortran 90 is a good test-bench for this point of view. It is not the perfect language for parallel programming. But it is *a* language, and it is the only cross-platform *standard* language now available. The committee that developed the language between 1978 and 1991 (known technically as X3J3) had strong representation from both a traditional "vectorization" viewpoint (e.g., from the Cray XMP and YMP series of computers), and also from the "data parallel" or "SIMD" viewpoints of parallel machines like the CM-2 and CM-5 from Thinking Machines, Inc. Language compromises were made, and a few (in our view) almost essential features were left out (see §22.5). But, by and large, the necessary tools are there: If you learn to think parallel in Fortran 90, you will easily be able to transfer the skill to future parallel standards, whether they are Fortran-based, C-based, or other.

CITED REFERENCES AND FURTHER READING:
Metcalf, M., and Reid, J. 1996, *Fortran 90/95 Explained* (Oxford: Oxford University Press).

22.1 Fortran 90 Data Parallelism: Arrays and Intrinsics

The underlying model for parallel computation in Fortran 90 is *data parallelism*, implemented by the use of arrays of data, and by the provision of operations and intrinsic functions that act on those arrays in parallel, in a manner optimized by the compiler for each particular hardware architecture. We will not try to draw a fine definitional distinction between "data parallelism" and so-called SIMD (single instruction multiple data) programming. For our purposes the two terms mean about the same thing: The programmer writes a single operation, "+" say, and the compiler causes it to be carried out on multiple pieces of data in as parallel a manner as the underlying hardware allows.

Any kind of parallel computing that is not SIMD is generally called MIMD (multiple instruction multiple data). A parallel programming language with MIMD features might allow, for example, several different subroutines — acting on different parts of the data — to be called into execution simultaneously. Fortran 90 has few, if any, MIMD constructions. A Fortran 90 compiler might, on some machines, execute MIMD code in implementing some Fortran 90 intrinsic functions (`pack` or `unpack`, e.g.), but this will be hidden from the Fortran 90 user. Some extensions of Fortran 90, like HPF, do implement MIMD features explicitly; but we will not consider these in this book. Fortran 95's `forall` and `PURE` extensions (see §21.6) will allow some significantly greater access to MIMD features (see §22.5).

Array Parallel Operations

We have already met the most basic, and most important, parallel facility of Fortran 90, namely, the ability to use whole arrays in expressions and assignments, with the indicated operations being effected in parallel across the array. Suppose, for example, we have the two-dimensional matrices `a`, `b`, and `c`,

```
REAL, DIMENSION(30,30) :: a,b,c
```

Then, instead of the serial construction,

```
do j=1,30
   do k=1,30
      c(j,k)=a(j,k)+b(j,k)
   end do
end do
```

which is of course perfectly valid Fortran 90 code, we can simply write

```
c=a+b
```

The compiler deduces from the declaration statement that a, b, and c are matrices, and what their bounding dimensions are.

Let us dwell for a moment on the conceptual differences between the serial code and parallel code for the above matrix addition. Although one is perhaps used to seeing the nested do-loops as simply an idiom for "do-the-enclosed-on-all-components," it in fact, according to the rules of Fortran, specifies a very particular time-ordering for the desired operations. The matrix elements are added by rows, in order (j=1,30), and within each row, by columns, in order (k=1,30).

In fact, the serial code above *overspecifies* the desired task, since it is guaranteed by the laws of mathematics that the order in which the element operations are done is of no possible relevance. Over the 50 year lifetime of serial von Neuman computers, we programmers have been brainwashed to break up all problems into single executable streams *in the time dimension only*. Indeed, the major design problem for supercomputer compilers for the last 20 years has been to *undo* such serial constructions and recover the underlying "parallel thoughts," for execution in vector or parallel processors. Now, rather than taking this expensive detour into and out of serial-land, we are asked simply to say what we mean in the first place, c=a+b.

The essence of parallel programming is *not* to force "into the time dimension" (i.e., to serialize) operations that naturally extend across a span of data, that is, "in the space dimension." If it were not for 50-year-old collective habits, and the languages designed to support them, parallel programming would probably strike us as more natural than its serial counterpart.

Broadcasts and Dimensional Expansion: SSP vs. MMP

We have previously mentioned the Fortran 90 rule that a scalar variable is conformable with any shape array. Thus, we can implement a calculation such as

$$y_i = x_i + s, \qquad i = 1, \ldots, n \tag{22.1.1}$$

with code like

```
y=x+s
```

where we of course assume previous declarations like

```
REAL(SP) :: s
REAL(SP), DIMENSION(n) :: x,y
```

with n a compile-time constant or dummy argument. (Hereafter, we will omit the declarations in examples that are this simple.)

This seemingly simple construction actually hides an important underlying parallel capability, namely, that of *broadcast*. The sums in y=x+s are done in parallel on different CPUs, each CPU accessing different components of x and y. Yet, they all must access the same scalar value s. If the hardware has local memory for each CPU, the value of s must be replicated and transferred to each CPU's local memory. On the other hand, if the hardware implements a single, global memory space, it is vital to do something that mitigates the traffic jam potentially caused by all the CPUs trying to access the same memory location at the same time. (We will use the term "broadcast" to refer equally to both cases.) Although hidden from the user, Fortran 90's ability to do broadcasts is an essential feature of it as a parallel language.

Broadcasts can be more complicated than the above simple example. Consider, for example, the calculation

$$w_i = \sum_{j=1}^{n} |x_i + x_j|, \qquad i = 1, \ldots, n \tag{22.1.2}$$

Here, we are doing n^2 operations: For each of n values of i there is a sum over n values of j.

Serial code for this calculation might be

```
do i=1,n
   w(i)=0.
   do j=1,n
      w(i)=w(i)+abs(x(i)+x(j))
   end do
end do
```

The obvious immediate parallelization in Fortran 90 uses the sum intrinsic function to eliminate the inner do-loop. This would be a suitable amount of parallelization for a small-scale parallel machine, with a few processors:

```
do i=1,n
   w(i)=sum(abs(x(i)+x))
end do
```

Notice that the conformability rule implies that a new value of x(i), a scalar, is being broadcast to all the processors involved in the abs and sum, with each iteration of the loop over i.

What about the outer do-loop? Do we need, or want, to eliminate it, too? That depends on the architecture of your computer, and on the tradeoff between time and memory in your problem (a common feature of all computing, no less so parallel computing). Here is an implementation that is free of all do-loops, in principle capable of being executed in a small number (independent of n) of parallel operations:

```
REAL(SP), DIMENSION(n,n) :: a
...
a = spread(x,dim=2,ncopies=n)+spread(x,dim=1,ncopies=n)
w = sum(abs(a),dim=1)
```

This is an example of what we call *dimensional expansion*, as implemented by the spread intrinsic. Although the above may strike you initially as quite a cryptic construction, it is easy to learn to read it. In the first assignment line, a matrix is constructed with all possible values of x(i)+x(j). In the second assignment line, this matrix is collapsed back to a vector by applying the sum operation to the absolute value of its elements, across one of its dimensions.

More explicitly, the first line creates a matrix `a` by adding two matrices each constructed via `spread`. In `spread`, the `dim` argument specifies which argument is *duplicated*, so that the first term *varies* across its first (row) dimension, and vice versa for the second term:

$$a_{ij} = x_i + x_j$$

$$= \begin{pmatrix} x_1 & x_1 & x_1 & \cdots \\ x_2 & x_2 & x_2 & \cdots \\ x_3 & x_3 & x_3 & \cdots \\ \vdots & \vdots & \vdots & \ddots \end{pmatrix} + \begin{pmatrix} x_1 & x_2 & x_3 & \cdots \\ x_1 & x_2 & x_3 & \cdots \\ x_1 & x_2 & x_3 & \cdots \\ \vdots & \vdots & \vdots & \ddots \end{pmatrix} \tag{22.1.3}$$

Since equation (22.1.2) above is symmetric in i and j, it doesn't really matter what value of `dim` we put in the `sum` construction, but the value `dim=1` corresponds to summing across the rows, that is, down each column of equation (22.1.3).

Be sure that you understand that the `spread` construction changed an $O(n)$ memory requirement into an $O(n^2)$ one! If your values of n are large, this is an impossible burden, and the previous implementation with a single do-loop remains the only practical one. On the other hand, if you are working on a massively parallel machine, whose number of processors is comparable to n^2 (or at least much larger than n), then the `spread` construction, and the underlying broadcast capability that it invokes, leads to a big win: All n^2 operations can be done in parallel. This distinction between small-scale parallel machines — which we will hereafter refer to as *SSP machines* — and massively multiprocessor machines — which we will refer to as *MMP machines* — is an important one. A main goal of parallelism is to saturate the available number of processors, and algorithms for doing so are often different in the SSP and MMP opposite limits. Dimensional expansion is one method for saturating processors in the MMP case.

Masks and "Index Loss"

An instructive extension of the above example is the following case of a product that omits one term (the diagonal one):

$$w_i = \prod_{\substack{j=1 \\ j \neq i}}^{n} (x_j - x_i), \qquad i = 1, \ldots, n \tag{22.1.4}$$

Formulas like equation (22.1.4) frequently occur in the context of interpolation, where all the x_i's are known to be distinct, so let us for the moment assume that this is the case.

Serial code for equation (22.1.4) could be

```
do i=1,n
   w(i)=1.0_sp
   do j=1,n
      if (j /= i) w(i)=w(i)*(x(j)-x(i))
   end do
end do
```

Parallel code for SSP machines, or for large enough n on MMP machines, could be

```
do i=1,n
   w(i)=product( x-x(i), mask=(x/=x(i)) )
end do
```

Here, the `mask` argument in the `product` intrinsic function causes the diagonal term to be omitted from the product, as we desire. There are some features of this code, however, that bear commenting on.

First, notice that, according to the rules of conformability, the expression `x/=x(i)` broadcasts the scalar `x(i)` and generates a logical array of length n, suitable for use as a `mask` in the `product` intrinsic. It is quite common in Fortran 90 to generate masks "on the fly" in this way, particularly if the mask is to be used only once.

Second, notice that the `j` index has disappeared completely. It is now implicit in the two occurrences of `x` (equivalent to `x(1:n)`) on the right-hand side. With the disappearance of the `j` index, we also lose the ability to do the test on `i` and `j`, but must use, in essence, `x(i)` and `x(j)` instead! That is a very general feature in Fortran 90: when an operation is done in parallel across an array, there is *no associated index* available within the operation. This "index loss," as we will see in later discussion, can sometimes be quite an annoyance.

A language construction present in CM [Connection Machine] Fortran, the so-called `forall`, which would have allowed access to an associated index in many cases, was eliminated from Fortran 90 by the X3J3 committee, in a controversial decision. Such a construction will come into the language in Fortran 95.

What about code for an MMP machine, where we are willing to use dimensional expansion to achieve greater parallelism? Here, we can write,

```
a = spread(x,dim=2,ncopies=n)-spread(x,dim=1,ncopies=n)
w = product(a,dim=1,mask=(a/=0.))
```

This time it does matter that the value of `dim` in the `product` intrinsic is 1 rather than 2. If you write out the analog of equation (22.1.3) for the present example, you'll see that the above fragment is the right way around. The problem of index loss is still with us: we have to construct a mask from the array `a`, not from its indices, *both* of which are now lost to us!

In most cases, there are workarounds (more, or less, awkward as they may be) for the problem of index loss. In the worst cases, which are quite rare, you have to create objects to hold, and thus bring back into play, the lost indices. For example,

```
INTEGER(I4B), DIMENSION(n) :: jj
...
jj = (/ (i,i=1,n) /)
do i=1,n
   w(i)=product( x-x(i), mask=(jj/=i) )
end do
```

Now the array `jj` is filled with the "lost" j index, so that it is available for use in the mask. A similar technique, involving spreads of `jj`, can be used in the above MMP code fragment, which used dimensional expansion. (Fortran 95's `forall` construction will make index loss much less of a problem. See §21.6.)

Incidentally, the above Fortran 90 construction, `(/ (i,i=1,n) /)`, is called an *array constructor with implied do list*. For reasons to be explained in §22.2, we almost never use this construction, in most cases substituting a Numerical Recipes utility function for generating arithmetical progressions, which we call `arth`.

Interprocessor Communication Costs

It is both a blessing and a curse that Fortran 90 completely hides from the user the underlying machinery of interprocessor communication, that is, the way that data values computed by (or stored locally near) one CPU make their way to a different CPU that might need them next. The blessing is that, by and large, the Fortran 90 programmer need not be concerned with how this machinery works. If you write

```
a(1:10,1:10) = b(1:10,1:10) + c(10:1:-1,10:1:-1)
```

the required upside-down-and-backwards values of the array `c` are just *there*, no matter that a great deal of routing and switching may have taken place. An ancillary blessing is that this book, unlike so many other (more highly technical) books on parallel programming (see references below) need not be filled with complex and subtle discussions of CPU connectivity, topology, routing algorithms, and so on.

The curse is, just as you might expect, that the Fortran 90 programmer can't control the interprocessor communication, even when it is desirable to do so. A few regular communication patterns are "known" to the compiler through Fortran 90 intrinsic functions, for example `b=transpose(a)`. These, presumably, are done in an optimal way. However, many other regular patterns of communication, which might also allow highly optimized implementations, don't have corresponding intrinsic functions. (An obvious example is the "butterfly" pattern of communication that occurs in fast Fourier transforms.) These, if coded in Fortran 90 by using general vector subscripts (e.g., `barr=arr(iarr)` or `barr(jarr)=arr`, where `iarr` and `jarr` are integer arrays), lose all possibility of being optimized. The compiler can't distinguish a communication step with regular structure from one with general structure, so it must assume the worst case, potentially resulting in very slow execution.

About the only thing a Fortran 90 programmer can do is to start with a general awareness of the kind of apparently parallel constructions that *might* be quite slow on his/her parallel machine, and then to refine that awareness by actual experience and experiment. Here is our list of constructions most likely to cause interprocessor communication bottlenecks:

- vector subscripts, like `barr=arr(iarr)` or `barr(jarr)=arr` (that is, general gather/scatter operations)
- the `pack` and `unpack` intrinsic functions
- mixing positive strides and negative strides in a single expression (as in the above `b(1:10,1:10)+c(10:1:-1,10:1:-1)`)
- the `reshape` intrinsic when used with the `order` argument
- possibly, the `cshift` and `eoshift` extrinsics, especially for nonsmall values of the shift.

On the other hand, the fact is that these constructions *are* parallel, and *are* there for you to use. If the alternative to using them is strictly serial code, you should almost always give them a try.

Linear Algebra

You should be alert for opportunities to use combinations of the `matmul`, `spread`, and `dot_product` intrinsics to perform complicated linear algebra calculations. One useful intrinsic that is not provided in Fortran 90 is the *outer product* of two vectors,

$$c_{ij} = a_i b_j \tag{22.1.5}$$

We already know how to implement this (cf. equation 22.1.3):

```
c = spread(a,dim=2,ncopies=size(b))*spread(b,dim=1,ncopies=size(a))
```

In fact, this operation occurs frequently enough to justify making it a utility function, outerprod, which we will do in Chapter 23. There we also define other "outer" operations between vectors, where the multiplication in the outer product is replaced by another binary operation, such as addition or division.

Here is an example of using these various functions: Many linear algebra routines require that a submatrix be updated according to a formula like

$$a_{jk} = a_{jk} + b_i a_{ji} \sum_{p=i}^{m} a_{pi} a_{pk}, \qquad j = i, \dots, m, \quad k = l, \dots, n \tag{22.1.6}$$

where i, m, l, and n are fixed values. Using an array slice like a(:,i) to turn a_{pi} into a vector indexed by p, we can code the sum with a matmul, yielding a vector indexed by k:

```
temp(l:n)=b(i)*matmul(a(i:m,i),a(i:m,l:n))
```

Here we have also included the multiplication by b_i, a scalar for fixed i. The vector temp, along with the vector a_{ji} = a(:,i), is then turned into a matrix by the outerprod utility and used to increment a_{jk}:

```
a(i:m,l:n)=a(i:m,l:n)+outerprod(a(i:m,i),temp(l:n))
```

Sometimes the update formula is similar to (22.1.6), but with a slight permutation of the indices. Such cases can be coded as above if you are careful about the order of the quantities in the matmul and the outerprod.

CITED REFERENCES AND FURTHER READING:

Akl, S.G. 1989, *The Design and Analysis of Parallel Algorithms* (Englewood Cliffs, NJ: Prentice Hall).

Bertsekas, D.P., and Tsitsiklis, J.N. 1989, *Parallel and Distributed Computation: Numerical Methods* (Englewood Cliffs, NJ: Prentice Hall).

Carey, G.F. 1989, *Parallel Supercomputing: Methods, Algorithms, and Applications* (New York: Wiley).

Fountain, T.J. 1994, *Parallel Computing: Principles and Practice* (New York: Cambridge University Press).

Golub, G., and Ortega, J.M. 1993, *Scientific Computing: An Introduction with Parallel Computing* (San Diego, CA: Academic Press).

Fox, G.C., et al. 1988, *Solving Problems on Concurrent Processors*, Volume I (Englewood Cliffs, NJ: Prentice Hall).

Hockney, R.W., and Jesshope, C.R. 1988, *Parallel Computers 2* (Bristol and Philadelphia: Adam Hilger).

Kumar, V., et al. 1994, *Introduction to Parallel Computing: Design and Analysis of Parallel Algorithms* (Redwood City, CA: Benjamin/Cummings).

Lewis, T.G., and El-Rewini, H. 1992, *Introduction to Parallel Computing* (Englewood Cliffs, NJ: Prentice Hall).

Modi, J.J. 1988, *Parallel Algorithms and Matrix Computation* (New York: Oxford University Press).
Smith, J.R. 1993, *The Design and Analysis of Parallel Algorithms* (New York: Oxford University Press).
Van de Velde, E. 1994, *Concurrent Scientific Computing* (New York: Springer-Verlag).

22.2 Linear Recurrence and Related Calculations

We have already seen that Fortran 90's *array constructor with implied do list* can be used to generate simple series of integers, like `(/ (i,i=1,n) /)`. Slightly more generally, one might want to generate an arithmetic progression, by the formula

$$v_j = b + (j-1)a, \qquad j = 1, \ldots, n \tag{22.2.1}$$

This is readily coded as

```
v(1:n) = (/ (b+(j-1)*a, j=1,n) /)
```

Although it is concise, and valid, *we don't like this coding.* The reason is that it violates the fundamental rule of "thinking parallel": it turns a parallel operation across a data vector into a serial do-loop over the components of that vector. Yes, we know that the compiler might be smart enough to generate parallel code for implied do lists; but it also might *not* be smart enough, here or in more complicated examples.

Equation (22.2.1) is also the simplest example of a *linear recurrence relation.* It can be rewritten as

$$v_1 = b, \qquad v_j = v_{j-1} + a, \qquad j = 2, \ldots, n \tag{22.2.2}$$

In this form (assuming that, in more complicated cases, one doesn't know an explicit solution like equation 22.2.1) one can't write an explicit array constructor. Code like

```
v(1) = b
v(2:n) = (/ (v(j-1)+a,j=2,n) /)        ! wrong
```

is legal Fortran 90 syntax, but illegal semantics; it does *not* do the desired recurrence! (The rules of Fortran 90 require that all the components of v on the right-hand side be evaluated before any of the components on the left-hand side are set.) Yet, as we shall see, techniques for accomplishing the evaluation in parallel are available.

With this as our starting point, we now survey some particular tricks of the (parallel) trade.

Subvector Scaling: Arithmetic and Geometric Progressions

For explicit arithmetic progressions like equation (22.2.1), the simplest parallel technique is *subvector scaling* [1]. The idea is to work your way through the desired vector in larger and larger parallel chunks:

$$
\begin{aligned}
v_1 &= b \\
v_2 &= b + a \\
v_{3\ldots4} &= v_{1\ldots2} + 2a \\
v_{5\ldots8} &= v_{1\ldots4} + 4a \\
v_{9\ldots16} &= v_{1\ldots8} + 8a
\end{aligned} \tag{22.2.3}
$$

And so on, until you reach the length of your vector. (The last step will not necessarily go all the way to the next power of 2, therefore.) The powers of 2, times a, can of course be obtained by successive doublings, rather than the explicit multiplications shown above.

You can see that subvector scaling requires about $\log_2 n$ parallel steps to process a vector of length n. Equally important for serial machines, or SSP machines, the scalar operation count for subvector scaling is no worse than entirely serial code: each new component v_i is produced by a single addition.

If addition is replaced by multiplication, the identical algorithm will produce geometric progressions, instead of arithmetic progressions. In Chapter 23, we will use subvector scaling to implement our utility functions `arth` and `geop` for these two progressions. (You can then call one of these functions instead of recoding equation 22.2.3 every time you need it.)

Vector Reduction: Evaluation of Polynomials

Logically related to subvector scaling is the case where a calculation can be parallelized across a vector that *shrinks* by a factor of 2 in each iteration, until a desired *scalar* result is reached. A good example of this is the parallel evaluation of a polynomial [2]

$$
P(x) = \sum_{j=0}^{N} c_j x^j \tag{22.2.4}
$$

For clarity we take the special case of $N = 5$. Start with the vector of coefficients (imagining appended zeros, as shown):

$$
c_0, \quad c_1, \quad c_2, \quad c_3, \quad c_4, \quad c_5, \quad 0, \quad \ldots
$$

Now, add the elements by pairs, multiplying the second of each pair by x:

$$
c_0 + c_1 x, \quad c_2 + c_3 x, \quad c_4 + c_5 x, \quad 0, \quad \ldots
$$

Now, the same operation, but with the multiplier x^2:

$$
(c_0 + c_1 x) + (c_2 + c_3 x) x^2, \quad (c_4 + c_5 x) + (0) x^2, \quad 0, \quad \ldots
$$

And a final time, with multiplier x^4:

$$[(c_0 + c_1 x) + (c_2 + c_3 x)x^2] + [(c_4 + c_5 x) + (0)x^2]x^4, \quad 0, \quad \dots$$

We are left with a vector of (active) length 1, whose value is the desired polynomial evaluation. (You can see that the zeros are just a bookkeeping device for taking account of the case where the active subvector has odd length.) The key point is that the combining by pairs is a parallel operation at each stage.

As in subvector scaling, there are about $\log_2 n$ parallel stages. Also as in subvector scaling, our total operations count is only negligibly different from purely scalar code: We do one add and one multiply for each original coefficient c_j. The only extra operations are $\log_2 n$ successive squarings of x; but this comes with the extra benefit of better roundoff properties than the standard scalar coding. In Chapter 23 we use vector reduction to implement our utility function `poly` for polynomial evaluation.

Recursive Doubling: Linear Recurrence Relations

Please don't confuse our use of the word "recurrence" (as in "recurrence relation," "linear recurrence," or equation 22.2.2) with the words "recursion" and "recursive," which both refer to the idea of a subroutine calling itself to obtain an efficient or concise algorithm. There are ample grounds for confusion, because recursive algorithms are in fact a good way of obtaining parallel solutions to linear recurrence relations, as we shall now see!

Consider the general first order linear recurrence relation

$$u_j = a_j + b_{j-1} u_{j-1}, \qquad j = 2, 3, \dots, n \tag{22.2.5}$$

with initial value $u_1 = a_1$. On a serial machine, we evaluate such a recurrence with a simple do-loop. To parallelize the recurrence, we can employ the powerful general strategy of *recursive doubling*. Write down equation (22.2.5) for $2j$ and for $2j - 1$:

$$u_{2j} = a_{2j} + b_{2j-1} u_{2j-1} \tag{22.2.6}$$

$$u_{2j-1} = a_{2j-1} + b_{2j-2} u_{2j-2} \tag{22.2.7}$$

Substitute equation (22.2.7) in equation (22.2.6) to eliminate u_{2j-1} and get

$$u_{2j} = (a_{2j} + a_{2j-1} b_{2j-1}) + (b_{2j-2} b_{2j-1}) u_{2j-2} \tag{22.2.8}$$

This is a new recurrence of the same form as (22.2.5) but over only the even u_j, and hence involving only $n/2$ terms. Clearly we can continue this process recursively, halving the number of terms in the recurrence at each stage, until we are left with a recurrence of length 1 or 2 that we can do explicitly. Each time we finish a subpart of the recursion, we fill in the odd terms in the recurrence, using equation (22.2.7). In practice, it's even easier than it sounds. Turn to Chapter B5 to see a straightforward implementation of this algorithm as the recipe `recur1`.

On a machine with more processors than n, all the arithmetic at each stage of the recursion can be done simultaneously. Since there are of order $\log n$ stages in the recursion, the execution time is $O(\log n)$. The total number of operations carried out is of order $n + n/2 + n/4 + \cdots = O(n)$, the same as for the obvious serial do-loop.

In the utility routines of Chapter 23, we will use recursive doubling to implement the routines poly_term, cumsum, and cumprod. We *could* use recursive doubling to implement parallel versions of arth and geop (arithmetic and geometric progressions), and zroots_unity (complex nth roots of unity), but these can be done slightly more efficiently by subvector scaling, as discussed above.

Cyclic Reduction: Linear Recurrence Relations

There is a variant of recursive doubling, called *cyclic reduction*, that can be implemented with a straightforward iteration loop, instead of a recursive procedure call. [3] Here we start by writing down the recurrence (22.2.5) for *all* adjacent terms u_j and u_{j-1} (not just the even ones, as before). Eliminating u_{j-1}, just as in equation (22.2.8), gives

$$u_j = (a_j + a_{j-1}b_{j-1}) + (b_{j-2}b_{j-1})u_{j-2} \tag{22.2.9}$$

which is a first order recurrence with new coefficients a'_j and b'_j. Repeating this process gives successive formulas for u_j in terms of u_{j-2}, u_{j-4}, u_{j-8}.... The procedure terminates when we reach u_{j-n} (for n a power of 2), which is zero for all j. Thus the last step gives u_j equal to the last set of a'_j's.

Here is a code fragment that implements cyclic reduction by direct iteration. The quantities a'_j are stored in the variable recur1.

```
recur1=a
bb=b
j=1
do
   if (j >= n) exit
   recur1(j+1:n)=recur1(j+1:n)+bb(j:n-1)*recur1(1:n-j)
   bb(2*j:n-1)=bb(2*j:n-1)*bb(j:n-j-1)
   j=2*j
enddo
```

In cyclic reduction the length of the vector u_j that is updated at each stage does *not* decrease by a factor of 2 at each stage, but rather only decreases from $\sim n$ to $\sim n/2$ during all $\log_2 n$ stages. Thus the total number of operations carried out is $O(n \log n)$, as opposed to $O(n)$ for recursive doubling. For a serial machine or SSP machine, therefore, cyclic reduction is rarely superior to recursive doubling when the latter can be used. For an MMP machine, however, the issue is less clear cut, because the pattern of communication in cyclic reduction is quite different (and, for some parallel architectures, possibly more favorable) than that of recursive doubling.

Second Order Recurrence Relations

Consider the second order recurrence relation

$$y_j = a_j + b_{j-2}y_{j-1} + c_{j-2}y_{j-2}, \qquad j = 3, 4, \ldots, n \tag{22.2.10}$$

with initial values

$$y_1 = a_1, \qquad y_2 = a_2 \tag{22.2.11}$$

Our labeling of subscripts is designed to make it easy to enter the coefficients in a computer program: You need to supply $a_1, \ldots, a_n$, $b_1, \ldots, b_{n-2}$, and $c_1, \ldots, c_{n-2}$. Rewrite the recurrence relation in the form ([3])

$$\begin{pmatrix} y_j \\ y_{j+1} \end{pmatrix} = \begin{pmatrix} 0 \\ a_{j+1} \end{pmatrix} + \begin{pmatrix} 0 & 1 \\ c_{j-1} & b_{j-1} \end{pmatrix} \begin{pmatrix} y_{j-1} \\ y_j \end{pmatrix}, \qquad j = 2, \ldots, n-1 \tag{22.2.12}$$

that is,

$$\mathbf{u}_j = \mathbf{a}_j + \mathbf{b}_{j-1} \cdot \mathbf{u}_{j-1}, \qquad j = 2, \ldots, n-1 \tag{22.2.13}$$

where

$$\mathbf{u}_j = \begin{pmatrix} y_j \\ y_{j+1} \end{pmatrix}, \quad \mathbf{a}_j = \begin{pmatrix} 0 \\ a_{j+1} \end{pmatrix}, \quad \mathbf{b}_{j-1} = \begin{pmatrix} 0 & 1 \\ c_{j-1} & b_{j-1} \end{pmatrix}, \quad j = 2, \ldots, n-1 \tag{22.2.14}$$

and

$$\mathbf{u}_1 = \mathbf{a}_1 = \begin{pmatrix} y_1 \\ y_2 \end{pmatrix} = \begin{pmatrix} a_1 \\ a_2 \end{pmatrix} \tag{22.2.15}$$

This is a first order recurrence relation for the vectors $\mathbf{u}_j$, and can be solved by the algorithm described above (and implemented in the recipe `recur1`). The only difference is that the multiplications are matrix multiplications with the 2×2 matrices $\mathbf{b}_j$. After the first recursive call, the zeros in **a** and **b** are lost, so we have to write the routine for general two-dimensional vectors and matrices.

Note that this algorithm does not avoid the potential instability problems associated with second order recurrences that are discussed in §5.5 of Volume 1. Also note that the algorithm generalizes in the obvious way to higher-order recurrences: An nth order recurrence can be written as a first order recurrence involving n-dimensional vectors and matrices.

Parallel Solution of Tridiagonal Systems

Closely related to recurrence relations, recursive doubling, and cyclic reduction is the parallel solution of tridiagonal systems. Since Fortran 90 vectors "know their own size," it is most logical to number the components of both the sub- and super-diagonals of the tridiagonal matrix from 1 to $N - 1$. Thus equation (2.4.1), here written in the special case of $N = 7$, becomes (blank elements denoting zero),

$$\begin{bmatrix} b_1 & c_1 & & & & & \\ a_1 & b_2 & c_2 & & & & \\ & a_2 & b_3 & c_3 & & & \\ & & a_3 & b_4 & c_4 & & \\ & & & a_4 & b_5 & c_5 & \\ & & & & a_5 & b_6 & c_6 \\ & & & & & a_6 & b_7 \end{bmatrix} \cdot \begin{bmatrix} u_1 \\ u_2 \\ u_3 \\ u_4 \\ u_5 \\ u_6 \\ u_7 \end{bmatrix} = \begin{bmatrix} r_1 \\ r_2 \\ r_3 \\ r_4 \\ r_5 \\ r_6 \\ r_7 \end{bmatrix} \tag{22.2.16}$$

The basic idea for solving equation (22.2.16) on a parallel computer is to partition the problem into even and odd elements, recurse to solve the former, and then solve the latter in parallel. Specifically, we first rewrite (22.2.16), by permuting its rows and columns, as

$$\begin{bmatrix} b_1 & & & & c_1 & & \\ & b_3 & & & a_2 & c_3 & \\ & & b_5 & & & a_4 & c_5 \\ & & & b_7 & & & a_6 \\ a_1 & c_2 & & & b_2 & & \\ & a_3 & c_4 & & & b_4 & \\ & & a_5 & c_6 & & & b_6 \end{bmatrix} \cdot \begin{bmatrix} u_1 \\ u_3 \\ u_5 \\ u_7 \\ u_2 \\ u_4 \\ u_6 \end{bmatrix} = \begin{bmatrix} r_1 \\ r_3 \\ r_5 \\ r_7 \\ r_2 \\ r_4 \\ r_6 \end{bmatrix} \tag{22.2.17}$$

Now observe that, by row operations that subtract multiples of the first four rows from each of the last three rows, we can eliminate all nonzero elements in the lower-left quadrant. The price we pay is bringing some new elements into the lower-right quadrant, whose nonzero elements we now call x's, y's, and z's. We call the modified right-hand sides q. The transformed problem is now

$$\begin{bmatrix} b_1 & & & & c_1 & & \\ & b_3 & & & a_2 & c_3 & \\ & & b_5 & & & a_4 & c_5 \\ & & & b_7 & & & a_6 \\ & & & & y_1 & z_1 & \\ & & & & x_1 & y_2 & z_2 \\ & & & & & x_2 & y_3 \end{bmatrix} \cdot \begin{bmatrix} u_1 \\ u_3 \\ u_5 \\ u_7 \\ u_2 \\ u_4 \\ u_6 \end{bmatrix} = \begin{bmatrix} r_1 \\ r_3 \\ r_5 \\ r_7 \\ q_1 \\ q_2 \\ q_3 \end{bmatrix} \tag{22.2.18}$$

Notice that the last three rows form a new, smaller, tridiagonal problem, which we can solve simply by recursing! Once its solution is known, the first four rows can be solved by a simple, parallelizable, substitution. This algorithm is implemented in `tridag` in Chapter B2.

The above method is essentially cyclic reduction, but in the case of the tridiagonal problem, it does not "unwind" into a simple iteration; on the contrary, a recursive subroutine is required. For discussion of this and related methods for parallelizing tridiagonal systems, and references to the literature, see Hockney and Jesshope [3].

Recursive doubling can also be used to solve tridiagonal systems, the method requiring the parallel solution (as above) of both a first order recurrence and a second order recurrence [3,4]. For tridiagonal systems, however, cyclic reduction is usually more efficient than recursive doubling.

CITED REFERENCES AND FURTHER READING:

Van Loan, C.F. 1992, *Computational Frameworks for the Fast Fourier Transform* (Philadelphia: S.I.A.M.) §1.4.2. [1]

Estrin, G. 1960, quoted in Knuth, D.E. 1981, *Seminumerical Algorithms*, volume 2 of *The Art of Computer Programming* (Reading, MA: Addison-Wesley), §4.6.4. [2]

Hockney, R.W., and Jesshope, C.R. 1988, *Parallel Computers 2: Architecture, Programming, and Algorithms* (Bristol and Philadelphia: Adam Hilger), §5.2.4 (cyclic reduction); §5.4.2 (second order recurrences); §5.4 (tridiagonal systems). [3]

Stone, H.S. 1973, *Journal of the ACM*, vol. 20, pp. 27–38; 1975, *ACM Transactions on Mathematical Software*, vol. 1, pp. 289–307. [4]

22.3 Parallel Synthetic Division and Related Algorithms

There are several techniques for parallelization that relate to synthetic division but that can also find application in wider contexts, as we shall see.

Cumulants of a Polynomial

Suppose we have a polynomial

$$P(x) = \sum_{j=0}^{N} c_j x^{N-j} \tag{22.3.1}$$

(Note that, here, the c_j's are indexed from highest degree to lowest, the reverse of the usual convention.) Then we can define the *cumulants* of the polynomial to be partial sums that occur in the polynomial's usual, serial evaluation,

$$\begin{aligned} P_0 &= c_0 \\ P_1 &= c_0 x + c_1 \\ &\cdots \\ P_N &= c_0 x^N + \cdots + c_N = P(x) \end{aligned} \tag{22.3.2}$$

Evidently, the cumulants satisfy a simple, linear first order recurrence relation,

$$P_0 = c_0, \qquad P_j = c_j + x P_{j-1}, \qquad j = 2, \dots, N \tag{22.3.3}$$

This is slightly simpler than the general first order recurrence, because the value of x does not depend on j. We already know, from §22.2's discussion of recursive doubling, how to parallelize equation (22.3.3) via a recursive subroutine. In Chapter 23, the utility routine `poly_term` will implement just such a procedure. An example of a routine that calls `poly_term` to evaluate a recurrence equivalent to equation (22.3.3) is `eulsum` in Chapter B5.

Notice that while we could use equation (22.3.3), parallelized by recursive doubling, simply to evaluate the polynomial $P(x) = P_N$, this is likely somewhat slower than the alternative technique of vector reduction, also discussed in §22.2, and implemented in the utility function `poly`. Equation (22.3.3) should be saved for cases where the rest of the P_j's (not just P_N) can be put to good use.

Synthetic Division by a Monomial

We now show that evaluation of the cumulants of a polynomial is equivalent to synthetic division of the polynomial by a monomial, also called *deflation* (see §9.5 in Volume 1). To review briefly, and by example, here is a standard tableau from high school algebra for the (long) division of a polynomial $2x^3 - 7x^2 + x + 3$ by the monomial factor $x - 3$.

$$
\begin{array}{r|l}
 & 2x^2 - \ x \ - 2 \\
\cline{2-2}
x-3 & 2x^3 - 7x^2 + \ x + 3 \\
 & 2x^3 - \underline{6x^2} \\
 & \qquad -x^2 + x \\
 & \qquad -x^2 \ \underline{+3x} \\
 & \qquad\qquad -2x + 3 \\
 & \qquad\qquad -2x \ \underline{+6} \\
 & \qquad\qquad\qquad -3 \text{ (remainder)}
\end{array}
\tag{22.3.4}
$$

Now, here is the same calculation written as a *synthetic division*, really the same procedure as tableau (22.3.4), but with unnecessary notational baggage omitted (and also a changed sign for the monomial's constant, so that subtractions become additions):

$$
\begin{array}{r|rrrr}
 & & 6 & -3 & -6 \\
3 & 2 & -7 & +1 & +3 \\
\hline
 & 2 & -1 & -2 & -3
\end{array}
\tag{22.3.5}
$$

If we substitute symbols for the above quantities with the correspondence

$$
\begin{array}{r|cccc}
x & c_0 & c_1 & c_2 & c_3 \\
\hline
 & P_0 & P_1 & P_2 & P_3
\end{array}
\tag{22.3.6}
$$

then it is immediately clear that the P_j's in equation (22.3.6) are simply the P_j's of equation (22.3.3); the calculation is thus parallelizable by recursive doubling. In this context, the utility routine `poly_term` is used by the routine `zroots` in Chapter B9.

Repeated Synthetic Division

It is well known from high-school algebra that repeated synthetic division of a polynomial yields, as the remainders that occur, first the value of the polynomial, next the value of its first derivative, and then (up to multiplication by the factorial of an integer) the values of higher derivatives.

If you want to parallelize the calculation of the value of a polynomial and one or two of its derivatives, it is not unreasonable to evaluate equation (22.3.3), parallelized by recursive doubling, two or three times. Our routine `ddpoly` in Chapter B5 is meant for such use, and it does just this, as does the routine `laguer` in Chapter B9.

There are other cases, however, for which you want to perform repeated synthetic division and "go all the way," until only a constant remains. For example, this is the preferred way of "shifting a polynomial," that is, evaluating the coefficients of a polynomial in a variable y that differs from the original variable x by an additive constant. (The recipe `pcshft` has this as its assigned task.) By way of example, consider the polynomial $3x^3 + x^2 + 4x + 7$, and let us perform repeated synthetic division by a general monomial $x - a$. The conventional calculation then proceeds according to the following tableau, reading it in conventional lexical order (left-to-right and top-to-bottom):

$$
\begin{array}{ccccccc}
3 & & 1 & & 4 & & 7 \\
\downarrow & & \downarrow & & \downarrow & & \downarrow \\
3 & \xrightarrow{a} & 3a+1 & \xrightarrow{a} & 3a^2+a+4 & \xrightarrow{a} & 3a^3+a^2+4a+7 \\
\downarrow & & \downarrow & & \downarrow & & \\
3 & \xrightarrow{a} & 6a+1 & \xrightarrow{a} & 9a^2+2a+4 & & \\
\downarrow & & \downarrow & & & & \\
3 & \xrightarrow{a} & 9a+1 & & & & \\
\downarrow & & & & & & \\
3 & & & & & &
\end{array}
\tag{22.3.7}
$$

Here, each row (after the first) shows a synthetic division or, equivalently, evaluation of the cumulants of the polynomial whose coefficients are the preceding row. The results at the right edge of the rows are the values of the polynomial and (up to integer factorials) its three nonzero derivatives, or (equivalently, without factorials) coefficients of the shifted polynomial.

We could parallelize the calculation of each row of tableau (22.3.7) by recursive doubling. That is a lot of recursion, which incurs a nonnegligible overhead. A much better way of doing the calculation is to deform tableau (22.3.7) into the following equivalent tableau,

$$
\begin{array}{ccccccccc}
3 \longrightarrow & 3 & & & & & & & \\
 & {}^{a}\!\downarrow & \searrow & & & & & & \\
1 \longrightarrow & 3a+1 & & 3 & & & & & \\
 & {}^{a}\!\downarrow & \searrow & {}^{a}\!\downarrow & \searrow & & & & \\
4 \longrightarrow & 3a^2+a+4 & & 6a+1 & & 3 & & & \\
 & {}^{a}\!\downarrow & \searrow & {}^{a}\!\downarrow & \searrow & {}^{a}\!\downarrow & \searrow & & \\
7 \longrightarrow & 3a^3+a^2+4a+7 & & 9a^2+2a+4 & & 9a+1 & & 3 &
\end{array}
\tag{22.3.8}
$$

Now each row explicitly depends on only the previous row (and the given first column), so the rows can be calculated in turn by an explicit parallel expression, with no recursive calls needed. An example of coding (22.3.8) in Fortran 90 can be found in the routine `pcshft` in Chapter B5. (It is also possible to eliminate most of the multiplications in (22.3.8), at the expense of a much smaller number of divisions. We have not done this because of the necessity for then treating all possible divisions by zero as special cases. See [1] for details and references.)

Actually, the deformation of (22.3.7) into (22.3.8) is the same trick as was used in Volume 1, p. 167, for evaluating a polynomial and its derivative simultaneously, also generalized in the Fortran 77 implementation of the routine `ddpoly` (Chapter 5). In the Fortran 90 implementation of `ddpoly` (Chapter B5) we *don't* use this trick, but instead use `poly_term`, because, there, we want to parallelize over the length of the polynomial, not over the number of desired derivatives.

Don't confuse the cases of *iterated* synthetic division, discussed here, with the simpler case of doing many simultaneous synthetic divisions. In the latter case, you can simply implement equation (22.3.3) serially, exactly as written, but with each operation being data-parallel across your problem set. (This case occurs in our routine polcoe in Chapter B3.)

Polynomial Coefficients from Roots

A parallel calculation algorithmically very similar to (22.3.7) or (22.3.8) occurs when we want to find the coefficients of a polynomial $P(x)$ from its roots $r_1, \ldots, r_N$. For this, the tableau is

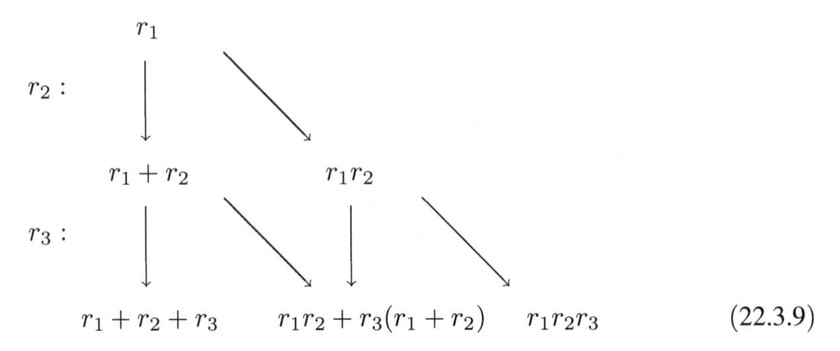

(22.3.9)

As before, the rows are computed consecutively, from top to bottom. Each row is computed via a single parallel expression. Note that values moving on vertical arrows are simply added in, while values moving on diagonal arrows are multiplied by a new root before adding. Examples of coding (22.3.9) in Fortran 90 can be found in the routines vander (Chapter B2) and polcoe (Chapter B3).

An equivalent deformation of (22.3.9) is

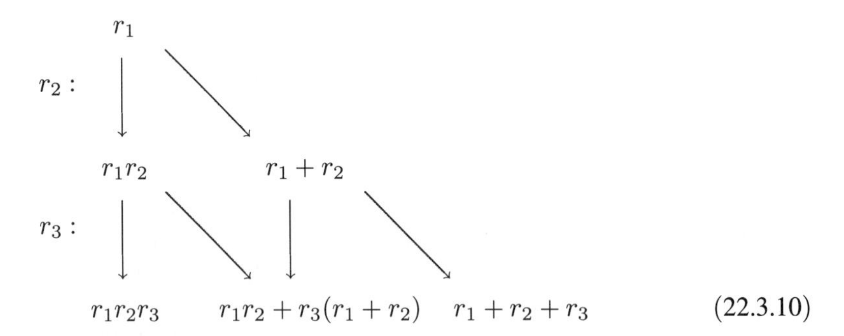

(22.3.10)

Here the diagonal arrows are simple additions, while the vertical arrows represent multiplication by a root value. Note that the coefficient answers in (22.3.10) come out in the opposite order from (22.3.9). An example of coding (22.3.10) in Fortran 90 can be found in the routine fixrts in Chapter B13.

CITED REFERENCES AND FURTHER READING:

Knuth, D.E. 1981, *Seminumerical Algorithms*, 2nd ed., vol. 2 of *The Art of Computer Programming* (Reading, MA: Addison-Wesley), §4.6.4, p. 470. [1]

22.4 Fast Fourier Transforms

Fast Fourier transforms are beloved by computer scientists, especially those who are interested in parallel algorithms, because the FFT's hierarchical structure generates a complicated, but analyzable, set of requirements for interprocessor communication on MMPs. Thus, almost all books on parallel algorithms (e.g., [1–3]) have a chapter on FFTs.

Unfortunately, the resulting algorithms are highly specific to particular parallel architectures, and therefore of little use to us in writing general purpose code in an architecture-independent parallel language like Fortran 90.

Luckily there is a good alternative that covers almost all cases of both serial and parallel machines. If, for a one-dimensional FFT of size N, one is satisfied with parallelism of order $\sqrt{N}$, then there is a good, general way of achieving a parallel FFT with *quite minimal* interprocessor communication; and the communication required is simply the matrix transpose operation, which Fortran 90 implements as an intrinsic. That is the approach that we discuss in this section, and implement in Chapter B12.

For a machine with M processors, this approach will saturate the processors (the desirable condition where none are idle) when the size of a one-dimensional Fourier transform, N, is large enough: $N > M^2$. Smaller N's will not achieve maximum parallelism. But such N's are in fact so small for one-dimensional problems that they are unlikely to be the rate-determining step in scientific calculations. If they are, it is usually because you are doing many such transforms independently, and you should recover "outer parallelism" by doing them all at once.

For two or more dimensions, the adopted approach will saturate M processors when *each* dimension of the problem is larger than M.

Column- and Row-Parallel FFTs

The basic building block that we assume (and implement in Chapter B12) is a routine for simultaneously taking the FFT of each *row* of a two-dimensional matrix. The method is exactly that of Volume 1's `four1` routine, but with array sections like `data(:,j)` replacing scalars like `data(j)`. Chapter B12's implementation of this is a routine called `fourrow`. If all the data for one column (that is, all the values `data(i,:)`, for some `i`) are local to a single processor, then the parallelism involves no interprocessor communication at all: The independent FFTs simply proceed, data parallel and in lockstep. This is architecture-independent parallelism with a vengeance.

We will also need to take the FFT of each *column* of a two-dimensional matrix. One way to do this is to take the transpose (a Fortran 90 intrinsic that hides a lot of interprocessor communication), then take the FFT of the rows using `fourrow`, then take the transpose again. An alternative method is to recode the `four1` routine with array sections in the other dimension (`data(j,:)`) replacing `four1`'s scalars (`data(j)`). This scheme, in Chapter B12, is a routine called `fourcol`. In this case, good parallelism will be achieved only if the values `data(:,i)`, for some `i`, are local to a single processor. Of course, Fortran 90 does not give the user direct control over how data are distributed over the machine; but extensions such as HPF are designed to give just such control.

On a serial machine, you might think that `fourrow` and `fourcol` should have identical timings (acting on a square matrix, say). The two routines do exactly the same operations, after all. Not so! On modern serial computers, `fourrow` and `fourcol` can have timings that differ by a factor of 2 or more, even when their detailed arithmetic is made identical (by giving to one a data array that is the transpose of the data array given to the other). This effect is due to the multilevel cache architecture of most computer memories, and the fact that serial Fortran always stores matrices by columns (first index changing most rapidly). On our workstations, `fourrow` is significantly faster than `fourcol`, and this is likely the generic behavior. However, we do not exclude the possibility that some machines, and some sizes of matrices, are the other way around.

One-Dimensional FFTs

Turn now to the problem of how to do a single, one-dimensional, FFT. We are given a complex array f of length N, an integer power of 2. The basic idea is to address the input array as if it were a two-dimensional array of size $m \times M$, where m and M are each integer powers of 2. Then the components of f can be addressed as

$$f(Jm + j), \qquad 0 \le j < m,\ 0 \le J < M \tag{22.4.1}$$

where the j index changes more rapidly, the J index more slowly, and parentheses denote Fortran-style subscripts.

Now, suppose we had some magical (parallel) method to compute the discrete Fourier transform

$$F(kM + K) \equiv \sum_{j,J} e^{2\pi i(kM+K)(Jm+j)/(Mm)} f(Jm + j),$$
$$0 \le k < m,\ 0 \le K < M \tag{22.4.2}$$

Then, you can see that the indices k and K would address the desired result (FFT of the original array), with K varying more rapidly.

Starting with equation (22.4.2) it is easy to verify the following identity,

$$F(kM + K) = \sum_j \left[e^{2\pi ijk/m} \left(e^{2\pi ijK/(Mm)} \left[\sum_J e^{2\pi iJK/M} f(Jm + j) \right] \right) \right] \tag{22.4.3}$$

But this, reading it from the innermost operation outward, is just the magical method that we need:

- Reshape the original array to $m \times M$ in Fortran normal order (storage by columns).
- FFT on the second (column) index for all values of the first (row) index, using the routine `fourrow`.
- Multiply each component by a phase factor $\exp[2\pi ijK/(Mm)]$.
- Transpose.

- Again FFT on the second (column) index for all values of the first (row) index, using the routine `fourrow`.
- Reshape the two-dimensional array back into one-dimensional output.

The above scheme uses `fourrow` exclusively, on the assumption that it is faster than its sibling `fourcol`. When that is the case (as we typically find), it is likely that the above method, implemented as `four1` in Chapter B12, is faster, even on scalar machines, than Volume 1's scalar version of `four1` (Chapter 12). The reason, as already mentioned, is that `fourrow`'s parallelism is taking better advantage of cache memory locality.

If `fourrow` is *not* faster than `fourcol` on your machine, then you should instead try the following alternative scheme, using `fourcol` only:

- Reshape the original array to $m \times M$ in Fortran normal order (storage by columns).
- Transpose.
- FFT on the first (row) index for all values of the second (column) index, using the routine `fourcol`.
- Multiply each component by a phase factor $\exp[2\pi ijK/(Mm)]$.
- Transpose.
- Again FFT on the first (row) index for all values of the second (column) index, using the routine `fourcol`.
- Transpose.
- Reshape the two-dimensional array back into one-dimensional output.

In Chapter B12, this scheme is implemented as `four1_alt`. You might wonder why `four1_alt` has three transpose operations, while `four1` had only one. Shouldn't there be a symmetry here? No. Fortran makes the arbitrary, but consistent, choice of storing two-dimensional arrays by columns, and this choice favors `four1` in terms of transposes. Luckily, at least on our serial workstations, `fourrow` (used by `four1`) is faster than `fourcol` (used by `four1_alt`), so it is a double win.

For further discussion and references on the ideas behind `four1` and `four1_alt` see [4], where these algorithms are called the four-step and six-step frameworks, respectively.

CITED REFERENCES AND FURTHER READING:

Fox, G.C., et al. 1988, *Solving Problems on Concurrent Processors*, Volume I (Englewood Cliffs, NJ: Prentice Hall), Chapter 11. [1]

Akl, S.G. 1989, *The Design and Analysis of Parallel Algorithms* (Englewood Cliffs, NJ: Prentice Hall), Chapter 9. [2]

Hockney, R.W., and Jesshope, C.R. 1988, *Parallel Computers 2* (Bristol and Philadelphia: Adam Hilger), §5.5. [3]

Van Loan, C. 1992, *Computational Frameworks for the Fast Fourier Transform* (Philadelphia: S.I.A.M.), §3.3. [4]

22.5 Missing Language Features

A few facilities that are fairly important to parallel programming are missing from the Fortran 90 language standard. On scalar machines this lack is not a problem, since one can readily program the missing features by using do-loops. On parallel machines, both SSP machines and MMP machines, one must hope that hardware manufacturers provide library routines, callable from Fortran 90, that provide access to the necessary facilities, or use extensions of Fortran 90, such as High Performance Fortran (HPF).

Scatter-with-Combine Functions

Fortran 90 allows the use of *vector subscripts* for so-called *gather* and *scatter* operations. For example, with the setup

```
REAL(SP), DIMENSION(6) :: arr,barr,carr
INTEGER(I4B), DIMENSION(6) :: iarr,jarr
...
iarr = (/ 1,3,5,2,4,6 /)
jarr = (/ 3,2,3,2,1,1 /)
```

Fortran 90 allows both the *one-to-one* gather and the *one-to-many* gather,

```
barr=arr(iarr)
carr=arr(jarr)
```

It also allows the one-to-one scatter,

```
barr(iarr)=carr
```

where the elements of `carr` are "scattered" into `barr` under the direction of the vector subscript `iarr`.

Fortran 90 does *not* allow the *many-to-one* scatter

```
barr(jarr)=carr          ! illegal for this jarr
```

because the repeated values in `jarr` try to assign different components of `carr` to the same location in `barr`. The result would not be deterministic.

Sometimes, however, one would in fact like a many-to-one construction, where the colliding elements get combined by a (commutative and associative) operation, like + or *, or `max()`. These so-called *scatter-with-combine* functions are readily implemented on serial machines by a do-loop, for example,

```
barr=0.
do j=1,size(carr)
   barr(jarr(j))=barr(jarr(j))+carr(j)
end do
```

Fortran 90 unfortunately provides no means for effecting scatter-with-combine functions in parallel. Luckily, almost all parallel machines do provide such a facility as a library program, as does HPF, where the above facility is called `SUM_SCATTER`. In Chapter 23 we will define utility routines `scatter_add` and `scatter_max` for scatter-with-combine functionalities, but the implementation given in Fortran 90 will be strictly serial, with a do-loop.

Skew Sections

Fortran 90 provides no good, parallel way to access the diagonal elements of a matrix, either to read them or to set them. Do-loops will obviously serve this need on serial machines. In principle, a construction like the following bizarre fragment could also be utilized,

```
REAL(SP), DIMENSION(n,n) :: mat
REAL(SP), DIMENSION(n*n) :: arr
REAL(SP), DIMENSION(n) :: diag
...
arr = reshape(mat,shape(arr))
diag = arr(1:n*n:n+1)
```

which extracts every $(n+1)$st element from a one-dimensional array derived by reshaping the input matrix. However, it is unlikely that any foreseeable parallel compiler will implement the above fragment without a prohibitive amount of unnecessary data movement; and code like the above is also exceedingly slow on all serial machines that we have tried.

In Chapter 23 we will define utility routines `get_diag`, `put_diag`, `diagadd`, `diagmult`, and `unit_matrix` to manipulate diagonal elements, but the implementation given in Fortran 90 will again be strictly serial, with do-loops.

Fortran 95 (see §21.6) will essentially fix Fortran 90's skew sections deficiency. For example, using its `forall` construction, the diagonal elements of an array can be accessed by a statement like

```
forall (j=1:20) diag(j) = arr(j,j)
```

SIMD vs. MIMD

Recall that we use "SIMD" (single-instruction, multiple data) and "data parallel" as interchangeable terms, and that "MIMD" (multiple-instruction, multiple data) is a more general programming model. (See §22.1.)

You should not be too quick to jump to the conclusion that Fortran 90's data parallel or SIMD model is "bad," and that MIMD features, absent in Fortran 90, are therefore "good." On the contrary, Fortran 90's basic data-parallel paradigm has a lot going for it. As we discussed in §22.1, most scientific problems naturally have a "data dimension" across which the time ordering of the calculation is irrelevant. Parallelism across this dimension, which is by nature most often SIMD, frees the mind to think clearly about the computational steps in an algorithm that actually need to be sequential. SIMD code has advantages of clarity and predictability that should not be taken lightly. The general MIMD model of "lots of different things all going on at the same time and communicating data with each other" is a programming and debugging nightmare.

Having said this, we must at the same time admit that a few MIMD features — most notably the ability to go through different logical branches for calculating different data elements in a data-parallel computation — are badly needed in certain programming situations. Fortran 90 is quite weak in this area.

Note that the `where...elsewhere...end where` construction is *not* a MIMD construction. Fortran 90 requires that the `where` clause be executed completely before the `elsewhere` is started. (This allows the results of any calculations in the former clause to be available for use in the latter.) So, this construction cannot be used to allow two logical branches to be calculated in parallel.

Special functions, where one would like to calculate function values for an array of input quantities, are a particularly compelling example of the need for some MIMD access. Indeed, you will find that Chapter B6 contains a number of intricate, and in a few cases truly bizarre, workarounds, using allowed combinations of `merge`, `where`, and `CONTAINS` (the latter, for separating different logical branches into formally different subprograms).

Fortran 95's `ELEMENTAL` and `PURE` constructions, and to some extent also `forall` (whose body will be able to include `PURE` function calls), will go a long way towards providing exactly the kind of MIMD constructions that are most needed. Once Fortran 95 becomes available and widespread, you can expect to see a new version of this volume, with a much-improved Chapter B6.

Conversely, the number of routines outside of Chapter B6 that can be significantly improved by the use of MIMD features is relatively small; this illustrates the underlying viability of the basic data-parallel SIMD model, even in a future language version with useful MIMD features.

Chapter 23. Numerical Recipes Utility Functions for Fortran 90

23.0 Introduction and Summary Listing

This chapter describes and summarizes the Numerical Recipes utility routines that are used throughout the rest of this volume. A complete implementation of these routines in Fortran 90 is listed in Appendix C1.

Why do we need utility routines? Aren't there already enough of them built into the language as Fortran 90 intrinsics? The answers lie in this volume's dual purpose: to implement the Numerical Recipes routines in Fortran 90 code that runs efficiently on fast serial machines, *and* to implement them, wherever possible, with efficient parallel code for multiprocessor machines that will become increasingly common in the future. We have found three kinds of situations where additional utility routines seem desirable:

1. Fortran 90 is a big language, with many high-level constructs — single statements that actually result in a lot of computing. We like this; it gives the language the potential for expressing algorithms very readably, getting them "out of the mud" of microscopic coding. In coding the 350+ Recipes for this volume, we kept a systematic watch for bits of microscopic coding that were repeated in many routines, and that seemed to be at a lower level of coding than that aspired to by good Fortran 90 style. Once these bits were identified, we pulled them out and substituted calls to new utility routines. These are the utilities that arguably ought to be new language intrinsics, equally useful for serial and parallel machines. (A prime example is `swap`.)

2. Fortran 90 contains many highly parallelizable language constructions. But, as we have seen in §22.5, it is also missing a few important constructions. Most parallel machines will provide these missing elements as machine-coded library subroutines. Some of our utility routines are provided simply as a standard interface to these common, but nonstandard, functionalities. Note that it is the nature of these routines that our specific implementation, in Appendix C1, will be serial, and therefore inefficient on parallel machines. If you have a parallel machine, you will need to recode these; this often involves no more than substituting a one-line library function call for the body of our implementation. Utilities in this category will likely become unnecessary over time, either as machine-dependent libraries converge to standard interfaces, or as the utilities get added to future Fortran

versions. (Indeed, some routines in this category will be unnecessary in Fortran 95, once it is available; see §23.7.)

3. Some tasks should just be done differently in serial, versus parallel, implementation. Linear recurrence relations are a good example (§22.2). These are trivially coded with a do-loop on serial machines, but require a fairly elaborate recursive construction for good parallelization. Rather than provide separate serial and parallel versions of the Numerical Recipes, we have chosen to pull out of the Recipes, and into utility routines, some identifiable tasks of this kind. These are cases where some recoding of our implementation in Appendix C1 might result in improved performance on your particular hardware. Unfortunately, it is not so simple as providing a single "serial implementation" and another single "parallel implementation," because even the seemingly simple word "serial" hides, at the hardware level, a variety of different degrees of pipelining, wide instructions, and so on. Appendix C1 therefore provides only a single implementation, although with some adjustable parameters that you can customize (by experiment) to maximize performance on your hardware.

The above three cases are not really completely distinct, and it is therefore not possible to assign any single utility routine to exactly one situation. Instead, we organize the rest of this chapter as follows: first, an alphabetical list, with short summary, of all the new utility routines; next, a series of short sections, grouped by functionality, that contain the detailed descriptions of the routines.

Alphabetical Listing

The following list gives an abbreviated mnemonic for the type, rank, and/or shape of the returned values (as in §21.4), the routine's calling sequence (optional arguments shown in italics), and a brief, often incomplete, description. The complete description of the routine is given in the later section shown in square brackets.

For each entry, the number shown in parentheses is the approximate number of distinct Recipes in this book that make use of that particular utility function, and is thus a rough guide to that utility's importance. (There may be multiple invocations of the utility in each such Recipe.) Where this number is small or zero, it is usually because the utility routine is a member of a related family of routines whose total usage was deemed significant enough to include, and we did not want users to have to "guess" which family members were instantiated.

`call array_copy(src,dest,n_copied,n_not_copied)`
Copy one-dimensional array (whose size is not necessarily known). [23.1] (9)

[Arr] `arth(first,increment,n)`
Return an arithmetic progression as an array. [23.4] (55)

`call assert(n1,`*`n2,...,`*`string)`
Exit with error message if any logical arguments are false. [23.3] (50)

[Int] `assert_eq(n1,`*`n2,...,`*`string)`
Exit with error message if all integer arguments are not equal; otherwise return common value. [23.3] (133)

[argTS] `cumprod(arr,`*`seed`*`)`

Cumulative products of one-dimensional array, with optional seed value. [23.4] (3)

[argTS] `cumsum(arr,seed)`
Cumulative sums of one-dimensional array, with optional seed value. [23.4] (9)

`call diagadd(mat,diag)`
Adds vector to diagonal of a matrix. [23.7] (4)

`call diagmult(mat,diag)`
Multiplies vector into diagonal of a matrix. [23.7] (2)

[Arr] `geop(first,factor,n)`
Return a geometrical progression as an array. [23.4] (7)

[Arr] `get_diag(mat)`
Gets diagonal of a matrix. [23.7] (2)

[Int] `ifirstloc(arr)`
Location of first true value in a logical array, returned as an integer. [23.2] (3)

[Int] `imaxloc(arr)`
Location of array maximum, returned as an integer. [23.2] (11)

[Int] `iminloc(arr)`
Location of array minimum, returned as an integer. [23.2] (8)

[Mat] `lower_triangle(j,k,extra)`
Returns a lower triangular logical mask. [23.7] (1)

`call nrerror(string)`
Exit with error message. [23.3] (96)

[Mat] `outerand(a,b)`
Returns the outer logical and of two vectors. [23.5] (1)

[Mat] `outerdiff(a,b)`
Returns the outer difference of two vectors. [23.5] (4)

[Mat] `outerdiv(a,b)`
Returns the outer quotient of two vectors. [23.5] (0)

[Mat] `outerprod(a,b)`
Returns the outer product of two vectors. [23.5] (14)

[Mat] `outersum(a,b)`
Returns the outer sum of two vectors. [23.5] (0)

[argTS] `poly(x,coeffs,mask)`
Evaluate a polynomial $P(x)$ for one or more values x, with optional mask. [23.4] (15)

[argTS] `poly_term(a,x)`
Returns partial cumulants of a polynomial, equivalent to synthetic

division. [23.4] (4)

`call put_diag(diag,mat)`
Sets diagonal of a matrix. [23.7] (0)

[Ptr] `reallocate(p,n,m,...)`
Reallocate pointer to new size, preserving its contents. [23.1] (5)

`call scatter_add(dest,source,dest_index)`
Scatter-adds source vector to specified components of destination vector. [23.6] (2)

`call scatter_max(dest,source,dest_index)`
Scatter-max source vector to specified components of destination vector. [23.6] (0)

`call swap(a,b,mask)`
Swap corresponding elements of a and b. [23.1] (24)

`call unit_matrix(mat)`
Sets matrix to be a unit matrix. [23.7] (6)

[Mat] `upper_triangle(j,k,extra)`
Returns an upper triangular logical mask. [23.7] (4)

[Real] `vabs(v)`
Length of a vector in L_2 norm. [23.8] (6)

[CArr] `zroots_unity(n,nn)`
Returns nn consecutive powers of the complex nth root of unity. [23.4] (4)

Comment on Relative Frequencies of Use

We find it interesting to compare our frequency of using the `nrutil` utility routines, with our most used language intrinsics (see §21.4). On this basis, the following routines are as useful to us as the *top 10* language intrinsics: `arth`, `assert`, `assert_eq`, `outerprod`, `poly`, and `swap`. We strongly recommend that the X3J3 standards committee, as well as individual compiler library implementors, consider the inclusion of new language intrinsics (or library routines) that subsume the capabilities of at least these routines. In the next tier of importance, we would put some further cumulative operations (`geop`, `cumsum`), some other "outer" operations on vectors (e.g., `outerdiff`), basic operations on the diagonals of matrices (`get_diag`, `put_diag`, `diag_add`), and some means of access to an array of unknown size (`array_copy`).

23.1 Routines That Move Data

To describe our utility routines, we introduce two items of Fortran 90 pseudocode: We use the symbol **T** to denote some type and rank declaration (including scalar rank, i.e., zero); and when we append a colon to a type specification, as in INTEGER(I4B)(:), for example, we denote an array of the given type.

The routines swap, array_copy, and reallocate simply move data around in useful ways.

⋆ ⋆ ⋆

swap (swaps corresponding elements)

User interface (or, "USE nrutil"):

```
SUBROUTINE swap(a,b,mask)
T, INTENT(INOUT) :: a,b
LOGICAL(LGT), INTENT(IN), OPTIONAL :: mask
END SUBROUTINE swap
```

Applicable types and ranks:

T ≡ *any type, any rank*

Types and ranks implemented (overloaded) in nrutil*:*

T ≡ INTEGER(I4B), REAL(SP), REAL(SP)(:), REAL(DP), COMPLEX(SPC), COMPLEX(SPC)(:), COMPLEX(SPC)(:,:), COMPLEX(DPC), COMPLEX(DPC)(:), COMPLEX(DPC)(:,:)

Action:

Swaps the corresponding elements of a and b. If mask is present, performs the swap only where mask is true. (Following code is the unmasked case. For speed at run time, the masked case is implemented by overloading, not by testing for the optional argument.)

Reference implementation:

```
T :: dum
dum=a
a=b
b=dum
```

⋆ ⋆ ⋆

array_copy (copy one-dimensional array)

User interface (or, "USE nrutil"):

```
SUBROUTINE array_copy(src,dest,n_copied,n_not_copied)
T, INTENT(IN) :: src
T, INTENT(OUT) :: dest
INTEGER(I4B), INTENT(OUT) :: n_copied, n_not_copied
END SUBROUTINE array_copy
```

Applicable types and ranks:

T ≡ *any type, rank 1*

Types and ranks implemented (overloaded) in nrutil*:*

T ≡ INTEGER(I4B)(:), REAL(SP)(:), REAL(DP)(:)

Action:

Copies to a destination array dest the one-dimensional array src, or as much of src as will fit in dest. Returns the number of components copied as n_copied, and the number of components not copied as n_not_copied.

The main use of this utility is where src is an expression that returns an array whose size is not known in advance, for example, the value returned by the pack intrinsic.

Reference implementation:

```
n_copied=min(size(src),size(dest))
n_not_copied=size(src)-n_copied
dest(1:n_copied)=src(1:n_copied)
```

⋆ ⋆ ⋆

reallocate (reallocate a pointer, preserving contents)

User interface (or, "`USE nrutil`*"):*

```
FUNCTION reallocate(p,n[,m,...])
T, POINTER :: p, reallocate
INTEGER(I4B), INTENT(IN) :: n[,m,...]
END FUNCTION reallocate
```

Applicable types and ranks:

T ≡ *any type, rank 1 or greater*

Types and ranks implemented (overloaded) in `nrutil`*:*

T ≡ `INTEGER(I4B)(:)`, `INTEGER(I4B)(:,:)`, `REAL(SP)(:)`, `REAL(SP)(:,:)`, `CHARACTER(1)(:)`

Action:

Allocates storage for a new array with shape specified by the integer(s) `n`, `m`, ... (equal in number to the rank of pointer `p`). Then, copies the contents of `p`'s target (or as much as will fit) into the new storage. Then, deallocates `p` and returns a pointer to the new storage.

The typical use is `p=reallocate(p,n[,m,...])`, which has the effect of changing the allocated size of `p` while preserving the contents.

The reference implementation, below, shows only the case of rank 1.

Reference implementation:

```
INTEGER(I4B) :: nold,ierr
allocate(reallocate(n),stat=ierr)
if (ierr /= 0) call &
   nrerror('reallocate: problem in attempt to allocate memory')
if (.not. associated(p)) RETURN
nold=size(p)
reallocate(1:min(nold,n))=p(1:min(nold,n))
deallocate(p)
```

23.2 Routines Returning a Location

Fortran 90's intrinsics `maxloc` and `minloc` return rank-one arrays. When, in the most frequent usage, their argument is a one-dimensional array, the answer comes back in the inconvenient form of an array containing a single component, which cannot be itself used in a subscript calculation. While there are workaround tricks (e.g., use of the `sum` intrinsic to convert the array to a scalar), it seems clearer to define routines `imaxloc` and `iminloc` that return integers directly.

The routine `ifirstloc` adds a related facility missing among the intrinsics: Return the first true location in a logical array.

⋆ ⋆ ⋆

imaxloc (location of array maximum as an integer)

User interface (or, "`USE nrutil`*"):*

```
FUNCTION imaxloc(arr)
T, INTENT(IN) :: arr
INTEGER(I4B) :: imaxloc
END FUNCTION imaxloc
```

Applicable types and ranks:

T ≡ *any integer or real type, rank 1*

Types and ranks implemented (overloaded) in `nrutil`*:*

T ≡ `INTEGER(I4B)(:), REAL(SP)(:)`

Action:

For one-dimensional arrays, identical to the `maxloc` intrinsic, except returns its answer as an integer rather than as `maxloc`'s somewhat awkward rank-one array containing a single component.

Reference implementation:

```
INTEGER(I4B), DIMENSION(1) :: imax
imax=maxloc(arr(:))
imaxloc=imax(1)
```

⋆ ⋆ ⋆

iminloc (location of array minimum as an integer)

User interface (or, "`USE nrutil`*"):*

```
FUNCTION iminloc(arr)
T, INTENT(IN) :: arr
INTEGER(I4B) :: iminloc
END FUNCTION iminloc
```

Applicable types and ranks:

T ≡ *any integer or real type, rank 1*

Types and ranks implemented (overloaded) in `nrutil`*:*

T ≡ `REAL(SP)(:)`

Action:

For one-dimensional arrays, identical to the `minloc` intrinsic, except returns its answer as an integer rather than as `minloc`'s somewhat awkward rank-one array containing a single component.

Reference implementation:

```
INTEGER(I4B), DIMENSION(1) :: imin
imin=minloc(arr(:))
iminloc=imin(1)
```

⋆ ⋆ ⋆

ifirstloc (returns location of first "true" in a logical vector)

User interface (or, "`USE nrutil`*"):*

```
FUNCTION ifirstloc(mask)
T, INTENT(IN) :: mask
INTEGER(I4B) :: ifirstloc
END FUNCTION ifirstloc
```

Applicable types and ranks:

T ≡ *any logical type, rank 1*

Types and ranks implemented (overloaded) in `nrutil`*:*

T ≡ `LOGICAL(LGT)`

Action:

Returns the index (subscript value) of the first location, in a one-dimensional logical `mask`, that has the value `.TRUE.`, or returns `size(mask)+1` if all components of `mask` are `.FALSE.`

Note that while the reference implementation uses a do-loop, the function is parallelized in `nrutil` by instead using the `merge` and `maxloc` intrinsics.

Reference implementation:

```
INTEGER(I4B) :: i
do i=1,size(mask)
   if (mask(i)) then
      ifirstloc=i
      return
   end if
end do
ifirstloc=i
```

23.3 Argument Checking and Error Handling

It is good programming practice for a routine to check the assumptions ("assertions") that it makes about the sizes of input arrays, allowed range of numerical arguments, and so forth. The routines `assert` and `assert_eq` are meant for this kind of use. The routine `nrerror` is our default error reporting routine.

⋆ ⋆ ⋆

assert (exit with error message if any assertion is false)

User interface (or, "`USE nrutil`*"):*

```
SUBROUTINE assert(n1,n2,...,string)
CHARACTER(LEN=*), INTENT(IN) :: string
LOGICAL, INTENT(IN) :: n1,n2,...
END SUBROUTINE assert
```

Action:

Embedding program dies gracefully with an error message if any of the logical arguments are false. Typical use is with logical expressions as the actual arguments. `nrutil` implements and overloads forms with 1, 2, 3, and 4 logical arguments, plus a form with a vector logical argument,

```
LOGICAL, DIMENSION(:), INTENT(IN) :: n
```

that is checked by the `all(n)` intrinsic.

Reference implementation:

```
if (.not. (n1.and.n2.and...)) then
   write (*,*) 'nrerror: an assertion failed with this tag:', string
   STOP 'program terminated by assert'
end if
```

⋆ ⋆ ⋆

assert_eq (exit with error message if integer arguments not all equal)

User interface (or, "USE nrutil"):

```
FUNCTION assert_eq(n1,n2,n3,...,string)
CHARACTER(LEN=*), INTENT(IN) :: string
INTEGER, INTENT(IN) :: n1,n2,n3,...
INTEGER :: assert_eq
END FUNCTION assert_eq
```

Action:

Embedding program dies gracefully with an error message if any of the integer arguments are not equal to the first. Otherwise, return the value of the first argument. Typical use is for enforcing equality on the sizes of arrays passed to a subprogram. `nrutil` implements and overloads forms with 1, 2, 3, and 4 integer arguments, plus a form with a vector integer argument,

```
INTEGER, DIMENSION(:), INTENT(IN) :: n
```

that is checked by the conditional `if (all(nn(2:)==nn(1)))`.

Reference implementation:

```
if (n1==n2.and.n2==n3.and...) then
   assert_eq=n1
else
   write (*,*) 'nrerror: an assert_eq failed with this tag:', string
   STOP 'program terminated by assert_eq'
end if
```

⋆ ⋆ ⋆

nrerror (report error message and stop)

User interface (or, "USE nrutil"):

```
SUBROUTINE nrerror(string)
CHARACTER(LEN=*), INTENT(IN) :: string
END SUBROUTINE nrerror
```

Action:

This is the minimal error handler used in this book. In applications of any complexity, it is intended only as a placeholder for a user's more complicated error handling strategy.

Reference implementation:

```
write (*,*) 'nrerror: ',string
STOP 'program terminated by nrerror'
```

23.4 Routines for Polynomials and Recurrences

Apart from programming convenience, these routines are designed to allow for nontrivial parallel implementations, as discussed in §22.2 and §22.3.

⋆ ⋆ ⋆

arth (returns arithmetic progression as an array)

User interface (or, "USE nrutil"):

```
FUNCTION arth(first,increment,n)
T, INTENT(IN) :: first,increment
INTEGER(I4B), INTENT(IN) :: n
T, DIMENSION(n) [or, 1 rank higher than T]:: arth
END FUNCTION arth
```

Applicable types and ranks:

T ≡ *any numerical type, any rank*

Types and ranks implemented (overloaded) in `nrutil`*:*

T ≡ `INTEGER(I4B), REAL(SP), REAL(DP)`

Action:

Returns an array of length `n` containing an arithmetic progression whose first value is `first` and whose increment is `increment`. If `first` and `increment` have rank greater than zero, returns an array of one larger rank, with the last subscript having size `n` and indexing the progressions. Note that the following reference implementation (for the scalar case) is definitional only, and neither parallelized nor optimized for roundoff error. See §22.2 and Appendix C1 for implementation by subvector scaling.

Reference implementation:

```
INTEGER(I4B) :: k
if (n > 0) arth(1)=first
do k=2,n
   arth(k)=arth(k-1)+increment
end do
```

⋆ ⋆ ⋆

geop (returns geometric progression as an array)

User interface (or, "USE nrutil"):

```
FUNCTION geop(first,factor,n)
T, INTENT(IN) :: first,factor
INTEGER(I4B), INTENT(IN) :: n
T, DIMENSION(n) [or, 1 rank higher than T]:: geop
END FUNCTION geop
```

Applicable types and ranks:

T ≡ *any numerical type, any rank*

Types and ranks implemented (overloaded) in `nrutil`*:*

T ≡ `INTEGER(I4B), REAL(SP), REAL(DP), REAL(DP)(:), COMPLEX(SPC)`

Action:

Returns an array of length n containing a geometric progression whose first value is first and whose multiplier is factor. If first and factor have rank greater than zero, returns an array of one larger rank, with the last subscript having size n and indexing the progression. Note that the following reference implementation (for the scalar case) is definitional only, and neither parallelized nor optimized for roundoff error. See §22.2 and Appendix C1 for implementation by subvector scaling.

Reference implementation:

```
INTEGER(I4B) :: k
if (n > 0) geop(1)=first
do k=2,n
   geop(k)=geop(k-1)*factor
end do
```

⋆ ⋆ ⋆

cumsum (cumulative sum on an array, with optional additive seed)

User interface (or, "USE nrutil"):

```
FUNCTION cumsum(arr,seed)
T, DIMENSION(:), INTENT(IN) :: arr
T, OPTIONAL, INTENT(IN) :: seed
T, DIMENSION(size(arr)), INTENT(OUT) :: cumsum
END FUNCTION cumsum
```

Applicable types and ranks:

T ≡ *any numerical type*

Types and ranks implemented (overloaded) in nrutil*:*

T ≡ INTEGER(I4B), REAL(SP)

Action:

Given the rank 1 array arr of type **T**, returns an array of identical type and size containing the cumulative sums of arr. If the optional argument seed is present, it is added to the first component (and therefore, by cumulation, all components) of the result. See §22.2 for parallelization ideas.

Reference implementation:

```
INTEGER(I4B) :: n,j
T :: sd
n=size(arr)
if (n == 0) return
sd=0.0
if (present(seed)) sd=seed
cumsum(1)=arr(1)+sd
do j=2,n
   cumsum(j)=cumsum(j-1)+arr(j)
end do
```

⋆ ⋆ ⋆

cumprod (cumulative prod on an array, with optional multiplicative seed)

User interface (or, "USE nrutil"):

```
FUNCTION cumprod(arr,seed)
T, DIMENSION(:), INTENT(IN) :: arr
T, OPTIONAL, INTENT(IN) :: seed
T, DIMENSION(size(arr)), INTENT(OUT) :: cumprod
END FUNCTION cumprod
```

Applicable types and ranks:
T ≡ *any numerical type*

Types and ranks implemented (overloaded) in `nrutil`*:*
T ≡ `REAL(SP)`

Action:
Given the rank 1 array `arr` of type **T**, returns an array of identical type and size containing the cumulative products of `arr`. If the optional argument `seed` is present, it is multiplied into the first component (and therefore, by cumulation, into all components) of the result. See §22.2 for parallelization ideas.

Reference implementation:

```
INTEGER(I4B) :: n,j
T :: sd
n=size(arr)
if (n == 0) return
sd=1.0
if (present(seed)) sd=seed
cumprod(1)=arr(1)*sd
do j=2,n
   cumprod(j)=cumprod(j-1)*arr(j)
end do
```

⋆ ⋆ ⋆

poly (polynomial evaluation)

User interface (or, "`USE nrutil`*"):*

```
FUNCTION poly(x,coeffs,mask)
T,, DIMENSION(:,...), INTENT(IN) :: x
T, DIMENSION(:), INTENT(IN) :: coeffs
LOGICAL(LGT), DIMENSION(:,...), OPTIONAL, INTENT(IN) :: mask
T :: poly
END FUNCTION poly
```

Applicable types and ranks:
T ≡ *any numerical type (*`x` *may be scalar or have any rank;* `x` *and* `coeffs` *may have different numerical types)*

Types and ranks implemented (overloaded) in `nrutil`*:*
T ≡ *various combinations of* `REAL(SP)`, `REAL(SP)(:)`, `REAL(DP)`, `REAL(DP)(:)`, `COMPLEX(SPC)` *(see Appendix C1 for details)*

Action:
Returns a scalar value or array with the same type and shape as `x`, containing the result of evaluating the polynomial with one-dimensional coefficient vector `coeffs` on each component of `x`. The optional argument `mask`, if present, has the same shape as `x`, and suppresses evaluation of the polynomial where its components are `.false.`. The following reference code shows the case where `mask` is not present. (The other case can be included by overloading.)

Reference implementation:

```
INTEGER(I4B) :: i,n
n=size(coeffs)
if (n <= 0) then
   poly=0.0
else
   poly=coeffs(n)
   do i=n-1,1,-1
      poly=x*poly+coeffs(i)
   end do
end if
```

⋆ ⋆ ⋆

poly_term (partial cumulants of a polynomial)

User interface (or, "`USE nrutil`*"):*

```
FUNCTION poly_term(a,x)
T, DIMENSION(:), INTENT(IN) :: a
T, INTENT(IN) :: x
T, DIMENSION(size(a)) :: poly_term
END FUNCTION poly_term
```

Applicable types and ranks:

T ≡ *any numerical type*

Types and ranks implemented (overloaded) in `nrutil`*:*

T ≡ `REAL(SP), COMPLEX(SPC)`

Action:

Returns an array of type and size the same as the one-dimensional array `a`, containing the partial cumulants of the polynomial with coefficients `a` (arranged from highest-order to lowest-order coefficients, n.b.) evaluated at `x`. This is equivalent to synthetic division, and can be parallelized. See §22.3. Note that the order of arguments is reversed in `poly` and `poly_term` — each routine returns a value with the size and shape of the *first* argument, the usual Fortran 90 convention.

Reference implementation:

```
INTEGER(I4B) :: n,j
n=size(a)
if (n <= 0) return
poly_term(1)=a(1)
do j=2,n
   poly_term(j)=a(j)+x*poly_term(j-1)
end do
```

⋆ ⋆ ⋆

zroots_unity (returns powers of complex nth root of unity)

User interface (or, "`USE nrutil`*"):*

```
FUNCTION zroots_unity(n,nn)
INTEGER(I4B), INTENT(IN) :: n,nn
COMPLEX(SPC), DIMENSION(nn) :: zroots_unity
END FUNCTION zroots_unity
```

Action:

Returns a complex array containing nn consecutive powers of the nth complex root of unity. Note that the following reference implementation is definitional only, and neither parallelized nor optimized for roundoff error. See Appendix C1 for implementation by subvector scaling.

Reference implementation:

```
INTEGER(I4B) :: k
REAL(SP) :: theta
if (nn==0) return
zroots_unity(1)=1.0
if (nn==1) return
theta=TWOPI/n
zroots_unity(2)=cmplx(cos(theta),sin(theta))
do k=3,nn
   zroots_unity(k)=zroots_unity(k-1)*zroots_unity(2)
end do
```

23.5 Routines for Outer Operations on Vectors

Outer operations on vectors take two vectors as input, and return a matrix as output. One dimension of the matrix is the size of the first vector, the other is the size of the second vector. Our convention is always the standard one,

$$\texttt{result(i,j)} = \texttt{first\_operand(i)}\ (op)\ \texttt{second\_operand(j)}$$

where (op) is any of addition, subtraction, multiplication, division, and logical and. The reason for coding these as utility routines is that Fortran 90's native construction, with two spreads (cf. §22.1), is difficult to read and thus prone to programmer errors.

⋆ ⋆ ⋆

outerprod (outer product)

User interface (or, "USE nrutil"):

```
FUNCTION outerprod(a,b)
T, DIMENSION(:), INTENT(IN) :: a,b
T, DIMENSION(size(a),size(b)) :: outerprod
END FUNCTION outerprod
```

Applicable types and ranks:

T ≡ *any numerical type*

Types and ranks implemented (overloaded) in nrutil*:*

T ≡ REAL(SP), REAL(DP)

Action:

Returns a matrix that is the outer product of two vectors.

Reference implementation:

```
outerprod = spread(a,dim=2,ncopies=size(b)) * &
   spread(b,dim=1,ncopies=size(a))
```

⋆ ⋆ ⋆

outerdiv (outer quotient)

User interface (or, "USE nrutil"):

```
FUNCTION outerdiv(a,b)
T, DIMENSION(:), INTENT(IN) :: a,b
T, DIMENSION(size(a),size(b)) :: outerdiv
END FUNCTION outerdiv
```

Applicable types and ranks:

T ≡ *any numerical type*

Types and ranks implemented (overloaded) in `nrutil`*:*

T ≡ `REAL(SP)`

Action:

Returns a matrix that is the outer quotient of two vectors.

Reference implementation:

```
outerdiv = spread(a,dim=2,ncopies=size(b)) / &
   spread(b,dim=1,ncopies=size(a))
```

⋆ ⋆ ⋆

outersum (outer sum)

User interface (or, "USE nrutil"):

```
FUNCTION outersum(a,b)
T, DIMENSION(:), INTENT(IN) :: a,b
T, DIMENSION(size(a),size(b)) :: outersum
END FUNCTION outersum
```

Applicable types and ranks:

T ≡ *any numerical type*

Types and ranks implemented (overloaded) in `nrutil`*:*

T ≡ `REAL(SP)`

Action:

Returns a matrix that is the outer sum of two vectors.

Reference implementation:

```
outersum = spread(a,dim=2,ncopies=size(b)) + &
   spread(b,dim=1,ncopies=size(a))
```

⋆ ⋆ ⋆

outerdiff (outer difference)

User interface (or, "USE nrutil"):

```
FUNCTION outerdiff(a,b)
T, DIMENSION(:), INTENT(IN) :: a,b
T, DIMENSION(size(a),size(b)) :: outerdiff
END FUNCTION outerdiff
```

Applicable types and ranks:

T ≡ *any numerical type*

Types and ranks implemented (overloaded) in `nrutil`*:*

T ≡ `INTEGER(I4B), REAL(SP), REAL(DP)`

Action:

Returns a matrix that is the outer difference of two vectors.

Reference implementation:

```
outerdiff = spread(a,dim=2,ncopies=size(b)) - &
   spread(b,dim=1,ncopies=size(a))
```

⋆ ⋆ ⋆

outerand (outer logical and)

User interface (or, "`USE nrutil`*"):*

```
FUNCTION outerand(a,b)
LOGICAL(LGT), DIMENSION(:), INTENT(IN) :: a,b
LOGICAL(LGT), DIMENSION(size(a),size(b)) :: outerand
END FUNCTION outerand
```

Applicable types and ranks:

T ≡ *any logical type*

Types and ranks implemented (overloaded) in `nrutil`*:*

T ≡ `LOGICAL(LGT)`

Action:

Returns a matrix that is the outer logical and of two vectors.

Reference implementation:

```
outerand = spread(a,dim=2,ncopies=size(b)) .and. &
   spread(b,dim=1,ncopies=size(a))
```

23.6 Routines for Scatter with Combine

These are common parallel functions that Fortran 90 simply doesn't provide a means for implementing. If you have a parallel machine, you should substitute library routines specific to your hardware.

⋆ ⋆ ⋆

scatter_add (scatter-add source to specified components of destination)

User interface (or, "`USE nrutil`*"):*

```
SUBROUTINE scatter_add(dest,source,dest_index)
T, DIMENSION(:), INTENT(OUT) :: dest
T, DIMENSION(:), INTENT(IN) :: source
INTEGER(I4B), DIMENSION(:), INTENT(IN) :: dest_index
END SUBROUTINE scatter_add
```

Applicable types and ranks:

T ≡ *any numerical type*

Types and ranks implemented (overloaded) in `nrutil`*:*

T ≡ `REAL(SP), REAL(DP)`

Action:

Adds each component of the array `source` into a component of `dest` specified by the index array `dest_index`. (The user will usually have zeroed `dest` before the call to this routine.) Note that `dest_index` has the size of `source`, but must contain values in the range from 1 to `size(dest)`, inclusive. Out-of-range values are ignored. There is no parallel implementation of this routine accessible from Fortran 90; most parallel machines supply an implementation as a library routine.

Reference implementation:

```
INTEGER(I4B) :: m,n,j,i
n=assert_eq(size(source),size(dest_index),'scatter_add')
m=size(dest)
do j=1,n
   i=dest_index(j)
   if (i > 0 .and. i <= m) dest(i)=dest(i)+source(j)
end do
```

⋆ ⋆ ⋆

scatter_max (scatter-max source to specified components of destination)

User interface (or, "`USE nrutil`*"):*

```
SUBROUTINE scatter_max(dest,source,dest_index)
T, DIMENSION(:), INTENT(OUT) :: dest
T, DIMENSION(:), INTENT(IN) :: source
INTEGER(I4B), DIMENSION(:), INTENT(IN) :: dest_index
END SUBROUTINE scatter_max
```

Applicable types and ranks:

T ≡ *any integer or real type*

Types and ranks implemented (overloaded) in `nrutil`*:*

T ≡ `REAL(SP), REAL(DP)`

Action:

Takes the `max` operation between each component of the array `source` and a component of `dest` specified by the index array `dest_index`, replacing that component of `dest` with the value obtained ("maxing into" operation). (The user will often want to fill the array `dest` with the value −`huge` before the call to this routine.) Note that `dest_index` has the size of `source`, but must contain values in the range from 1 to `size(dest)`, inclusive. Out-of-range values are ignored. There is no parallel implementation of this routine accessible from Fortran 90; most parallel machines supply an implementation as a library routine.

Reference implementation:

```
INTEGER(I4B) :: m,n,j,i
n=assert_eq(size(source),size(dest_index),'scatter_max')
m=size(dest)
do j=1,n
   i=dest_index(j)
   if (i > 0 .and. i <= m) dest(i)=max(dest(i),source(j))
end do
```

23.7 Routines for Skew Operations on Matrices

These are also missing parallel capabilities in Fortran 90. In Appendix C1 they are coded serially, with one or more do-loops.

⋆ ⋆ ⋆

diagadd (adds vector to diagonal of a matrix)

User interface (or, "USE nrutil"):

```
SUBROUTINE diagadd(mat,diag)
T, DIMENSION(:,:), INTENT(INOUT) :: mat
T, DIMENSION(:), INTENT(IN) :: diag
END SUBROUTINE diagadd
```

Applicable types and ranks:

T ≡ *any numerical type*

Types and ranks implemented (overloaded) in `nrutil`*:*

T ≡ `REAL(SP)`

Action:

The argument `diag`, either a scalar or else a vector whose size must be the smaller of the two dimensions of matrix `mat`, is added to the diagonal of the matrix `mat`. The following shows an implementation where `diag` is a vector; the scalar case can be overloaded (see Appendix C1).

Reference implementation:

```
INTEGER(I4B) :: j,n
n = assert_eq(size(diag),min(size(mat,1),size(mat,2)),'diagadd')
do j=1,n
   mat(j,j)=mat(j,j)+diag(j)
end do
```

⋆ ⋆ ⋆

diagmult (multiplies vector into diagonal of a matrix)

User interface (or, "USE nrutil"):

```
SUBROUTINE diagmult(mat,diag)
T, DIMENSION(:,:), INTENT(INOUT) :: mat
T, DIMENSION(:), INTENT(IN) :: diag
END SUBROUTINE diagmult
```

Applicable types and ranks:

T ≡ *any numerical type*

Types and ranks implemented (overloaded) in `nrutil`*:*

T ≡ `REAL(SP)`

Action:

The argument `diag`, either a scalar or else a vector whose size must be the smaller of the two dimensions of matrix `mat`, is multiplied onto the diagonal of the matrix `mat`. The following shows an implementation where `diag` is a vector; the scalar case can be overloaded (see Appendix C1).

Reference implementation:

```
INTEGER(I4B) :: j,n
n = assert_eq(size(diag),min(size(mat,1),size(mat,2)),'diagmult')
do j=1,n
   mat(j,j)=mat(j,j)*diag(j)
end do
```

⋆ ⋆ ⋆

get_diag (gets diagonal of matrix)

User interface (or, "USE nrutil"):

```
FUNCTION get_diag(mat)
T, DIMENSION(:,:), INTENT(IN) :: mat
T, DIMENSION(min(size(mat,1),size(mat,2))) :: get_diag
END FUNCTION get_diag
```

Applicable types and ranks:

T $\equiv$ *any type*

Types and ranks implemented (overloaded) in nrutil*:*

T $\equiv$ REAL(SP), REAL(DP)

Action:

Returns a vector containing the diagonal values of the matrix mat.

Reference implementation:

```
INTEGER(I4B) :: j
do j=1,min(size(mat,1),size(mat,2))
   get_diag(j)=mat(j,j)
end do
```

⋆ ⋆ ⋆

put_diag (sets the diagonal elements of a matrix)

User interface (or, "USE nrutil"):

```
SUBROUTINE put_diag(diag,mat)
T, DIMENSION(:), INTENT(IN) :: diag
T, DIMENSION(:,:), INTENT(INOUT) :: mat
END SUBROUTINE put_diag
```

Applicable types and ranks:

T $\equiv$ *any type*

Types and ranks implemented (overloaded) in nrutil*:*

T $\equiv$ REAL(SP)

Action:

Sets the diagonal of matrix mat equal to the argument diag, either a scalar or else a vector whose size must be the smaller of the two dimensions of matrix mat. The following shows an implementation where diag is a vector; the scalar case can be overloaded (see Appendix C1).

Reference implementation:

```
INTEGER(I4B) :: j,n
n=assert_eq(size(diag),min(size(mat,1),size(mat,2)),'put_diag')
do j=1,n
   mat(j,j)=diag(j)
end do
```

⋆ ⋆ ⋆

unit_matrix (returns a unit matrix)

User interface (or, "USE nrutil"):

```
SUBROUTINE unit_matrix(mat)
T, DIMENSION(:,:), INTENT(OUT) :: mat
END SUBROUTINE unit_matrix
```

Applicable types and ranks:

T ≡ *any numerical type*

Types and ranks implemented (overloaded) in nrutil*:*

T ≡ REAL(SP)

Action:

Sets the diagonal components of mat to unity, all other components to zero. When mat is square, this will be the unit matrix; otherwise, a unit matrix with appended rows or columns of zeros.

Reference implementation:

```
INTEGER(I4B) :: i,n
n=min(size(mat,1),size(mat,2))
mat(:,:)=0.0
do i=1,n
   mat(i,i)=1.0
end do
```

⋆ ⋆ ⋆

upper_triangle (returns an upper triangular mask)

User interface (or, "USE nrutil"):

```
FUNCTION upper_triangle(j,k,extra)
INTEGER(I4B), INTENT(IN) :: j,k
INTEGER(I4B), OPTIONAL, INTENT(IN) :: extra
LOGICAL(LGT), DIMENSION(j,k) :: upper_triangle
END FUNCTION upper_triangle
```

Action:

When the optional argument extra is zero or absent, returns a logical mask of shape (j, k) whose values are true above and to the right of the diagonal, false elsewhere (including on the diagonal). When extra is present and positive, a corresponding number of additional (sub-)diagonals are returned as true. (extra = 1 makes the main diagonal return true.) When extra is present and negative, it suppresses a corresponding number of superdiagonals.

Reference implementation:

```
INTEGER(I4B) :: n,jj,kk
n=0
if (present(extra)) n=extra
do jj=1,j
   do kk=1,k
      upper_triangle(jj,kk)= (jj-kk < n)
   end do
end do
```

⋆ ⋆ ⋆

lower_triangle (returns a lower triangular mask)

User interface (or, "USE nrutil"):

```
FUNCTION lower_triangle(j,k,extra)
INTEGER(I4B), INTENT(IN) :: j,k
INTEGER(I4B), OPTIONAL, INTENT(IN) :: extra
LOGICAL(LGT), DIMENSION(j,k) :: lower_triangle
END FUNCTION lower_triangle
```

Action:

When the optional argument extra is zero or absent, returns a logical mask of shape (j,k) whose values are true below and to the left of the diagonal, false elsewhere (including on the diagonal). When extra is present and positive, a corresponding number of additional (super-)diagonals are returned as true. (extra = 1 makes the main diagonal return true.) When extra is present and negative, it suppresses a corresponding number of subdiagonals.

Reference implementation:

```
INTEGER(I4B) :: n,jj,kk
n=0
if (present(extra)) n=extra
do jj=1,j
   do kk=1,k
      lower_triangle(jj,kk)= (kk-jj < n)
   end do
end do
```

Fortran 95's forall construction will make the parallel implementation of all our skew operations utilities extremely simple. For example, the do-loop in diagadd will collapse to

```
forall (j=1:n) mat(j,j)=mat(j,j)+diag(j)
```

In fact, this implementation is so simple as to raise the question of whether a separate utility like diagadd will be needed at all. There are valid arguments on both sides of this question: The "con" argument, against a routine like diagadd, is that it is just another reserved name that you have to remember (if you want to use it). The "pro" argument is that a separate routine avoids the "index pollution" (the opposite disease from "index loss" discussed in §22.1) of introducing a superfluous variable j, and that a separate utility allows for additional error checking on the sizes and compatibility of its arguments. We expect that different programmers will have differing tastes.

The argument for keeping a routine like upper_triangle or lower_triangle, once Fortran 95's *masked* forall constructions become available, is less persuasive. We recommend that you consider these two routines as placeholders for "remember to recode this in Fortran 95, someday."

23.8 Other Routine(s)

You might argue that we don't really need a routine for the idiom

```
sqrt(dot_product(v,v))
```

You might be right. The ability to overload the complex case, with its additional complex conjugate, is an argument in its favor, however.

⋆ ⋆ ⋆

vabs (L_2 norm of a vector)

User interface (or, "`USE nrutil`*"):*

```
FUNCTION vabs(v)
T, DIMENSION(:), INTENT(IN) :: v
T :: vabs
END FUNCTION vabs
```

Applicable types and ranks:

T ≡ *any real or complex type*

Types and ranks implemented (overloaded) in `nrutil`*:*

T ≡ `REAL(SP)`

Action:

Returns the length of a vector `v` in L_2 norm, that is, the square root of the sum of the squares of the components. (For complex types, the `dot_product` should be between the vector and its complex conjugate.)

Reference implementation:

```
vabs=sqrt(dot_product(v,v))
```

Fortran 90 Code Chapters B1–B20

Fortran 90 versions of all the Numerical Recipes routines appear in the following Chapters B1 through B20, numbered in correspondence with Chapters 1 through 20 in Volume 1. Within each chapter, the routines appear in the same order as in Volume 1, but not broken out separately by section number within Volume 1's chapters.

There are commentaries accompanying many of the routines, generally following the printed listing of the routine to which they apply. These are of two kinds: issues related to parallelizing the algorithm in question, and issues related to the Fortran 90 implementation. To distinguish between these two, rather different, kinds of discussions, we use the two icons,

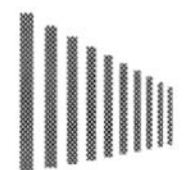

the left icon (above) indicating a "parallel note," and the right icon denoting a "Fortran 90 tip." Specific code segments of the routine that are discussed in these commentaries are singled out by reproducing some of the code as an "index line" next to the icon, or at the beginning of subsequent paragraphs if there are several items that are commented on.

`d=merge(FPMIN,d,abs(d)<FPMIN)` This would be the start of a discussion of code that begins at the line in the listing containing the indicated code fragment.

⋆ ⋆ ⋆

A row of stars, like the above, is used between unrelated routines, or at the beginning and end of related groups of routines.

Some chapters contain discussions that are more general than commentary on individual routines, but that were deemed too specific for inclusion in Chapters 21 through 23. Here are some highlights of this additional material:

- Approximations to roots of orthogonal polynomials for parallel computation of Gaussian quadrature formulas (Chapter B4)
- Difficulty of, and tricks for, parallel calculation of special function values in a SIMD model of computation (Chapter B6)
- Parallel random number generation (Chapter B7)
- Fortran 90 tricks for dealing with ties in sorted arrays, counting things in boxes, etc. (Chapter B14)
- Use of recursion in implementing multigrid elliptic PDE solvers (Chapter B19)

Chapter B1. Preliminaries

```
SUBROUTINE flmoon(n,nph,jd,frac)
USE nrtype; USE nrutil, ONLY : nrerror
IMPLICIT NONE
INTEGER(I4B), INTENT(IN) :: n,nph
INTEGER(I4B), INTENT(OUT) :: jd
REAL(SP), INTENT(OUT) :: frac
```

Our programs begin with an introductory comment summarizing their purpose and explaining their calling sequence. This routine calculates the phases of the moon. Given an integer n and a code nph for the phase desired (nph $= 0$ for new moon, 1 for first quarter, 2 for full, 3 for last quarter), the routine returns the Julian Day Number jd, and the fractional part of a day frac to be added to it, of the nth such phase since January, 1900. Greenwich Mean Time is assumed.

```
REAL(SP), PARAMETER :: RAD=PI/180.0_sp
INTEGER(I4B) :: i
REAL(SP) :: am,as,c,t,t2,xtra
c=n+nph/4.0_sp                          This is how we comment an individual line.
t=c/1236.85_sp
t2=t**2
as=359.2242_sp+29.105356_sp*c           You aren't really intended to understand this al-
am=306.0253_sp+385.816918_sp*c+0.010730_sp*t2        gorithm, but it does work!
jd=2415020+28*n+7*nph
xtra=0.75933_sp+1.53058868_sp*c+(1.178e-4_sp-1.55e-7_sp*t)*t2
select case(nph)
    case(0,2)
        xtra=xtra+(0.1734_sp-3.93e-4_sp*t)*sin(RAD*as)-0.4068_sp*sin(RAD*am)
    case(1,3)
        xtra=xtra+(0.1721_sp-4.0e-4_sp*t)*sin(RAD*as)-0.6280_sp*sin(RAD*am)
    case default
        call nrerror('flmoon: nph is unknown')       This is how we will indicate error
end select                                               conditions.
i=int(merge(xtra,xtra-1.0_sp, xtra >= 0.0))
jd=jd+i
frac=xtra-i
END SUBROUTINE flmoon
```

f90 `select case(nph)...case(0,2)...end select` Fortran 90 includes a case construction that executes at most one of several blocks of code, depending on the value of an integer, logical, or character expression. Ideally, the case construction will execute more efficiently than a long sequence of cascaded `if...else if...else if...` constructions. C programmers should note that the Fortran 90 construction, perhaps mercifully, does not have C's "drop-through" feature.

`merge(xtra,xtra-1.0_sp, xtra >= 0.0)` The merge construction in Fortran 90, while intended primarily for use with vector arguments, is also a convenient way of generating conditional scalar expressions, that is, expressions with one value, or another, depending on the result of a logical test.

When the arguments of a merge are vectors, parallelization by the compiler is straightforward as an array parallel operation (see p. 964). Less obvious is how the scalar case, as above, is handled. For small-scale parallel (SSP) machines, the natural gain is via speculative evaluation of both of the first two arguments simultaneously with evaluation of the test.

A good compiler should not penalize a scalar machine for use of either the scalar or vector merge construction. The Fortran 90 standard states that "it is not necessary for a processor to evaluate all of the operands of an expression, or to evaluate entirely each operand, if the value of the expression can be determined otherwise." Therefore, for each test on a scalar machine, only one or the other of the first two argument components need be evaluated.

⋆ ⋆ ⋆

```
FUNCTION julday(mm,id,iyyy)
USE nrtype; USE nrutil, ONLY : nrerror
IMPLICIT NONE
INTEGER(I4B), INTENT(IN) :: mm,id,iyyy
INTEGER(I4B) :: julday
   In this routine julday returns the Julian Day Number that begins at noon of the calendar
   date specified by month mm, day id, and year iyyy, all integer variables. Positive year
   signifies A.D.; negative, B.C. Remember that the year after 1 B.C. was 1 A.D.
INTEGER(I4B), PARAMETER :: IGREG=15+31*(10+12*1582)     Gregorian Calendar adopted
INTEGER(I4B) :: ja,jm,jy                                      Oct. 15, 1582.
jy=iyyy
if (jy == 0) call nrerror('julday: there is no year zero')
if (jy < 0) jy=jy+1
if (mm > 2) then                                  Here is an example of a block IF-structure.
    jm=mm+1
else
    jy=jy-1
    jm=mm+13
end if
julday=int(365.25_sp*jy)+int(30.6001_sp*jm)+id+1720995
if (id+31*(mm+12*iyyy) >= IGREG) then             Test whether to change to Gregorian Cal-
    ja=int(0.01_sp*jy)                                endar.
    julday=julday+2-ja+int(0.25_sp*ja)
end if
END FUNCTION julday
```

⋆ ⋆ ⋆

```
PROGRAM badluk
USE nrtype
USE nr, ONLY : flmoon,julday
IMPLICIT NONE
INTEGER(I4B) :: ic,icon,idwk,ifrac,im,iyyy,jd,jday,n
INTEGER(I4B) :: iybeg=1900,iyend=2000      The range of dates to be searched.
REAL(SP) :: frac
REAL(SP), PARAMETER :: TIMZON=-5.0_sp/24.0_sp
  Time zone -5 is Eastern Standard Time.
write (*,'(1x,a,i5,a,i5)') 'Full moons on Friday the 13th from',&
    iybeg,' to',iyend
do iyyy=iybeg,iyend                          Loop over each year,
    do im=1,12                               and each month.
        jday=julday(im,13,iyyy)              Is the 13th a Friday?
        idwk=mod(jday+1,7)
```

```
        if (idwk == 5) then
            n=12.37_sp*(iyyy-1900+(im-0.5_sp)/12.0_sp)
              This value n is a first approximation to how many full moons have occurred
              since 1900. We will feed it into the phase routine and adjust it up or down until
              we determine that our desired 13th was or was not a full moon. The variable
              icon signals the direction of adjustment.
            icon=0
            do
                call flmoon(n,2,jd,frac)        Get date of full moon n.
                ifrac=nint(24.0_sp*(frac+TIMZON))     Convert to hours in correct time
                if (ifrac < 0) then                       zone.
                    jd=jd-1                     Convert from Julian Days beginning at noon
                    ifrac=ifrac+24                 to civil days beginning at midnight.
                end if
                if (ifrac > 12) then
                    jd=jd+1
                    ifrac=ifrac-12
                else
                    ifrac=ifrac+12
                end if
                if (jd == jday) then            Did we hit our target day?
                    write (*,'(/1x,i2,a,i2,a,i4)') im,'/',13,'/',iyyy
                    write (*,'(1x,a,i2,a)') 'Full moon ',ifrac,&
                        ' hrs after midnight (EST).'
                      Don't worry if you are unfamiliar with FORTRAN's esoteric input/output
                      statements; very few programs in this book do any input/output.
                    exit                        Part of the break-structure, case of a match.
                else                            Didn't hit it.
                    ic=isign(1,jday-jd)
                    if (ic == -icon) exit          Another break, case of no match.
                    icon=ic
                    n=n+ic
                end if
            end do
        end if
    end do
end do
END PROGRAM badluk
```

f90 `...IGREG=15+31*(10+12*1582)` (in julday), `...TIMZON=-5.0_sp/24.0_sp` (in badluk) These are two examples of initialization expressions for "named constants" (that is, PARAMETERs). Because the initialization expressions will generally be evaluated at compile time, Fortran 90 puts some restrictions on what kinds of intrinsic functions they can contain. Although the evaluation of a real expression like `-5.0_sp/24.0_sp` *ought* to give identical results at compile time and at execution time, all the way down to the least significant bit, in our opinion the conservative programmer shouldn't count on strict identity at the level of floating-point roundoff error. (In the special case of *cross*-compilers, such roundoff-level discrepancies between compile time and run time are almost inevitable.)

⋆ ⋆ ⋆

```
SUBROUTINE caldat(julian,mm,id,iyyy)
USE nrtype
IMPLICIT NONE
INTEGER(I4B), INTENT(IN) :: julian
INTEGER(I4B), INTENT(OUT) :: mm,id,iyyy
    Inverse of the function julday given above. Here julian is input as a Julian Day Number,
    and the routine outputs mm,id, and iyyy as the month, day, and year on which the specified
    Julian Day started at noon.
INTEGER(I4B) :: ja,jalpha,jb,jc,jd,je
INTEGER(I4B), PARAMETER :: IGREG=2299161
if (julian >= IGREG) then                    Cross-over to Gregorian Calendar produces this
    jalpha=int(((julian-1867216)-0.25_sp)/36524.25_sp)      correction.
    ja=julian+1+jalpha-int(0.25_sp*jalpha)
else if (julian < 0) then                    Make day number positive by adding integer num-
    ja=julian+36525*(1-julian/36525)              ber of Julian centuries, then subtract them
else                                              off at the end.
    ja=julian
end if
jb=ja+1524
jc=int(6680.0_sp+((jb-2439870)-122.1_sp)/365.25_sp)
jd=365*jc+int(0.25_sp*jc)
je=int((jb-jd)/30.6001_sp)
id=jb-jd-int(30.6001_sp*je)
mm=je-1
if (mm > 12) mm=mm-12
iyyy=jc-4715
if (mm > 2) iyyy=iyyy-1
if (iyyy <= 0) iyyy=iyyy-1
if (julian < 0) iyyy=iyyy-100*(1-julian/36525)
END SUBROUTINE caldat
```

Chapter B2. Solution of Linear Algebraic Equations

```
SUBROUTINE gaussj(a,b)
USE nrtype; USE nrutil, ONLY : assert_eq,nrerror,outerand,outerprod,swap
IMPLICIT NONE
REAL(SP), DIMENSION(:,:), INTENT(INOUT) :: a,b
```

Linear equation solution by Gauss-Jordan elimination, equation (2.1.1). a is an $N \times N$ input coefficient matrix. b is an $N \times M$ input matrix containing M right-hand-side vectors. On output, a is replaced by its matrix inverse, and b is replaced by the corresponding set of solution vectors.

```
INTEGER(I4B), DIMENSION(size(a,1)) :: ipiv,indxr,indxc
```

These arrays are used for bookkeeping on the pivoting.

```
LOGICAL(LGT), DIMENSION(size(a,1)) :: lpiv
REAL(SP) :: pivinv
REAL(SP), DIMENSION(size(a,1)) :: dumc
INTEGER(I4B), TARGET :: irc(2)
INTEGER(I4B) :: i,l,n
INTEGER(I4B), POINTER :: irow,icol
n=assert_eq(size(a,1),size(a,2),size(b,1),'gaussj')
irow => irc(1)
icol => irc(2)
ipiv=0
do i=1,n                          Main loop over columns to be reduced.
    lpiv = (ipiv == 0)            Begin search for a pivot element.
    irc=maxloc(abs(a),outerand(lpiv,lpiv))
    ipiv(icol)=ipiv(icol)+1
    if (ipiv(icol) > 1) call nrerror('gaussj: singular matrix (1)')
```

We now have the pivot element, so we interchange rows, if needed, to put the pivot element on the diagonal. The columns are not physically interchanged, only relabeled: indxc(i), the column of the ith pivot element, is the ith column that is reduced, while indxr(i) is the row in which that pivot element was originally located. If indxr(i) $\neq$ indxc(i) there is an implied column interchange. With this form of bookkeeping, the solution b's will end up in the correct order, and the inverse matrix will be scrambled by columns.

```
    if (irow /= icol) then
        call swap(a(irow,:),a(icol,:))
        call swap(b(irow,:),b(icol,:))
    end if
    indxr(i)=irow                 We are now ready to divide the pivot row by the pivot
    indxc(i)=icol                     element, located at irow and icol.
    if (a(icol,icol) == 0.0) &
        call nrerror('gaussj: singular matrix (2)')
    pivinv=1.0_sp/a(icol,icol)
    a(icol,icol)=1.0
    a(icol,:)=a(icol,:)*pivinv
    b(icol,:)=b(icol,:)*pivinv
    dumc=a(:,icol)                Next, we reduce the rows, except for the pivot one, of
    a(:,icol)=0.0                     course.
```

```
      a(icol,icol)=pivinv
      a(1:icol-1,:)=a(1:icol-1,:)-outerprod(dumc(1:icol-1),a(icol,:))
      b(1:icol-1,:)=b(1:icol-1,:)-outerprod(dumc(1:icol-1),b(icol,:))
      a(icol+1:,:)=a(icol+1:,:)-outerprod(dumc(icol+1:),a(icol,:))
      b(icol+1:,:)=b(icol+1:,:)-outerprod(dumc(icol+1:),b(icol,:))
end do
```

It only remains to unscramble the solution in view of the column interchanges. We do this by interchanging pairs of columns in the reverse order that the permutation was built up.

```
do l=n,1,-1
      call swap(a(:,indxr(l)),a(:,indxc(l)))
end do
END SUBROUTINE gaussj
```

f90 `irow => irc(1) ... icol => irc(2)` The `maxloc` intrinsic returns the location of the maximum value of an array as an integer array, in this case of size 2. Pre-pointing pointer variables to components of the array that will be thus set makes possible convenient references to the desired row and column positions.

`irc=maxloc(abs(a),outerand(lpiv,lpiv))` The combination of `maxloc` and one of the `outer...` routines from `nrutil` allows for a very concise formulation. If this task is done with loops, it becomes the ungainly "flying vee,"

```
aa=0.0
do i=1,n
   if (lpiv(i)) then
      do j=1,n
         if (lpiv(j)) then
            if (abs(a(i,j)) > aa) then
               aa=abs(a(i,j))
               irow=i
               icol=j
            endif
         endif
      end do
   end do
end do
```

`call swap(a(irow,:),a(icol,:))` The `swap` routine (in `nrutil`) is concise and convenient. Fortran 90's ability to overload multiple routines onto a single name is vital here: Much of the convenience would vanish if we had to remember variant routine names for each variable type and rank of object that might be swapped.

Even better, here, than overloading would be if Fortran 90 allowed user-written *elemental* procedures (procedures with unspecified or arbitrary rank and shape), like the intrinsic elemental procedures built into the language. Fortran 95 will, but Fortran 90 doesn't.

One quick (if superficial) test for how much parallelism is achieved in a Fortran 90 routine is to count its do-loops, and compare that number to the number of do-loops in the Fortran 77 version of the same routine. Here, in `gaussj`, 13 do-loops are reduced to 2.

`a(1:icol-1,:)=... b(1:icol-1,:)=...`
`a(icol+1:,:)=... b(icol+1:,:)=...`

Here the same operation is applied to every row of a, and to every row of b, *except* row number icol. On a massively multiprocessor (MMP) machine it would be better to use a logical mask and do all of a in a single statement, all of b in another one. For a small-scale parallel (SSP) machine, the lines as written should saturate the machine's concurrency, and they avoid the additional overhead of testing the mask.

This would be a good place to point out, however, that linear algebra routines written in Fortran 90 are likely *never* to be competitive with the hand-coded library routines that are generally supplied as part of MMP programming environments. If you are using our routines instead of library routines written specifically for your architecture, you are wasting cycles!

⋆ ⋆ ⋆

```
SUBROUTINE ludcmp(a,indx,d)
USE nrtype; USE nrutil, ONLY : assert_eq,imaxloc,nrerror,outerprod,swap
IMPLICIT NONE
REAL(SP), DIMENSION(:,:), INTENT(INOUT) :: a
INTEGER(I4B), DIMENSION(:), INTENT(OUT) :: indx
REAL(SP), INTENT(OUT) :: d
    Given an N × N input matrix a, this routine replaces it by the LU decomposition of a
    rowwise permutation of itself. On output, a is arranged as in equation (2.3.14); indx is an
    output vector of length N that records the row permutation effected by the partial pivoting;
    d is output as ±1 depending on whether the number of row interchanges was even or odd,
    respectively. This routine is used in combination with lubksb to solve linear equations or
    invert a matrix.
REAL(SP), DIMENSION(size(a,1)) :: vv          vv stores the implicit scaling of each row.
REAL(SP), PARAMETER :: TINY=1.0e-20_sp        A small number.
INTEGER(I4B) :: j,n,imax
n=assert_eq(size(a,1),size(a,2),size(indx),'ludcmp')
d=1.0                                         No row interchanges yet.
vv=maxval(abs(a),dim=2)                       Loop over rows to get the implicit scaling
if (any(vv == 0.0)) call nrerror('singular matrix in ludcmp')      information.
  There is a row of zeros.
vv=1.0_sp/vv                                  Save the scaling.
do j=1,n
    imax=(j-1)+imaxloc(vv(j:n)*abs(a(j:n,j)))        Find the pivot row.
    if (j /= imax) then                       Do we need to interchange rows?
        call swap(a(imax,:),a(j,:))           Yes, do so...
        d=-d                                  ...and change the parity of d.
        vv(imax)=vv(j)                        Also interchange the scale factor.
    end if
    indx(j)=imax
    if (a(j,j) == 0.0) a(j,j)=TINY
      If the pivot element is zero the matrix is singular (at least to the precision of the al-
      gorithm). For some applications on singular matrices, it is desirable to substitute TINY
      for zero.
    a(j+1:n,j)=a(j+1:n,j)/a(j,j)              Divide by the pivot element.
    a(j+1:n,j+1:n)=a(j+1:n,j+1:n)-outerprod(a(j+1:n,j),a(j,j+1:n))
      Reduce remaining submatrix.
end do
END SUBROUTINE ludcmp
```

f90 vv=maxval(abs(a),dim=2) A single statement finds the maximum absolute value in each row. Fortran 90 intrinsics like maxval generally "do their thing" *in* the dimension specified by *dim* and return a result with a shape corresponding to the *other* dimensions. Thus, here, vv's size is that of the *first* dimension of a.

`imax=(j-1)+imaxloc(vv(j:n)*abs(a(j:n,j)))` Here we see why the `nrutil` routine `imaxloc` is handy: We want the index, in the range `1:n` of a quantity to be searched for only in the limited range `j:n`. Using `imaxloc`, we just add back the proper offset of `j-1`. (Using only Fortran 90 intrinsics, we could write `imax=(j-1)+sum(maxloc(vv(j:n)*abs(a(j:n,j))))`, but the use of `sum` just to turn an array of length 1 into a scalar seems sufficiently confusing as to be avoided.)

`a(j+1:n,j+1:n)=a(j+1:n,j+1:n)-outerprod(a(j+1:n,j),a(j,j+1:n))` The Fortran 77 version of `ludcmp`, using Crout's algorithm for the reduction, does not parallelize well: The elements are updated by $O(N^2)$ separate dot product operations in a particular order. Here we use a slightly different reduction, termed "outer product Gaussian elimination" by Golub and Van Loan [1], that requires just N steps of matrix-parallel reduction. (See their §3.2.3 and §3.2.9 for the algorithm, and their §3.4.1 to understand how the pivoting is performed.)

We use `nrutil`'s routine `outerprod` instead of the more cumbersome pure Fortran 90 construction:

```
spread(a(j+1:n,j),dim=2,ncopies=n-j)*spread(a(j,j+1:n),dim=1,ncopies=n-j)
```

```
SUBROUTINE lubksb(a,indx,b)
USE nrtype; USE nrutil, ONLY : assert_eq
IMPLICIT NONE
REAL(SP), DIMENSION(:,:), INTENT(IN) :: a
INTEGER(I4B), DIMENSION(:), INTENT(IN) :: indx
REAL(SP), DIMENSION(:), INTENT(INOUT) :: b
    Solves the set of N linear equations A · X = B. Here the N × N matrix a is input, not
    as the original matrix A, but rather as its LU decomposition, determined by the routine
    ludcmp. indx is input as the permutation vector of length N returned by ludcmp. b is
    input as the right-hand-side vector B, also of length N, and returns with the solution vector
    X. a and indx are not modified by this routine and can be left in place for successive calls
    with different right-hand sides b. This routine takes into account the possibility that b will
    begin with many zero elements, so it is efficient for use in matrix inversion.
INTEGER(I4B) :: i,n,ii,ll
REAL(SP) :: summ
n=assert_eq(size(a,1),size(a,2),size(indx),'lubksb')
ii=0                              When ii is set to a positive value, it will become the in-
do i=1,n                              dex of the first nonvanishing element of b. We now do
    ll=indx(i)                        the forward substitution, equation (2.3.6). The only new
    summ=b(ll)                        wrinkle is to unscramble the permutation as we go.
    b(ll)=b(i)
    if (ii /= 0) then
        summ=summ-dot_product(a(i,ii:i-1),b(ii:i-1))
    else if (summ /= 0.0) then
        ii=i                      A nonzero element was encountered, so from now on we will
    end if                            have to do the dot product above.
    b(i)=summ
end do
do i=n,1,-1                       Now we do the backsubstitution, equation (2.3.7).
    b(i) = (b(i)-dot_product(a(i,i+1:n),b(i+1:n)))/a(i,i)
end do
END SUBROUTINE lubksb
```

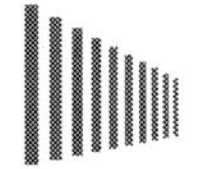

Conceptually, the search for the first nonvanishing element of b (index `ii`) should be moved out of the first do-loop. However, in practice, the need to unscramble the permutation, and also considerations of performance on scalar machines, cause us to write this very scalar-looking code. The performance penalty on parallel machines should be minimal.

⋆ ⋆ ⋆

Serial and parallel algorithms for tridiagonal problems are quite different. We therefore provide separate routines tridag_ser and tridag_par. In the MODULE nr interface file, one or the other of these (your choice) is given the generic name tridag. Of course, *either* version will work correctly on any computer; it is only a question of efficiency. See §22.2 for the numbering of the equation coefficients, and for a description of the parallel algorithm.

```
SUBROUTINE tridag_ser(a,b,c,r,u)
USE nrtype; USE nrutil, ONLY : assert_eq,nrerror
IMPLICIT NONE
REAL(SP), DIMENSION(:), INTENT(IN) :: a,b,c,r
REAL(SP), DIMENSION(:), INTENT(OUT) :: u
   Solves for a vector u of size N the tridiagonal linear set given by equation (2.4.1) using a
   serial algorithm. Input vectors b (diagonal elements) and r (right-hand sides) have size N,
   while a and c (off-diagonal elements) are size N − 1.
REAL(SP), DIMENSION(size(b)) :: gam          One vector of workspace, gam is needed.
INTEGER(I4B) :: n,j
REAL(SP) :: bet
n=assert_eq((/size(a)+1,size(b),size(c)+1,size(r),size(u)/),'tridag_ser')
bet=b(1)
if (bet == 0.0) call nrerror('tridag_ser: Error at code stage 1')
  If this happens then you should rewrite your equations as a set of order N − 1, with u2
  trivially eliminated.
u(1)=r(1)/bet
do j=2,n                                     Decomposition and forward substitution.
    gam(j)=c(j-1)/bet
    bet=b(j)-a(j-1)*gam(j)
    if (bet == 0.0) &                        Algorithm fails; see below routine in Vol. 1.
        call nrerror('tridag_ser: Error at code stage 2')
    u(j)=(r(j)-a(j-1)*u(j-1))/bet
end do
do j=n-1,1,-1                                Backsubstitution.
    u(j)=u(j)-gam(j+1)*u(j+1)
end do
END SUBROUTINE tridag_ser
```

```
RECURSIVE SUBROUTINE tridag_par(a,b,c,r,u)
USE nrtype; USE nrutil, ONLY : assert_eq,nrerror
USE nr, ONLY : tridag_ser
IMPLICIT NONE
REAL(SP), DIMENSION(:), INTENT(IN) :: a,b,c,r
REAL(SP), DIMENSION(:), INTENT(OUT) :: u
   Solves for a vector u of size N the tridiagonal linear set given by equation (2.4.1) using a
   parallel algorithm. Input vectors b (diagonal elements) and r (right-hand sides) have size
   N, while a and c (off-diagonal elements) are size N − 1.
INTEGER(I4B), PARAMETER :: NPAR_TRIDAG=4     Determines when serial algorithm is in-
INTEGER(I4B) :: n,n2,nm,nx                       voked.
REAL(SP), DIMENSION(size(b)/2) :: y,q,piva
REAL(SP), DIMENSION(size(b)/2-1) :: x,z
REAL(SP), DIMENSION(size(a)/2) :: pivc
n=assert_eq((/size(a)+1,size(b),size(c)+1,size(r),size(u)/),'tridag_par')
if (n < NPAR_TRIDAG) then
    call tridag_ser(a,b,c,r,u)
else
    if (maxval(abs(b(1:n))) == 0.0) &        Algorithm fails; see below routine in Vol. 1.
```

```
        call nrerror('tridag_par: possible singular matrix')
    n2=size(y)
    nm=size(pivc)
    nx=size(x)
    piva = a(1:n-1:2)/b(1:n-1:2)             Zero the odd a's and even c's, giving x,
    pivc = c(2:n-1:2)/b(3:n:2)                   y, z, q.
    y(1:nm) = b(2:n-1:2)-piva(1:nm)*c(1:n-2:2)-pivc*a(2:n-1:2)
    q(1:nm) = r(2:n-1:2)-piva(1:nm)*r(1:n-2:2)-pivc*r(3:n:2)
    if (nm < n2) then
        y(n2) = b(n)-piva(n2)*c(n-1)
        q(n2) = r(n)-piva(n2)*r(n-1)
    end if
    x = -piva(2:n2)*a(2:n-2:2)
    z = -pivc(1:nx)*c(3:n-1:2)
    call tridag_par(x,y,z,q,u(2:n:2))        Recurse and get even u's.
    u(1) = (r(1)-c(1)*u(2))/b(1)             Substitute and get odd u's.
    u(3:n-1:2) = (r(3:n-1:2)-a(2:n-2:2)*u(2:n-2:2) &
        -c(3:n-1:2)*u(4:n:2))/b(3:n-1:2)
    if (nm == n2) u(n)=(r(n)-a(n-1)*u(n-1))/b(n)
end if
END SUBROUTINE tridag_par
```

f90 The serial version `tridag_ser` is called when the routine has recursed its way down to sufficiently small subproblems. The point at which this occurs is determined by the parameter `NPAR_TRIDAG` whose optimal value is likely machine-dependent. Notice that `tridag_ser` must here be called by its specific name, not by the generic `tridag` (which might itself be overloaded with either `tridag_ser` or `tridag_par`).

⋆ ⋆ ⋆

```
SUBROUTINE banmul(a,m1,m2,x,b)
USE nrtype; USE nrutil, ONLY : assert_eq,arth
IMPLICIT NONE
REAL(SP), DIMENSION(:,:), INTENT(IN) :: a
INTEGER(I4B), INTENT(IN) :: m1,m2
REAL(SP), DIMENSION(:), INTENT(IN) :: x
REAL(SP), DIMENSION(:), INTENT(OUT) :: b
```

Matrix multiply $\mathbf{b} = \mathbf{A} \cdot \mathbf{x}$, where $\mathbf{A}$ is band diagonal with `m1` rows below the diagonal and `m2` rows above. If the input vector $\mathbf{x}$ and output vector $\mathbf{b}$ are of length N, then the array `a(1:`N`,1:m1+m2+1)` stores $\mathbf{A}$ as follows: The diagonal elements are in `a(1:`N`,m1+1)`. Subdiagonal elements are in `a(`j`:`N`,1:m1)` (with $j > 1$ appropriate to the number of elements on each subdiagonal). Superdiagonal elements are in `a(1:`j`,m1+2:m1+m2+1)` with $j < N$ appropriate to the number of elements on each superdiagonal.

```
INTEGER(I4B) :: m,n
n=assert_eq(size(a,1),size(b),size(x),'banmul: n')
m=assert_eq(size(a,2),m1+m2+1,'banmul: m')
b=sum(a*eoshift(spread(x,dim=2,ncopies=m), &
    dim=1,shift=arth(-m1,1,m)),dim=2)
END SUBROUTINE banmul
```

f90
```
b=sum(a*eoshift(spread(x,dim=2,ncopies=m), &
    dim=1,shift=arth(-m1,1,m)),dim=2)
```

This is a good example of Fortran 90 at both its best and its worst: best, because it allows quite subtle combinations of fully parallel operations to be built up; worst, because the resulting code is virtually incomprehensible!

What is going on becomes clearer if we imagine a temporary array y with a declaration like `REAL(SP), DIMENSION(size(a,1),size(a,2)) :: y`. Then, the above single line decomposes into

```
y=spread(x,dim=2,ncopies=m)              [Duplicate x into columns of y.]
y=eoshift(y,dim=1,shift=arth(-m1,1,m))   [Shift columns by a linear progression.]
b=sum(a*y,dim=2)                         [Multiply by the band-diagonal elements,
                                              and sum.]
```

We use here a relatively rare subcase of the `eoshift` intrinsic, using a vector value for the `shift` argument to accomplish the simultaneous shifting of a bunch of columns, by different amounts (here specified by the linear progression returned by `arth`).

If you still don't see how this accomplishes the multiplication of a band diagonal matrix by a vector, work through a simple example by hand.

```
SUBROUTINE bandec(a,m1,m2,al,indx,d)
USE nrtype; USE nrutil, ONLY : assert_eq,imaxloc,swap,arth
IMPLICIT NONE
REAL(SP), DIMENSION(:,:), INTENT(INOUT) :: a
INTEGER(I4B), INTENT(IN) :: m1,m2
REAL(SP), DIMENSION(:,:), INTENT(OUT) :: al
INTEGER(I4B), DIMENSION(:), INTENT(OUT) :: indx
REAL(SP), INTENT(OUT) :: d
REAL(SP), PARAMETER :: TINY=1.0e-20_sp
```

Given an $N \times N$ band diagonal matrix $\mathbf{A}$ with `m1` subdiagonal rows and `m2` superdiagonal rows, compactly stored in the array `a(1:`N`,1:m1+m2+1)` as described in the comment for routine `banmul`, this routine constructs an LU decomposition of a rowwise permutation of $\mathbf{A}$. The upper triangular matrix replaces `a`, while the lower triangular matrix is returned in `al(1:`N`,1:m1)`. `indx` is an output vector of length N that records the row permutation effected by the partial pivoting; `d` is output as ± 1 depending on whether the number of row interchanges was even or odd, respectively. This routine is used in combination with `banbks` to solve band-diagonal sets of equations.

```
INTEGER(I4B) :: i,k,l,mdum,mm,n
REAL(SP) :: dum
n=assert_eq(size(a,1),size(al,1),size(indx),'bandec: n')
mm=assert_eq(size(a,2),m1+m2+1,'bandec: mm')
mdum=assert_eq(size(al,2),m1,'bandec: mdum')
a(1:m1,:)=eoshift(a(1:m1,:),dim=2,shift=arth(m1,-1,m1))     Rearrange the storage a
d=1.0                                                            bit.
do k=1,n                                  For each row...
    l=min(m1+k,n)
    i=imaxloc(abs(a(k:l,1)))+k-1          Find the pivot element.
    dum=a(i,1)
    if (dum == 0.0) a(k,1)=TINY
```

Matrix is algorithmically singular, but proceed anyway with TINY pivot (desirable in some applications).

```
    indx(k)=i
    if (i /= k) then                      Interchange rows.
        d=-d
        call swap(a(k,1:mm),a(i,1:mm))
    end if
    do i=k+1,l                            Do the elimination.
        dum=a(i,1)/a(k,1)
        al(k,i-k)=dum
        a(i,1:mm-1)=a(i,2:mm)-dum*a(k,2:mm)
        a(i,mm)=0.0
    end do
end do
END SUBROUTINE bandec
```

`a(1:m1,:)=eoshift(a(1:m1,:),...` See similar discussion of `eoshift` for `banmul`, just above.

`i=imaxloc(abs(a(k:l,1)))+k-1` See discussion of `imaxloc` on p. 1017.

Notice that the above is *not* well parallelized for MMP machines: the outer do-loop is done N times, where N, the diagonal length, is potentially the largest dimension in the problem. Small-scale parallel (SSP) machines, and scalar machines, are not disadvantaged, because the parallelism of order `mm=m1+m2+1` in the inner loops can be enough to saturate their concurrency.

We don't know of an N-parallel algorithm for decomposing band diagonal matrices, at least one that has any reasonably concise expression in Fortran 90. Conceptually, one can view a band diagonal matrix as a *block tridiagonal* matrix, and then apply the same recursive strategy as was used in `tridag_par`. However, the implementation details of this are daunting. (We would welcome a user-contributed routine, clear, concise, and with parallelism of order N.)

```
SUBROUTINE banbks(a,m1,m2,al,indx,b)
USE nrtype; USE nrutil, ONLY : assert_eq,swap
IMPLICIT NONE
REAL(SP), DIMENSION(:,:), INTENT(IN) :: a,al
INTEGER(I4B), INTENT(IN) :: m1,m2
INTEGER(I4B), DIMENSION(:), INTENT(IN) :: indx
REAL(SP), DIMENSION(:), INTENT(INOUT) :: b
    Given the arrays a, al, and indx as returned from bandec, and given a right-hand-side
    vector b, solves the band diagonal linear equations A·x = b. The solution vector x overwrites
    b. The other input arrays are not modified, and can be left in place for successive calls with
    different right-hand sides.
INTEGER(I4B) :: i,k,l,mdum,mm,n
n=assert_eq(size(a,1),size(al,1),size(b),size(indx),'banbks: n')
mm=assert_eq(size(a,2),m1+m2+1,'banbks: mm')
mdum=assert_eq(size(al,2),m1,'banbks: mdum')
do k=1,n                         Forward substitution, unscrambling the permuted rows as we
    l=min(n,m1+k)                    go.
    i=indx(k)
    if (i /= k) call swap(b(i),b(k))
    b(k+1:l)=b(k+1:l)-al(k,1:l-k)*b(k)
end do
do i=n,1,-1                      Backsubstitution.
    l=min(mm,n-i+1)
    b(i)=(b(i)-dot_product(a(i,2:l),b(1+i:i+l-1)))/a(i,1)
end do
END SUBROUTINE banbks
```

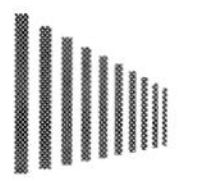

As for `bandec`, the routine `banbks` is not parallelized on the large dimension N, though it does give the compiler the opportunity for ample small-scale parallelization inside the loops.

⋆ ⋆ ⋆

```
SUBROUTINE mprove(a,alud,indx,b,x)
USE nrtype; USE nrutil, ONLY : assert_eq
USE nr, ONLY : lubksb
IMPLICIT NONE
REAL(SP), DIMENSION(:,:), INTENT(IN) :: a,alud
INTEGER(I4B), DIMENSION(:), INTENT(IN) :: indx
REAL(SP), DIMENSION(:), INTENT(IN) :: b
REAL(SP), DIMENSION(:), INTENT(INOUT) :: x
    Improves a solution vector x of the linear set of equations A·X = B. The N×N matrix a
    and the N-dimensional vectors b and x are input. Also input is alud, the LU decomposition
    of a as returned by ludcmp, and the N-dimensional vector indx also returned by that
    routine. On output, only x is modified, to an improved set of values.
INTEGER(I4B) :: ndum
REAL(SP), DIMENSION(size(a,1)) :: r
ndum=assert_eq((/size(a,1),size(a,2),size(alud,1),size(alud,2),size(b),&
    size(x),size(indx)/),'mprove')
r=matmul(real(a,dp),real(x,dp))-real(b,dp)
  Calculate the right-hand side, accumulating the residual in double precision.
call lubksb(alud,indx,r)          Solve for the error term,
x=x-r                             and subtract it from the old solution.
END SUBROUTINE mprove
```

f90 `assert_eq((/.../),'mprove')` This overloaded version of the `nrutil` routine `assert_eq` makes use of a trick for passing a variable number of scalar arguments to a routine: Put them into an array constructor, `(/.../)`, and pass the array. The receiving routine can use the `size` intrinsic to count them. The technique has some obvious limitations: All the arguments in the array must be of the same type; and the arguments are passed, in effect, by *value*, not by address, so they must be, in effect, `INTENT(IN)`.

`r=matmul(real(a,dp),real(x,dp))-real(b,dp)` Since Fortran 90's elemental intrinsics operate with the type of their arguments, we can use the `real(...,dp)`'s to force the `matmul` matrix multiplication to be done in double precision, which is what we want. In Fortran 77, we would have to do the matrix multiplication with temporary double precision variables, both inconvenient and (since Fortran 77 has no dynamic memory allocation) a waste of memory.

* * *

```
SUBROUTINE svbksb_sp(u,w,v,b,x)
USE nrtype; USE nrutil, ONLY : assert_eq
REAL(SP), DIMENSION(:,:), INTENT(IN) :: u,v
REAL(SP), DIMENSION(:), INTENT(IN) :: w,b
REAL(SP), DIMENSION(:), INTENT(OUT) :: x
    Solves A·X = B for a vector X, where A is specified by the arrays u, v, w as returned
    by svdcmp. Here u is M×N, v is N×N, and w is of length N. b is the M-dimensional
    input right-hand side. x is the N-dimensional output solution vector. No input quantities
    are destroyed, so the routine may be called sequentially with different b's.
INTEGER(I4B) :: mdum,ndum
REAL(SP), DIMENSION(size(x)) :: tmp
mdum=assert_eq(size(u,1),size(b),'svbksb_sp: mdum')
ndum=assert_eq((/size(u,2),size(v,1),size(v,2),size(w),size(x)/),&
    'svbksb_sp: ndum')
where (w /= 0.0)
    tmp=matmul(b,u)/w           Calculate diag(1/w_j)U^T B,
elsewhere
    tmp=0.0                     but replace 1/w_j by zero if w_j = 0.
end where
```

```
x=matmul(v,tmp)                Matrix multiply by V to get answer.
END SUBROUTINE svbksb_sp
```

f90 `where (w /= 0.0)...tmp=...elsewhere...tmp=` Normally, when a `where` `...elsewhere` construction is used to set a variable (here `tmp`) to one or another value, we like to replace it with a `merge` expression. Here, however, the `where` is required to guarantee that a division by zero doesn't occur. The rule is that `where` will *never* evaluate expressions that are excluded by the mask in the `where` line, but other constructions, like `merge`, *might* perform speculative evaluation of more than one possible outcome before selecting the applicable one.

Because singular value decomposition is something that one often wants to do in double precision, we include a double-precision version. In `nr`, the single- and double-precision versions are overloaded onto the name `svbksb`.

```
SUBROUTINE svbksb_dp(u,w,v,b,x)
USE nrtype; USE nrutil, ONLY : assert_eq
REAL(DP), DIMENSION(:,:), INTENT(IN) :: u,v
REAL(DP), DIMENSION(:), INTENT(IN) :: w,b
REAL(DP), DIMENSION(:), INTENT(OUT) :: x
INTEGER(I4B) :: mdum,ndum
REAL(DP), DIMENSION(size(x)) :: tmp
mdum=assert_eq(size(u,1),size(b),'svbksb_dp: mdum')
ndum=assert_eq((/size(u,2),size(v,1),size(v,2),size(w),size(x)/),&
    'svbksb_dp: ndum')
where (w /= 0.0)
    tmp=matmul(b,u)/w
elsewhere
    tmp=0.0
end where
x=matmul(v,tmp)
END SUBROUTINE svbksb_dp
```

```
SUBROUTINE svdcmp_sp(a,w,v)
USE nrtype; USE nrutil, ONLY : assert_eq,nrerror,outerprod
USE nr, ONLY : pythag
IMPLICIT NONE
REAL(SP), DIMENSION(:,:), INTENT(INOUT) :: a
REAL(SP), DIMENSION(:), INTENT(OUT) :: w
REAL(SP), DIMENSION(:,:), INTENT(OUT) :: v
    Given an M × N matrix a, this routine computes its singular value decomposition, A =
    U · W · V^T. The matrix U replaces a on output. The diagonal matrix of singular values
    W is output as the N-dimensional vector w. The N × N matrix V (not the transpose V^T)
    is output as v.
INTEGER(I4B) :: i,its,j,k,l,m,n,nm
REAL(SP) :: anorm,c,f,g,h,s,scale,x,y,z
REAL(SP), DIMENSION(size(a,1)) :: tempm
REAL(SP), DIMENSION(size(a,2)) :: rv1,tempn
m=size(a,1)
n=assert_eq(size(a,2),size(v,1),size(v,2),size(w),'svdcmp_sp')
g=0.0
scale=0.0
do i=1,n                                  Householder reduction to bidiagonal form.
    l=i+1
    rv1(i)=scale*g
    g=0.0
    scale=0.0
    if (i <= m) then
```

```
        scale=sum(abs(a(i:m,i)))
        if (scale /= 0.0) then
            a(i:m,i)=a(i:m,i)/scale
            s=dot_product(a(i:m,i),a(i:m,i))
            f=a(i,i)
            g=-sign(sqrt(s),f)
            h=f*g-s
            a(i,i)=f-g
            tempn(l:n)=matmul(a(i:m,i),a(i:m,l:n))/h
            a(i:m,l:n)=a(i:m,l:n)+outerprod(a(i:m,i),tempn(l:n))
            a(i:m,i)=scale*a(i:m,i)
        end if
    end if
    w(i)=scale*g
    g=0.0
    scale=0.0
    if ((i <= m) .and. (i /= n)) then
        scale=sum(abs(a(i,l:n)))
        if (scale /= 0.0) then
            a(i,l:n)=a(i,l:n)/scale
            s=dot_product(a(i,l:n),a(i,l:n))
            f=a(i,l)
            g=-sign(sqrt(s),f)
            h=f*g-s
            a(i,l)=f-g
            rv1(l:n)=a(i,l:n)/h
            tempm(l:m)=matmul(a(l:m,l:n),a(i,l:n))
            a(l:m,l:n)=a(l:m,l:n)+outerprod(tempm(l:m),rv1(l:n))
            a(i,l:n)=scale*a(i,l:n)
        end if
    end if
end do
anorm=maxval(abs(w)+abs(rv1))
do i=n,1,-1                              Accumulation of right-hand transformations.
    if (i < n) then
        if (g /= 0.0) then
            v(l:n,i)=(a(i,l:n)/a(i,l))/g       Double division to avoid possible under-
            tempn(l:n)=matmul(a(i,l:n),v(l:n,l:n))       flow.
            v(l:n,l:n)=v(l:n,l:n)+outerprod(v(l:n,i),tempn(l:n))
        end if
        v(i,l:n)=0.0
        v(l:n,i)=0.0
    end if
    v(i,i)=1.0
    g=rv1(i)
    l=i
end do
do i=min(m,n),1,-1                       Accumulation of left-hand transformations.
    l=i+1
    g=w(i)
    a(i,l:n)=0.0
    if (g /= 0.0) then
        g=1.0_sp/g
        tempn(l:n)=(matmul(a(l:m,i),a(l:m,l:n))/a(i,i))*g
        a(i:m,l:n)=a(i:m,l:n)+outerprod(a(i:m,i),tempn(l:n))
        a(i:m,i)=a(i:m,i)*g
    else
        a(i:m,i)=0.0
    end if
    a(i,i)=a(i,i)+1.0_sp
end do
do k=n,1,-1                              Diagonalization of the bidiagonal form: Loop over
    do its=1,30                              singular values, and over allowed iterations.
        do l=k,1,-1                      Test for splitting.
```

```
            nm=l-1
            if ((abs(rv1(l))+anorm) == anorm) exit
              Note that rv1(1) is always zero, so can never fall through bottom of loop.
            if ((abs(w(nm))+anorm) == anorm) then
                c=0.0                       Cancellation of rv1(l), if l > 1.
                s=1.0
                do i=l,k
                    f=s*rv1(i)
                    rv1(i)=c*rv1(i)
                    if ((abs(f)+anorm) == anorm) exit
                    g=w(i)
                    h=pythag(f,g)
                    w(i)=h
                    h=1.0_sp/h
                    c= (g*h)
                    s=-(f*h)
                    tempm(1:m)=a(1:m,nm)
                    a(1:m,nm)=a(1:m,nm)*c+a(1:m,i)*s
                    a(1:m,i)=-tempm(1:m)*s+a(1:m,i)*c
                end do
                exit
            end if
        end do
        z=w(k)
        if (l == k) then                    Convergence.
            if (z < 0.0) then               Singular value is made nonnegative.
                w(k)=-z
                v(1:n,k)=-v(1:n,k)
            end if
            exit
        end if
        if (its == 30) call nrerror('svdcmp_sp: no convergence in svdcmp')
        x=w(l)                              Shift from bottom 2-by-2 minor.
        nm=k-1
        y=w(nm)
        g=rv1(nm)
        h=rv1(k)
        f=((y-z)*(y+z)+(g-h)*(g+h))/(2.0_sp*h*y)
        g=pythag(f,1.0_sp)
        f=((x-z)*(x+z)+h*((y/(f+sign(g,f)))-h))/x
        c=1.0                               Next QR transformation:
        s=1.0
        do j=l,nm
            i=j+1
            g=rv1(i)
            y=w(i)
            h=s*g
            g=c*g
            z=pythag(f,h)
            rv1(j)=z
            c=f/z
            s=h/z
            f= (x*c)+(g*s)
            g=-(x*s)+(g*c)
            h=y*s
            y=y*c
            tempn(1:n)=v(1:n,j)
            v(1:n,j)=v(1:n,j)*c+v(1:n,i)*s
            v(1:n,i)=-tempn(1:n)*s+v(1:n,i)*c
            z=pythag(f,h)
            w(j)=z                          Rotation can be arbitrary if z = 0.
            if (z /= 0.0) then
                z=1.0_sp/z
                c=f*z
```

```
                s=h*z
            end if
            f= (c*g)+(s*y)
            x=-(s*g)+(c*y)
            tempm(1:m)=a(1:m,j)
            a(1:m,j)=a(1:m,j)*c+a(1:m,i)*s
            a(1:m,i)=-tempm(1:m)*s+a(1:m,i)*c
        end do
        rv1(l)=0.0
        rv1(k)=f
        w(k)=x
    end do
end do
END SUBROUTINE svdcmp_sp
```

The SVD algorithm implemented above does not parallelize very well. There are two parts to the algorithm. The first, reduction to bidiagonal form, can be parallelized. The second, the iterative diagonalization of the bidiagonal form, uses QR transformations that are intrinsically serial. There have been proposals for parallel SVD algorithms [2], but we do not have sufficient experience with them yet to recommend them over the well-established serial algorithm.

`tempn(l:n)=matmul...a(i:m,l:n)=...outerprod...` Here is an example of an update as in equation (22.1.6). In this case b_i is independent of i: It is simply `1/h`. The lines beginning `tempm(l:m)=matmul` about 16 lines down are of a similar form, but with the terms in the opposite order in the `matmul`.

As with `svbksb`, single- and double-precision versions of the routines are overloaded onto the name `svdcmp` in `nr`.

```
SUBROUTINE svdcmp_dp(a,w,v)
USE nrtype; USE nrutil, ONLY : assert_eq,nrerror,outerprod
USE nr, ONLY : pythag
IMPLICIT NONE
REAL(DP), DIMENSION(:,:), INTENT(INOUT) :: a
REAL(DP), DIMENSION(:), INTENT(OUT) :: w
REAL(DP), DIMENSION(:,:), INTENT(OUT) :: v
INTEGER(I4B) :: i,its,j,k,l,m,n,nm
REAL(DP) :: anorm,c,f,g,h,s,scale,x,y,z
REAL(DP), DIMENSION(size(a,1)) :: tempm
REAL(DP), DIMENSION(size(a,2)) :: rv1,tempn
m=size(a,1)
n=assert_eq(size(a,2),size(v,1),size(v,2),size(w),'svdcmp_dp')
g=0.0
scale=0.0
do i=1,n
    l=i+1
    rv1(i)=scale*g
    g=0.0
    scale=0.0
    if (i <= m) then
        scale=sum(abs(a(i:m,i)))
        if (scale /= 0.0) then
            a(i:m,i)=a(i:m,i)/scale
            s=dot_product(a(i:m,i),a(i:m,i))
            f=a(i,i)
            g=-sign(sqrt(s),f)
```

```
                h=f*g-s
                a(i,i)=f-g
                tempn(l:n)=matmul(a(i:m,i),a(i:m,l:n))/h
                a(i:m,l:n)=a(i:m,l:n)+outerprod(a(i:m,i),tempn(l:n))
                a(i:m,i)=scale*a(i:m,i)
            end if
        end if
        w(i)=scale*g
        g=0.0
        scale=0.0
        if ((i <= m) .and. (i /= n)) then
            scale=sum(abs(a(i,l:n)))
            if (scale /= 0.0) then
                a(i,l:n)=a(i,l:n)/scale
                s=dot_product(a(i,l:n),a(i,l:n))
                f=a(i,l)
                g=-sign(sqrt(s),f)
                h=f*g-s
                a(i,l)=f-g
                rv1(l:n)=a(i,l:n)/h
                tempm(l:m)=matmul(a(l:m,l:n),a(i,l:n))
                a(l:m,l:n)=a(l:m,l:n)+outerprod(tempm(l:m),rv1(l:n))
                a(i,l:n)=scale*a(i,l:n)
            end if
        end if
    end do
    anorm=maxval(abs(w)+abs(rv1))
    do i=n,1,-1
        if (i < n) then
            if (g /= 0.0) then
                v(l:n,i)=(a(i,l:n)/a(i,l))/g
                tempn(l:n)=matmul(a(i,l:n),v(l:n,l:n))
                v(l:n,l:n)=v(l:n,l:n)+outerprod(v(l:n,i),tempn(l:n))
            end if
            v(i,l:n)=0.0
            v(l:n,i)=0.0
        end if
        v(i,i)=1.0
        g=rv1(i)
        l=i
    end do
    do i=min(m,n),1,-1
        l=i+1
        g=w(i)
        a(i,l:n)=0.0
        if (g /= 0.0) then
            g=1.0_dp/g
            tempn(l:n)=(matmul(a(l:m,i),a(l:m,l:n))/a(i,i))*g
            a(i:m,l:n)=a(i:m,l:n)+outerprod(a(i:m,i),tempn(l:n))
            a(i:m,i)=a(i:m,i)*g
        else
            a(i:m,i)=0.0
        end if
        a(i,i)=a(i,i)+1.0_dp
    end do
    do k=n,1,-1
        do its=1,30
            do l=k,1,-1
                nm=l-1
                if ((abs(rv1(l))+anorm) == anorm) exit
                if ((abs(w(nm))+anorm) == anorm) then
                    c=0.0
                    s=1.0
                    do i=l,k
```

```
                    f=s*rv1(i)
                    rv1(i)=c*rv1(i)
                    if ((abs(f)+anorm) == anorm) exit
                    g=w(i)
                    h=pythag(f,g)
                    w(i)=h
                    h=1.0_dp/h
                    c= (g*h)
                    s=-(f*h)
                    tempm(1:m)=a(1:m,nm)
                    a(1:m,nm)=a(1:m,nm)*c+a(1:m,i)*s
                    a(1:m,i)=-tempm(1:m)*s+a(1:m,i)*c
                end do
                exit
            end if
        end do
        z=w(k)
        if (l == k) then
            if (z < 0.0) then
                w(k)=-z
                v(1:n,k)=-v(1:n,k)
            end if
            exit
        end if
        if (its == 30) call nrerror('svdcmp_dp: no convergence in svdcmp')
        x=w(l)
        nm=k-1
        y=w(nm)
        g=rv1(nm)
        h=rv1(k)
        f=((y-z)*(y+z)+(g-h)*(g+h))/(2.0_dp*h*y)
        g=pythag(f,1.0_dp)
        f=((x-z)*(x+z)+h*((y/(f+sign(g,f)))-h))/x
        c=1.0
        s=1.0
        do j=l,nm
            i=j+1
            g=rv1(i)
            y=w(i)
            h=s*g
            g=c*g
            z=pythag(f,h)
            rv1(j)=z
            c=f/z
            s=h/z
            f= (x*c)+(g*s)
            g=-(x*s)+(g*c)
            h=y*s
            y=y*c
            tempn(1:n)=v(1:n,j)
            v(1:n,j)=v(1:n,j)*c+v(1:n,i)*s
            v(1:n,i)=-tempn(1:n)*s+v(1:n,i)*c
            z=pythag(f,h)
            w(j)=z
            if (z /= 0.0) then
                z=1.0_dp/z
                c=f*z
                s=h*z
            end if
            f= (c*g)+(s*y)
            x=-(s*g)+(c*y)
            tempm(1:m)=a(1:m,j)
            a(1:m,j)=a(1:m,j)*c+a(1:m,i)*s
            a(1:m,i)=-tempm(1:m)*s+a(1:m,i)*c
```

```
            end do
            rv1(l)=0.0
            rv1(k)=f
            w(k)=x
        end do
    end do
    END SUBROUTINE svdcmp_dp
```

```
FUNCTION pythag_sp(a,b)
USE nrtype
IMPLICIT NONE
REAL(SP), INTENT(IN) :: a,b
REAL(SP) :: pythag_sp
```

Computes $(a^2 + b^2)^{1/2}$ without destructive underflow or overflow.

```
REAL(SP) :: absa,absb
absa=abs(a)
absb=abs(b)
if (absa > absb) then
    pythag_sp=absa*sqrt(1.0_sp+(absb/absa)**2)
else
    if (absb == 0.0) then
        pythag_sp=0.0
    else
        pythag_sp=absb*sqrt(1.0_sp+(absa/absb)**2)
    end if
end if
END FUNCTION pythag_sp
```

```
FUNCTION pythag_dp(a,b)
USE nrtype
IMPLICIT NONE
REAL(DP), INTENT(IN) :: a,b
REAL(DP) :: pythag_dp
REAL(DP) :: absa,absb
absa=abs(a)
absb=abs(b)
if (absa > absb) then
    pythag_dp=absa*sqrt(1.0_dp+(absb/absa)**2)
else
    if (absb == 0.0) then
        pythag_dp=0.0
    else
        pythag_dp=absb*sqrt(1.0_dp+(absa/absb)**2)
    end if
end if
END FUNCTION pythag_dp
```

⋆ ⋆ ⋆

```
SUBROUTINE cyclic(a,b,c,alpha,beta,r,x)
USE nrtype; USE nrutil, ONLY : assert,assert_eq
USE nr, ONLY : tridag
IMPLICIT NONE
REAL(SP), DIMENSION(:), INTENT(IN):: a,b,c,r
REAL(SP), INTENT(IN) :: alpha,beta
REAL(SP), DIMENSION(:), INTENT(OUT):: x
   Solves the "cyclic" set of linear equations given by equation (2.7.9). a, b, c, and r are
   input vectors, while x is the output solution vector, all of the same size. alpha and beta
   are the corner entries in the matrix. The input is not modified.
INTEGER(I4B) :: n
REAL(SP) :: fact,gamma
REAL(SP), DIMENSION(size(x)) :: bb,u,z
n=assert_eq((/size(a),size(b),size(c),size(r),size(x)/),'cyclic')
call assert(n > 2, 'cyclic arg')
gamma=-b(1)                                  Avoid subtraction error in forming bb(1).
bb(1)=b(1)-gamma                             Set up the diagonal of the modified tridiag-
bb(n)=b(n)-alpha*beta/gamma                      onal system.
bb(2:n-1)=b(2:n-1)
call tridag(a(2:n),bb,c(1:n-1),r,x)          Solve A · x = r.
u(1)=gamma                                   Set up the vector u.
u(n)=alpha
u(2:n-1)=0.0
call tridag(a(2:n),bb,c(1:n-1),u,z)          Solve A · z = u.
fact=(x(1)+beta*x(n)/gamma)/(1.0_sp+z(1)+beta*z(n)/gamma)      Form v·x/(1+v·z).
x=x-fact*z                                   Now get the solution vector x.
END SUBROUTINE cyclic
```

The parallelism in cyclic is in tridag. Users with multiprocessor machines will want to be sure that, in nrutil, they have set the name tridag to be overloaded with tridag_par instead of tridag_ser.

⋆ ⋆ ⋆

The routines sprsin, sprsax, sprstx, sprstp, and sprsdiag give roughly equivalent functionality to the corresponding Fortran 77 routines, but they are *not* plug compatible. Instead, they take advantage of (and illustrate) several Fortran 90 features that are not present in Fortran 77.

In the module nrtype we define a TYPE sprs2_sp for two-dimensional sparse, square, matrices, in single precision, as follows

```
TYPE sprs2_sp
   INTEGER(I4B) :: n,len
   REAL(SP), DIMENSION(:), POINTER :: val
   INTEGER(I4B), DIMENSION(:), POINTER :: irow
   INTEGER(I4B), DIMENSION(:), POINTER :: jcol
END TYPE sprs2_sp
```

This has much less structure to it than the "row-indexed sparse storage mode" used in Volume 1. Here, a sparse matrix is just a list of values, and corresponding lists giving the row and column number that each value is in. Two integers n and len give, respectively, the underlying size (number of rows or columns) in the full matrix, and the number of stored nonzero values. While the previously used row-indexed scheme can be somewhat more efficient for serial machines, it does not parallelize conveniently, while this one does (though with some caveats; see below).

```
SUBROUTINE sprsin_sp(a,thresh,sa)
USE nrtype; USE nrutil, ONLY : arth,assert_eq
IMPLICIT NONE
REAL(SP), DIMENSION(:,:), INTENT(IN) :: a
REAL(SP), INTENT(IN) :: thresh
TYPE(sprs2_sp), INTENT(OUT) :: sa
   Converts a square matrix a to sparse storage format as sa. Only elements of a with mag-
   nitude ≥ thresh are retained.
INTEGER(I4B) :: n,len
LOGICAL(LGT), DIMENSION(size(a,1),size(a,2)) :: mask
n=assert_eq(size(a,1),size(a,2),'sprsin_sp')
mask=abs(a)>thresh
len=count(mask)                  How many elements to store?
allocate(sa%val(len),sa%irow(len),sa%jcol(len))
sa%n=n
sa%len=len
sa%val=pack(a,mask)              Grab the values, row, and column numbers.
sa%irow=pack(spread(arth(1,1,n),2,n),mask)
sa%jcol=pack(spread(arth(1,1,n),1,n),mask)
END SUBROUTINE sprsin_sp
```

```
SUBROUTINE sprsin_dp(a,thresh,sa)
USE nrtype; USE nrutil, ONLY : arth,assert_eq
IMPLICIT NONE
REAL(DP), DIMENSION(:,:), INTENT(IN) :: a
REAL(DP), INTENT(IN) :: thresh
TYPE(sprs2_dp), INTENT(OUT) :: sa
INTEGER(I4B) :: n,len
LOGICAL(LGT), DIMENSION(size(a,1),size(a,2)) :: mask
n=assert_eq(size(a,1),size(a,2),'sprsin_dp')
mask=abs(a)>thresh
len=count(mask)
allocate(sa%val(len),sa%irow(len),sa%jcol(len))
sa%n=n
sa%len=len
sa%val=pack(a,mask)
sa%irow=pack(spread(arth(1,1,n),2,n),mask)
sa%jcol=pack(spread(arth(1,1,n),1,n),mask)
END SUBROUTINE sprsin_dp
```

f90 Note that the routines sprsin_sp and sprsin_dp — single and double precision versions of the same algorithm — are overloaded onto the name sprsin in module nr. We supply both forms because the routine linbcg, below, works in double precision.

sa%irow=pack(spread(arth(1,1,n),2,n),mask) The trick here is to use the same mask, abs(a)>thresh, in three consecutive pack expressions, thus guaranteeing that the corresponding elements of the array argument get selected for packing. The first time, we get the desired matrix element values. The second time (above code fragment), we construct a matrix with each element having the value of its *row* number. The third time, we construct a matrix with each element having the value of its *column* number.

```
SUBROUTINE sprsax_sp(sa,x,b)
USE nrtype; USE nrutil, ONLY : assert_eq,scatter_add
IMPLICIT NONE
TYPE(sprs2_sp), INTENT(IN) :: sa
REAL(SP), DIMENSION (:), INTENT(IN) :: x
REAL(SP), DIMENSION (:), INTENT(OUT) :: b
   Multiply a matrix sa in sparse matrix format by a vector x, giving a vector b.
INTEGER(I4B) :: ndum
ndum=assert_eq(sa%n,size(x),size(b),'sprsax_sp')
b=0.0_sp
call scatter_add(b,sa%val*x(sa%jcol),sa%irow)
  Each sparse matrix entry adds a term to some component of b.
END SUBROUTINE sprsax_sp
```

```
SUBROUTINE sprsax_dp(sa,x,b)
USE nrtype; USE nrutil, ONLY : assert_eq,scatter_add
IMPLICIT NONE
TYPE(sprs2_dp), INTENT(IN) :: sa
REAL(DP), DIMENSION (:), INTENT(IN) :: x
REAL(DP), DIMENSION (:), INTENT(OUT) :: b
INTEGER(I4B) :: ndum
ndum=assert_eq(sa%n,size(x),size(b),'sprsax_dp')
b=0.0_dp
call scatter_add(b,sa%val*x(sa%jcol),sa%irow)
END SUBROUTINE sprsax_dp
```

`call scatter_add(b,sa%val*x(sa%jcol),sa%irow)` Since more than one component of the middle vector argument will, in general, need to be added into the same component of `b`, we must resort to a call to the `nrutil` routine `scatter_add` to achieve parallelism. *However,* this parallelism is achieved only if a parallel version of `scatter_add` is available! As we have discussed previously (p. 984), Fortran 90 does not provide any scatter-with-combine (here, scatter-with-add) facility, insisting instead that indexed operations yield non-colliding addresses. Luckily, almost all parallel machines do provide such a facility as a library program. In HPF, for example, the equivalent of `scatter_add` is `SUM_SCATTER`.

The call to `scatter_add` above is equivalent to the do-loop

```
b=0.0
do k=1,sa%len
   b(sa%irow(k))=b(sa%irow(k))+sa%val(k)*x(sa%jcol(k))
end do
```

```
SUBROUTINE sprstx_sp(sa,x,b)
USE nrtype; USE nrutil, ONLY : assert_eq,scatter_add
IMPLICIT NONE
TYPE(sprs2_sp), INTENT(IN) :: sa
REAL(SP), DIMENSION (:), INTENT(IN) :: x
REAL(SP), DIMENSION (:), INTENT(OUT) :: b
   Multiply the transpose of a matrix sa in sparse matrix format by a vector x, giving a vector b.
INTEGER(I4B) :: ndum
ndum=assert_eq(sa%n,size(x),size(b),'sprstx_sp')
b=0.0_sp
call scatter_add(b,sa%val*x(sa%irow),sa%jcol)
  Each sparse matrix entry adds a term to some component of b.
END SUBROUTINE sprstx_sp
```

```
SUBROUTINE sprstx_dp(sa,x,b)
USE nrtype; USE nrutil, ONLY : assert_eq,scatter_add
IMPLICIT NONE
TYPE(sprs2_dp), INTENT(IN) :: sa
REAL(DP), DIMENSION (:), INTENT(IN) :: x
REAL(DP), DIMENSION (:), INTENT(OUT) :: b
INTEGER(I4B) :: ndum
ndum=assert_eq(sa%n,size(x),size(b),'sprstx_dp')
b=0.0_dp
call scatter_add(b,sa%val*x(sa%irow),sa%jcol)
END SUBROUTINE sprstx_dp
```

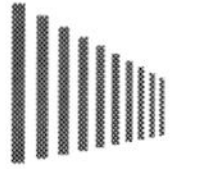

Precisely the same comments as for sprsax apply to sprstx. The call to scatter_add is here equivalent to

```
b=0.0
do k=1,sa%len
   b(sa%jcol(k))=b(sa%jcol(k))+sa%val(k)*x(sa%irow(k))
end do
```

```
SUBROUTINE sprstp(sa)
USE nrtype
IMPLICIT NONE
TYPE(sprs2_sp), INTENT(INOUT) :: sa
   Replaces sa, in sparse matrix format, by its transpose.
INTEGER(I4B), DIMENSION(:), POINTER :: temp
temp=>sa%irow                    We need only swap the row and column pointers.
sa%irow=>sa%jcol
sa%jcol=>temp
END SUBROUTINE sprstp
```

```
SUBROUTINE sprsdiag_sp(sa,b)
USE nrtype; USE nrutil, ONLY : array_copy,assert_eq
IMPLICIT NONE
TYPE(sprs2_sp), INTENT(IN) :: sa
REAL(SP), DIMENSION(:), INTENT(OUT) :: b
   Extracts the diagonal of a matrix sa in sparse matrix format into a vector b.
REAL(SP), DIMENSION(size(b)) :: val
INTEGER(I4B) :: k,l,ndum,nerr
INTEGER(I4B), DIMENSION(size(b)) :: i
LOGICAL(LGT), DIMENSION(:), ALLOCATABLE :: mask
ndum=assert_eq(sa%n,size(b),'sprsdiag_sp')
l=sa%len
allocate(mask(l))
mask = (sa%irow(1:l) == sa%jcol(1:l))        Find diagonal elements.
call array_copy(pack(sa%val(1:l),mask),val,k,nerr)      Grab the values...
i(1:k)=pack(sa%irow(1:l),mask)               ...and their locations.
deallocate(mask)
b=0.0                                        Zero b because zero values not stored in sa.
b(i(1:k))=val(1:k)                           Scatter values into correct slots.
END SUBROUTINE sprsdiag_sp
```

```
SUBROUTINE sprsdiag_dp(sa,b)
USE nrtype; USE nrutil, ONLY : array_copy,assert_eq
IMPLICIT NONE
TYPE(sprs2_dp), INTENT(IN) :: sa
REAL(DP), DIMENSION(:), INTENT(OUT) :: b
REAL(DP), DIMENSION(size(b)) :: val
INTEGER(I4B) :: k,l,ndum,nerr
INTEGER(I4B), DIMENSION(size(b)) :: i
LOGICAL(LGT), DIMENSION(:), ALLOCATABLE :: mask
ndum=assert_eq(sa%n,size(b),'sprsdiag_dp')
l=sa%len
allocate(mask(l))
mask = (sa%irow(1:l) == sa%jcol(1:l))
call array_copy(pack(sa%val(1:l),mask),val,k,nerr)
i(1:k)=pack(sa%irow(1:l),mask)
deallocate(mask)
b=0.0
b(i(1:k))=val(1:k)
END SUBROUTINE sprsdiag_dp
```

f90 `call array_copy(pack(sa%val(1:l),mask),val,k,nerr)` We use the nrutil routine array_copy because we don't know in advance how many nonzero diagonal elements will be selected by mask. Of course we could count them with a count(mask), but this is an extra step, and inefficient on scalar machines.

`i(1:k)=pack(sa%irow(1:l),mask)` Using the same mask, we pick out the corresponding locations of the diagonal elements. No need to use array_copy now, since we know the value of k.

`b(i(1:k))=val(1:k)` Finally, we can put each element in the right place. Notice that if the sparse matrix is ill-formed, with more than one value stored for the same diagonal element (which should not happen!) then the vector subscript i(1:k) is a "many-one section" and its use on the left-hand side is illegal.

⋆ ⋆ ⋆

```
SUBROUTINE linbcg(b,x,itol,tol,itmax,iter,err)
USE nrtype; USE nrutil, ONLY : assert_eq,nrerror
USE nr, ONLY : atimes,asolve,snrm
IMPLICIT NONE
REAL(DP), DIMENSION(:), INTENT(IN) :: b          Double precision is a good idea in this
REAL(DP), DIMENSION(:), INTENT(INOUT) :: x          routine.
INTEGER(I4B), INTENT(IN) :: itol,itmax
REAL(DP), INTENT(IN) :: tol
INTEGER(I4B), INTENT(OUT) :: iter
REAL(DP), INTENT(OUT) :: err
REAL(DP), PARAMETER :: EPS=1.0e-14_dp
```

Solves $\mathbf{A} \cdot \mathbf{x} = \mathbf{b}$ for x, given b of the same length, by the iterative biconjugate gradient method. On input x should be set to an initial guess of the solution (or all zeros); itol is 1,2,3, or 4, specifying which convergence test is applied (see text); itmax is the maximum number of allowed iterations; and tol is the desired convergence tolerance. On output, x is reset to the improved solution, iter is the number of iterations actually taken, and err is the estimated error. The matrix $\mathbf{A}$ is referenced only through the user-supplied routines atimes, which computes the product of either $\mathbf{A}$ or its transpose on a vector; and asolve, which solves $\widetilde{\mathbf{A}} \cdot \mathbf{x} = \mathbf{b}$ or $\widetilde{\mathbf{A}}^T \cdot \mathbf{x} = \mathbf{b}$ for some preconditioner matrix $\widetilde{\mathbf{A}}$ (possibly the trivial diagonal part of $\mathbf{A}$).

```
INTEGER(I4B) :: n
REAL(DP) :: ak,akden,bk,bkden,bknum,bnrm,dxnrm,xnrm,zm1nrm,znrm
REAL(DP), DIMENSION(size(b)) :: p,pp,r,rr,z,zz
n=assert_eq(size(b),size(x),'linbcg')
iter=0
call atimes(x,r,0)                    Calculate initial residual. Input to atimes is
r=b-r                                     x(1:n), output is r(1:n); the final 0
rr=r                                      indicates that the matrix (not its trans-
! call atimes(r,rr,0)                     pose) is to be used.
  Uncomment this line to get the "minimum residual" variant of the algorithm.
select case(itol)                     Calculate norms for use in stopping criterion,
    case(1)                               and initialize z.
        bnrm=snrm(b,itol)
        call asolve(r,z,0)            Input to asolve is r(1:n), output is z(1:n);
    case(2)                               the final 0 indicates that the matrix Ã
        call asolve(b,z,0)                (not its transpose) is to be used.
        bnrm=snrm(z,itol)
        call asolve(r,z,0)
    case(3:4)
        call asolve(b,z,0)
        bnrm=snrm(z,itol)
        call asolve(r,z,0)
        znrm=snrm(z,itol)
    case default
        call nrerror('illegal itol in linbcg')
end select
do                                    Main loop.
    if (iter > itmax) exit
    iter=iter+1
    call asolve(rr,zz,1)              Final 1 indicates use of transpose matrix Ã^T.
    bknum=dot_product(z,rr)           Calculate coefficient bk and direction vectors
    if (iter == 1) then                   p and pp.
        p=z
        pp=zz
    else
        bk=bknum/bkden
        p=bk*p+z
        pp=bk*pp+zz
    end if
    bkden=bknum                       Calculate coefficient ak, new iterate x, and
    call atimes(p,z,0)                    new residuals r and rr.
    akden=dot_product(z,pp)
    ak=bknum/akden
    call atimes(pp,zz,1)
    x=x+ak*p
    r=r-ak*z
    rr=rr-ak*zz
    call asolve(r,z,0)                Solve Ã·z = r and check stopping criterion.
    select case(itol)
        case(1)
            err=snrm(r,itol)/bnrm
        case(2)
            err=snrm(z,itol)/bnrm
        case(3:4)
            zm1nrm=znrm
            znrm=snrm(z,itol)
            if (abs(zm1nrm-znrm) > EPS*znrm) then
                dxnrm=abs(ak)*snrm(p,itol)
                err=znrm/abs(zm1nrm-znrm)*dxnrm
            else
                err=znrm/bnrm         Error may not be accurate, so loop again.
                cycle
```

```
            end if
            xnrm=snrm(x,itol)
            if (err <= 0.5_dp*xnrm) then
                err=err/xnrm
            else
                err=znrm/bnrm             Error may not be accurate, so loop again.
                cycle
            end if
    end select
    write (*,*) ' iter=',iter,' err=',err
    if (err <= tol) exit
end do
END SUBROUTINE linbcg
```

`case default...call nrerror('illegal itol in linbcg')` It's *always* a good idea to trap errors when the value of a case construction is supplied externally to the routine, as here.

```
FUNCTION snrm(sx,itol)
USE nrtype
IMPLICIT NONE
REAL(DP), DIMENSION(:), INTENT(IN) :: sx
INTEGER(I4B), INTENT(IN) :: itol
REAL(DP) :: snrm
    Compute one of two norms for a vector sx, as signaled by itol. Used by linbcg.
if (itol <= 3) then
    snrm=sqrt(dot_product(sx,sx))         Vector magnitude norm.
else
    snrm=maxval(abs(sx))                  Largest component norm.
end if
END FUNCTION snrm
```

```
SUBROUTINE atimes(x,r,itrnsp)
USE nrtype; USE nrutil, ONLY : assert_eq
USE nr, ONLY : sprsax,sprstx        DOUBLE PRECISION versions of sprsax and sprstx.
USE xlinbcg_data                    The matrix is accessed through this module.
REAL(DP), DIMENSION(:), INTENT(IN) :: x
REAL(DP), DIMENSION(:), INTENT(OUT) :: r
INTEGER(I4B), INTENT(IN) :: itrnsp
INTEGER(I4B) :: n
n=assert_eq(size(x),size(r),'atimes')
if (itrnsp == 0) then
    call sprsax(sa,x,r)
else
    call sprstx(sa,x,r)
end if
END SUBROUTINE atimes
```

```
SUBROUTINE asolve(b,x,itrnsp)
USE nrtype; USE nrutil, ONLY : assert_eq,nrerror
USE nr, ONLY : sprsdiag          DOUBLE PRECISION version of sprsdiag.
USE xlinbcg_data                 The matrix is accessed through this module.
REAL(DP), DIMENSION(:), INTENT(IN) :: b
REAL(DP), DIMENSION(:), INTENT(OUT) :: x
INTEGER(I4B), INTENT(IN) :: itrnsp
INTEGER(I4B) :: ndum
ndum=assert_eq(size(b),size(x),'asolve')
call sprsdiag(sa,x)
```

The matrix $\widetilde{\mathbf{A}}$ is taken to be the diagonal part of $\mathbf{A}$. Since the transpose matrix has the same diagonal, the flag `itrnsp` is not used.

```
if (any(x == 0.0)) call nrerror('asolve: singular diagonal matrix')
x=b/x
END SUBROUTINE asolve
```

f90 The routines `atimes` and `asolve` are examples of user-supplied routines that interface `linbcg` to a user-supplied method for multiplying the user's sparse matrix by a vector, and for solving the preconditioner matrix equation. Here, we have used these routines to connect `linbcg` to the sparse matrix machinery developed above. If we were instead using the different sparse matrix machinery of Volume 1, we would modify `atimes` and `asolve` accordingly.

`USE xlinbcg_data` This user-supplied module is assumed to have `sa` (the sparse matrix) in it.

⋆ ⋆ ⋆

```
FUNCTION vander(x,q)
USE nrtype; USE nrutil, ONLY : assert_eq,outerdiff
IMPLICIT NONE
REAL(DP), DIMENSION(:), INTENT(IN) :: x,q
REAL(DP), DIMENSION(size(x)) :: vander
```

Solves the Vandermonde linear system $\sum_{i=1}^{N} x_i^{k-1} w_i = q_k$ $(k = 1,\ldots,N)$. Input consists of the vectors `x` and `q` of length N. The solution w (also of length N) is returned in `vander`.

```
REAL(DP), DIMENSION(size(x)) :: c
REAL(DP), DIMENSION(size(x),size(x)) :: a
INTEGER(I4B) :: i,n
n=assert_eq(size(x),size(q),'vander')
if (n == 1) then
    vander(1)=q(1)
else
    c(:)=0.0                            Initialize array.
    c(n)=-x(1)                          Coefficients of the master polynomial are found
    do i=2,n                               by recursion.
        c(n+1-i:n-1)=c(n+1-i:n-1)-x(i)*c(n+2-i:n)
        c(n)=c(n)-x(i)
    end do
    a(:,:)=outerdiff(x,x)               Make vector w_j = prod_{n/=j}(x_j - x_n).
    vander(:)=product(a,dim=2,mask=(a /= 0.0))
```

Now do synthetic division by $x - x_j$. The division for all x_j can be done in parallel (on a parallel machine), since the : in the loop below is over j.

```
    a(:,1)=-c(1)/x(:)
    do i=2,n
        a(:,i)=-(c(i)-a(:,i-1))/x(:)
    end do
    vander(:)=matmul(a,q)/vander(:)     Solve linear system and supply denomina-
end if                                     tor.
END FUNCTION vander
```

f90 `a=outerdiff...w=product...` Here is an example of the coding of equation (22.1.4). Since in this case the product is over the second index (n in $x_j - x_n$), we have `dim=2` in the `product`.

```
FUNCTION toeplz(r,y)
USE nrtype; USE nrutil, ONLY : assert_eq,nrerror
IMPLICIT NONE
REAL(SP), DIMENSION(:), INTENT(IN) :: r,y
REAL(SP), DIMENSION(size(y)) :: toeplz
```

Solves the Toeplitz system $\sum_{j=1}^{N} R_{(N+i-j)} x_j = y_i$ $(i = 1, \ldots, N)$. The Toeplitz matrix need not be symmetric. `y` (of length N) and `r` (of length $2N-1$) are input arrays; the solution x (of length N) is returned in `toeplz`.

```
INTEGER(I4B) :: m,m1,n,ndum
REAL(SP) :: sd,sgd,sgn,shn,sxn
REAL(SP), DIMENSION(size(y)) :: g,h,t
n=size(y)
ndum=assert_eq(2*n-1,size(r),'toeplz: ndum')
if (r(n) == 0.0) call nrerror('toeplz: initial singular minor')
toeplz(1)=y(1)/r(n)                          Initialize for the recursion.
if (n == 1) RETURN
g(1)=r(n-1)/r(n)
h(1)=r(n+1)/r(n)
do m=1,n                                     Main loop over the recursion.
    m1=m+1
    sxn=-y(m1)+dot_product(r(n+1:n+m),toeplz(m:1:-1))
      Compute numerator and denominator for x,
    sd=-r(n)+dot_product(r(n+1:n+m),g(1:m))
    if (sd == 0.0) exit
    toeplz(m1)=sxn/sd                        whence x.
    toeplz(1:m)=toeplz(1:m)-toeplz(m1)*g(m:1:-1)
    if (m1 == n) RETURN
    sgn=-r(n-m1)+dot_product(r(n-m:n-1),g(1:m))     Compute numerator and denom-
    shn=-r(n+m1)+dot_product(r(n+m:n+1:-1),h(1:m))    inator for G and H,
    sgd=-r(n)+dot_product(r(n-m:n-1),h(m:1:-1))
    if (sd == 0.0 .or. sgd == 0.0) exit
    g(m1)=sgn/sgd                            whence G and H.
    h(m1)=shn/sd
    t(1:m)=g(1:m)
    g(1:m)=g(1:m)-g(m1)*h(m:1:-1)
    h(1:m)=h(1:m)-h(m1)*t(m:1:-1)
end do                                       Back for another recurrence.
if (m > n) call nrerror('toeplz: sanity check failed in routine')
call nrerror('toeplz: singular principal minor')
END FUNCTION toeplz
```

* * *

```
SUBROUTINE choldc(a,p)
USE nrtype; USE nrutil, ONLY : assert_eq,nrerror
IMPLICIT NONE
REAL(SP), DIMENSION(:,:), INTENT(INOUT) :: a
REAL(SP), DIMENSION(:), INTENT(OUT) :: p
```

Given an $N \times N$ positive-definite symmetric matrix `a`, this routine constructs its Cholesky decomposition, $\mathbf{A} = \mathbf{L} \cdot \mathbf{L}^T$. On input, only the upper triangle of `a` need be given; it is not modified. The Cholesky factor $\mathbf{L}$ is returned in the lower triangle of `a`, except for its diagonal elements, which are returned in `p`, a vector of length N.

```
INTEGER(I4B) :: i,n
REAL(SP) :: summ
n=assert_eq(size(a,1),size(a,2),size(p),'choldc')
do i=1,n
```

```
    summ=a(i,i)-dot_product(a(i,1:i-1),a(i,1:i-1))
    if (summ <= 0.0) call nrerror('choldc failed')      a, with rounding errors, is
    p(i)=sqrt(summ)                                      not positive definite.
    a(i+1:n,i)=(a(i,i+1:n)-matmul(a(i+1:n,1:i-1),a(i,1:i-1)))/p(i)
end do
END SUBROUTINE choldc
```

```
SUBROUTINE cholsl(a,p,b,x)
USE nrtype; USE nrutil, ONLY : assert_eq
IMPLICIT NONE
REAL(SP), DIMENSION(:,:), INTENT(IN) :: a
REAL(SP), DIMENSION(:), INTENT(IN) :: p,b
REAL(SP), DIMENSION(:), INTENT(INOUT) :: x
```

Solves the set of N linear equations $\mathbf{A} \cdot \mathbf{x} = \mathbf{b}$, where a is a positive-definite symmetric matrix. a ($N \times N$) and p (of length N) are input as the output of the routine choldc. Only the lower triangle of a is accessed. b is the input right-hand-side vector, of length N. The solution vector, also of length N, is returned in x. a and p are not modified and can be left in place for successive calls with different right-hand sides b. b is not modified unless you identify b and x in the calling sequence, which is allowed.

```
INTEGER(I4B) :: i,n
n=assert_eq((/size(a,1),size(a,2),size(p),size(b),size(x)/),'cholsl')
do i=1,n                          Solve L · y = b, storing y in x.
    x(i)=(b(i)-dot_product(a(i,1:i-1),x(1:i-1)))/p(i)
end do
do i=n,1,-1                       Solve L^T · x = y.
    x(i)=(x(i)-dot_product(a(i+1:n,i),x(i+1:n)))/p(i)
end do
END SUBROUTINE cholsl
```

⋆ ⋆ ⋆

```
SUBROUTINE qrdcmp(a,c,d,sing)
USE nrtype; USE nrutil, ONLY : assert_eq,outerprod,vabs
IMPLICIT NONE
REAL(SP), DIMENSION(:,:), INTENT(INOUT) :: a
REAL(SP), DIMENSION(:), INTENT(OUT) :: c,d
LOGICAL(LGT), INTENT(OUT) :: sing
```

Constructs the QR decomposition of the $n \times n$ matrix a. The upper triangular matrix $\mathbf{R}$ is returned in the upper triangle of a, except for the diagonal elements of $\mathbf{R}$, which are returned in the n-dimensional vector d. The orthogonal matrix $\mathbf{Q}$ is represented as a product of $n-1$ Householder matrices $\mathbf{Q}_1 \dots \mathbf{Q}_{n-1}$, where $\mathbf{Q}_j = \mathbf{1} - \mathbf{u}_j \otimes \mathbf{u}_j / c_j$. The ith component of $\mathbf{u}_j$ is zero for $i = 1, \dots, j-1$ while the nonzero components are returned in a(i,j) for $i = j, \dots, n$. sing returns as true if singularity is encountered during the decomposition, but the decomposition is still completed in this case.

```
INTEGER(I4B) :: k,n
REAL(SP) :: scale,sigma
n=assert_eq(size(a,1),size(a,2),size(c),size(d),'qrdcmp')
sing=.false.
do k=1,n-1
    scale=maxval(abs(a(k:n,k)))
    if (scale == 0.0) then            Singular case.
        sing=.true.
        c(k)=0.0
        d(k)=0.0
    else                              Form Q_k and Q_k · A.
        a(k:n,k)=a(k:n,k)/scale
        sigma=sign(vabs(a(k:n,k)),a(k,k))
        a(k,k)=a(k,k)+sigma
        c(k)=sigma*a(k,k)
```

```
        d(k)=-scale*sigma
        a(k:n,k+1:n)=a(k:n,k+1:n)-outerprod(a(k:n,k),&
            matmul(a(k:n,k),a(k:n,k+1:n)))/c(k)
    end if
end do
d(n)=a(n,n)
if (d(n) == 0.0) sing=.true.
END SUBROUTINE qrdcmp
```

`a(k:n,k+1:n)=a(k:n,k+1:n)-outerprod...matmul...` See discussion of equation (22.1.6).

```
SUBROUTINE qrsolv(a,c,d,b)
USE nrtype; USE nrutil, ONLY : assert_eq
USE nr, ONLY : rsolv
IMPLICIT NONE
REAL(SP), DIMENSION(:,:), INTENT(IN) :: a
REAL(SP), DIMENSION(:), INTENT(IN) :: c,d
REAL(SP), DIMENSION(:), INTENT(INOUT) :: b
```

Solves the set of n linear equations $\mathbf{A} \cdot \mathbf{x} = \mathbf{b}$. The $n \times n$ matrix a and the n-dimensional vectors c and d are input as the output of the routine qrdcmp and are not modified. b is input as the right-hand-side vector of length n, and is overwritten with the solution vector on output.

```
INTEGER(I4B) :: j,n
REAL(SP) :: tau
n=assert_eq((/size(a,1),size(a,2),size(b),size(c),size(d)/),'qrsolv')
do j=1,n-1                          Form Q^T . b.
    tau=dot_product(a(j:n,j),b(j:n))/c(j)
    b(j:n)=b(j:n)-tau*a(j:n,j)
end do
call rsolv(a,d,b)                   Solve R . x = Q^T . b.
END SUBROUTINE qrsolv
```

```
SUBROUTINE rsolv(a,d,b)
USE nrtype; USE nrutil, ONLY : assert_eq
IMPLICIT NONE
REAL(SP), DIMENSION(:,:), INTENT(IN) :: a
REAL(SP), DIMENSION(:), INTENT(IN) :: d
REAL(SP), DIMENSION(:), INTENT(INOUT) :: b
```

Solves the set of n linear equations $\mathbf{R} \cdot \mathbf{x} = \mathbf{b}$, where $\mathbf{R}$ is an upper triangular matrix stored in a and d. The $n \times n$ matrix a and the vector d of length n are input as the output of the routine qrdcmp and are not modified. b is input as the right-hand-side vector of length n, and is overwritten with the solution vector on output.

```
INTEGER(I4B) :: i,n
n=assert_eq(size(a,1),size(a,2),size(b),size(d),'rsolv')
b(n)=b(n)/d(n)
do i=n-1,1,-1
    b(i)=(b(i)-dot_product(a(i,i+1:n),b(i+1:n)))/d(i)
end do
END SUBROUTINE rsolv
```

⋆ ⋆ ⋆

```
SUBROUTINE qrupdt(r,qt,u,v)
USE nrtype; USE nrutil, ONLY : assert_eq,ifirstloc
USE nr, ONLY : rotate,pythag
IMPLICIT NONE
REAL(SP), DIMENSION(:,:), INTENT(INOUT) :: r,qt
REAL(SP), DIMENSION(:), INTENT(INOUT) :: u
REAL(SP), DIMENSION(:), INTENT(IN) :: v
```

Given the QR decomposition of some $n \times n$ matrix, calculates the QR decomposition of the matrix $\mathbf{Q} \cdot (\mathbf{R} + \mathbf{u} \otimes \mathbf{v})$. Here `r` and `qt` are $n \times n$ matrices, `u` and `v` are n-dimensional vectors. Note that $\mathbf{Q}^T$ is input and returned in `qt`.

```
INTEGER(I4B) :: i,k,n
n=assert_eq((/size(r,1),size(r,2),size(qt,1),size(qt,2),size(u),&
    size(v)/),'qrupdt')
k=n+1-ifirstloc(u(n:1:-1) /= 0.0)           Find largest k such that u(k) ≠ 0.
if (k < 1) k=1
do i=k-1,1,-1                               Transform R + u ⊗ v to upper Hessenberg.
    call rotate(r,qt,i,u(i),-u(i+1))
    u(i)=pythag(u(i),u(i+1))
end do
r(1,:)=r(1,:)+u(1)*v
do i=1,k-1                                  Transform upper Hessenberg matrix to upper
    call rotate(r,qt,i,r(i,i),-r(i+1,i))       triangular.
end do
END SUBROUTINE qrupdt
```

f90 `k=n+1-ifirstloc(u(n:1:-1) /= 0.0)` The function `ifirstloc` in `nrutil` returns the first occurrence of `.true.` in a logical vector. See the discussion of the analogous routine `imaxloc` on p. 1017.

```
SUBROUTINE rotate(r,qt,i,a,b)
USE nrtype; USE nrutil, ONLY : assert_eq
IMPLICIT NONE
REAL(SP), DIMENSION(:,:), TARGET, INTENT(INOUT) :: r,qt
INTEGER(I4B), INTENT(IN) :: i
REAL(SP), INTENT(IN) :: a,b
```

Given $n \times n$ matrices `r` and `qt`, carry out a Jacobi rotation on rows `i` and `i`$+1$ of each matrix. `a` and `b` are the parameters of the rotation: $\cos\theta = a/\sqrt{a^2+b^2}$, $\sin\theta = b/\sqrt{a^2+b^2}$.

```
REAL(SP), DIMENSION(size(r,1)) :: temp
INTEGER(I4B) :: n
REAL(SP) :: c,fact,s
n=assert_eq(size(r,1),size(r,2),size(qt,1),size(qt,2),'rotate')
if (a == 0.0) then                  Avoid unnecessary overflow or underflow.
    c=0.0
    s=sign(1.0_sp,b)
else if (abs(a) > abs(b)) then
    fact=b/a
    c=sign(1.0_sp/sqrt(1.0_sp+fact**2),a)
    s=fact*c
else
    fact=a/b
    s=sign(1.0_sp/sqrt(1.0_sp+fact**2),b)
    c=fact*s
end if
temp(i:n)=r(i,i:n)                  Premultiply r by Jacobi rotation.
r(i,i:n)=c*temp(i:n)-s*r(i+1,i:n)
r(i+1,i:n)=s*temp(i:n)+c*r(i+1,i:n)
temp=qt(i,:)                        Premultiply qt by Jacobi rotation.
qt(i,:)=c*temp-s*qt(i+1,:)
qt(i+1,:)=s*temp+c*qt(i+1,:)
END SUBROUTINE rotate
```

CITED REFERENCES AND FURTHER READING:

Golub, G.H., and Van Loan, C.F. 1989, *Matrix Computations*, 2nd ed. (Baltimore: Johns Hopkins University Press). [1]

Gu, M., Demmel, J., and Dhillon, I. 1994, LAPACK Working Note #88 (Computer Science Department, University of Tennessee at Knoxville, Preprint UT-CS-94-257; available from Netlib, or as http://www.cs.utk.edu/∼library/TechReports/1994/ut-cs-94-257.ps.Z). [2] See also discussion after `tqli` in Chapter B11.

Chapter B3. Interpolation and Extrapolation

```
SUBROUTINE polint(xa,ya,x,y,dy)
USE nrtype; USE nrutil, ONLY : assert_eq,iminloc,nrerror
IMPLICIT NONE
REAL(SP), DIMENSION(:), INTENT(IN) :: xa,ya
REAL(SP), INTENT(IN) :: x
REAL(SP), INTENT(OUT) :: y,dy
```

Given arrays xa and ya of length N, and given a value x, this routine returns a value y, and an error estimate dy. If $P(x)$ is the polynomial of degree $N-1$ such that $P(\mathtt{xa}_i) = \mathtt{ya}_i, i = 1, \ldots, N$, then the returned value $\mathtt{y} = P(\mathtt{x})$.

```
INTEGER(I4B) :: m,n,ns
REAL(SP), DIMENSION(size(xa)) :: c,d,den,ho
n=assert_eq(size(xa),size(ya),'polint')
c=ya                                        Initialize the tableau of c's and d's.
d=ya
ho=xa-x
ns=iminloc(abs(x-xa))                       Find index ns of closest table entry.
y=ya(ns)                                    This is the initial approximation to y.
ns=ns-1
do m=1,n-1                                  For each column of the tableau,
    den(1:n-m)=ho(1:n-m)-ho(1+m:n)          we loop over the current c's and d's and up-
    if (any(den(1:n-m) == 0.0)) &               date them.
        call nrerror('polint: calculation failure')
          This error can occur only if two input xa's are (to within roundoff) identical.
    den(1:n-m)=(c(2:n-m+1)-d(1:n-m))/den(1:n-m)
    d(1:n-m)=ho(1+m:n)*den(1:n-m)           Here the c's and d's are updated.
    c(1:n-m)=ho(1:n-m)*den(1:n-m)
    if (2*ns < n-m) then                After each column in the tableau is completed, we decide
        dy=c(ns+1)                          which correction, c or d, we want to add to our accu-
    else                                    mulating value of y, i.e., which path to take through
        dy=d(ns)                            the tableau—forking up or down. We do this in such a
        ns=ns-1                             way as to take the most "straight line" route through the
    end if                                  tableau to its apex, updating ns accordingly to keep track
    y=y+dy                                  of where we are. This route keeps the partial approxima-
end do                                      tions centered (insofar as possible) on the target x. The
END SUBROUTINE polint                       last dy added is thus the error indication.
```

```
SUBROUTINE ratint(xa,ya,x,y,dy)
USE nrtype; USE nrutil, ONLY : assert_eq,iminloc,nrerror
IMPLICIT NONE
REAL(SP), DIMENSION(:), INTENT(IN) :: xa,ya
REAL(SP), INTENT(IN) :: x
REAL(SP), INTENT(OUT) :: y,dy
```

Given arrays xa and ya of length N, and given a value of x, this routine returns a value of y and an accuracy estimate dy. The value returned is that of the diagonal rational function, evaluated at x, that passes through the N points $(\mathtt{xa}_i, \mathtt{ya}_i)$, $i = 1 \ldots N$.

```
INTEGER(I4B) :: m,n,ns
REAL(SP), DIMENSION(size(xa)) :: c,d,dd,h,t
```

```
REAL(SP), PARAMETER :: TINY=1.0e-25_sp                A small number.
n=assert_eq(size(xa),size(ya),'ratint')
h=xa-x
ns=iminloc(abs(h))
y=ya(ns)
if (x == xa(ns)) then
    dy=0.0
    RETURN
end if
c=ya
d=ya+TINY                                             The TINY part is needed to prevent
                                                          a rare zero-over-zero condition.
ns=ns-1
do m=1,n-1
    t(1:n-m)=(xa(1:n-m)-x)*d(1:n-m)/h(1+m:n)          h will never be zero, since this was
                                                          tested in the initializing loop.
    dd(1:n-m)=t(1:n-m)-c(2:n-m+1)
    if (any(dd(1:n-m) == 0.0)) &
        call nrerror('failure in ratint')             This error condition indicates that
                                                          the interpolating function has a
                                                          pole at the requested value of
                                                          x.
    dd(1:n-m)=(c(2:n-m+1)-d(1:n-m))/dd(1:n-m)
    d(1:n-m)=c(2:n-m+1)*dd(1:n-m)
    c(1:n-m)=t(1:n-m)*dd(1:n-m)
    if (2*ns < n-m) then
        dy=c(ns+1)
    else
        dy=d(ns)
        ns=ns-1
    end if
    y=y+dy
end do
END SUBROUTINE ratint
```

⋆ ⋆ ⋆

```
SUBROUTINE spline(x,y,yp1,ypn,y2)
USE nrtype; USE nrutil, ONLY : assert_eq
USE nr, ONLY : tridag
IMPLICIT NONE
REAL(SP), DIMENSION(:), INTENT(IN) :: x,y
REAL(SP), INTENT(IN) :: yp1,ypn
REAL(SP), DIMENSION(:), INTENT(OUT) :: y2
```

Given arrays x and y of length N containing a tabulated function, i.e., $y_i = f(x_i)$, with $x_1 < x_2 < \ldots < x_N$, and given values yp1 and ypn for the first derivative of the interpolating function at points 1 and N, respectively, this routine returns an array y2 of length N that contains the second derivatives of the interpolating function at the tabulated points x_i. If yp1 and/or ypn are equal to 1×10^{30} or larger, the routine is signaled to set the corresponding boundary condition for a natural spline, with zero second derivative on that boundary.

```
INTEGER(I4B) :: n
REAL(SP), DIMENSION(size(x)) :: a,b,c,r
n=assert_eq(size(x),size(y),size(y2),'spline')
c(1:n-1)=x(2:n)-x(1:n-1)                  Set up the tridiagonal equations.
r(1:n-1)=6.0_sp*((y(2:n)-y(1:n-1))/c(1:n-1))
r(2:n-1)=r(2:n-1)-r(1:n-2)
a(2:n-1)=c(1:n-2)
b(2:n-1)=2.0_sp*(c(2:n-1)+a(2:n-1))
b(1)=1.0
b(n)=1.0
if (yp1 > 0.99e30_sp) then                The lower boundary condition is set either to be "nat-
                                              ural"
    r(1)=0.0
    c(1)=0.0
else                                      or else to have a specified first derivative.
    r(1)=(3.0_sp/(x(2)-x(1)))*((y(2)-y(1))/(x(2)-x(1))-yp1)
```

```
    c(1)=0.5
end if                                  The upper boundary condition is set either to be
if (ypn > 0.99e30_sp) then                  "natural"
    r(n)=0.0
    a(n)=0.0
else                                    or else to have a specified first derivative.
    r(n)=(-3.0_sp/(x(n)-x(n-1)))*((y(n)-y(n-1))/(x(n)-x(n-1))-ypn)
    a(n)=0.5
end if
call tridag(a(2:n),b(1:n),c(1:n-1),r(1:n),y2(1:n))
END SUBROUTINE spline
```

```
FUNCTION splint(xa,ya,y2a,x)
USE nrtype; USE nrutil, ONLY : assert_eq,nrerror
USE nr, ONLY: locate
IMPLICIT NONE
REAL(SP), DIMENSION(:), INTENT(IN) :: xa,ya,y2a
REAL(SP), INTENT(IN) :: x
REAL(SP) :: splint
```

Given the arrays xa and ya, which tabulate a function (with the xa_i's in increasing or decreasing order), and given the array y2a, which is the output from spline above, and given a value of x, this routine returns a cubic-spline interpolated value. The arrays xa, ya and y2a are all of the same size.

```
INTEGER(I4B) :: khi,klo,n
REAL(SP) :: a,b,h
n=assert_eq(size(xa),size(ya),size(y2a),'splint')
klo=max(min(locate(xa,x),n-1),1)
```

We will find the right place in the table by means of locate's bisection algorithm. This is optimal if sequential calls to this routine are at random values of x. If sequential calls are in order, and closely spaced, one would do better to store previous values of klo and khi and test if they remain appropriate on the next call.

```
khi=klo+1                          klo and khi now bracket the input value of x.
h=xa(khi)-xa(klo)
if (h == 0.0) call nrerror('bad xa input in splint')     The xa's must be distinct.
a=(xa(khi)-x)/h                    Cubic spline polynomial is now evaluated.
b=(x-xa(klo))/h
splint=a*ya(klo)+b*ya(khi)+((a**3-a)*y2a(klo)+(b**3-b)*y2a(khi))*(h**2)/6.0_sp
END FUNCTION splint
```

f90 klo=max(min(locate(xa,x),n-1),1) In the Fortran 77 version of splint, there is in-line code to find the location in the table by bisection. Here we prefer an explicit call to locate, which performs the bisection. On some massively multiprocessor (MMP) machines, one might substitute a different, more parallel algorithm (see next note).

⋆ ⋆ ⋆

```
FUNCTION locate(xx,x)
USE nrtype
IMPLICIT NONE
REAL(SP), DIMENSION(:), INTENT(IN) :: xx
REAL(SP), INTENT(IN) :: x
INTEGER(I4B) :: locate
```

Given an array xx(1:N), and given a value x, returns a value j such that x is between xx(j) and xx($j+1$). xx must be monotonic, either increasing or decreasing. $j=0$ or $j=N$ is returned to indicate that x is out of range.

```
INTEGER(I4B) :: n,jl,jm,ju
LOGICAL :: ascnd
```

```
n=size(xx)
ascnd = (xx(n) >= xx(1))                True if ascending order of table, false otherwise.
jl=0                                    Initialize lower
ju=n+1                                  and upper limits.
do
    if (ju-jl <= 1) exit                Repeat until this condition is satisfied.
    jm=(ju+jl)/2                        Compute a midpoint,
    if (ascnd .eqv. (x >= xx(jm))) then
        jl=jm                           and replace either the lower limit
    else
        ju=jm                           or the upper limit, as appropriate.
    end if
end do
if (x == xx(1)) then                    Then set the output, being careful with the endpoints.
    locate=1
else if (x == xx(n)) then
    locate=n-1
else
    locate=jl
end if
END FUNCTION locate
```

The use of bisection is perhaps a sin on a genuinely parallel machine, but (since the process takes only logarithmically many sequential steps) it is at most a *small* sin. One can imagine a "fully parallel" implementation like,

```
k=iminloc(abs(x-xx))
if ((x < xx(k)) .eqv. (xx(1) < xx(n))) then
        locate=k-1
else
        locate=k
end if
```

Problem is, unless the number of *physical* (not logical) processors participating in the `iminloc` is larger than N, the length of the array, this "parallel" code turns a $\log N$ algorithm into one scaling as N, quite an unacceptable inefficiency. So we prefer to be small sinners and bisect.

```
SUBROUTINE hunt(xx,x,jlo)
USE nrtype
IMPLICIT NONE
INTEGER(I4B), INTENT(INOUT) :: jlo
REAL(SP), INTENT(IN) :: x
REAL(SP), DIMENSION(:), INTENT(IN) :: xx
    Given an array xx(1:N), and given a value x, returns a value jlo such that x is between
    xx(jlo) and xx(jlo+1). xx must be monotonic, either increasing or decreasing. jlo = 0
    or jlo = N is returned to indicate that x is out of range. jlo on input is taken as the
    initial guess for jlo on output.
INTEGER(I4B) :: n,inc,jhi,jm
LOGICAL :: ascnd
n=size(xx)
ascnd = (xx(n) >= xx(1))                    True if ascending order of table, false otherwise.
if (jlo <= 0 .or. jlo > n) then             Input guess not useful. Go immediately to bisec-
    jlo=0                                       tion.
    jhi=n+1
else
    inc=1                                   Set the hunting increment.
    if (x >= xx(jlo) .eqv. ascnd) then          Hunt up:
        do
            jhi=jlo+inc
            if (jhi > n) then               Done hunting, since off end of table.
```

```
                jhi=n+1
                exit
            else
                if (x < xx(jhi) .eqv. ascnd) exit
                jlo=jhi                     Not done hunting,
                inc=inc+inc                 so double the increment
            end if
        end do                              and try again.
    else                                    Hunt down:
        jhi=jlo
        do
            jlo=jhi-inc
            if (jlo < 1) then               Done hunting, since off end of table.
                jlo=0
                exit
            else
                if (x >= xx(jlo) .eqv. ascnd) exit
                jhi=jlo                     Not done hunting,
                inc=inc+inc                 so double the increment
            end if
        end do                              and try again.
    end if
end if                                      Done hunting, value bracketed.
do                                          Hunt is done, so begin the final bisection phase:
    if (jhi-jlo <= 1) then
        if (x == xx(n)) jlo=n-1
        if (x == xx(1)) jlo=1
        exit
    else
        jm=(jhi+jlo)/2
        if (x >= xx(jm) .eqv. ascnd) then
            jlo=jm
        else
            jhi=jm
        end if
    end if
end do
END SUBROUTINE hunt
```

⋆ ⋆ ⋆

```
FUNCTION polcoe(x,y)
USE nrtype; USE nrutil, ONLY : assert_eq,outerdiff
IMPLICIT NONE
REAL(SP), DIMENSION(:), INTENT(IN) :: x,y
REAL(SP), DIMENSION(size(x)) :: polcoe
```

Given same-size arrays x and y containing a tabulated function $y_i = f(x_i)$, this routine returns a same-size array of coefficients c_j, such that $y_i = \sum_j c_j x_i^{j-1}$.

```
INTEGER(I4B) :: i,k,n
REAL(SP), DIMENSION(size(x)) :: s
REAL(SP), DIMENSION(size(x),size(x)) :: a
n=assert_eq(size(x),size(y),'polcoe')
s=0.0                               Coefficients s_i of the master polynomial P(x) are found by
s(n)=-x(1)                              recurrence.
do i=2,n
    s(n+1-i:n-1)=s(n+1-i:n-1)-x(i)*s(n+2-i:n)
    s(n)=s(n)-x(i)
end do
a=outerdiff(x,x)                    Make vector w_j = prod_{j/=n}(x_j - x_n), using polcoe for tempo-
polcoe=product(a,dim=2,mask=a /= 0.0)       rary storage.
```

```
    Now do synthetic division by x - x_j. The division for all x_j can be done in parallel (on a
    parallel machine), since the : in the loop below is over j.
a(:,1)=-s(1)/x(:)
do k=2,n
    a(:,k)=-(s(k)-a(:,k-1))/x(:)
end do
s=y/polcoe
polcoe=matmul(s,a)              Solve linear system.
END FUNCTION polcoe
```

For a description of the coding here, see §22.3, especially equation (22.3.9). You might also want to compare the coding here with the Fortran 77 version, and also look at the description of the method on p. 84 in Volume 1. The Fortran 90 implementation here is in fact much closer to that description than is the Fortran 77 method, which goes through some acrobatics to roll the synthetic division and matrix multiplication into a single set of two nested loops. The price we pay, here, is storage for the matrix a. Since the degree of any useful polynomial is not a very large number, this is essentially no penalty.

Also worth noting is the way that parallelism is brought to the required synthetic division. For a *single* such synthetic division (e.g., as accomplished by the `nrutil` routine `poly_term`), parallelism can be obtained only by recursion. Here things are much simpler, because we need a whole bunch of simultaneous and independent synthetic divisions; so we can just do them in the obvious, data-parallel, way.

```
FUNCTION polcof(xa,ya)
USE nrtype; USE nrutil, ONLY : assert_eq,iminloc
USE nr, ONLY : polint
IMPLICIT NONE
REAL(SP), DIMENSION(:), INTENT(IN) :: xa,ya
REAL(SP), DIMENSION(size(xa)) :: polcof
    Given same-size arrays xa and ya containing a tabulated function ya_i = f(xa_i), this routine
    returns a same-size array of coefficients c_j such that ya_i = sum_j c_j xa_i^(j-1).
INTEGER(I4B) :: j,k,m,n
REAL(SP) :: dy
REAL(SP), DIMENSION(size(xa)) :: x,y
n=assert_eq(size(xa),size(ya),'polcof')
x=xa
y=ya
do j=1,n
    m=n+1-j
    call polint(x(1:m),y(1:m),0.0_sp,polcof(j),dy)
        Use the polynomial interpolation routine of §3.1 to extrapolate to x = 0.
    k=iminloc(abs(x(1:m)))                  Find the remaining x_k of smallest absolute value,
    where (x(1:m) /= 0.0) y(1:m)=(y(1:m)-polcof(j))/x(1:m)      reduce all the terms,
    y(k:m-1)=y(k+1:m)                       and eliminate x_k.
    x(k:m-1)=x(k+1:m)
end do
END FUNCTION polcof
```

⋆ ⋆ ⋆

```
SUBROUTINE polin2(x1a,x2a,ya,x1,x2,y,dy)
USE nrtype; USE nrutil, ONLY : assert_eq
USE nr, ONLY : polint
IMPLICIT NONE
REAL(SP), DIMENSION(:), INTENT(IN) :: x1a,x2a
REAL(SP), DIMENSION(:,:), INTENT(IN) :: ya
REAL(SP), INTENT(IN) :: x1,x2
REAL(SP), INTENT(OUT) :: y,dy
```

Given arrays `x1a` of length M and `x2a` of length N of independent variables, and an $M \times N$ array of function values `ya`, tabulated at the grid points defined by `x1a` and `x2a`, and given values `x1` and `x2` of the independent variables, this routine returns an interpolated function value `y`, and an accuracy indication `dy` (based only on the interpolation in the `x1` direction, however).

```
INTEGER(I4B) :: j,m,ndum
REAL(SP), DIMENSION(size(x1a)) :: ymtmp
REAL(SP), DIMENSION(size(x2a)) :: yntmp
m=assert_eq(size(x1a),size(ya,1),'polin2: m')
ndum=assert_eq(size(x2a),size(ya,2),'polin2: ndum')
do j=1,m                                        Loop over rows.
    yntmp=ya(j,:)                               Copy row into temporary storage.
    call polint(x2a,yntmp,x2,ymtmp(j),dy)       Interpolate answer into temporary stor-
end do                                              age.
call polint(x1a,ymtmp,x1,y,dy)                  Do the final interpolation.
END SUBROUTINE polin2
```

⋆ ⋆ ⋆

```
SUBROUTINE bcucof(y,y1,y2,y12,d1,d2,c)
USE nrtype
IMPLICIT NONE
REAL(SP), INTENT(IN) :: d1,d2
REAL(SP), DIMENSION(4), INTENT(IN) :: y,y1,y2,y12
REAL(SP), DIMENSION(4,4), INTENT(OUT) :: c
```

Given arrays `y`, `y1`, `y2`, and `y12`, each of length 4, containing the function, gradients, and cross derivative at the four grid points of a rectangular grid cell (numbered counterclockwise from the lower left), and given `d1` and `d2`, the length of the grid cell in the 1- and 2-directions, this routine returns the 4×4 table `c` that is used by routine `bcuint` for bicubic interpolation.

```
REAL(SP), DIMENSION(16) :: x
REAL(SP), DIMENSION(16,16) :: wt
DATA wt /1,0,-3,2,4*0,-3,0,9,-6,2,0,-6,4,&
    8*0,3,0,-9,6,-2,0,6,-4,10*0,9,-6,2*0,-6,4,2*0,3,-2,6*0,-9,6,&
    2*0,6,-4,4*0,1,0,-3,2,-2,0,6,-4,1,0,-3,2,8*0,-1,0,3,-2,1,0,-3,&
    2,10*0,-3,2,2*0,3,-2,6*0,3,-2,2*0,-6,4,2*0,3,-2,0,1,-2,1,5*0,&
    -3,6,-3,0,2,-4,2,9*0,3,-6,3,0,-2,4,-2,10*0,-3,3,2*0,2,-2,2*0,&
    -1,1,6*0,3,-3,2*0,-2,2,5*0,1,-2,1,0,-2,4,-2,0,1,-2,1,9*0,-1,2,&
    -1,0,1,-2,1,10*0,1,-1,2*0,-1,1,6*0,-1,1,2*0,2,-2,2*0,-1,1/
x(1:4)=y                                   Pack a temporary vector x.
x(5:8)=y1*d1
x(9:12)=y2*d2
x(13:16)=y12*d1*d2
x=matmul(wt,x)                             Matrix multiply by the stored table.
c=reshape(x,(/4,4/),order=(/2,1/))         Unpack the result into the output table.
END SUBROUTINE bcucof
```

f90 `x=matmul(wt,x) ... c=reshape(x,(/4,4/),order=(/2,1/))` It is a powerful technique to combine the `matmul` intrinsic with `reshape`'s of the input or output. The idea is to use `matmul` whenever the calculation can be cast into the form of a linear mapping between input and output objects. Here the `order=(/2,1/)` parameter specifies that we want the packing to be by rows, not by Fortran's default of columns. (In this two-dimensional case, it's the equivalent of applying `transpose`.)

```
SUBROUTINE bcuint(y,y1,y2,y12,x1l,x1u,x2l,x2u,x1,x2,ansy,ansy1,ansy2)
USE nrtype; USE nrutil, ONLY : nrerror
USE nr, ONLY : bcucof
IMPLICIT NONE
REAL(SP), DIMENSION(4), INTENT(IN) :: y,y1,y2,y12
REAL(SP), INTENT(IN) :: x1l,x1u,x2l,x2u,x1,x2
REAL(SP), INTENT(OUT) :: ansy,ansy1,ansy2
    Bicubic interpolation within a grid square. Input quantities are y,y1,y2,y12 (as described
    in bcucof); x1l and x1u, the lower and upper coordinates of the grid square in the 1-
    direction; x2l and x2u likewise for the 2-direction; and x1,x2, the coordinates of the
    desired point for the interpolation. The interpolated function value is returned as ansy,
    and the interpolated gradient values as ansy1 and ansy2. This routine calls bcucof.
INTEGER(I4B) :: i
REAL(SP) :: t,u
REAL(SP), DIMENSION(4,4) :: c
call bcucof(y,y1,y2,y12,x1u-x1l,x2u-x2l,c)          Get the c's.
if (x1u == x1l .or. x2u == x2l) call &
    nrerror('bcuint: problem with input values - boundary pair equal?')
t=(x1-x1l)/(x1u-x1l)                                 Equation (3.6.4).
u=(x2-x2l)/(x2u-x2l)
ansy=0.0
ansy2=0.0
ansy1=0.0
do i=4,1,-1                                          Equation (3.6.6).
    ansy=t*ansy+((c(i,4)*u+c(i,3))*u+c(i,2))*u+c(i,1)
    ansy2=t*ansy2+(3.0_sp*c(i,4)*u+2.0_sp*c(i,3))*u+c(i,2)
    ansy1=u*ansy1+(3.0_sp*c(4,i)*t+2.0_sp*c(3,i))*t+c(2,i)
end do
ansy1=ansy1/(x1u-x1l)
ansy2=ansy2/(x2u-x2l)
END SUBROUTINE bcuint
```

⋆ ⋆ ⋆

```
SUBROUTINE splie2(x1a,x2a,ya,y2a)
USE nrtype; USE nrutil, ONLY : assert_eq
USE nr, ONLY : spline
IMPLICIT NONE
REAL(SP), DIMENSION(:), INTENT(IN) :: x1a,x2a
REAL(SP), DIMENSION(:,:), INTENT(IN) :: ya
REAL(SP), DIMENSION(:,:), INTENT(OUT) :: y2a
    Given an M × N tabulated function ya, and N tabulated independent variables x2a, this
    routine constructs one-dimensional natural cubic splines of the rows of ya and returns the
    second derivatives in the M × N array y2a. (The array x1a is included in the argument
    list merely for consistency with routine splin2.)
INTEGER(I4B) :: j,m,ndum
m=assert_eq(size(x1a),size(ya,1),size(y2a,1),'splie2: m')
ndum=assert_eq(size(x2a),size(ya,2),size(y2a,2),'splie2: ndum')
do j=1,m
    call spline(x2a,ya(j,:),1.0e30_sp,1.0e30_sp,y2a(j,:))
```

```
        Values 1 × 10^30 signal a natural spline.
end do
END SUBROUTINE splie2
```

```
FUNCTION splin2(x1a,x2a,ya,y2a,x1,x2)
USE nrtype; USE nrutil, ONLY : assert_eq
USE nr, ONLY : spline,splint
IMPLICIT NONE
REAL(SP), DIMENSION(:), INTENT(IN) :: x1a,x2a
REAL(SP), DIMENSION(:,:), INTENT(IN) :: ya,y2a
REAL(SP), INTENT(IN) :: x1,x2
REAL(SP) :: splin2
    Given x1a, x2a, ya as described in splie2 and y2a as produced by that routine; and given
    a desired interpolating point x1,x2; this routine returns an interpolated function value by
    bicubic spline interpolation.
INTEGER(I4B) :: j,m,ndum
REAL(SP), DIMENSION(size(x1a)) :: yytmp,y2tmp2
m=assert_eq(size(x1a),size(ya,1),size(y2a,1),'splin2: m')
ndum=assert_eq(size(x2a),size(ya,2),size(y2a,2),'splin2: ndum')
do j=1,m
    yytmp(j)=splint(x2a,ya(j,:),y2a(j,:),x2)
        Perform m evaluations of the row splines constructed by splie2, using the one-dimensional
        spline evaluator splint.
end do
call spline(x1a,yytmp,1.0e30_sp,1.0e30_sp,y2tmp2)
  Construct the one-dimensional column spline and evaluate it.
splin2=splint(x1a,yytmp,y2tmp2,x1)
END FUNCTION splin2
```

Chapter B4. Integration of Functions

```
SUBROUTINE trapzd(func,a,b,s,n)
USE nrtype; USE nrutil, ONLY : arth
IMPLICIT NONE
REAL(SP), INTENT(IN) :: a,b
REAL(SP), INTENT(INOUT) :: s
INTEGER(I4B), INTENT(IN) :: n
INTERFACE
    FUNCTION func(x)
    USE nrtype
    REAL(SP), DIMENSION(:), INTENT(IN) :: x
    REAL(SP), DIMENSION(size(x)) :: func
    END FUNCTION func
END INTERFACE
```

This routine computes the `n`th stage of refinement of an extended trapezoidal rule. `func` is input as the name of the function to be integrated between limits `a` and `b`, also input. When called with `n`=1, the routine returns as `s` the crudest estimate of $\int_a^b f(x)dx$. Subsequent calls with `n`=2,3,... (in that sequential order) will improve the accuracy of `s` by adding 2^{n-2} additional interior points. `s` should not be modified between sequential calls.

```
REAL(SP) :: del,fsum
INTEGER(I4B) :: it
if (n == 1) then
    s=0.5_sp*(b-a)*sum(func( (/ a,b /) ))
else
    it=2**(n-2)
    del=(b-a)/it                       This is the spacing of the points to be added.
    fsum=sum(func(arth(a+0.5_sp*del,del,it)))
    s=0.5_sp*(s+del*fsum)              This replaces s by its refined value.
end if
END SUBROUTINE trapzd
```

f90 While most of the quadrature routines in this chapter are coded as functions, `trapzd` is a subroutine because the argument `s` that returns the function value must also be supplied as an input parameter. We could change the subroutine into a function by declaring `s` to be a local variable with the SAVE attribute. However, this would prevent us from being able to use the routine recursively to do multidimensional quadrature (see quad3d on p. 1065). When `s` is left as an argument, a fresh copy is created on each recursive call. As a SAVE'd variable, by contrast, its value would get overwritten on each call, and the code would not be properly "re-entrant."

`s=0.5_sp*(b-a)*sum(func( (/ a,b /) ))` Note how we use the `(/.../)` construct to supply a set of scalar arguments to a vector function.

⋆ ⋆ ⋆

```
FUNCTION qtrap(func,a,b)
USE nrtype; USE nrutil, ONLY : nrerror
USE nr, ONLY : trapzd
IMPLICIT NONE
REAL(SP), INTENT(IN) :: a,b
REAL(SP) :: qtrap
INTERFACE
    FUNCTION func(x)
    USE nrtype
    REAL(SP), DIMENSION(:), INTENT(IN) :: x
    REAL(SP), DIMENSION(size(x)) :: func
    END FUNCTION func
END INTERFACE
INTEGER(I4B), PARAMETER :: JMAX=20
REAL(SP), PARAMETER :: EPS=1.0e-6_sp, UNLIKELY=-1.0e30_sp
```

Returns the integral of the function `func` from `a` to `b`. The parameter `EPS` should be set to the desired fractional accuracy and `JMAX` so that 2 to the power `JMAX-1` is the maximum allowed number of steps. Integration is performed by the trapezoidal rule.

```
REAL(SP) :: olds
INTEGER(I4B) :: j
olds=UNLIKELY                       Any number that is unlikely to be the average of the
do j=1,JMAX                             function at its endpoints will do here.
    call trapzd(func,a,b,qtrap,j)
    if (j > 5) then                 Avoid spurious early convergence.
        if (abs(qtrap-olds) < EPS*abs(olds) .or. &
            (qtrap == 0.0 .and. olds == 0.0)) RETURN
    end if
    olds=qtrap
end do
call nrerror('qtrap: too many steps')
END FUNCTION qtrap
```

⋆ ⋆ ⋆

```
FUNCTION qsimp(func,a,b)
USE nrtype; USE nrutil, ONLY : nrerror
USE nr, ONLY : trapzd
IMPLICIT NONE
REAL(SP), INTENT(IN) :: a,b
REAL(SP) :: qsimp
INTERFACE
    FUNCTION func(x)
    USE nrtype
    REAL(SP), DIMENSION(:), INTENT(IN) :: x
    REAL(SP), DIMENSION(size(x)) :: func
    END FUNCTION func
END INTERFACE
INTEGER(I4B), PARAMETER :: JMAX=20
REAL(SP), PARAMETER :: EPS=1.0e-6_sp, UNLIKELY=-1.0e30_sp
```

Returns the integral of the function `func` from `a` to `b`. The parameter `EPS` should be set to the desired fractional accuracy and `JMAX` so that 2 to the power `JMAX-1` is the maximum allowed number of steps. Integration is performed by Simpson's rule.

```
INTEGER(I4B) :: j
REAL(SP) :: os,ost,st
ost=UNLIKELY
os= UNLIKELY
do j=1,JMAX
    call trapzd(func,a,b,st,j)
    qsimp=(4.0_sp*st-ost)/3.0_sp        Compare equation (4.2.4).
    if (j > 5) then                     Avoid spurious early convergence.
        if (abs(qsimp-os) < EPS*abs(os) .or. &
            (qsimp == 0.0 .and. os == 0.0)) RETURN
    end if
    os=qsimp
```

```
    ost=st
end do
call nrerror('qsimp: too many steps')
END FUNCTION qsimp
```

⋆ ⋆ ⋆

```
FUNCTION qromb(func,a,b)
USE nrtype; USE nrutil, ONLY : nrerror
USE nr, ONLY : polint,trapzd
IMPLICIT NONE
REAL(SP), INTENT(IN) :: a,b
REAL(SP) :: qromb
INTERFACE
    FUNCTION func(x)
    USE nrtype
    REAL(SP), DIMENSION(:), INTENT(IN) :: x
    REAL(SP), DIMENSION(size(x)) :: func
    END FUNCTION func
END INTERFACE
INTEGER(I4B), PARAMETER :: JMAX=20,JMAXP=JMAX+1,K=5,KM=K-1
REAL(SP), PARAMETER :: EPS=1.0e-6_sp
```

Returns the integral of the function `func` from `a` to `b`. Integration is performed by Romberg's method of order 2K, where, e.g., K=2 is Simpson's rule.

Parameters: `EPS` is the fractional accuracy desired, as determined by the extrapolation error estimate; `JMAX` limits the total number of steps; `K` is the number of points used in the extrapolation.

```
REAL(SP), DIMENSION(JMAXP) :: h,s
REAL(SP) :: dqromb
INTEGER(I4B) :: j
h(1)=1.0
do j=1,JMAX
    call trapzd(func,a,b,s(j),j)
    if (j >= K) then
        call polint(h(j-KM:j),s(j-KM:j),0.0_sp,qromb,dqromb)
        if (abs(dqromb) <= EPS*abs(qromb)) RETURN
    end if
    s(j+1)=s(j)
    h(j+1)=0.25_sp*h(j)
end do
call nrerror('qromb: too many steps')
END FUNCTION qromb
```

(`h,s`:) These store the successive trapezoidal approximations and their relative stepsizes.

(`h(j+1)=0.25_sp*h(j)`:) This is a key step: The factor is 0.25 even though the stepsize is decreased by only 0.5. This makes the extrapolation a polynomial in h^2 as allowed by equation (4.2.1), not just a polynomial in h.

⋆ ⋆ ⋆

```
SUBROUTINE midpnt(func,a,b,s,n)
USE nrtype; USE nrutil, ONLY : arth
IMPLICIT NONE
REAL(SP), INTENT(IN) :: a,b
REAL(SP), INTENT(INOUT) :: s
INTEGER(I4B), INTENT(IN) :: n
INTERFACE
    FUNCTION func(x)
    USE nrtype
    REAL(SP), DIMENSION(:), INTENT(IN) :: x
    REAL(SP), DIMENSION(size(x)) :: func
    END FUNCTION func
END INTERFACE
```

This routine computes the `n`th stage of refinement of an extended midpoint rule. `func` is input as the name of the function to be integrated between limits `a` and `b`, also input. When

```
    called with n=1, the routine returns as s the crudest estimate of ∫_a^b f(x)dx. Subsequent
    calls with n=2,3,... (in that sequential order) will improve the accuracy of s by adding
    (2/3) × 3^(n-1) additional interior points. s should not be modified between sequential calls.
REAL(SP) :: del
INTEGER(I4B) :: it
REAL(SP), DIMENSION(2*3**(n-2)) :: x
if (n == 1) then
    s=(b-a)*sum(func( (/0.5_sp*(a+b)/) ))
else
    it=3**(n-2)
    del=(b-a)/(3.0_sp*it)                         The added points alternate in spacing between
    x(1:2*it-1:2)=arth(a+0.5_sp*del,3.0_sp*del,it)    del and 2*del.
    x(2:2*it:2)=x(1:2*it-1:2)+2.0_sp*del
    s=s/3.0_sp+del*sum(func(x))                   The new sum is combined with the old integral
end if                                                to give a refined integral.
END SUBROUTINE midpnt
```

f90 `midpnt` is a subroutine and not a function for the same reasons as `trapzd`. This is also true for the other `mid...` routines below.

`s=(b-a)*sum(func( (/0.5_sp*(a+b)/) ))` Here we use `(/.../)` to pass a single scalar argument to a vector function.

⋆ ⋆ ⋆

```
FUNCTION qromo(func,a,b,choose)
USE nrtype; USE nrutil, ONLY : nrerror
USE nr, ONLY : polint
IMPLICIT NONE
REAL(SP), INTENT(IN) :: a,b
REAL(SP) :: qromo
INTERFACE
    FUNCTION func(x)
    USE nrtype
    IMPLICIT NONE
    REAL(SP), DIMENSION(:), INTENT(IN) :: x
    REAL(SP), DIMENSION(size(x)) :: func
    END FUNCTION func

    SUBROUTINE choose(funk,aa,bb,s,n)
    USE nrtype
    IMPLICIT NONE
    REAL(SP), INTENT(IN) :: aa,bb
    REAL(SP), INTENT(INOUT) :: s
    INTEGER(I4B), INTENT(IN) :: n
    INTERFACE
        FUNCTION funk(x)
        USE nrtype
        IMPLICIT NONE
        REAL(SP), DIMENSION(:), INTENT(IN) :: x
        REAL(SP), DIMENSION(size(x)) :: funk
        END FUNCTION funk
    END INTERFACE
    END SUBROUTINE choose
END INTERFACE
INTEGER(I4B), PARAMETER :: JMAX=14,JMAXP=JMAX+1,K=5,KM=K-1
REAL(SP), PARAMETER :: EPS=1.0e-6
    Romberg integration on an open interval. Returns the integral of the function func from a
    to b, using any specified integrating subroutine choose and Romberg's method. Normally
    choose will be an open formula, not evaluating the function at the endpoints. It is assumed
    that choose triples the number of steps on each call, and that its error series contains only
```

even powers of the number of steps. The routines midpnt, midinf, midsql, midsqu, and midexp are possible choices for choose. The parameters have the same meaning as in qromb.

```
REAL(SP), DIMENSION(JMAXP) :: h,s
REAL(SP) :: dqromo
INTEGER(I4B) :: j
h(1)=1.0
do j=1,JMAX
    call choose(func,a,b,s(j),j)
    if (j >= K) then
        call polint(h(j-KM:j),s(j-KM:j),0.0_sp,qromo,dqromo)
        if (abs(dqromo) <= EPS*abs(qromo)) RETURN
    end if
    s(j+1)=s(j)
    h(j+1)=h(j)/9.0_sp              This is where the assumption of step tripling and an even
end do                                  error series is used.
call nrerror('qromo: too many steps')
END FUNCTION qromo
```

⋆ ⋆ ⋆

```
SUBROUTINE midinf(funk,aa,bb,s,n)
USE nrtype; USE nrutil, ONLY : arth,assert
IMPLICIT NONE
REAL(SP), INTENT(IN) :: aa,bb
REAL(SP), INTENT(INOUT) :: s
INTEGER(I4B), INTENT(IN) :: n
INTERFACE
    FUNCTION funk(x)
    USE nrtype
    REAL(SP), DIMENSION(:), INTENT(IN) :: x
    REAL(SP), DIMENSION(size(x)) :: funk
    END FUNCTION funk
END INTERFACE
```

This routine is an exact replacement for midpnt, i.e., returns as s the nth stage of refinement of the integral of funk from aa to bb, except that the function is evaluated at evenly spaced points in $1/x$ rather than in x. This allows the upper limit bb to be as large and positive as the computer allows, or the lower limit aa to be as large and negative, but not both. aa and bb must have the same sign.

```
REAL(SP) :: a,b,del
INTEGER(I4B) :: it
REAL(SP), DIMENSION(2*3**(n-2)) :: x
call assert(aa*bb > 0.0, 'midinf args')
b=1.0_sp/aa                         These two statements change the limits of integration ac-
a=1.0_sp/bb                             cordingly.
if (n == 1) then                    From this point on, the routine is exactly identical to midpnt.
    s=(b-a)*sum(func( (/0.5_sp*(a+b)/) ))
else
    it=3**(n-2)
    del=(b-a)/(3.0_sp*it)
    x(1:2*it-1:2)=arth(a+0.5_sp*del,3.0_sp*del,it)
    x(2:2*it:2)=x(1:2*it-1:2)+2.0_sp*del
    s=s/3.0_sp+del*sum(func(x))
end if
CONTAINS

    FUNCTION func(x)                This internal function effects the change of variable.
    REAL(SP), DIMENSION(:), INTENT(IN) :: x
    REAL(SP), DIMENSION(size(x)) :: func
    func=funk(1.0_sp/x)/x**2
    END FUNCTION func
END SUBROUTINE midinf
```

f90 `FUNCTION func(x)` The change of variable could have been effected by a statement function in `midinf` itself. However, the statement function is a Fortran 77 feature that is deprecated in Fortran 90 because it does not allow the benefits of having an explicit interface, i.e., a complete set of specification statements. Statement functions can always be coded as internal subprograms instead.

```
SUBROUTINE midsql(funk,aa,bb,s,n)
USE nrtype; USE nrutil, ONLY : arth
IMPLICIT NONE
REAL(SP), INTENT(IN) :: aa,bb
REAL(SP), INTENT(INOUT) :: s
INTEGER(I4B), INTENT(IN) :: n
INTERFACE
    FUNCTION funk(x)
    USE nrtype
    REAL(SP), DIMENSION(:), INTENT(IN) :: x
    REAL(SP), DIMENSION(size(x)) :: funk
    END FUNCTION funk
END INTERFACE
    This routine is an exact replacement for midpnt, i.e., returns as s the nth stage of refinement
    of the integral of funk from aa to bb, except that it allows for an inverse square-root
    singularity in the integrand at the lower limit aa.
REAL(SP) :: a,b,del
INTEGER(I4B) :: it
REAL(SP), DIMENSION(2*3**(n-2)) :: x
b=sqrt(bb-aa)                    These two statements change the limits of integration ac-
a=0.0                                cordingly.
if (n == 1) then                 From this point on, the routine is exactly identical to midpnt.
    s=(b-a)*sum(func( (/0.5_sp*(a+b)/) ))
else
    it=3**(n-2)
    del=(b-a)/(3.0_sp*it)
    x(1:2*it-1:2)=arth(a+0.5_sp*del,3.0_sp*del,it)
    x(2:2*it:2)=x(1:2*it-1:2)+2.0_sp*del
    s=s/3.0_sp+del*sum(func(x))
end if
CONTAINS

    FUNCTION func(x)             This internal function effects the change of variable.
    REAL(SP), DIMENSION(:), INTENT(IN) :: x
    REAL(SP), DIMENSION(size(x)) :: func
    func=2.0_sp*x*funk(aa+x**2)
    END FUNCTION func
END SUBROUTINE midsql
```

```
SUBROUTINE midsqu(funk,aa,bb,s,n)
USE nrtype; USE nrutil, ONLY : arth
IMPLICIT NONE
REAL(SP), INTENT(IN) :: aa,bb
REAL(SP), INTENT(INOUT) :: s
INTEGER(I4B), INTENT(IN) :: n
INTERFACE
    FUNCTION funk(x)
    USE nrtype
    REAL(SP), DIMENSION(:), INTENT(IN) :: x
    REAL(SP), DIMENSION(size(x)) :: funk
    END FUNCTION funk
END INTERFACE
    This routine is an exact replacement for midpnt, i.e., returns as s the nth stage of refinement
    of the integral of funk from aa to bb, except that it allows for an inverse square-root
    singularity in the integrand at the upper limit bb.
REAL(SP) :: a,b,del
```

```
INTEGER(I4B) :: it
REAL(SP), DIMENSION(2*3**(n-2)) :: x
b=sqrt(bb-aa)                  These two statements change the limits of integration ac-
a=0.0                              cordingly.
if (n == 1) then               From this point on, the routine is exactly identical to midpnt.
    s=(b-a)*sum(func( (/0.5_sp*(a+b)/) ))
else
    it=3**(n-2)
    del=(b-a)/(3.0_sp*it)
    x(1:2*it-1:2)=arth(a+0.5_sp*del,3.0_sp*del,it)
    x(2:2*it:2)=x(1:2*it-1:2)+2.0_sp*del
    s=s/3.0_sp+del*sum(func(x))
end if
CONTAINS
    FUNCTION func(x)           This internal function effects the change of variable.
    REAL(SP), DIMENSION(:), INTENT(IN) :: x
    REAL(SP), DIMENSION(size(x)) :: func
    func=2.0_sp*x*funk(bb-x**2)
    END FUNCTION func
END SUBROUTINE midsqu
```

```
SUBROUTINE midexp(funk,aa,bb,s,n)
USE nrtype; USE nrutil, ONLY : arth
IMPLICIT NONE
REAL(SP), INTENT(IN) :: aa,bb
REAL(SP), INTENT(INOUT) :: s
INTEGER(I4B), INTENT(IN) :: n
INTERFACE
    FUNCTION funk(x)
    USE nrtype
    REAL(SP), DIMENSION(:), INTENT(IN) :: x
    REAL(SP), DIMENSION(size(x)) :: funk
    END FUNCTION funk
END INTERFACE
```

This routine is an exact replacement for `midpnt`, i.e., returns as `s` the `n`th stage of refinement of the integral of `funk` from `aa` to `bb`, except that `bb` is assumed to be infinite (value passed not actually used). It is assumed that the function `funk` decreases exponentially rapidly at infinity.

```
REAL(SP) :: a,b,del
INTEGER(I4B) :: it
REAL(SP), DIMENSION(2*3**(n-2)) :: x
b=exp(-aa)                     These two statements change the limits of integration ac-
a=0.0                              cordingly.
if (n == 1) then               From this point on, the routine is exactly identical to midpnt.
    s=(b-a)*sum(func( (/0.5_sp*(a+b)/) ))
else
    it=3**(n-2)
    del=(b-a)/(3.0_sp*it)
    x(1:2*it-1:2)=arth(a+0.5_sp*del,3.0_sp*del,it)
    x(2:2*it:2)=x(1:2*it-1:2)+2.0_sp*del
    s=s/3.0_sp+del*sum(func(x))
end if
CONTAINS
    FUNCTION func(x)           This internal function effects the change of variable.
    REAL(SP), DIMENSION(:), INTENT(IN) :: x
    REAL(SP), DIMENSION(size(x)) :: func
    func=funk(-log(x))/x
    END FUNCTION func
END SUBROUTINE midexp
```

⋆ ⋆ ⋆

```
SUBROUTINE gauleg(x1,x2,x,w)
USE nrtype; USE nrutil, ONLY : arth,assert_eq,nrerror
IMPLICIT NONE
REAL(SP), INTENT(IN) :: x1,x2
REAL(SP), DIMENSION(:), INTENT(OUT) :: x,w
REAL(DP), PARAMETER :: EPS=3.0e-14_dp
```

Given the lower and upper limits of integration x1 and x2, this routine returns arrays x and w of length N containing the abscissas and weights of the Gauss-Legendre N-point quadrature formula. The parameter EPS is the relative precision. Note that internal computations are done in double precision.

```
INTEGER(I4B) :: its,j,m,n
INTEGER(I4B), PARAMETER :: MAXIT=10
REAL(DP) :: xl,xm
REAL(DP), DIMENSION((size(x)+1)/2) :: p1,p2,p3,pp,z,z1
LOGICAL(LGT), DIMENSION((size(x)+1)/2) :: unfinished
n=assert_eq(size(x),size(w),'gauleg')
m=(n+1)/2                                     The roots are symmetric in the interval,
xm=0.5_dp*(x2+x1)                                 so we only have to find half of them.
xl=0.5_dp*(x2-x1)
z=cos(PI_D*(arth(1,1,m)-0.25_dp)/(n+0.5_dp))      Initial approximations to the roots.
unfinished=.true.
do its=1,MAXIT                                Newton's method carried out simultane-
    where (unfinished)                            ously on the roots.
        p1=1.0
        p2=0.0
    end where
    do j=1,n                                  Loop up the recurrence relation to get
        where (unfinished)                        the Legendre polynomials evaluated
            p3=p2                                 at z.
            p2=p1
            p1=((2.0_dp*j-1.0_dp)*z*p2-(j-1.0_dp)*p3)/j
        end where
    end do
```

p1 now contains the desired Legendre polynomials. We next compute pp, the derivatives, by a standard relation involving also p2, the polynomials of one lower order.

```
    where (unfinished)
        pp=n*(z*p1-p2)/(z*z-1.0_dp)
        z1=z
        z=z1-p1/pp                            Newton's method.
        unfinished=(abs(z-z1) > EPS)
    end where
    if (.not. any(unfinished)) exit
end do
if (its == MAXIT+1) call nrerror('too many iterations in gauleg')
x(1:m)=xm-xl*z                                Scale the root to the desired interval,
x(n:n-m+1:-1)=xm+xl*z                         and put in its symmetric counterpart.
w(1:m)=2.0_dp*xl/((1.0_dp-z**2)*pp**2)        Compute the weight
w(n:n-m+1:-1)=w(1:m)                          and its symmetric counterpart.
END SUBROUTINE gauleg
```

f90 Often we have an iterative procedure that has to be applied until all components of a vector have satisfied a convergence criterion. Some components of the vector might converge sooner than others, and it is inefficient on a small-scale parallel (SSP) machine to continue iterating on those components. The general structure we use for such an iteration is exemplified by the following lines from gauleg:

```
LOGICAL(LGT), DIMENSION((size(x)+1)/2) :: unfinished
    ...
unfinished=.true.
do its=1,MAXIT
```

```
      where (unfinished)
          ...
         unfinished=(abs(z-z1) > EPS)
      end where
      if (.not. any(unfinished)) exit
   end do
   if (its == MAXIT+1) call nrerror('too many iterations in gauleg')
```

We use the logical mask `unfinished` to control which vector components are processed inside the `where`. The mask gets updated on each iteration by testing whether any further vector components have converged. When all have converged, we exit the iteration loop. Finally, we check the value of `its` to see whether the maximum allowed number of iterations was exceeded before all components converged.

The logical expression controlling the `where` block (in this case `unfinished`) gets evaluated completely on entry into the `where`, and it is then perfectly fine to modify it inside the block. The modification affects only the *next* execution of the `where`.

On a strictly *serial* machine, there is of course some penalty associated with the above scheme: after a vector component converges, its corresponding component in `unfinished` is redundantly tested on each further iteration, until the slowest-converging component is done. If the number of iterations required does not vary too greatly from component to component, this is a minor, often negligible, penalty. However, one should be on the alert against algorithms whose worst-case convergence could differ from typical convergence by orders of magnitude. For these, one would need to implement a more complicated packing-unpacking scheme. (See discussion in Chapter B6, especially introduction, p. 1083, and notes for `factrl`, p. 1087.)

```
SUBROUTINE gaulag(x,w,alf)
USE nrtype; USE nrutil, ONLY : arth,assert_eq,nrerror
USE nr, ONLY : gammln
IMPLICIT NONE
REAL(SP), INTENT(IN) :: alf
REAL(SP), DIMENSION(:), INTENT(OUT) :: x,w
REAL(DP), PARAMETER :: EPS=3.0e-13_dp
```

Given `alf`, the parameter α of the Laguerre polynomials, this routine returns arrays `x` and `w` of length N containing the abscissas and weights of the N-point Gauss-Laguerre quadrature formula. The abscissas are returned in ascending order. The parameter EPS is the relative precision. Note that internal computations are done in double precision.

```
INTEGER(I4B) :: its,j,n
INTEGER(I4B), PARAMETER :: MAXIT=10
REAL(SP) :: anu
REAL(SP), PARAMETER :: C1=9.084064e-01_sp,C2=5.214976e-02_sp,&
    C3=2.579930e-03_sp,C4=3.986126e-03_sp
REAL(SP), DIMENSION(size(x)) :: rhs,r2,r3,theta
REAL(DP), DIMENSION(size(x)) :: p1,p2,p3,pp,z,z1
LOGICAL(LGT), DIMENSION(size(x)) :: unfinished
n=assert_eq(size(x),size(w),'gaulag')
anu=4.0_sp*n+2.0_sp*alf+2.0_sp          Initial approximations to the roots go into z.
rhs=arth(4*n-1,-4,n)*PI/anu
r3=rhs**(1.0_sp/3.0_sp)
r2=r3**2
theta=r3*(C1+r2*(C2+r2*(C3+r2*C4)))
z=anu*cos(theta)**2
unfinished=.true.
do its=1,MAXIT                            Newton's method carried out simultaneously on
    where (unfinished)                        the roots.
```

```
        p1=1.0
        p2=0.0
    end where
    do j=1,n                                Loop up the recurrence relation to get the La-
        where (unfinished)                      guerre polynomials evaluated at z.
            p3=p2
            p2=p1
            p1=((2.0_dp*j-1.0_dp+alf-z)*p2-(j-1.0_dp+alf)*p3)/j
        end where
    end do
      p1 now contains the desired Laguerre polynomials. We next compute pp, the derivatives,
      by a standard relation involving also p2, the polynomials of one lower order.
    where (unfinished)
        pp=(n*p1-(n+alf)*p2)/z
        z1=z
        z=z1-p1/pp                          Newton's formula.
        unfinished=(abs(z-z1) > EPS*z)
    end where
    if (.not. any(unfinished)) exit
end do
if (its == MAXIT+1) call nrerror('too many iterations in gaulag')
x=z                                         Store the root and the weight.
w=-exp(gammln(alf+n)-gammln(real(n,sp)))/(pp*n*p2)
END SUBROUTINE gaulag
```

The key difficulty in parallelizing this routine starting from the Fortran 77 version is that the initial guesses for the roots of the Laguerre polynomials were given in terms of previously determined roots. This prevents one from finding all the roots simultaneously. The solution is to come up with a new approximation to the roots that is a simple explicit formula, like the formula we used for the Legendre roots in gauleg.

We start with the approximation to $L_n^\alpha(x)$ given in equation (10.15.8) of [1]. We keep only the first term and ask when it is zero. This gives the following prescription for the kth root x_k of $L_n^\alpha(x)$: Solve for θ the equation

$$2\theta - \sin 2\theta = \frac{4n - 4k + 3}{4n + 2\alpha + 2}\pi \tag{B4.1}$$

Since $1 \le k \le n$ and $\alpha > -1$, we can always find a value such that $0 < \theta < \pi/2$. Then the approximation to the root is

$$x_k = (4n + 2\alpha + 2)\cos^2\theta \tag{B4.2}$$

This typically gives 3-digit accuracy, more than enough for the Newton iteration to be able to refine the root. Unfortunately equation (B4.1) is not an explicit formula for θ. (You may recognize it as being of the same form as Kepler's equation in mechanics.) If we call the right-hand side of (B4.1) y, then we can get an explicit formula by working out the power series for $y^{1/3}$ near $\theta = 0$ (using a computer algebra program). Next invert the series to give θ as a function of $y^{1/3}$. Finally, economize the series (see §5.11). The result is the concise approximation

$$\theta = 0.9084064y^{1/3} + 5.214976 \times 10^{-2}y + 2.579930 \times 10^{-3}y^{5/3} + 3.986126 \times 10^{-3}y^{7/3} \tag{B4.3}$$

```
SUBROUTINE gauher(x,w)
USE nrtype; USE nrutil, ONLY : arth,assert_eq,nrerror
IMPLICIT NONE
REAL(SP), DIMENSION(:), INTENT(OUT) :: x,w
REAL(DP), PARAMETER :: EPS=3.0e-13_dp,PIM4=0.7511255444649425_dp
```

This routine returns arrays x and w of length N containing the abscissas and weights of the N-point Gauss-Hermite quadrature formula. The abscissas are returned in descending order. Note that internal computations are done in double precision.
Parameters: EPS is the relative precision, PIM4 $= 1/\pi^{1/4}$.

```
INTEGER(I4B) :: its,j,m,n
INTEGER(I4B), PARAMETER :: MAXIT=10
REAL(SP) :: anu
REAL(SP), PARAMETER :: C1=9.084064e-01_sp,C2=5.214976e-02_sp,&
    C3=2.579930e-03_sp,C4=3.986126e-03_sp
REAL(SP), DIMENSION((size(x)+1)/2) :: rhs,r2,r3,theta
REAL(DP), DIMENSION((size(x)+1)/2) :: p1,p2,p3,pp,z,z1
LOGICAL(LGT), DIMENSION((size(x)+1)/2) :: unfinished
n=assert_eq(size(x),size(w),'gauher')
m=(n+1)/2                        The roots are symmetric about the origin, so we have to
anu=2.0_sp*n+1.0_sp                  find only half of them.
rhs=arth(3,4,m)*PI/anu
r3=rhs**(1.0_sp/3.0_sp)
r2=r3**2
theta=r3*(C1+r2*(C2+r2*(C3+r2*C4)))
z=sqrt(anu)*cos(theta)           Initial approximations to the roots.
unfinished=.true.
do its=1,MAXIT                   Newton's method carried out simultaneously on the roots.
    where (unfinished)
        p1=PIM4
        p2=0.0
    end where
    do j=1,n                     Loop up the recurrence relation to get the Hermite poly-
        where (unfinished)           nomials evaluated at z.
            p3=p2
            p2=p1
            p1=z*sqrt(2.0_dp/j)*p2-sqrt(real(j-1,dp)/real(j,dp))*p3
        end where
    end do
```

p1 now contains the desired Hermite polynomials. We next compute pp, the derivatives, by the relation (4.5.21) using p2, the polynomials of one lower order.

```
    where (unfinished)
        pp=sqrt(2.0_dp*n)*p2
        z1=z
        z=z1-p1/pp               Newton's formula.
        unfinished=(abs(z-z1) > EPS)
    end where
    if (.not. any(unfinished)) exit
end do
if (its == MAXIT+1) call nrerror('too many iterations in gauher')
x(1:m)=z                         Store the root
x(n:n-m+1:-1)=-z                 and its symmetric counterpart.
w(1:m)=2.0_dp/pp**2              Compute the weight
w(n:n-m+1:-1)=w(1:m)             and its symmetric counterpart.
END SUBROUTINE gauher
```

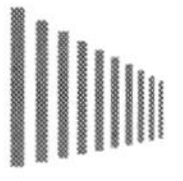

Once again we need an explicit approximation for the polynomial roots, this time for $H_n(x)$. We can use the same approximation scheme as for $L_n^\alpha(x)$, since

$$H_{2m}(x) \propto L_m^{-1/2}(x^2), \qquad H_{2m+1}(x) \propto x L_m^{1/2}(x^2) \tag{B4.4}$$

Equations (B4.1) and (B4.2) become

$$2\theta - \sin 2\theta = \frac{4k-1}{2n+1}\pi$$
$$x_k = \sqrt{2n+1}\cos\theta \qquad \text{(B4.5)}$$

Here $k = 1, 2, \ldots, m$ where $m = [(n+1)/2]$, and $k = 1$ is the largest root. The negative roots follow from symmetry. The root at $x = 0$ for odd n is included in this approximation.

```
SUBROUTINE gaujac(x,w,alf,bet)
USE nrtype; USE nrutil, ONLY : arth,assert_eq,nrerror
USE nr, ONLY : gammln
IMPLICIT NONE
REAL(SP), INTENT(IN) :: alf,bet
REAL(SP), DIMENSION(:), INTENT(OUT) :: x,w
REAL(DP), PARAMETER :: EPS=3.0e-14_dp
    Given alf and bet, the parameters α and β of the Jacobi polynomials, this routine returns
    arrays x and w of length N containing the abscissas and weights of the N-point Gauss-
    Jacobi quadrature formula. The abscissas are returned in descending order. The parameter
    EPS is the relative precision. Note that internal computations are done in double precision.
INTEGER(I4B) :: its,j,n
INTEGER(I4B), PARAMETER :: MAXIT=10
REAL(DP) :: alfbet,a,c,temp
REAL(DP), DIMENSION(size(x)) :: b,p1,p2,p3,pp,z,z1
LOGICAL(LGT), DIMENSION(size(x)) :: unfinished
n=assert_eq(size(x),size(w),'gaujac')
alfbet=alf+bet                      Initial approximations to the roots go into z.
z=cos(PI*(arth(1,1,n)-0.25_dp+0.5_dp*alf)/(n+0.5_dp*(alfbet+1.0_dp)))
unfinished=.true.
do its=1,MAXIT                      Newton's method carried out simultaneously on the roots.
    temp=2.0_dp+alfbet
    where (unfinished)              Start the recurrence with P0 and P1 to avoid a division
        p1=(alf-bet+temp*z)/2.0_dp      by zero when α + β = 0 or −1.
        p2=1.0
    end where
    do j=2,n                        Loop up the recurrence relation to get the Jacobi poly-
        a=2*j*(j+alfbet)*temp           nomials evaluated at z.
        temp=temp+2.0_dp
        c=2.0_dp*(j-1.0_dp+alf)*(j-1.0_dp+bet)*temp
        where (unfinished)
            p3=p2
            p2=p1
            b=(temp-1.0_dp)*(alf*alf-bet*bet+temp*&
                (temp-2.0_dp)*z)
            p1=(b*p2-c*p3)/a
        end where
    end do
      p1 now contains the desired Jacobi polynomials. We next compute pp, the derivatives,
      by a standard relation involving also p2, the polynomials of one lower order.
    where (unfinished)
        pp=(n*(alf-bet-temp*z)*p1+2.0_dp*(n+alf)*&
            (n+bet)*p2)/(temp*(1.0_dp-z*z))
        z1=z
        z=z1-p1/pp                  Newton's formula.
        unfinished=(abs(z-z1) > EPS)
    end where
    if (.not. any(unfinished)) exit
end do
if (its == MAXIT+1) call nrerror('too many iterations in gaujac')
x=z                                 Store the root and the weight.
```

```
w=exp(gammln(alf+n)+gammln(bet+n)-gammln(n+1.0_sp)-&
    gammln(n+alf+bet+1.0_sp))*temp*2.0_sp**alfbet/(pp*p2)
END SUBROUTINE gaujac
```

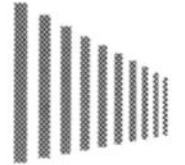

Now we need an explicit approximation for the roots of the Jacobi polynomials $P_n^{(\alpha,\beta)}(x)$. We start with the asymptotic expansion (10.14.10) of [1]. Setting this to zero gives the formula

$$x = \cos\left[\frac{k - 1/4 + \alpha/2}{n + (\alpha + \beta + 1)/2}\pi\right] \tag{B4.6}$$

This is better than the formula (22.16.1) in [2], especially at small and moderate n.

⋆ ⋆ ⋆

```
SUBROUTINE gaucof(a,b,amu0,x,w)
USE nrtype; USE nrutil, ONLY : assert_eq,unit_matrix
USE nr, ONLY : eigsrt,tqli
IMPLICIT NONE
REAL(SP), DIMENSION(:), INTENT(INOUT) :: a,b
REAL(SP), INTENT(IN) :: amu0
REAL(SP), DIMENSION(:), INTENT(OUT) :: x,w
```

Computes the abscissas and weights for a Gaussian quadrature formula from the Jacobi matrix. On input, a and b of length N are the coefficients of the recurrence relation for the set of monic orthogonal polynomials. The quantity $\mu_0 \equiv \int_a^b W(x)\,dx$ is input as amu0. The abscissas are returned in descending order in array x of length N, with the corresponding weights in w, also of length N. The arrays a and b are modified. Execution can be speeded up by modifying tqli and eigsrt to compute only the first component of each eigenvector.

```
REAL(SP), DIMENSION(size(a),size(a)) :: z
INTEGER(I4B) :: n
n=assert_eq(size(a),size(b),size(x),size(w),'gaucof')
b(2:n)=sqrt(b(2:n))             Set up superdiagonal of Jacobi matrix.
call unit_matrix(z)             Set up identity matrix for tqli to compute eigenvectors.
call tqli(a,b,z)
call eigsrt(a,z)                Sort eigenvalues into descending order.
x=a
w=amu0*z(1,:)**2                Equation (4.5.12).
END SUBROUTINE gaucof
```

⋆ ⋆ ⋆

```
SUBROUTINE orthog(anu,alpha,beta,a,b)
USE nrtype; USE nrutil, ONLY : assert_eq
IMPLICIT NONE
REAL(SP), DIMENSION(:), INTENT(IN) :: anu,alpha,beta
REAL(SP), DIMENSION(:), INTENT(OUT) :: a,b
```

Computes the coefficients a_j and b_j, $j = 0, \ldots N-1$, of the recurrence relation for monic orthogonal polynomials with weight function $W(x)$ by Wheeler's algorithm. On input, alpha and beta contain the $2N-1$ coefficients α_j and β_j, $j = 0, \ldots 2N-2$, of the recurrence relation for the chosen basis of orthogonal polynomials. The $2N$ modified moments ν_j are input in anu for $j = 0, \ldots 2N-1$. The first N coefficients are returned in a and b.

```
INTEGER(I4B) :: k,n,ndum
REAL(SP), DIMENSION(2*size(a)+1,2*size(a)+1) :: sig
n=assert_eq(size(a),size(b),'orthog: n')
ndum=assert_eq(2*n,size(alpha)+1,size(anu),size(beta)+1,'orthog: ndum')
sig(1,3:2*n)=0.0                  Initialization, Equation (4.5.33).
sig(2,2:2*n+1)=anu(1:2*n)
a(1)=alpha(1)+anu(2)/anu(1)
b(1)=0.0
do k=3,n+1                        Equation (4.5.34).
    sig(k,k:2*n-k+3)=sig(k-1,k+1:2*n-k+4)+(alpha(k-1:2*n-k+2) &
        -a(k-2))*sig(k-1,k:2*n-k+3)-b(k-2)*sig(k-2,k:2*n-k+3) &
        +beta(k-1:2*n-k+2)*sig(k-1,k-1:2*n-k+2)
    a(k-1)=alpha(k-1)+sig(k,k+1)/sig(k,k)-sig(k-1,k)/sig(k-1,k-1)
    b(k-1)=sig(k,k)/sig(k-1,k-1)
end do
END SUBROUTINE orthog
```

⋆ ⋆ ⋆

f90 As discussed in Volume 1, multidimensional quadrature can be performed by calling a one-dimensional quadrature routine along each dimension. If the same routine is used for all such calls, then the calls are recursive. The file quad3d.f90 contains two modules, quad3d_qgaus_mod and quad3d_qromb_mod. In the first, the basic one-dimensional quadrature routine is a 10-point Gaussian quadrature routine called qgaus and three-dimensional quadrature is performed by calling quad3d_qgaus. In the second, the basic one-dimensional routine is qromb of §4.3 and the three-dimensional routine is quad3d_qromb. The Gaussian quadrature is simpler but its accuracy is not controllable. The Romberg integration lets you specify an accuracy, but is apt to be very slow if you try for too much accuracy. The only difference between the stand-alone version of trapzd and the version included here is that we have to add the keyword RECURSIVE. The only changes from the stand-alone version of qromb are: We have to add RECURSIVE; we remove trapzd from the list of routines in USE nr; we increase EPS to 3×10^{-6}. Even this value could be too ambitious for difficult functions. You may want to set JMAX to a smaller value than 20 to avoid burning up a lot of computer time. Some people advocate using a smaller EPS on the inner quadrature (over z in our routine) than on the outer quadratures (over x or y). That strategy would require separate copies of qromb.

```
MODULE quad3d_qgaus_mod
USE nrtype
PRIVATE                           Hide all names from the outside,
PUBLIC quad3d_qgaus               except quad3d itself.
REAL(SP) :: xsav,ysav
INTERFACE                         User-supplied functions.
    FUNCTION func(x,y,z)          The three-dimensional function to be integrated.
    USE nrtype
    REAL(SP), INTENT(IN) :: x,y
    REAL(SP), DIMENSION(:), INTENT(IN) :: z
    REAL(SP), DIMENSION(size(z)) :: func
    END FUNCTION func

    FUNCTION y1(x)
    USE nrtype
```

```
        REAL(SP), INTENT(IN) :: x
        REAL(SP) :: y1
        END FUNCTION y1

        FUNCTION y2(x)
        USE nrtype
        REAL(SP), INTENT(IN) :: x
        REAL(SP) :: y2
        END FUNCTION y2

        FUNCTION z1(x,y)
        USE nrtype
        REAL(SP), INTENT(IN) :: x,y
        REAL(SP) :: z1
        END FUNCTION z1

        FUNCTION z2(x,y)
        USE nrtype
        REAL(SP), INTENT(IN) :: x,y
        REAL(SP) :: z2
        END FUNCTION z2
    END INTERFACE
```

The routine quad3d_qgaus returns as ss the integral of a user-supplied function func over a three-dimensional region specified by the limits x1, x2, and by the user-supplied functions y1, y2, z1, and z2, as defined in (4.6.2). Integration is performed by calling qgaus recursively.

```
CONTAINS

FUNCTION h(x)                          This is H of eq. (4.6.5).
REAL(SP), DIMENSION(:), INTENT(IN) :: x
REAL(SP), DIMENSION(size(x)) :: h
INTEGER(I4B) :: i
do i=1,size(x)
    xsav=x(i)
    h(i)=qgaus(g,y1(xsav),y2(xsav))
end do
END FUNCTION h

FUNCTION g(y)                          This is G of eq. (4.6.4).
REAL(SP), DIMENSION(:), INTENT(IN) :: y
REAL(SP), DIMENSION(size(y)) :: g
INTEGER(I4B) :: j
do j=1,size(y)
    ysav=y(j)
    g(j)=qgaus(f,z1(xsav,ysav),z2(xsav,ysav))
end do
END FUNCTION g

FUNCTION f(z)                          The integrand f(x,y,z) evaluated at fixed x and y.
REAL(SP), DIMENSION(:), INTENT(IN) :: z
REAL(SP), DIMENSION(size(z)) :: f
f=func(xsav,ysav,z)
END FUNCTION f

RECURSIVE FUNCTION qgaus(func,a,b)
REAL(SP), INTENT(IN) :: a,b
REAL(SP) :: qgaus
INTERFACE
    FUNCTION func(x)
    USE nrtype
    REAL(SP), DIMENSION(:), INTENT(IN) :: x
    REAL(SP), DIMENSION(size(x)) :: func
    END FUNCTION func
END INTERFACE
REAL(SP) :: xm,xr
REAL(SP), DIMENSION(5) :: dx, w = (/ 0.2955242247_sp,0.2692667193_sp,&
    0.2190863625_sp,0.1494513491_sp,0.0666713443_sp /),&
    x = (/ 0.1488743389_sp,0.4333953941_sp,0.6794095682_sp,&
```

```
    0.8650633666_sp,0.9739065285_sp /)
xm=0.5_sp*(b+a)
xr=0.5_sp*(b-a)
dx(:)=xr*x(:)
qgaus=xr*sum(w(:)*(func(xm+dx)+func(xm-dx)))
END FUNCTION qgaus

SUBROUTINE quad3d_qgaus(x1,x2,ss)
REAL(SP), INTENT(IN) :: x1,x2
REAL(SP), INTENT(OUT) :: ss
ss=qgaus(h,x1,x2)
END SUBROUTINE quad3d_qgaus
END MODULE quad3d_qgaus_mod
```

f90 `PRIVATE...PUBLIC quad3d_qgaus` By default, all module entities are accessible by a routine that uses the module (unless we restrict the `USE` statement with `ONLY`). In this module, the user needs access only to the routine `quad3d_qgaus`; the variables `xsav`, `ysav` and the procedures `f`, `g`, `h`, and `qgaus` are purely internal. It is good programming practice to prevent duplicate name conflicts or data overwriting by limiting access to only the desired entities. Here the `PRIVATE` statement with no variable names resets the default from `PUBLIC`. Then we include in the `PUBLIC` statement only the function name we want to be accessible.

`REAL(SP) :: xsav,ysav` In Fortran 90, we generally avoid declaring global variables in `COMMON` blocks. Instead, we give them complete specifications in a module. A deficiency of Fortran 90 is that it does not allow pointers to functions. So here we have to use the fixed-name function `func` for the function to be integrated over. If we could have a pointer to a function as a global variable, then we would just set the pointer to point to the user function (of any name) in the calling program. Similarly the functions `y1`, `y2`, `z1`, and `z2` could also have any name.

`CONTAINS` Here follow the internal subprograms `f`, `g`, `h`, `qgaus`, and `quad3d_qgaus`. Note that such internal subprograms are all "visible" to each other, i.e., their interfaces are mutually explicit, and do not require `INTERFACE` statements.

`RECURSIVE SUBROUTINE qgaus(func,a,b,ss)` The `RECURSIVE` keyword is required for the compiler to process correctly any procedure that is invoked again in its body before the return from the first call has been completed. While some compilers may let you get away without explicitly informing them that a routine is recursive, don't count on it!

```
MODULE quad3d_qromb_mod
    Alternative to quad3d_qgaus_mod that uses qromb to perform each one-dimensional in-
    tegration.
USE nrtype
PRIVATE
PUBLIC quad3d_qromb
REAL(SP) :: xsav,ysav
INTERFACE
    FUNCTION func(x,y,z)
    USE nrtype
    REAL(SP), INTENT(IN) :: x,y
    REAL(SP), DIMENSION(:), INTENT(IN) :: z
    REAL(SP), DIMENSION(size(z)) :: func
    END FUNCTION func

    FUNCTION y1(x)
```

```
    USE nrtype
    REAL(SP), INTENT(IN) :: x
    REAL(SP) :: y1
    END FUNCTION y1

    FUNCTION y2(x)
    USE nrtype
    REAL(SP), INTENT(IN) :: x
    REAL(SP) :: y2
    END FUNCTION y2

    FUNCTION z1(x,y)
    USE nrtype
    REAL(SP), INTENT(IN) :: x,y
    REAL(SP) :: z1
    END FUNCTION z1

    FUNCTION z2(x,y)
    USE nrtype
    REAL(SP), INTENT(IN) :: x,y
    REAL(SP) :: z2
    END FUNCTION z2
END INTERFACE
CONTAINS

FUNCTION h(x)
REAL(SP), DIMENSION(:), INTENT(IN) :: x
REAL(SP), DIMENSION(size(x)) :: h
INTEGER(I4B) :: i
do i=1,size(x)
    xsav=x(i)
    h(i)=qromb(g,y1(xsav),y2(xsav))
end do
END FUNCTION h

FUNCTION g(y)
REAL(SP), DIMENSION(:), INTENT(IN) :: y
REAL(SP), DIMENSION(size(y)) :: g
INTEGER(I4B) :: j
do j=1,size(y)
    ysav=y(j)
    g(j)=qromb(f,z1(xsav,ysav),z2(xsav,ysav))
end do
END FUNCTION g

FUNCTION f(z)
REAL(SP), DIMENSION(:), INTENT(IN) :: z
REAL(SP), DIMENSION(size(z)) :: f
f=func(xsav,ysav,z)
END FUNCTION f

RECURSIVE FUNCTION qromb(func,a,b)
USE nrtype; USE nrutil, ONLY : nrerror
USE nr, ONLY : polint
IMPLICIT NONE
REAL(SP), INTENT(IN) :: a,b
REAL(SP) :: qromb
INTERFACE
    FUNCTION func(x)
    USE nrtype
    REAL(SP), DIMENSION(:), INTENT(IN) :: x
    REAL(SP), DIMENSION(size(x)) :: func
    END FUNCTION func
END INTERFACE
INTEGER(I4B), PARAMETER :: JMAX=20,JMAXP=JMAX+1,K=5,KM=K-1
REAL(SP), PARAMETER :: EPS=3.0e-6_sp
REAL(SP), DIMENSION(JMAXP) :: h,s
REAL(SP) :: dqromb
```

```
INTEGER(I4B) :: j
h(1)=1.0
do j=1,JMAX
    call trapzd(func,a,b,s(j),j)
    if (j >= K) then
        call polint(h(j-KM:j),s(j-KM:j),0.0_sp,qromb,dqromb)
        if (abs(dqromb) <= EPS*abs(qromb)) RETURN
    end if
    s(j+1)=s(j)
    h(j+1)=0.25_sp*h(j)
end do
call nrerror('qromb: too many steps')
END FUNCTION qromb

RECURSIVE SUBROUTINE trapzd(func,a,b,s,n)
USE nrtype; USE nrutil, ONLY : arth
IMPLICIT NONE
REAL(SP), INTENT(IN) :: a,b
REAL(SP), INTENT(INOUT) :: s
INTEGER(I4B), INTENT(IN) :: n
INTERFACE
    FUNCTION func(x)
    USE nrtype
    REAL(SP), DIMENSION(:), INTENT(IN) :: x
    REAL(SP), DIMENSION(size(x)) :: func
    END FUNCTION func
END INTERFACE
REAL(SP) :: del,fsum
INTEGER(I4B) :: it
if (n == 1) then
    s=0.5_sp*(b-a)*sum(func( (/ a,b /) ))
else
    it=2**(n-2)
    del=(b-a)/it
    fsum=sum(func(arth(a+0.5_sp*del,del,it)))
    s=0.5_sp*(s+del*fsum)
end if
END SUBROUTINE trapzd

SUBROUTINE quad3d_qromb(x1,x2,ss)
REAL(SP), INTENT(IN) :: x1,x2
REAL(SP), INTENT(OUT) :: ss
ss=qromb(h,x1,x2)
END SUBROUTINE quad3d_qromb
END MODULE quad3d_qromb_mod
```

`MODULE quad3d_qromb_mod` The only difference between this module and the previous one is that all calls to `qgaus` are replaced by calls to `qromb` and that the routine `qgaus` is replaced by `qromb` and `trapzd`.

CITED REFERENCES AND FURTHER READING:

Erdélyi, A., Magnus, W., Oberhettinger, F., and Tricomi, F.G. 1953, *Higher Transcendental Functions*, Volume II (New York: McGraw-Hill). [1]

Abramowitz, M., and Stegun, I.A. 1964, *Handbook of Mathematical Functions*, Applied Mathematics Series, Volume 55 (Washington: National Bureau of Standards; reprinted 1968 by Dover Publications, New York). [2]

Chapter B5. Evaluation of Functions

```
SUBROUTINE eulsum(sum,term,jterm)
USE nrtype; USE nrutil, ONLY : poly_term,reallocate
IMPLICIT NONE
REAL(SP), INTENT(INOUT) :: sum
REAL(SP), INTENT(IN) :: term
INTEGER(I4B), INTENT(IN) :: jterm
```

Incorporates into `sum` the `jterm`'th term, with value `term`, of an alternating series. `sum` is input as the previous partial sum, and is output as the new partial sum. The first call to this routine, with the first `term` in the series, should be with `jterm=1`. On the second call, `term` should be set to the second term of the series, with sign opposite to that of the first call, and `jterm` should be 2. And so on.

```
REAL(SP), DIMENSION(:), POINTER, SAVE :: wksp
INTEGER(I4B), SAVE :: nterm                      Number of saved differences in wksp.
LOGICAL(LGT), SAVE :: init=.true.
if (init) then                                   Initialize.
    init=.false.
    nullify(wksp)
end if
if (jterm == 1) then
    nterm=1
    wksp=>reallocate(wksp,100)
    wksp(1)=term
    sum=0.5_sp*term                              Return first estimate.
else
    if (nterm+1 > size(wksp)) wksp=>reallocate(wksp,2*size(wksp))
    wksp(2:nterm+1)=0.5_sp*wksp(1:nterm)         Update saved quantities by van Wijn-
    wksp(1)=term                                     gaarden's algorithm.
    wksp(1:nterm+1)=poly_term(wksp(1:nterm+1),0.5_sp)
    if (abs(wksp(nterm+1)) <= abs(wksp(nterm))) then     Favorable to increase p,
        sum=sum+0.5_sp*wksp(nterm+1)
        nterm=nterm+1                            and the table becomes longer.
    else                                         Favorable to increase n,
        sum=sum+wksp(nterm+1)                    the table doesn't become longer.
    end if
end if
END SUBROUTINE eulsum
```

f90 This routine uses the function `reallocate` in `nrutil` to define a temporary workspace and then, if necessary, enlarge the workspace without destroying the earlier contents. The pointer `wksp` is declared with the `SAVE` attribute. Since Fortran 90 pointers are born "in limbo," we cannot immediately test whether they are associated or not. Hence the code `if (init)...nullify(wksp)`. Then the line `wksp=>reallocate(wksp,100)` allocates an array of length 100 and points `wksp` to it. On subsequent calls to `eulsum`, if `nterm` ever gets bigger than the size of `wksp`, the call to `reallocate` doubles the size of `wksp` and copies the old contents into the new storage.

You could achieve the same effect as the code `if (init)...nullify(wksp)... wksp=>reallocate(wksp,100)` with a simple `allocate(wksp,100)`. You would then use reallocate only for increasing the storage if necessary. Don't! The advantage of the above scheme becomes clear if you consider what happens if eulsum is invoked *twice* by the calling program to evaluate two different sums. On the second invocation, when jterm $= 1$ again, you would be allocating an already allocated pointer. This does not generate an error — it simply leaves the original target inaccessible. Using reallocate instead not only allocates a new array of length 100, but also detects that wksp had already been associated. It dutifully (and wastefully) copies the first 100 elements of the old wksp into the new storage, and, more importantly, deallocates the old wksp, reclaiming its storage. While only two invocations of eulsum without intervening deallocation of memory would not cause a problem, many such invocations might well. We believe that, as a general rule, the potential for catastrophe from reckless use of allocate is great enough that you should *always* deallocate whenever storage is no longer required.

The unnecessary copying of 100 elements when eulsum is invoked a second time could be avoided by making init an argument. It hardly seems worth it to us.

For Fortran 90 neophytes, note that unlike in C you have to do nothing special to get the contents of the storage a pointer is addressing. The compiler figures out from the context whether you mean the contents, such as wksp(1:nterm), or the address, such as both occurrences of wksp in wksp=>reallocate(wksp,100).

wksp(1:nterm+1)=poly_term(wksp(1:nterm+1),0.5_sp) The poly_term function in nrutil tabulates the partial sums of a polynomial, or, equivalently, performs the synthetic division of a polynomial by a monomial.

Small-scale parallelism in eulsum is achieved straightforwardly by the use of vector constructions and poly_term (which parallelizes recursively). The routine is not written to take advantage of data parallelism in the (infrequent) case of wanting to sum many different series simultaneously; nor, since wksp is a SAVEd variable, can it be used in many simultaneous instances on a MIMD machine. (You can easily recode these generalizations if you need them.)

* * *

```
SUBROUTINE ddpoly(c,x,pd)
USE nrtype; USE nrutil, ONLY : arth,cumprod,poly_term
IMPLICIT NONE
REAL(SP), INTENT(IN) :: x
REAL(SP), DIMENSION(:), INTENT(IN) :: c
REAL(SP), DIMENSION(:), INTENT(OUT) :: pd
    Given the coefficients of a polynomial of degree Nc − 1 as an array c(1:Nc) with c(1)
    being the constant term, and given a value x, this routine returns the polynomial evaluated
    at x as pd(1) and Nd − 1 derivatives as pd(2:Nd).
INTEGER(I4B) :: i,nc,nd
REAL(SP), DIMENSION(size(pd)) :: fac
REAL(SP), DIMENSION(size(c)) :: d
nc=size(c)
nd=size(pd)
d(nc:1:-1)=poly_term(c(nc:1:-1),x)
do i=2,min(nd,nc)
    d(nc:i:-1)=poly_term(d(nc:i:-1),x)
end do
pd=d(1:nd)
fac=cumprod(arth(1.0_sp,1.0_sp,nd))         After the first derivative, factorial constants
pd(3:nd)=fac(2:nd-1)*pd(3:nd)                  come in.
END SUBROUTINE ddpoly
```

f90 `d(nc:1:-1)=poly_term(c(nc:1:-1),x)` The `poly_term` function in `nrutil` tabulates the partial sums of a polynomial, or, equivalently, performs synthetic division. See §22.3 for a discussion of why `ddpoly` is coded this way.

`fac=cumprod(arth(1.0_sp,1.0_sp,nd))` Here the function `arth` from `nrutil` generates the sequence $1, 2, 3\ldots$. The function `cumprod` then tabulates the cumulative products, thus making a table of factorials.

Notice that `ddpoly` doesn't need an argument to pass N_d, the number of output terms desired by the user: It gets that information from the length of the array `pd` that the user provides for it to fill. It is a minor curiosity that `pd`, declared as `INTENT(OUT)`, can thus be used, on the sly, to pass some `INTENT(IN)` information. (A Fortran 90 brain teaser could be: A subroutine with only `INTENT(OUT)` arguments can be called to print any specified integer. How is this done?)

```
SUBROUTINE poldiv(u,v,q,r)
USE nrtype; USE nrutil, ONLY : assert_eq
IMPLICIT NONE
REAL(SP), DIMENSION(:), INTENT(IN) :: u,v
REAL(SP), DIMENSION(:), INTENT(OUT) :: q,r
```

Given the N coefficients of a polynomial in `u`, and the N_v coefficients of another polynomial in `v`, divide the polynomial `u` by the polynomial `v` ("u"/"v") giving a quotient polynomial whose coefficients are returned in `q`, and a remainder polynomial whose coefficients are returned in `r`. The arrays `q` and `r` are of length N, but only the first $N - N_v + 1$ elements of `q` and the first $N_v - 1$ elements of `r` are used. The remaining elements are returned as zero.

```
INTEGER(I4B) :: i,n,nv
n=assert_eq(size(u),size(q),size(r),'poldiv')
nv=size(v)
r(:)=u(:)
q(:)=0.0
do i=n-nv,0,-1
    q(i+1)=r(nv+i)/v(nv)
    r(i+1:nv+i-1)=r(i+1:nv+i-1)-q(i+1)*v(1:nv-1)
end do
r(nv:n)=0.0
END SUBROUTINE poldiv
```

⋆ ⋆ ⋆

```
FUNCTION ratval_s(x,cof,mm,kk)
USE nrtype; USE nrutil, ONLY : poly
IMPLICIT NONE
REAL(DP), INTENT(IN) :: x
```

Note precision! Change to `REAL(SP)` if desired.

```
INTEGER(I4B), INTENT(IN) :: mm,kk
REAL(DP), DIMENSION(mm+kk+1), INTENT(IN) :: cof
REAL(DP) :: ratval_s
```

Given `mm`, `kk`, and `cof(1:mm+kk+1)`, evaluate and return the rational function $(\mathtt{cof(1)} + \mathtt{cof(2)}x + \cdots + \mathtt{cof(mm+1)}x^{\mathtt{mm}})/(1 + \mathtt{cof(mm+2)}x + \cdots + \mathtt{cof(mm+kk+1)}x^{\mathtt{kk}})$.

```
ratval_s=poly(x,cof(1:mm+1))/(1.0_dp+x*poly(x,cof(mm+2:mm+kk+1)))
END FUNCTION ratval_s
```

f90 This simple routine uses the function `poly` from `nrutil` to evaluate the numerator and denominator polynomials. Single- and double-precision versions, `ratval_s` and `ratval_v`, are overloaded onto the name `ratval` when the module `nr` is used.

```
FUNCTION ratval_v(x,cof,mm,kk)
USE nrtype; USE nrutil, ONLY : poly
IMPLICIT NONE
REAL(DP), DIMENSION(:), INTENT(IN) :: x
INTEGER(I4B), INTENT(IN) :: mm,kk
REAL(DP), DIMENSION(mm+kk+1), INTENT(IN) :: cof
REAL(DP), DIMENSION(size(x)) :: ratval_v
ratval_v=poly(x,cof(1:mm+1))/(1.0_dp+x*poly(x,cof(mm+2:mm+kk+1)))
END FUNCTION ratval_v
```

⋆ ⋆ ⋆

The routines `recur1` and `recur2` are new in this volume, and do not have Fortran 77 counterparts. First- and second-order linear recurrences are implemented as trivial do-loops on strictly serial machines. On parallel machines, however, they pose different, and quite interesting, programming challenges. Since many calculations can be decomposed into recurrences, it is useful to have general, parallelizable routines available. The algorithms behind `recur1` and `recur2` are discussed in §22.2.

```
RECURSIVE FUNCTION recur1(a,b) RESULT(u)
USE nrtype; USE nrutil, ONLY : assert_eq
IMPLICIT NONE
REAL(SP), DIMENSION(:), INTENT(IN) :: a,b
REAL(SP), DIMENSION(size(a)) :: u
INTEGER(I4B), PARAMETER :: NPAR_RECUR1=8
```

Given vectors a of size n and b of size $n-1$, returns a vector u that satisfies the first order linear recurrence $u_1 = a_1$, $u_j = a_j + b_{j-1}u_{j-1}$, for $j = 2, \ldots, n$. Parallelization is via a recursive evaluation.

```
INTEGER(I4B) :: n,j
n=assert_eq(size(a),size(b)+1,'recur1')
u(1)=a(1)
if (n < NPAR_RECUR1) then                  Do short vectors as a loop.
    do j=2,n
        u(j)=a(j)+b(j-1)*u(j-1)
    end do
else
```

Otherwise, combine coefficients and recurse on the even components, then evaluate all the odd components in parallel.

```
    u(2:n:2)=recur1(a(2:n:2)+a(1:n-1:2)*b(1:n-1:2), &
            b(3:n-1:2)*b(2:n-2:2))
    u(3:n:2)=a(3:n:2)+b(2:n-1:2)*u(2:n-1:2)
end if
END FUNCTION recur1
```

f90 `RECURSIVE FUNCTION recur1(a,b) RESULT(u)` When a recursive function invokes itself only indirectly through a sequence of function calls, then the function name can be used for the result just as in a nonrecursive function. When the function invokes itself directly, however, as in `recur1`, then another name must be used for the result. If you are hazy on the syntax for RESULT, see the discussion of recursion in §21.5.

⋆ ⋆ ⋆

```
FUNCTION recur2(a,b,c)
USE nrtype; USE nrutil, ONLY : assert_eq
IMPLICIT NONE
REAL(SP), DIMENSION(:), INTENT(IN) :: a,b,c
REAL(SP), DIMENSION(size(a)) :: recur2
```

Given vectors a of size n and b and c of size $n-2$, returns a vector u that satisfies the second order linear recurrence $u_1 = a_1$, $u_2 = a_2$, $u_j = a_j + b_{j-2}u_{j-1} + c_{j-2}u_{j-2}$, for $j = 3, \ldots, n$. Parallelization is via conversion to a first order recurrence for a two-dimensional vector.

```
INTEGER(I4B) :: n
REAL(SP), DIMENSION(size(a)-1) :: a1,a2,u1,u2
REAL(SP), DIMENSION(size(a)-2) :: b11,b12,b21,b22
n=assert_eq(size(a),size(b)+2,size(c)+2,'recur2')
a1(1)=a(1)                          Set up vector a.
a2(1)=a(2)
a1(2:n-1)=0.0
a2(2:n-1)=a(3:n)
b11(1:n-2)=0.0                      Set up matrix b.
b12(1:n-2)=1.0
b21(1:n-2)=c(1:n-2)
b22(1:n-2)=b(1:n-2)
call recur1_v(a1,a2,b11,b12,b21,b22,u1,u2)
recur2(1:n-1)=u1(1:n-1)
recur2(n)=u2(n-1)
CONTAINS

RECURSIVE SUBROUTINE recur1_v(a1,a2,b11,b12,b21,b22,u1,u2)
USE nrtype; USE nrutil, ONLY : assert_eq
IMPLICIT NONE
REAL(SP), DIMENSION(:), INTENT(IN) :: a1,a2,b11,b12,b21,b22
REAL(SP), DIMENSION(:), INTENT(OUT) :: u1,u2
INTEGER(I4B), PARAMETER :: NPAR_RECUR2=8
```

Used by recur2 to evaluate first order vector recurrence. Routine is a two-dimensional vector version of recur1, with matrix multiplication replacing scalar multiplication.

```
INTEGER(I4B) :: n,j,nn,nn1
REAL(SP), DIMENSION(size(a1)/2) :: aa1,aa2
REAL(SP), DIMENSION(size(a1)/2-1) :: bb11,bb12,bb21,bb22
n=assert_eq((/size(a1),size(a2),size(b11)+1,size(b12)+1,size(b21)+1,&
    size(b22)+1,size(u1),size(u2)/),'recur1_v')
u1(1)=a1(1)
u2(1)=a2(1)
if (n < NPAR_RECUR2) then           Do short vectors as a loop.
    do j=2,n
        u1(j)=a1(j)+b11(j-1)*u1(j-1)+b12(j-1)*u2(j-1)
        u2(j)=a2(j)+b21(j-1)*u1(j-1)+b22(j-1)*u2(j-1)
    end do
else
```

Otherwise, combine coefficients and recurse on the even components, then evaluate all the odd components in parallel.

```
    nn=n/2
    nn1=nn-1
    aa1(1:nn)=a1(2:n:2)+b11(1:n-1:2)*a1(1:n-1:2)+&
        b12(1:n-1:2)*a2(1:n-1:2)
    aa2(1:nn)=a2(2:n:2)+b21(1:n-1:2)*a1(1:n-1:2)+&
            b22(1:n-1:2)*a2(1:n-1:2)
    bb11(1:nn1)=b11(3:n-1:2)*b11(2:n-2:2)+&
            b12(3:n-1:2)*b21(2:n-2:2)
    bb12(1:nn1)=b11(3:n-1:2)*b12(2:n-2:2)+&
            b12(3:n-1:2)*b22(2:n-2:2)
    bb21(1:nn1)=b21(3:n-1:2)*b11(2:n-2:2)+&
            b22(3:n-1:2)*b21(2:n-2:2)
    bb22(1:nn1)=b21(3:n-1:2)*b12(2:n-2:2)+&
            b22(3:n-1:2)*b22(2:n-2:2)
    call recur1_v(aa1,aa2,bb11,bb12,bb21,bb22,u1(2:n:2),u2(2:n:2))
    u1(3:n:2)=a1(3:n:2)+b11(2:n-1:2)*u1(2:n-1:2)+&
```

```
            b12(2:n-1:2)*u2(2:n-1:2)
    u2(3:n:2)=a2(3:n:2)+b21(2:n-1:2)*u1(2:n-1:2)+&
            b22(2:n-1:2)*u2(2:n-1:2)
end if
END SUBROUTINE recur1_v
END FUNCTION recur2
```

⋆ ⋆ ⋆

```
FUNCTION dfridr(func,x,h,err)
USE nrtype; USE nrutil, ONLY : assert,geop,iminloc
IMPLICIT NONE
REAL(SP), INTENT(IN) :: x,h
REAL(SP), INTENT(OUT) :: err
REAL(SP) :: dfridr
INTERFACE
    FUNCTION func(x)
    USE nrtype
    IMPLICIT NONE
    REAL(SP), INTENT(IN) :: x
    REAL(SP) :: func
    END FUNCTION func
END INTERFACE
INTEGER(I4B),PARAMETER :: NTAB=10
REAL(SP), PARAMETER :: CON=1.4_sp,CON2=CON*CON,BIG=huge(x),SAFE=2.0
```

Returns the derivative of a function `func` at a point `x` by Ridders' method of polynomial extrapolation. The value `h` is input as an estimated initial stepsize; it need not be small, but rather should be an increment in `x` over which `func` changes *substantially*. An estimate of the error in the derivative is returned as `err`.

Parameters: Stepsize is decreased by `CON` at each iteration. Max size of tableau is set by `NTAB`. Return when error is `SAFE` worse than the best so far.

```
INTEGER(I4B) :: ierrmin,i,j
REAL(SP) :: hh
REAL(SP), DIMENSION(NTAB-1) :: errt,fac
REAL(SP), DIMENSION(NTAB,NTAB) :: a
call assert(h /= 0.0, 'dfridr arg')
hh=h
a(1,1)=(func(x+hh)-func(x-hh))/(2.0_sp*hh)
err=BIG
fac(1:NTAB-1)=geop(CON2,CON2,NTAB-1)
do i=2,NTAB                       Successive columns in the Neville tableau will go to smaller
    hh=hh/CON                        stepsizes and higher orders of extrapolation.
    a(1,i)=(func(x+hh)-func(x-hh))/(2.0_sp*hh)      Try new, smaller stepsize.
    do j=2,i
          Compute extrapolations of various orders, requiring no new function evaluations.
        a(j,i)=(a(j-1,i)*fac(j-1)-a(j-1,i-1))/(fac(j-1)-1.0_sp)
    end do
    errt(1:i-1)=max(abs(a(2:i,i)-a(1:i-1,i)),abs(a(2:i,i)-a(1:i-1,i-1)))
      The error strategy is to compare each new extrapolation to one order lower, both at the
      present stepsize and the previous one.
    ierrmin=iminloc(errt(1:i-1))
    if (errt(ierrmin) <= err) then        If error is decreased, save the improved an-
        err=errt(ierrmin)                     swer.
        dfridr=a(1+ierrmin,i)
    end if
    if (abs(a(i,i)-a(i-1,i-1)) >= SAFE*err) RETURN
      If higher order is worse by a significant factor SAFE, then quit early.
end do
END FUNCTION dfridr
```

`ierrmin=iminloc(errt(1:i-1))` The function `iminloc` in `nrutil` is useful when you need to know the index of the smallest element in an array.

⋆ ⋆ ⋆

```
FUNCTION chebft(a,b,n,func)
USE nrtype; USE nrutil, ONLY : arth,outerprod
IMPLICIT NONE
REAL(SP), INTENT(IN) :: a,b
INTEGER(I4B), INTENT(IN) :: n
REAL(SP), DIMENSION(n) :: chebft
INTERFACE
    FUNCTION func(x)
    USE nrtype
    IMPLICIT NONE
    REAL(SP), DIMENSION(:), INTENT(IN) :: x
    REAL(SP), DIMENSION(size(x)) :: func
    END FUNCTION func
END INTERFACE
```

Chebyshev fit: Given a function `func`, lower and upper limits of the interval [a,b], and a maximum degree `n`, this routine computes the `n` coefficients c_k such that `func`$(x) \approx [\sum_{k=1}^{\mathtt{n}} c_k T_{k-1}(y)] - c_1/2$, where y and x are related by (5.8.10). This routine is to be used with moderately large `n` (e.g., 30 or 50), the array of c's subsequently to be truncated at the smaller value m such that c_{m+1} and subsequent elements are negligible.

```
REAL(DP) :: bma,bpa
REAL(DP), DIMENSION(n) :: theta
bma=0.5_dp*(b-a)
bpa=0.5_dp*(b+a)
theta(:)=PI_D*arth(0.5_dp,1.0_dp,n)/n
chebft(:)=matmul(cos(outerprod(arth(0.0_dp,1.0_dp,n),theta)), &
    func(real(cos(theta)*bma+bpa,sp)))*2.0_dp/n
```

We evaluate the function at the `n` points required by (5.8.7). We accumulate the sum in double precision for safety.

```
END FUNCTION chebft
```

`chebft(:)=matmul(...)` Here again Fortran 90 produces a very concise parallelizable formulation that requires some effort to decode. Equation (5.8.7) is a product of the matrix of cosines, where the rows are indexed by j and the columns by k, with the vector of function values indexed by k. We use the `outerprod` function in `nrutil` to form the matrix of arguments for the cosine, and rely on the element-by-element application of `cos` to produce the matrix of cosines. `matmul` then takes care of the matrix product. A subtlety is that, while the calculation is being done in double precision to minimize roundoff, the function is assumed to be supplied in single precision. Thus `real(...,sp)` is used to convert the double precision argument to single precision.

```
FUNCTION chebev_s(a,b,c,x)
USE nrtype; USE nrutil, ONLY : nrerror
IMPLICIT NONE
REAL(SP), INTENT(IN) :: a,b,x
REAL(SP), DIMENSION(:), INTENT(IN) :: c
REAL(SP) :: chebev_s
```

Chebyshev evaluation: All arguments are input. `c` is an array of length M of Chebyshev coefficients, the first M elements of `c` output from `chebft` (which must have been called with the same a and b). The Chebyshev polynomial $\sum_{k=1}^{M} c_k T_{k-1}(y) - c_1/2$ is evaluated at a point $y = [x - (b + a)/2]/[(b - a)/2]$, and the result is returned as the function value.

```
INTEGER(I4B) :: j,m
REAL(SP) :: d,dd,sv,y,y2
if ((x-a)*(x-b) > 0.0) call nrerror('x not in range in chebev_s')
m=size(c)
d=0.0
dd=0.0
y=(2.0_sp*x-a-b)/(b-a)                Change of variable.
y2=2.0_sp*y
do j=m,2,-1                           Clenshaw's recurrence.
    sv=d
    d=y2*d-dd+c(j)
    dd=sv
end do
chebev_s=y*d-dd+0.5_sp*c(1)           Last step is different.
END FUNCTION chebev_s
```

```
FUNCTION chebev_v(a,b,c,x)
USE nrtype; USE nrutil, ONLY : nrerror
IMPLICIT NONE
REAL(SP), INTENT(IN) :: a,b
REAL(SP), DIMENSION(:), INTENT(IN) :: c,x
REAL(SP), DIMENSION(size(x)) :: chebev_v
INTEGER(I4B) :: j,m
REAL(SP), DIMENSION(size(x)) :: d,dd,sv,y,y2
if (any((x-a)*(x-b) > 0.0)) call nrerror('x not in range in chebev_v')
m=size(c)
d=0.0
dd=0.0
y=(2.0_sp*x-a-b)/(b-a)
y2=2.0_sp*y
do j=m,2,-1
    sv=d
    d=y2*d-dd+c(j)
    dd=sv
end do
chebev_v=y*d-dd+0.5_sp*c(1)
END FUNCTION chebev_v
```

f90 The name chebev is overloaded with scalar and vector versions. chebev_v is essentially identical to chebev_s except for the declarations of the variables. Fortran 90 does the appropriate scalar or vector arithmetic in the body of the routine, depending on the type of the variables.

⋆ ⋆ ⋆

```
FUNCTION chder(a,b,c)
USE nrtype; USE nrutil, ONLY : arth,cumsum
IMPLICIT NONE
REAL(SP), INTENT(IN) :: a,b
REAL(SP), DIMENSION(:), INTENT(IN) :: c
REAL(SP), DIMENSION(size(c)) :: chder
```

This routine returns an array of length N containing the Chebyshev coefficients of the derivative of the function whose coefficients are in the array c. Input are a,b,c, as output from routine chebft §5.8. The desired degree of approximation N is equal to the length of c supplied.

```
INTEGER(I4B) :: n
REAL(SP) :: con
REAL(SP), DIMENSION(size(c)) :: temp
n=size(c)
temp(1)=0.0
temp(2:n)=2.0_sp*arth(n-1,-1,n-1)*c(n:2:-1)
chder(n:1:-2)=cumsum(temp(1:n:2))              Equation (5.9.2).
chder(n-1:1:-2)=cumsum(temp(2:n:2))
con=2.0_sp/(b-a)
chder=chder*con                                Normalize to the interval b-a.
END FUNCTION chder
```

```
FUNCTION chint(a,b,c)
USE nrtype; USE nrutil, ONLY : arth
IMPLICIT NONE
REAL(SP), INTENT(IN) :: a,b
REAL(SP), DIMENSION(:), INTENT(IN) :: c
REAL(SP), DIMENSION(size(c)) :: chint
```

This routine returns an array of length N containing the Chebyshev coefficients of the integral of the function whose coefficients are in the array c. Input are a,b,c, as output from routine chebft §5.8. The desired degree of approximation N is equal to the length of c supplied. The constant of integration is set so that the integral vanishes at a.

```
INTEGER(I4B) :: n
REAL(SP) :: con
n=size(c)
con=0.25_sp*(b-a)                              Factor that normalizes to the interval b-a.
chint(2:n-1)=con*(c(1:n-2)-c(3:n))/arth(1,1,n-2)    Equation (5.9.1).
chint(n)=con*c(n-1)/(n-1)                      Special case of (5.9.1) for n.
chint(1)=2.0_sp*(sum(chint(2:n:2))-sum(chint(3:n:2)))    Set the constant of integration.
END FUNCTION chint
```

f90 If you look at equation (5.9.1) for the Chebyshev coefficients of the integral of a function, you will see c_{i-1} and c_{i+1} and be tempted to use eoshift. We think it is almost always better to use array sections instead, as in the code above, especially if your code will ever run on a serial machine.

⋆ ⋆ ⋆

```
FUNCTION chebpc(c)
USE nrtype
IMPLICIT NONE
REAL(SP), DIMENSION(:), INTENT(IN) :: c
REAL(SP), DIMENSION(size(c)) :: chebpc
```

Chebyshev polynomial coefficients. Given a coefficient array c of length N, this routine returns a coefficient array d of length N such that $\sum_{k=1}^{N} d_k y^{k-1} = \sum_{k=1}^{N} c_k T_{k-1}(y) - c_1/2$. The method is Clenshaw's recurrence (5.8.11), but now applied algebraically rather than arithmetically.

```
INTEGER(I4B) :: j,n
REAL(SP), DIMENSION(size(c)) :: dd,sv
n=size(c)
chebpc=0.0
dd=0.0
chebpc(1)=c(n)
do j=n-1,2,-1
    sv(2:n-j+1)=chebpc(2:n-j+1)
    chebpc(2:n-j+1)=2.0_sp*chebpc(1:n-j)-dd(2:n-j+1)
    dd(2:n-j+1)=sv(2:n-j+1)
```

```
        sv(1)=chebpc(1)
        chebpc(1)=-dd(1)+c(j)
        dd(1)=sv(1)
    end do
    chebpc(2:n)=chebpc(1:n-1)-dd(2:n)
    chebpc(1)=-dd(1)+0.5_sp*c(1)
    END FUNCTION chebpc
```

⋆ ⋆ ⋆

```
SUBROUTINE pcshft(a,b,d)
USE nrtype; USE nrutil, ONLY : geop
IMPLICIT NONE
REAL(SP), INTENT(IN) :: a,b
REAL(SP), DIMENSION(:), INTENT(INOUT) :: d
```

Polynomial coefficient shift. Given a coefficient array d of length N, this routine generates a coefficient array g of the same length such that $\sum_{k=1}^{N} \mathrm{d}_k y^{k-1} = \sum_{k=1}^{N} g_k x^{k-1}$, where x and y are related by (5.8.10), i.e., the interval $-1 < y < 1$ is mapped to the interval $\mathrm{a} < x < \mathrm{b}$. The array g is returned in d.

```
INTEGER(I4B) :: j,n
REAL(SP), DIMENSION(size(d)) :: dd
REAL(SP) :: x
n=size(d)
dd=d*geop(1.0_sp,2.0_sp/(b-a),n)
x=-0.5_sp*(a+b)
d(1)=dd(n)
d(2:n)=0.0
do j=n-1,1,-1
    d(2:n+1-j)=d(2:n+1-j)*x+d(1:n-j)
    d(1)=d(1)*x+dd(j)
end do
END SUBROUTINE pcshft
```

We accomplish the shift by synthetic division, that miracle of high-school algebra.

There is a subtle, but major, distinction between the synthetic division algorithm used in the Fortran 77 version of `pcshft` and that used above. In the Fortran 77 version, the synthetic division (translated to Fortran 90 notation) is

```
d(1:n)=dd(1:n)
do j=1,n-1
    do k=n-1,j,-1
        d(k)=x*d(k+1)+d(k)
    end do
end do
```

while, in Fortran 90, it is

```
d(1)=dd(n)
d(2:n)=0.0
do j=n-1,1,-1
    d(2:n+1-j)=d(2:n+1-j)*x+d(1:n-j)
    d(1)=d(1)*x+dd(j)
end do
```

As explained in §22.3, these are algebraically — but not algorithmically — equivalent. The inner loop in the Fortran 77 version does not parallelize, because each `k` value uses the result of the previous one. In fact, the `k` loop is a synthetic division, which can be parallelized *recursively* (as in the `nrutil` routine `poly_term`), but not simply vectorized. In the Fortran 90 version, since not one but n-1 successive synthetic divisions are to be performed (by the outer loop), it is possible to reorganize the calculation to allow vectorization.

⋆ ⋆ ⋆

```
FUNCTION pccheb(d)
USE nrtype; USE nrutil, ONLY : arth,cumprod,geop
IMPLICIT NONE
REAL(SP), DIMENSION(:), INTENT(IN) :: d
REAL(SP), DIMENSION(size(d)) :: pccheb
    Inverse of routine chebpc: given an array of polynomial coefficients d, returns an equivalent
    array of Chebyshev coefficients of the same length.
INTEGER(I4B) :: k,n
REAL(SP), DIMENSION(size(d)) :: denom,numer,pow
n=size(d)
pccheb(1)=2.0_sp*d(1)
pow=geop(1.0_sp,2.0_sp,n)                  Powers of 2.
numer(1)=1.0                               Combinatorial coefficients computed as numer/denom.
denom(1)=1.0
denom(2:(n+3)/2)=cumprod(arth(1.0_sp,1.0_sp,(n+1)/2))
pccheb(2:n)=0.0
do k=2,n                                   Loop over orders of x in the polynomial.
    numer(2:(k+3)/2)=cumprod(arth(k-1.0_sp,-1.0_sp,(k+1)/2))
    pccheb(k:1:-2)=pccheb(k:1:-2)+&
        d(k)/pow(k-1)*numer(1:(k+1)/2)/denom(1:(k+1)/2)
end do
END FUNCTION pccheb
```

⋆ ⋆ ⋆

```
SUBROUTINE pade(cof,resid)
USE nrtype
USE nr, ONLY : lubksb,ludcmp,mprove
IMPLICIT NONE
REAL(DP), DIMENSION(:), INTENT(INOUT) :: cof     DP for consistency with ratval.
REAL(SP), INTENT(OUT) :: resid
    Given cof(1:2N+1), the leading terms in the power series expansion of a function, solve
    the linear Padé equations to return the coefficients of a diagonal rational function approxi-
    mation to the same function, namely (cof(1) + cof(2)x + ··· + cof(N+1)x^N)/(1 +
    cof(N+2)x + ··· + cof(2N+1)x^N). The value resid is the norm of the residual
    vector; a small value indicates a well-converged solution.
INTEGER(I4B) :: k,n
INTEGER(I4B), DIMENSION((size(cof)-1)/2) :: indx
REAL(SP), PARAMETER :: BIG=1.0e30_sp       A big number.
REAL(SP) :: d,rr,rrold
REAL(SP), DIMENSION((size(cof)-1)/2) :: x,y,z
REAL(SP), DIMENSION((size(cof)-1)/2,(size(cof)-1)/2) :: q,qlu
n=(size(cof)-1)/2
x=cof(n+2:2*n+1)                           Set up matrix for solving.
y=x
do k=1,n
    q(:,k)=cof(n+2-k:2*n+1-k)
end do
qlu=q
call ludcmp(qlu,indx,d)                    Solve by LU decomposition and backsubsti-
call lubksb(qlu,indx,x)                        tution.
rr=BIG
do                                         Important to use iterative improvement, since
    rrold=rr                                   the Padé equations tend to be ill-conditioned.
```

```
    z=x
    call mprove(q,qlu,indx,y,x)
    rr=sum((z-x)**2)                    Calculate residual.
    if (rr >= rrold) exit               If it is no longer improving, call it quits.
end do
resid=sqrt(rrold)
do k=1,n                                Calculate the remaining coefficients.
    y(k)=cof(k+1)-dot_product(z(1:k),cof(k:1:-1))
end do
cof(2:n+1)=y                            Copy answers to output.
cof(n+2:2*n+1)=-z
END SUBROUTINE pade
```

⋆ ⋆ ⋆

```
SUBROUTINE ratlsq(func,a,b,mm,kk,cof,dev)
USE nrtype; USE nrutil, ONLY : arth,geop
USE nr, ONLY : ratval,svbksb,svdcmp
IMPLICIT NONE
REAL(DP), INTENT(IN) :: a,b
INTEGER(I4B), INTENT(IN) :: mm,kk
REAL(DP), DIMENSION(:), INTENT(OUT) :: cof
REAL(DP), INTENT(OUT) :: dev
INTERFACE
    FUNCTION func(x)
    USE nrtype
    REAL(DP), DIMENSION(:), INTENT(IN) :: x
    REAL(DP), DIMENSION(size(x)) :: func
    END FUNCTION func
END INTERFACE
INTEGER(I4B), PARAMETER :: NPFAC=8,MAXIT=5
REAL(DP), PARAMETER :: BIG=1.0e30_dp
```

Returns in cof(1:mm+kk+1) the coefficients of a rational function approximation to the function func in the interval (a, b). Input quantities mm and kk specify the order of the numerator and denominator, respectively. The maximum absolute deviation of the approximation (insofar as is known) is returned as dev. Note that double-precision versions of svdcmp and svbksb are called.

```
INTEGER(I4B) :: it,ncof,npt,npth
REAL(DP) :: devmax,e,theta
REAL(DP), DIMENSION((mm+kk+1)*NPFAC) :: bb,ee,fs,wt,xs
REAL(DP), DIMENSION(mm+kk+1) :: coff,w
REAL(DP), DIMENSION(mm+kk+1,mm+kk+1) :: v
REAL(DP), DIMENSION((mm+kk+1)*NPFAC,mm+kk+1) :: u,temp
ncof=mm+kk+1
npt=NPFAC*ncof                      Number of points where function is evaluated,
npth=npt/2                              i.e., fineness of the mesh.
dev=BIG
theta=PIO2_D/(npt-1)
xs(1:npth-1)=a+(b-a)*sin(theta*arth(0,1,npth-1))**2
```

Now fill arrays with mesh abscissas and function values. At each end, use formula that minimizes roundoff sensitivity in xs.

```
xs(npth:npt)=b-(b-a)*sin(theta*arth(npt-npth,-1,npt-npth+1))**2
fs=func(xs)
wt=1.0                              In later iterations we will adjust these weights to
ee=1.0                                  combat the largest deviations.
e=0.0
do it=1,MAXIT                       Loop over iterations.
    bb=wt*(fs+sign(e,ee))
```

Key idea here: Fit to $\text{fn}(x)+e$ where the deviation is positive, to $\text{fn}(x)-e$ where it is negative. Then e is supposed to become an approximation to the equal-ripple deviation.

```
    temp=geop(spread(1.0_dp,1,npt),xs,ncof)
```

```
            Note that vector form of geop (returning matrix) is being used.
        u(:,1:mm+1)=temp(:,1:mm+1)*spread(wt,2,mm+1)
            Set up the "design matrix" for the least squares fit.
        u(:,mm+2:ncof)=-temp(:,2:ncof-mm)*spread(bb,2,ncof-mm-1)
        call svdcmp(u,w,v)
            Singular Value Decomposition. In especially singular or difficult cases, one might here
            edit the singular values w(1:ncof), replacing small values by zero.
        call svbksb(u,w,v,bb,coff)
        ee=ratval(xs,coff,mm,kk)-fs         Tabulate the deviations and revise the weights.
        wt=abs(ee)                          Use weighting to emphasize most deviant points.
        devmax=maxval(wt)
        e=sum(wt)/npt                       Update e to be the mean absolute deviation.
        if (devmax <= dev) then             Save only the best coefficient set found.
            cof=coff
            dev=devmax
        end if
        write(*,10) it,devmax
    end do
10  format (' ratlsq iteration=',i2,' max error=',1p,e10.3)
    END SUBROUTINE ratlsq
```

`temp=geop(spread(1.0_dp,1,npt),xs,ncof)` The design matrix u_{ij} is defined for $i = 1, \ldots,$ npts by

$$u_{ij} = \begin{cases} w_i x_i^{j-1}, & j = 1, \ldots, m+1 \\ -b_i x_i^{j-m-2}, & j = m+2, \ldots, n \end{cases} \tag{B5.12}$$

The first case in equation (B5.12) is computed in parallel by constructing the matrix `temp` equal to

$$\begin{bmatrix} 1 & x_1 & x_1^2 & \cdots \\ 1 & x_2 & x_2^2 & \cdots \\ 1 & x_3 & x_3^2 & \cdots \\ \vdots & \vdots & \vdots & \ddots \end{bmatrix}$$

and then multiplying by the matrix `spread(wt,2,mm+1)`, which is just

$$\begin{bmatrix} w_1 & w_1 & w_1 & \cdots \\ w_2 & w_2 & w_2 & \cdots \\ w_3 & w_3 & w_3 & \cdots \\ \vdots & \vdots & \vdots & \ddots \end{bmatrix}$$

(Remember that multiplication using $*$ means element-by-element multiplication, not matrix multiplication.) A similar construction is used for the second part of the design matrix.

Chapter B6. Special Functions

f90 A Fortran 90 intrinsic function such as sin(x) is both *generic* and *elemental*. Generic means that the argument x can be any of multiple intrinsic data types and kind values (in the case of sin, any real or complex kind). Elemental means that x need not be a scalar, but can be an array of any rank and shape, in which case the calculation of sin is performed independently for each element.

Ideally, when we implement more complicated special functions in Fortran 90, as we do in this chapter, we would make them, too, both generic and elemental. Unfortunately, the language standard does not completely allow this. User-defined elemental functions are prohibited in Fortran 90, though they will be allowed in Fortran 95. And, there is no fully automatic way of providing for a single routine to allow arguments of multiple data types or kinds — nothing like C++'s "class templates," for example.

However, don't give up hope! Fortran 90 does provide a powerful mechanism for overloading, which can be used (perhaps not always with a maximum of convenience) to *simulate* both generic and elemental function features. In most cases, when we implement a special function with a scalar argument, gammln(x) say, we will also implement a corresponding vector-valued function of vector argument that evaluates the special function for each component of the vector argument. We will then overload the scalar and vector version of the function onto the same function name. For example, within the nr module are the lines

```
INTERFACE gammln
   FUNCTION gammln_s(xx)
   USE nrtype
   REAL(SP), INTENT(IN) :: xx
   REAL(SP) :: gammln_s
   END FUNCTION gammln_s

   FUNCTION gammln_v(xx)
   USE nrtype
   REAL(SP), DIMENSION(:), INTENT(IN) :: xx
   REAL(SP), DIMENSION(size(xx)) :: gammln_v
   END FUNCTION gammln_v
END INTERFACE
```

which can be included by a statement like "USE nr, ONLY: gammln," and then allow you to write gammln(x) without caring (or even thinking about) whether x is a scalar or a vector. If you want arguments of even higher rank (matrices, and so forth), you can provide these yourself, based on our models, and overload them, too.

That takes care of "elemental"; what about "generic"? Here, too, overloading provides an acceptable, if not perfect, solution. Where double-precision versions of special functions are needed, you can in many cases easily construct them from our provided routines by changing the variable kinds (and any necessary convergence parameters), and then additionally overload them onto the same generic function names. (In general, in the interest of brevity, we will not ourselves do this for the functions in this chapter.)

At first meeting, Fortran 90's overloading capability may seem trivial, or merely cosmetic, to the Fortran 77 programmer; but one soon comes to rely on it as an important conceptual simplification. Programming at a "higher level of abstraction" is usually more productive than spending time "bogged down in the mud." Furthermore, the use of overloading is generally fail-safe: If you invoke a generic name with arguments of shapes or types for which a specific routine has not been defined, the compiler tells you about it.

We won't reprint the module `nr`'s interface blocks for all the routines in this chapter. When you see routines named `something_s` and `something_v`, below, you can safely assume that the generic name `something` is defined in the module `nr` and overloaded with the two specific routine names. A full alphabetical listing of all the interface blocks in `nr` is given in Appendix C2.

Given our heavy investment, in this chapter, in overloadable vector-valued special function routines, it is worth discussing whether this effort is simply a stopgap measure for Fortran 90, soon to be made obsolete by Fortran 95's provision of user-definable `ELEMENTAL` procedures. The answer is "not necessarily," and takes us into some speculation about the future of SIMD, versus MIMD, computing.

Elemental procedures, while applying the same executable code to each element, do not insist that it be feasible to perform all the parallel calculations in lockstep. That is, elemental procedures can have tests and branches (`if-then-else` constructions) that result in different elements being calculated by totally different pieces of code, in a fashion that can only be determined at run time. For true 100% MIMD (multiple instruction, multiple data) machines, this is not a problem: individual processors do the individual element calculations asynchronously.

However, virtually none of today's (and likely tomorrow's) largest-scale parallel supercomputers are 100% MIMD in this way. While modern parallel supercomputers increasingly have MIMD features, they continue to reward the use of SIMD (single instruction, multiple data) code with greater computational speed, often because of hardware pipelining or vector processing features within the individual processors. The use of Fortran 90 (or, for that matter Fortran 95) in a data-parallel or SIMD mode is thus by no means superfluous, or obviated by Fortran 95's `ELEMENTAL` construction.

The problem we face is that parallel calculation of special function values often doesn't fit well into the SIMD mold: Since the calculation of the value of a special function typically requires the convergence of an iterative process, as well as possible branches for different values of arguments, it cannot *in general* be done efficiently with "lockstep" SIMD programming.

Luckily, in particular cases, including most (but not all) of the functions in this chapter, one can in fact make reasonably good parallel implementations with the SIMD tools provided by the language. We will in fact see a number of different tricks for accomplishing this in the code that follows.

We are interested in demonstrating SIMD techniques, but we are not completely impractical. None of the data-parallel implementations given below are too inefficient on a scalar machine, and some may in fact be faster than Fortran 95's `ELEMENTAL` alternative, or than do-loops over calls to the scalar version of the function. On a scalar machine, how can this be? We have already, above, hinted at the answer: (i) most modern scalar processors can overlap instructions to some degree, and data-parallel coding often provides compilers with the ability to accomplish this more efficiently; and (ii) data-parallel code can sometimes give better cache utilization.

⋆ ⋆ ⋆

```
FUNCTION gammln_s(xx)
USE nrtype; USE nrutil, ONLY : arth,assert
IMPLICIT NONE
REAL(SP), INTENT(IN) :: xx
REAL(SP) :: gammln_s
   Returns the value ln[Γ(xx)] for xx > 0.
REAL(DP) :: tmp,x
   Internal arithmetic will be done in double precision, a nicety that you can omit if five-figure
   accuracy is good enough.
REAL(DP) :: stp = 2.5066282746310005_dp
REAL(DP), DIMENSION(6) :: coef = (/76.18009172947146_dp,&
    -86.50532032941677_dp,24.01409824083091_dp,&
    -1.231739572450155_dp,0.1208650973866179e-2_dp,&
    -0.5395239384953e-5_dp/)
call assert(xx > 0.0, 'gammln_s arg')
x=xx
tmp=x+5.5_dp
tmp=(x+0.5_dp)*log(tmp)-tmp
gammln_s=tmp+log(stp*(1.000000000190015_dp+&
    sum(coef(:)/arth(x+1.0_dp,1.0_dp,size(coef))))/x)
END FUNCTION gammln_s
```

```
FUNCTION gammln_v(xx)
USE nrtype; USE nrutil, ONLY: assert
IMPLICIT NONE
INTEGER(I4B) :: i
REAL(SP), DIMENSION(:), INTENT(IN) :: xx
REAL(SP), DIMENSION(size(xx)) :: gammln_v
REAL(DP), DIMENSION(size(xx)) :: ser,tmp,x,y
REAL(DP) :: stp = 2.5066282746310005_dp
REAL(DP), DIMENSION(6) :: coef = (/76.18009172947146_dp,&
    -86.50532032941677_dp,24.01409824083091_dp,&
    -1.231739572450155_dp,0.1208650973866179e-2_dp,&
    -0.5395239384953e-5_dp/)
if (size(xx) == 0) RETURN
call assert(all(xx > 0.0), 'gammln_v arg')
x=xx
tmp=x+5.5_dp
tmp=(x+0.5_dp)*log(tmp)-tmp
ser=1.000000000190015_dp
y=x
do i=1,size(coef)
    y=y+1.0_dp
    ser=ser+coef(i)/y
end do
gammln_v=tmp+log(stp*ser/x)
END FUNCTION gammln_v
```

`call assert(xx > 0.0, 'gammln_s arg')` We use the `nrutil` routine `assert` for functions that have restrictions on the allowed range of arguments. One could instead have used an `if` statement with a call to `nrerror`; but we think that the uniformity of using `assert`, and the fact that its logical arguments read the "desired" way, not the "erroneous" way, make for a clearer programming style. In the vector version, the `assert` line is: `call assert(all(xx > 0.0), 'gammln_v arg')`

Notice that the scalar and vector versions achieve parallelism in quite different ways, something that we will see many times in this chapter. In the scalar case, parallelism (at least small-scale) is achieved through constructions like

```
sum(coef(:)/arth(x+1.0_dp,1.0_dp,size(coef)))
```

Here vector utilities construct the series $x + 1, x + 2, \ldots$ and then sum a series with these terms in the denominators and a vector of coefficients in the numerators. (This code may seem terse to Fortran 90 novices, but once you get used to it, it is quite clear to read.)

In the vector version, by contrast, parallelism is achieved across the components of the vector argument, and the above series is evaluated sequentially as a do-loop. Obviously the assumption is that the length of the vector argument is much longer than the very modest number (here, 6) of terms in the sum.

⋆ ⋆ ⋆

```
FUNCTION factrl_s(n)
USE nrtype; USE nrutil, ONLY : arth,assert,cumprod
USE nr, ONLY : gammln
IMPLICIT NONE
INTEGER(I4B), INTENT(IN) :: n
REAL(SP) :: factrl_s
   Returns the value n! as a floating-point number.
INTEGER(I4B), SAVE :: ntop=0
INTEGER(I4B), PARAMETER :: NMAX=32
REAL(SP), DIMENSION(NMAX), SAVE :: a        Table of stored values.
call assert(n >= 0, 'factrl_s arg')
if (n < ntop) then                          Already in table.
    factrl_s=a(n+1)
else if (n < NMAX) then                     Fill in table up to NMAX.
    ntop=NMAX
    a(1)=1.0
    a(2:NMAX)=cumprod(arth(1.0_sp,1.0_sp,NMAX-1))
    factrl_s=a(n+1)
else                                        Larger value than size of table is required.
    factrl_s=exp(gammln(n+1.0_sp))              Actually, this big a value is going to over-
end if                                          flow on many computers, but no harm in
END FUNCTION factrl_s                           trying.
```

`cumprod(arth(1.0_sp,1.0_sp,NMAX-1))` By now you should recognize this as an idiom for generating a vector of consecutive factorials. The routines `cumprod` and `arth`, both in `nrutil`, are both capable of being parallelized, e.g., by recursion, so this idiom is potentially faster than an in-line do-loop.

```
FUNCTION factrl_v(n)
USE nrtype; USE nrutil, ONLY : arth,assert,cumprod
USE nr, ONLY : gammln
IMPLICIT NONE
INTEGER(I4B), DIMENSION(:), INTENT(IN) :: n
REAL(SP), DIMENSION(size(n)) :: factrl_v
LOGICAL(LGT), DIMENSION(size(n)) :: mask
INTEGER(I4B), SAVE :: ntop=0
INTEGER(I4B), PARAMETER :: NMAX=32
REAL(SP), DIMENSION(NMAX), SAVE :: a
call assert(all(n >= 0), 'factrl_v arg')
if (ntop == 0) then
    ntop=NMAX
    a(1)=1.0
    a(2:NMAX)=cumprod(arth(1.0_sp,1.0_sp,NMAX-1))
end if
mask = (n >= NMAX)
factrl_v=unpack(exp(gammln(pack(n,mask)+1.0_sp)),mask,0.0_sp)
where (.not. mask) factrl_v=a(n+1)
END FUNCTION factrl_v
```

`unpack(exp(gammln(pack(n,mask)+1.0_sp)),mask,0.0_sp)` Here we meet the first of several solutions to a common problem: How shall we get answers, from an external vector-valued function, for just a *subset* of vector arguments, those defined by a mask? Here we use what we call the "pack-unpack" solution: Pack up all the arguments using the mask, send them to the function, and unpack the answers that come back. This packing and unpacking is not without cost (highly dependent on machine architecture, to be sure), but we hope to "earn it back" in the parallelism of the external function.

`where (.not. mask) factrl_v=a(n+1)` In some cases we might take care of the `.not.mask` case directly within the `unpack` construction, using its third ("`FIELD=`") argument to provide the not-unpacked values. However, there is no guarantee that the compiler won't evaluate all components of the "`FIELD=`" array, if it finds it efficient to do so. Here, since the index of `a(n+1)` would be out of range, we can't do it this way. Thus the separate `where` statement.

⋆ ⋆ ⋆

```
FUNCTION bico_s(n,k)
USE nrtype
USE nr, ONLY : factln
IMPLICIT NONE
INTEGER(I4B), INTENT(IN) :: n,k
REAL(SP) :: bico_s
    Returns the binomial coefficient (n k) as a floating-point number.
bico_s=nint(exp(factln(n)-factln(k)-factln(n-k)))
  The nearest-integer function cleans up roundoff error for smaller values of n and k.
END FUNCTION bico_s
```

```
FUNCTION bico_v(n,k)
USE nrtype; USE nrutil, ONLY : assert_eq
USE nr, ONLY : factln
IMPLICIT NONE
INTEGER(I4B), DIMENSION(:), INTENT(IN) :: n,k
REAL(SP), DIMENSION(size(n)) :: bico_v
INTEGER(I4B) :: ndum
ndum=assert_eq(size(n),size(k),'bico_v')
bico_v=nint(exp(factln(n)-factln(k)-factln(n-k)))
END FUNCTION bico_v
```

⋆ ⋆ ⋆

```
FUNCTION factln_s(n)
USE nrtype; USE nrutil, ONLY : arth,assert
USE nr, ONLY : gammln
IMPLICIT NONE
INTEGER(I4B), INTENT(IN) :: n
REAL(SP) :: factln_s
   Returns ln(n!).
INTEGER(I4B), PARAMETER :: TMAX=100
REAL(SP), DIMENSION(TMAX), SAVE :: a
LOGICAL(LGT), SAVE :: init=.true.
if (init) then                    Initialize the table.
    a(1:TMAX)=gammln(arth(1.0_sp,1.0_sp,TMAX))
    init=.false.
end if
call assert(n >= 0, 'factln_s arg')
if (n < TMAX) then                In range of the table.
    factln_s=a(n+1)
else                              Out of range of the table.
    factln_s=gammln(n+1.0_sp)
end if
END FUNCTION factln_s
```

```
FUNCTION factln_v(n)
USE nrtype; USE nrutil, ONLY : arth,assert
USE nr, ONLY : gammln
IMPLICIT NONE
INTEGER(I4B), DIMENSION(:), INTENT(IN) :: n
REAL(SP), DIMENSION(size(n)) :: factln_v
LOGICAL(LGT), DIMENSION(size(n)) :: mask
INTEGER(I4B), PARAMETER :: TMAX=100
REAL(SP), DIMENSION(TMAX), SAVE :: a
LOGICAL(LGT), SAVE :: init=.true.
if (init) then
    a(1:TMAX)=gammln(arth(1.0_sp,1.0_sp,TMAX))
    init=.false.
end if
call assert(all(n >= 0), 'factln_v arg')
mask = (n >= TMAX)
factln_v=unpack(gammln(pack(n,mask)+1.0_sp),mask,0.0_sp)
where (.not. mask) factln_v=a(n+1)
END FUNCTION factln_v
```

f90 `gammln(arth(1.0_sp,1.0_sp,TMAX))` Another example of the programming convenience of combining a function returning a vector (here, `arth`) with a special function whose generic name (here, `gammln`) has an overloaded vector version.

⋆ ⋆ ⋆

```
FUNCTION beta_s(z,w)
USE nrtype
USE nr, ONLY : gammln
IMPLICIT NONE
REAL(SP), INTENT(IN) :: z,w
REAL(SP) :: beta_s
   Returns the value of the beta function B(z,w).
beta_s=exp(gammln(z)+gammln(w)-gammln(z+w))
END FUNCTION beta_s
```

```
FUNCTION beta_v(z,w)
USE nrtype; USE nrutil, ONLY : assert_eq
USE nr, ONLY : gammln
IMPLICIT NONE
REAL(SP), DIMENSION(:), INTENT(IN) :: z,w
REAL(SP), DIMENSION(size(z)) :: beta_v
INTEGER(I4B) :: ndum
ndum=assert_eq(size(z),size(w),'beta_v')
beta_v=exp(gammln(z)+gammln(w)-gammln(z+w))
END FUNCTION beta_v
```

⋆ ⋆ ⋆

```
FUNCTION gammp_s(a,x)
USE nrtype; USE nrutil, ONLY : assert
USE nr, ONLY : gcf,gser
IMPLICIT NONE
REAL(SP), INTENT(IN) :: a,x
REAL(SP) :: gammp_s
   Returns the incomplete gamma function P(a,x).
call assert( x >= 0.0, a > 0.0, 'gammp_s args')
if (x<a+1.0_sp) then                Use the series representation.
   gammp_s=gser(a,x)
else                                Use the continued fraction representation
   gammp_s=1.0_sp-gcf(a,x)          and take its complement.
end if
END FUNCTION gammp_s
```

```
FUNCTION gammp_v(a,x)
USE nrtype; USE nrutil, ONLY : assert,assert_eq
USE nr, ONLY : gcf,gser
IMPLICIT NONE
REAL(SP), DIMENSION(:), INTENT(IN) :: a,x
REAL(SP), DIMENSION(size(x)) :: gammp_v
LOGICAL(LGT), DIMENSION(size(x)) :: mask
INTEGER(I4B) :: ndum
ndum=assert_eq(size(a),size(x),'gammp_v')
call assert( all(x >= 0.0), all(a > 0.0), 'gammp_v args')
mask = (x<a+1.0_sp)
gammp_v=merge(gser(a,merge(x,0.0_sp,mask)), &
   1.0_sp-gcf(a,merge(x,0.0_sp,.not. mask)),mask)
END FUNCTION gammp_v
```

`call assert( x >= 0.0,  a > 0.0, 'gammp_s args')` The generic routine `assert` in `nrutil` is overloaded with variants for more than one logical assertion, so you can make more than one assertion about argument ranges.

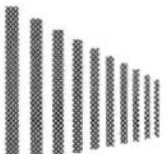

`gammp_v=merge(gser(a,merge(x,0.0_sp,mask)), &`
`1.0_sp-gcf(a,merge(x,0.0_sp,.not. mask)),mask)` Here we meet the *second* solution to the problem of getting masked values from an external vector function. (For the first solution, see note to `factrl`, above.) We call this one "merge with dummy values": Inappropriate values of the argument `x` (as determined by `mask`) are set to zero before `gser`, and later `gcf`, are called, and the supernumerary answers returned are discarded by a final `merge`. The assumption here is that the dummy value sent to the function (here, zero) is a special value that computes extremely fast, so that the overhead of computing and returning the supernumerary function values is outweighed by the parallelism achieved on the nontrivial components of `x`. Look at `gser_v` and `gcf_v` below to judge whether this assumption is realistic in this case.

```
FUNCTION gammq_s(a,x)
USE nrtype; USE nrutil, ONLY : assert
USE nr, ONLY : gcf,gser
IMPLICIT NONE
REAL(SP), INTENT(IN) :: a,x
REAL(SP) :: gammq_s
   Returns the incomplete gamma function Q(a,x) ≡ 1 − P(a,x).
call assert( x >= 0.0, a > 0.0, 'gammq_s args')
if (x<a+1.0_sp) then                   Use the series representation
   gammq_s=1.0_sp-gser(a,x)            and take its complement.
else                                   Use the continued fraction representation.
   gammq_s=gcf(a,x)
end if
END FUNCTION gammq_s
```

```
FUNCTION gammq_v(a,x)
USE nrtype; USE nrutil, ONLY : assert,assert_eq
USE nr, ONLY : gcf,gser
IMPLICIT NONE
REAL(SP), DIMENSION(:), INTENT(IN) :: a,x
REAL(SP), DIMENSION(size(a)) :: gammq_v
LOGICAL(LGT), DIMENSION(size(x)) :: mask
INTEGER(I4B) :: ndum
ndum=assert_eq(size(a),size(x),'gammq_v')
call assert( all(x >= 0.0), all(a > 0.0), 'gammq_v args')
mask = (x<a+1.0_sp)
gammq_v=merge(1.0_sp-gser(a,merge(x,0.0_sp,mask)), &
   gcf(a,merge(x,0.0_sp,.not. mask)),mask)
END FUNCTION gammq_v
```

```
FUNCTION gser_s(a,x,gln)
USE nrtype; USE nrutil, ONLY : nrerror
USE nr, ONLY : gammln
IMPLICIT NONE
REAL(SP), INTENT(IN) :: a,x
REAL(SP), OPTIONAL, INTENT(OUT) :: gln
REAL(SP) :: gser_s
INTEGER(I4B), PARAMETER :: ITMAX=100
REAL(SP), PARAMETER :: EPS=epsilon(x)
```

Returns the incomplete gamma function $P(a,x)$ evaluated by its series representation as gamser. Also optionally returns $\ln\Gamma(a)$ as gln.

```
INTEGER(I4B) :: n
REAL(SP) :: ap,del,summ
if (x == 0.0) then
    gser_s=0.0
    RETURN
end if
ap=a
summ=1.0_sp/a
del=summ
do n=1,ITMAX
    ap=ap+1.0_sp
    del=del*x/ap
    summ=summ+del
    if (abs(del) < abs(summ)*EPS) exit
end do
if (n > ITMAX) call nrerror('a too large, ITMAX too small in gser_s')
if (present(gln)) then
    gln=gammln(a)
    gser_s=summ*exp(-x+a*log(x)-gln)
else
    gser_s=summ*exp(-x+a*log(x)-gammln(a))
end if
END FUNCTION gser_s


FUNCTION gser_v(a,x,gln)
USE nrtype; USE nrutil, ONLY : assert_eq,nrerror
USE nr, ONLY : gammln
IMPLICIT NONE
REAL(SP), DIMENSION(:), INTENT(IN) :: a,x
REAL(SP), DIMENSION(:), OPTIONAL, INTENT(OUT) :: gln
REAL(SP), DIMENSION(size(a)) :: gser_v
INTEGER(I4B), PARAMETER :: ITMAX=100
REAL(SP), PARAMETER :: EPS=epsilon(x)
INTEGER(I4B) :: n
REAL(SP), DIMENSION(size(a)) :: ap,del,summ
LOGICAL(LGT), DIMENSION(size(a)) :: converged,zero
n=assert_eq(size(a),size(x),'gser_v')
zero=(x == 0.0)
where (zero) gser_v=0.0
ap=a
summ=1.0_sp/a
del=summ
converged=zero
do n=1,ITMAX
    where (.not. converged)
        ap=ap+1.0_sp
        del=del*x/ap
        summ=summ+del
        converged = (abs(del) < abs(summ)*EPS)
    end where
    if (all(converged)) exit
end do
if (n > ITMAX) call nrerror('a too large, ITMAX too small in gser_v')
if (present(gln)) then
    if (size(gln) < size(a)) call &
        nrerror('gser: Not enough space for gln')
    gln=gammln(a)
    where (.not. zero) gser_v=summ*exp(-x+a*log(x)-gln)
else
    where (.not. zero) gser_v=summ*exp(-x+a*log(x)-gammln(a))
```

```
end if
END FUNCTION gser_v
```

f90 `REAL(SP), OPTIONAL, INTENT(OUT) :: gln` Normally, an `OPTIONAL` argument will be `INTENT(IN)` and be used to provide a less-often-used extra input argument to a function. Here, the `OPTIONAL` argument is `INTENT(OUT)`, used to provide a useful value that is a byproduct of the main calculation.

Also note that although $x \geq 0$ is required, we omit our usual `call assert` check for this, because `gser` is supposed to be called only by `gammp` or `gammq` — and these routines supply the argument checking themselves.

`do n=1,ITMAX...end do...if (n > ITMAX)...` This is typical code in Fortran 90 for a loop with a maximum number of iterations, relying on Fortran 90's guarantee that the index of the do-loop will be available after normal completion of the loop with a predictable value, greater by one than the upper limit of the loop. If the `exit` statement within the loop is ever taken, the `if` statement is guaranteed to fail; if the loop goes all the way through `ITMAX` cycles, the `if` statement is guaranteed to succeed.

`zero=(x == 0.0)...where (zero) gser_v=0.0...converged=zero` This is the code that provides for very low overhead calculation of zero arguments, as is assumed by the merge-with-dummy-values strategy in `gammp` and `gammq`. Zero arguments are "pre-converged" and are never the holdouts in the convergence test.

```
FUNCTION gcf_s(a,x,gln)
USE nrtype; USE nrutil, ONLY : nrerror
USE nr, ONLY : gammln
IMPLICIT NONE
REAL(SP), INTENT(IN) :: a,x
REAL(SP), OPTIONAL, INTENT(OUT) :: gln
REAL(SP) :: gcf_s
INTEGER(I4B), PARAMETER :: ITMAX=100
REAL(SP), PARAMETER :: EPS=epsilon(x),FPMIN=tiny(x)/EPS
```

Returns the incomplete gamma function $Q(a,x)$ evaluated by its continued fraction representation as `gammcf`. Also optionally returns $\ln\Gamma(a)$ as `gln`.

Parameters: `ITMAX` is the maximum allowed number of iterations; `EPS` is the relative accuracy; `FPMIN` is a number near the smallest representable floating-point number.

```
INTEGER(I4B) :: i
REAL(SP) :: an,b,c,d,del,h
if (x == 0.0) then
    gcf_s=1.0
    RETURN
end if
b=x+1.0_sp-a                       Set up for evaluating continued fraction by mod-
c=1.0_sp/FPMIN                         ified Lentz's method (§5.2) with b0 = 0.
d=1.0_sp/b
h=d
do i=1,ITMAX                       Iterate to convergence.
    an=-i*(i-a)
    b=b+2.0_sp
    d=an*d+b
    if (abs(d) < FPMIN) d=FPMIN
    c=b+an/c
    if (abs(c) < FPMIN) c=FPMIN
```

```
    d=1.0_sp/d
    del=d*c
    h=h*del
    if (abs(del-1.0_sp) <= EPS) exit
end do
if (i > ITMAX) call nrerror('a too large, ITMAX too small in gcf_s')
if (present(gln)) then
    gln=gammln(a)
    gcf_s=exp(-x+a*log(x)-gln)*h          Put factors in front.
else
    gcf_s=exp(-x+a*log(x)-gammln(a))*h
end if
END FUNCTION gcf_s


FUNCTION gcf_v(a,x,gln)
USE nrtype; USE nrutil, ONLY : assert_eq,nrerror
USE nr, ONLY : gammln
IMPLICIT NONE
REAL(SP), DIMENSION(:), INTENT(IN) :: a,x
REAL(SP), DIMENSION(:), OPTIONAL, INTENT(OUT) :: gln
REAL(SP), DIMENSION(size(a)) :: gcf_v
INTEGER(I4B), PARAMETER :: ITMAX=100
REAL(SP), PARAMETER :: EPS=epsilon(x),FPMIN=tiny(x)/EPS
INTEGER(I4B) :: i
REAL(SP), DIMENSION(size(a)) :: an,b,c,d,del,h
LOGICAL(LGT), DIMENSION(size(a)) :: converged,zero
i=assert_eq(size(a),size(x),'gcf_v')
zero=(x == 0.0)
where (zero)
    gcf_v=1.0
elsewhere
    b=x+1.0_sp-a
    c=1.0_sp/FPMIN
    d=1.0_sp/b
    h=d
end where
converged=zero
do i=1,ITMAX
    where (.not. converged)
        an=-i*(i-a)
        b=b+2.0_sp
        d=an*d+b
        d=merge(FPMIN,d, abs(d)<FPMIN )
        c=b+an/c
        c=merge(FPMIN,c, abs(c)<FPMIN )
        d=1.0_sp/d
        del=d*c
        h=h*del
        converged = (abs(del-1.0_sp)<=EPS)
    end where
    if (all(converged)) exit
end do
if (i > ITMAX) call nrerror('a too large, ITMAX too small in gcf_v')
if (present(gln)) then
    if (size(gln) < size(a)) call &
        nrerror('gser: Not enough space for gln')
    gln=gammln(a)
    where (.not. zero) gcf_v=exp(-x+a*log(x)-gln)*h
else
    where (.not. zero) gcf_v=exp(-x+a*log(x)-gammln(a))*h
end if
END FUNCTION gcf_v
```

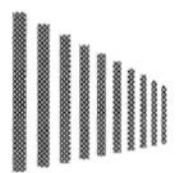

`zero=(x == 0.0)...where (zero) gcf_v=1.0...converged=zero` See note on `gser`. Here, too, we pre-converge the special value of zero.

⋆ ⋆ ⋆

```
FUNCTION erf_s(x)
USE nrtype
USE nr, ONLY : gammp
IMPLICIT NONE
REAL(SP), INTENT(IN) :: x
REAL(SP) :: erf_s
   Returns the error function erf(x).
erf_s=gammp(0.5_sp,x**2)
if (x < 0.0) erf_s=-erf_s
END FUNCTION erf_s
```

```
FUNCTION erf_v(x)
USE nrtype
USE nr, ONLY : gammp
IMPLICIT NONE
REAL(SP), DIMENSION(:), INTENT(IN) :: x
REAL(SP), DIMENSION(size(x)) :: erf_v
erf_v=gammp(spread(0.5_sp,1,size(x)),x**2)
where (x < 0.0) erf_v=-erf_v
END FUNCTION erf_v
```

`erf_v=gammp(spread(0.5_sp,1,size(x)),x**2)` Yes, we do have an overloaded vector version of `gammp`, but it is vectorized on *both* its arguments. Thus, in a case where we want to vectorize on only *one* argument, we need a `spread` construction. In many contexts, Fortran 90 automatically makes scalars conformable with arrays (i.e., it automatically spreads them to the shape of the array); but the language does *not* do so when trying to match a generic function or subroutine call to a specific overloaded name. Perhaps this is wise; it is safer to prevent "accidental" invocations of vector-specific functions. Or, perhaps it is an area where the language could be improved.

```
FUNCTION erfc_s(x)
USE nrtype
USE nr, ONLY : gammp,gammq
IMPLICIT NONE
REAL(SP), INTENT(IN) :: x
REAL(SP) :: erfc_s
   Returns the complementary error function erfc(x).
erfc_s=merge(1.0_sp+gammp(0.5_sp,x**2),gammq(0.5_sp,x**2), x < 0.0)
END FUNCTION erfc_s
```

`erfc_s=merge(1.0_sp+gammp(0.5_sp,x**2),gammq(0.5_sp,x**2), x < 0.0)` An example of our use of `merge` as an idiom for a conditional expression. Once you get used to these, you'll find them just as clear as the multiline `if...then...else` alternative.

```
FUNCTION erfc_v(x)
USE nrtype
USE nr, ONLY : gammp,gammq
IMPLICIT NONE
REAL(SP), DIMENSION(:), INTENT(IN) :: x
REAL(SP), DIMENSION(size(x)) :: erfc_v
LOGICAL(LGT), DIMENSION(size(x)) :: mask
mask = (x < 0.0)
erfc_v=merge(1.0_sp+gammp(spread(0.5_sp,1,size(x)), &
    merge(x,0.0_sp,mask)**2),gammq(spread(0.5_sp,1,size(x)), &
    merge(x,0.0_sp,.not. mask)**2),mask)
END FUNCTION erfc_v
```

f90 `erfc_v=merge(1.0_sp+...)` Another example of the "merge with dummy values" idiom described on p. 1090. Here positive values of x in the call to gammp, and negative values in the call to gammq, are first set to the dummy value zero. The value zero is a special argument that computes very fast. The unwanted dummy function values are then discarded by the final outer `merge`.

⋆ ⋆ ⋆

```
FUNCTION erfcc_s(x)
USE nrtype; USE nrutil, ONLY : poly
IMPLICIT NONE
REAL(SP), INTENT(IN) :: x
REAL(SP) :: erfcc_s
```

Returns the complementary error function erfc(x) with fractional error everywhere less than 1.2×10^{-7}.

```
REAL(SP) :: t,z
REAL(SP), DIMENSION(10) :: coef = (/-1.26551223_sp,1.00002368_sp,&
    0.37409196_sp,0.09678418_sp,-0.18628806_sp,0.27886807_sp,&
    -1.13520398_sp,1.48851587_sp,-0.82215223_sp,0.17087277_sp/)
z=abs(x)
t=1.0_sp/(1.0_sp+0.5_sp*z)
erfcc_s=t*exp(-z*z+poly(t,coef))
if (x < 0.0) erfcc_s=2.0_sp-erfcc_s
END FUNCTION erfcc_s
```

```
FUNCTION erfcc_v(x)
USE nrtype; USE nrutil, ONLY : poly
IMPLICIT NONE
REAL(SP), DIMENSION(:), INTENT(IN) :: x
REAL(SP), DIMENSION(size(x)) :: erfcc_v,t,z
REAL(SP), DIMENSION(10) :: coef = (/-1.26551223_sp,1.00002368_sp,&
    0.37409196_sp,0.09678418_sp,-0.18628806_sp,0.27886807_sp,&
    -1.13520398_sp,1.48851587_sp,-0.82215223_sp,0.17087277_sp/)
z=abs(x)
t=1.0_sp/(1.0_sp+0.5_sp*z)
erfcc_v=t*exp(-z*z+poly(t,coef))
where (x < 0.0) erfcc_v=2.0_sp-erfcc_v
END FUNCTION erfcc_v
```

f90 `erfcc_v=t*exp(-z*z+poly(t,coef))` The vector code is identical to the scalar, because the `nrutil` routine `poly` has overloaded cases for the evaluation of a polynomial at a single value of the independent variable, and at multiple values. One *could* also overload a version with a matrix of coefficients whose columns could be used for the simultaneous evaluation of different polynomials at different values of independent variable. The point is that as long as there are differences in the shapes of at least one argument, the intended version of `poly` can be discerned by the compiler.

⋆ ⋆ ⋆

```
FUNCTION expint(n,x)
USE nrtype; USE nrutil, ONLY : arth,assert,nrerror
IMPLICIT NONE
INTEGER(I4B), INTENT(IN) :: n
REAL(SP), INTENT(IN) :: x
REAL(SP) :: expint
INTEGER(I4B), PARAMETER :: MAXIT=100
REAL(SP), PARAMETER :: EPS=epsilon(x),BIG=huge(x)*EPS
    Evaluates the exponential integral En(x).
    Parameters: MAXIT is the maximum allowed number of iterations; EPS is the desired relative
    error, not smaller than the machine precision; BIG is a number near the largest representable
    floating-point number; EULER (in nrtype) is Euler's constant γ.
INTEGER(I4B) :: i,nm1
REAL(SP) :: a,b,c,d,del,fact,h
call assert(n >= 0, x >= 0.0, (x > 0.0 .or. n > 1), &
    'expint args')
if (n == 0) then                        Special case.
    expint=exp(-x)/x
    RETURN
end if
nm1=n-1
if (x == 0.0) then                      Another special case.
    expint=1.0_sp/nm1
else if (x > 1.0) then                  Lentz's algorithm (§5.2).
    b=x+n
    c=BIG
    d=1.0_sp/b
    h=d
    do i=1,MAXIT
        a=-i*(nm1+i)
        b=b+2.0_sp
        d=1.0_sp/(a*d+b)                Denominators cannot be zero.
        c=b+a/c
        del=c*d
        h=h*del
        if (abs(del-1.0_sp) <= EPS) exit
    end do
    if (i > MAXIT) call nrerror('expint: continued fraction failed')
    expint=h*exp(-x)
else                                    Evaluate series.
    if (nm1 /= 0) then                  Set first term.
        expint=1.0_sp/nm1
    else
        expint=-log(x)-EULER
    end if
    fact=1.0
    do i=1,MAXIT
        fact=-fact*x/i
        if (i /= nm1) then
            del=-fact/(i-nm1)
```

```
        else                         ψ(n) appears here.
            del=fact*(-log(x)-EULER+sum(1.0_sp/arth(1,1,nm1)))
        end if
        expint=expint+del
        if (abs(del) < abs(expint)*EPS) exit
    end do
    if (i > MAXIT) call nrerror('expint: series failed')
end if
END FUNCTION expint
```

expint does not readily parallelize, and we thus don't provide a vector version. For syntactic convenience you could make a vector version with a do-loop over calls to this scalar version; or, in Fortran 95, you can of course make the function ELEMENTAL.

⋆ ⋆ ⋆

```
FUNCTION ei(x)
USE nrtype; USE nrutil, ONLY : assert,nrerror
IMPLICIT NONE
REAL(SP), INTENT(IN) :: x
REAL(SP) :: ei
INTEGER(I4B), PARAMETER :: MAXIT=100
REAL(SP), PARAMETER :: EPS=epsilon(x),FPMIN=tiny(x)/EPS
   Computes the exponential integral Ei(x) for x > 0.
   Parameters: MAXIT is the maximum number of iterations allowed; EPS is the relative error,
   or absolute error near the zero of Ei at x = 0.3725; FPMIN is a number near the smallest
   representable floating-point number; EULER (in nrtype) is Euler's constant γ.
INTEGER(I4B) :: k
REAL(SP) :: fact,prev,sm,term
call assert(x > 0.0, 'ei arg')
if (x < FPMIN) then                    Special case: avoid failure of convergence test
    ei=log(x)+EULER                       because of underflow.
else if (x <= -log(EPS)) then          Use power series.
    sm=0.0
    fact=1.0
    do k=1,MAXIT
        fact=fact*x/k
        term=fact/k
        sm=sm+term
        if (term < EPS*sm) exit
    end do
    if (k > MAXIT) call nrerror('series failed in ei')
    ei=sm+log(x)+EULER
else                                   Use asymptotic series.
    sm=0.0                             Start with second term.
    term=1.0
    do k=1,MAXIT
        prev=term
        term=term*k/x
        if (term < EPS) exit           Since final sum is greater than one, term itself
        if (term < prev) then             approximates the relative error.
            sm=sm+term                 Still converging: add new term.
        else                           Diverging: subtract previous term and exit.
            sm=sm-prev
            exit
        end if
    end do
    if (k > MAXIT) call nrerror('asymptotic failed in ei')
    ei=exp(x)*(1.0_sp+sm)/x
end if
END FUNCTION ei
```

ei does not readily parallelize, and we thus don't provide a vector version. For syntactic convenience you could make a vector version with a do-loop over calls to this scalar version; or, in Fortran 95, you can of course make the function ELEMENTAL.

⋆ ⋆ ⋆

```
FUNCTION betai_s(a,b,x)
USE nrtype; USE nrutil, ONLY : assert
USE nr, ONLY : betacf,gammln
IMPLICIT NONE
REAL(SP), INTENT(IN) :: a,b,x
REAL(SP) :: betai_s
   Returns the incomplete beta function I_x(a,b).
REAL(SP) :: bt
call assert(x >= 0.0, x <= 1.0, 'betai_s arg')
if (x == 0.0 .or. x == 1.0) then
    bt=0.0
else                                              Factors in front of the continued frac-
    bt=exp(gammln(a+b)-gammln(a)-gammln(b)&          tion.
        +a*log(x)+b*log(1.0_sp-x))
end if
if (x < (a+1.0_sp)/(a+b+2.0_sp)) then             Use continued fraction directly.
    betai_s=bt*betacf(a,b,x)/a
else                                              Use continued fraction after making the
    betai_s=1.0_sp-bt*betacf(b,a,1.0_sp-x)/b         symmetry transformation.
end if
END FUNCTION betai_s
```

```
FUNCTION betai_v(a,b,x)
USE nrtype; USE nrutil, ONLY : assert,assert_eq
USE nr, ONLY : betacf,gammln
IMPLICIT NONE
REAL(SP), DIMENSION(:), INTENT(IN) :: a,b,x
REAL(SP), DIMENSION(size(a)) :: betai_v
REAL(SP), DIMENSION(size(a)) :: bt
LOGICAL(LGT), DIMENSION(size(a)) :: mask
INTEGER(I4B) :: ndum
ndum=assert_eq(size(a),size(b),size(x),'betai_v')
call assert(all(x >= 0.0), all(x <= 1.0), 'betai_v arg')
where (x == 0.0 .or. x == 1.0)
    bt=0.0
elsewhere
    bt=exp(gammln(a+b)-gammln(a)-gammln(b)&
        +a*log(x)+b*log(1.0_sp-x))
end where
mask=(x < (a+1.0_sp)/(a+b+2.0_sp))
betai_v=bt*betacf(merge(a,b,mask),merge(b,a,mask),&
    merge(x,1.0_sp-x,mask))/merge(a,b,mask)
where (.not. mask) betai_v=1.0_sp-betai_v
END FUNCTION betai_v
```

f90 Compare the scalar

```
if (x < (a+1.0_sp)/(a+b+2.0_sp)) then
   betai_s=bt*betacf(a,b,x)/a
else
   betai_s=1.0_sp-bt*betacf(b,a,1.0_sp-x)/b
end if
```

with the vector

```
mask=(x < (a+1.0_sp)/(a+b+2.0_sp))
betai_v=bt*betacf(merge(a,b,mask),merge(b,a,mask),&
   merge(x,1.0_sp-x,mask))/merge(a,b,mask)
where (.not. mask) betai_v=1.0_sp-betai_v
```

Here `merge` is used (several times) to evaluate all the required components in a single call to the vectorized `betacf`, notwithstanding that some components require one pattern of arguments, some a different pattern.

```
FUNCTION betacf_s(a,b,x)
USE nrtype; USE nrutil, ONLY : nrerror
IMPLICIT NONE
REAL(SP), INTENT(IN) :: a,b,x
REAL(SP) :: betacf_s
INTEGER(I4B), PARAMETER :: MAXIT=100
REAL(SP), PARAMETER :: EPS=epsilon(x), FPMIN=tiny(x)/EPS
   Used by betai: Evaluates continued fraction for incomplete beta function by modified
   Lentz's method (§5.2).
REAL(SP) :: aa,c,d,del,h,qab,qam,qap
INTEGER(I4B) :: m,m2
qab=a+b                                    These q's will be used in factors that occur
qap=a+1.0_sp                                  in the coefficients (6.4.6).
qam=a-1.0_sp
c=1.0                                      First step of Lentz's method.
d=1.0_sp-qab*x/qap
if (abs(d) < FPMIN) d=FPMIN
d=1.0_sp/d
h=d
do m=1,MAXIT
   m2=2*m
   aa=m*(b-m)*x/((qam+m2)*(a+m2))
   d=1.0_sp+aa*d                           One step (the even one) of the recurrence.
   if (abs(d) < FPMIN) d=FPMIN
   c=1.0_sp+aa/c
   if (abs(c) < FPMIN) c=FPMIN
   d=1.0_sp/d
   h=h*d*c
   aa=-(a+m)*(qab+m)*x/((a+m2)*(qap+m2))
   d=1.0_sp+aa*d                           Next step of the recurrence (the odd one).
   if (abs(d) < FPMIN) d=FPMIN
   c=1.0_sp+aa/c
   if (abs(c) < FPMIN) c=FPMIN
   d=1.0_sp/d
   del=d*c
   h=h*del
   if (abs(del-1.0_sp) <= EPS) exit        Are we done?
end do
if (m > MAXIT)&
   call nrerror('a or b too big, or MAXIT too small in betacf_s')
betacf_s=h
END FUNCTION betacf_s
```

```
FUNCTION betacf_v(a,b,x)
USE nrtype; USE nrutil, ONLY : assert_eq,nrerror
IMPLICIT NONE
REAL(SP), DIMENSION(:), INTENT(IN) :: a,b,x
REAL(SP), DIMENSION(size(x)) :: betacf_v
INTEGER(I4B), PARAMETER :: MAXIT=100
REAL(SP), PARAMETER :: EPS=epsilon(x), FPMIN=tiny(x)/EPS
REAL(SP), DIMENSION(size(x)) :: aa,c,d,del,h,qab,qam,qap
LOGICAL(LGT), DIMENSION(size(x)) :: converged
INTEGER(I4B) :: m
INTEGER(I4B), DIMENSION(size(x)) :: m2
m=assert_eq(size(a),size(b),size(x),'betacf_v')
qab=a+b
qap=a+1.0_sp
qam=a-1.0_sp
c=1.0
d=1.0_sp-qab*x/qap
where (abs(d) < FPMIN) d=FPMIN
d=1.0_sp/d
h=d
converged=.false.
do m=1,MAXIT
    where (.not. converged)
        m2=2*m
        aa=m*(b-m)*x/((qam+m2)*(a+m2))
        d=1.0_sp+aa*d
        d=merge(FPMIN,d, abs(d)<FPMIN )
        c=1.0_sp+aa/c
        c=merge(FPMIN,c, abs(c)<FPMIN )
        d=1.0_sp/d
        h=h*d*c
        aa=-(a+m)*(qab+m)*x/((a+m2)*(qap+m2))
        d=1.0_sp+aa*d
        d=merge(FPMIN,d, abs(d)<FPMIN )
        c=1.0_sp+aa/c
        c=merge(FPMIN,c, abs(c)<FPMIN )
        d=1.0_sp/d
        del=d*c
        h=h*del
        converged = (abs(del-1.0_sp) <= EPS)
    end where
    if (all(converged)) exit
end do
if (m > MAXIT)&
    call nrerror('a or b too big, or MAXIT too small in betacf_v')
betacf_v=h
END FUNCTION betacf_v
```

f90 `d=merge(FPMIN,d, abs(d)<FPMIN )` The scalar version does this with an `if`. Why does it become a `merge` here in the vector version, rather than a `where`? Because we are already inside a "`where (.not.converged)`" block, and Fortran 90 doesn't allow nested `where`'s! (Fortran 95 *will* allow nested `where`'s.)

* * *

```
FUNCTION bessj0_s(x)
USE nrtype; USE nrutil, ONLY : poly
IMPLICIT NONE
REAL(SP), INTENT(IN) :: x
REAL(SP) :: bessj0_s
    Returns the Bessel function J0(x) for any real x.
REAL(SP) :: ax,xx,z
REAL(DP) :: y                       We'll accumulate polynomials in double precision.
REAL(DP), DIMENSION(5) :: p = (/1.0_dp,-0.1098628627e-2_dp,&
    0.2734510407e-4_dp,-0.2073370639e-5_dp,0.2093887211e-6_dp/)
REAL(DP), DIMENSION(5) :: q = (/-0.1562499995e-1_dp,&
    0.1430488765e-3_dp,-0.6911147651e-5_dp,0.7621095161e-6_dp,&
    -0.934945152e-7_dp/)
REAL(DP), DIMENSION(6) :: r = (/57568490574.0_dp,-13362590354.0_dp,&
    651619640.7_dp,-11214424.18_dp,77392.33017_dp,&
    -184.9052456_dp/)
REAL(DP), DIMENSION(6) :: s = (/57568490411.0_dp,1029532985.0_dp,&
    9494680.718_dp,59272.64853_dp,267.8532712_dp,1.0_dp/)
if (abs(x) < 8.0) then              Direct rational function fit.
    y=x**2
    bessj0_s=poly(y,r)/poly(y,s)
else                                Fitting function (6.5.9).
    ax=abs(x)
    z=8.0_sp/ax
    y=z**2
    xx=ax-0.785398164_sp
    bessj0_s=sqrt(0.636619772_sp/ax)*(cos(xx)*&
        poly(y,p)-z*sin(xx)*poly(y,q))
end if
END FUNCTION bessj0_s
```

```
FUNCTION bessj0_v(x)
USE nrtype; USE nrutil, ONLY : poly
IMPLICIT NONE
REAL(SP), DIMENSION(:), INTENT(IN) :: x
REAL(SP), DIMENSION(size(x)) :: bessj0_v
REAL(SP), DIMENSION(size(x)) :: ax,xx,z
REAL(DP), DIMENSION(size(x)) :: y
LOGICAL(LGT), DIMENSION(size(x)) :: mask
REAL(DP), DIMENSION(5) :: p = (/1.0_dp,-0.1098628627e-2_dp,&
    0.2734510407e-4_dp,-0.2073370639e-5_dp,0.2093887211e-6_dp/)
REAL(DP), DIMENSION(5) :: q = (/-0.1562499995e-1_dp,&
    0.1430488765e-3_dp,-0.6911147651e-5_dp,0.7621095161e-6_dp,&
    -0.934945152e-7_dp/)
REAL(DP), DIMENSION(6) :: r = (/57568490574.0_dp,-13362590354.0_dp,&
    651619640.7_dp,-11214424.18_dp,77392.33017_dp,&
    -184.9052456_dp/)
REAL(DP), DIMENSION(6) :: s = (/57568490411.0_dp,1029532985.0_dp,&
    9494680.718_dp,59272.64853_dp,267.8532712_dp,1.0_dp/)
mask = (abs(x) < 8.0)
where (mask)
    y=x**2
    bessj0_v=poly(y,r,mask)/poly(y,s,mask)
elsewhere
    ax=abs(x)
    z=8.0_sp/ax
    y=z**2
    xx=ax-0.785398164_sp
    bessj0_v=sqrt(0.636619772_sp/ax)*(cos(xx)*&
        poly(y,p,.not. mask)-z*sin(xx)*poly(y,q,.not. mask))
end where
END FUNCTION bessj0_v
```

`where (mask)...bessj0_v=poly(y,r,mask)/poly(y,s,mask)` Here we meet the *third* solution to the problem of getting masked values from an external vector function. (For the other two solutions, see notes to `factrl`, p. 1087, and `gammp`, p. 1090.) Here we simply evade all responsibility and pass the mask into every routine that is supposed to be masked. Let it be somebody else's problem! That works here because your hardworking authors have overloaded the `nrutil` routine `poly` with a masked vector version. More typically, of course, it becomes *your* problem, and you have to remember to write masked versions of all the vector routines that you call in this way. (We'll meet examples of this later.)

⋆ ⋆ ⋆

```
FUNCTION bessy0_s(x)
USE nrtype; USE nrutil, ONLY : assert,poly
USE nr, ONLY : bessj0
IMPLICIT NONE
REAL(SP), INTENT(IN) :: x
REAL(SP) :: bessy0_s
   Returns the Bessel function Y0(x) for positive x.
REAL(SP) :: xx,z
REAL(DP) :: y                     We'll accumulate polynomials in double precision.
REAL(DP), DIMENSION(5) :: p = (/1.0_dp,-0.1098628627e-2_dp,&
    0.2734510407e-4_dp,-0.2073370639e-5_dp,0.2093887211e-6_dp/)
REAL(DP), DIMENSION(5) :: q = (/-0.1562499995e-1_dp,&
    0.1430488765e-3_dp,-0.6911147651e-5_dp,0.7621095161e-6_dp,&
    -0.934945152e-7_dp/)
REAL(DP), DIMENSION(6) :: r = (/-2957821389.0_dp,7062834065.0_dp,&
    -512359803.6_dp,10879881.29_dp,-86327.92757_dp,&
    228.4622733_dp/)
REAL(DP), DIMENSION(6) :: s = (/40076544269.0_dp,745249964.8_dp,&
    7189466.438_dp,47447.26470_dp,226.1030244_dp,1.0_dp/)
call assert(x > 0.0, 'bessy0_s arg')
if (abs(x) < 8.0) then            Rational function approximation of (6.5.8).
    y=x**2
    bessy0_s=(poly(y,r)/poly(y,s))+&
        0.636619772_sp*bessj0(x)*log(x)
else                              Fitting function (6.5.10).
    z=8.0_sp/x
    y=z**2
    xx=x-0.785398164_sp
    bessy0_s=sqrt(0.636619772_sp/x)*(sin(xx)*&
        poly(y,p)+z*cos(xx)*poly(y,q))
end if
END FUNCTION bessy0_s
```

```
FUNCTION bessy0_v(x)
USE nrtype; USE nrutil, ONLY : assert,poly
USE nr, ONLY : bessj0
IMPLICIT NONE
REAL(SP), DIMENSION(:), INTENT(IN) :: x
REAL(SP), DIMENSION(size(x)) :: bessy0_v
REAL(SP), DIMENSION(size(x)) :: xx,z
REAL(DP), DIMENSION(size(x)) :: y
LOGICAL(LGT), DIMENSION(size(x)) :: mask
REAL(DP), DIMENSION(5) :: p = (/1.0_dp,-0.1098628627e-2_dp,&
    0.2734510407e-4_dp,-0.2073370639e-5_dp,0.2093887211e-6_dp/)
REAL(DP), DIMENSION(5) :: q = (/-0.1562499995e-1_dp,&
    0.1430488765e-3_dp,-0.6911147651e-5_dp,0.7621095161e-6_dp,&
```

```
    -0.934945152e-7_dp/)
REAL(DP), DIMENSION(6) :: r = (/-2957821389.0_dp,7062834065.0_dp,&
    -512359803.6_dp,10879881.29_dp,-86327.92757_dp,&
    228.4622733_dp/)
REAL(DP), DIMENSION(6) :: s = (/40076544269.0_dp,745249964.8_dp,&
    7189466.438_dp,47447.26470_dp,226.1030244_dp,1.0_dp/)
call assert(all(x > 0.0), 'bessy0_v arg')
mask = (abs(x) < 8.0)
where (mask)
    y=x**2
    bessy0_v=(poly(y,r,mask)/poly(y,s,mask))+&
        0.636619772_sp*bessj0(x)*log(x)
elsewhere
    z=8.0_sp/x
    y=z**2
    xx=x-0.785398164_sp
    bessy0_v=sqrt(0.636619772_sp/x)*(sin(xx)*&
        poly(y,p,.not. mask)+z*cos(xx)*poly(y,q,.not. mask))
end where
END FUNCTION bessy0_v
```

⋆ ⋆ ⋆

```
FUNCTION bessj1_s(x)
USE nrtype; USE nrutil, ONLY : poly
IMPLICIT NONE
REAL(SP), INTENT(IN) :: x
REAL(SP) :: bessj1_s
   Returns the Bessel function J_1(x) for any real x.
REAL(SP) :: ax,xx,z
REAL(DP) :: y                         We'll accumulate polynomials in double precision.
REAL(DP), DIMENSION(6) :: r = (/72362614232.0_dp,&
    -7895059235.0_dp,242396853.1_dp,-2972611.439_dp,&
    15704.48260_dp,-30.16036606_dp/)
REAL(DP), DIMENSION(6) :: s = (/144725228442.0_dp,2300535178.0_dp,&
    18583304.74_dp,99447.43394_dp,376.9991397_dp,1.0_dp/)
REAL(DP), DIMENSION(5) :: p = (/1.0_dp,0.183105e-2_dp,&
    -0.3516396496e-4_dp,0.2457520174e-5_dp,-0.240337019e-6_dp/)
REAL(DP), DIMENSION(5) :: q = (/0.04687499995_dp,&
    -0.2002690873e-3_dp,0.8449199096e-5_dp,-0.88228987e-6_dp,&
    0.105787412e-6_dp/)
if (abs(x) < 8.0) then                Direct rational approximation.
    y=x**2
    bessj1_s=x*(poly(y,r)/poly(y,s))
else                                  Fitting function (6.5.9).
    ax=abs(x)
    z=8.0_sp/ax
    y=z**2
    xx=ax-2.356194491_sp
    bessj1_s=sqrt(0.636619772_sp/ax)*(cos(xx)*&
        poly(y,p)-z*sin(xx)*poly(y,q))*sign(1.0_sp,x)
end if
END FUNCTION bessj1_s
```

```
FUNCTION bessj1_v(x)
USE nrtype; USE nrutil, ONLY : poly
IMPLICIT NONE
REAL(SP), DIMENSION(:), INTENT(IN) :: x
REAL(SP), DIMENSION(size(x)) :: bessj1_v
REAL(SP), DIMENSION(size(x)) :: ax,xx,z
REAL(DP), DIMENSION(size(x)) :: y
LOGICAL(LGT), DIMENSION(size(x)) :: mask
REAL(DP), DIMENSION(6) :: r = (/72362614232.0_dp,&
    -7895059235.0_dp,242396853.1_dp,-2972611.439_dp,&
    15704.48260_dp,-30.16036606_dp/)
REAL(DP), DIMENSION(6) :: s = (/144725228442.0_dp,2300535178.0_dp,&
    18583304.74_dp,99447.43394_dp,376.9991397_dp,1.0_dp/)
REAL(DP), DIMENSION(5) :: p = (/1.0_dp,0.183105e-2_dp,&
    -0.3516396496e-4_dp,0.2457520174e-5_dp,-0.240337019e-6_dp/)
REAL(DP), DIMENSION(5) :: q = (/0.04687499995_dp,&
    -0.2002690873e-3_dp,0.8449199096e-5_dp,-0.88228987e-6_dp,&
    0.105787412e-6_dp/)
mask = (abs(x) < 8.0)
where (mask)
    y=x**2
    bessj1_v=x*(poly(y,r,mask)/poly(y,s,mask))
elsewhere
    ax=abs(x)
    z=8.0_sp/ax
    y=z**2
    xx=ax-2.356194491_sp
    bessj1_v=sqrt(0.636619772_sp/ax)*(cos(xx)*&
        poly(y,p,.not. mask)-z*sin(xx)*poly(y,q,.not. mask))*&
        sign(1.0_sp,x)
end where
END FUNCTION bessj1_v
```

⋆ ⋆ ⋆

```
FUNCTION bessy1_s(x)
USE nrtype; USE nrutil, ONLY : assert,poly
USE nr, ONLY : bessj1
IMPLICIT NONE
REAL(SP), INTENT(IN) :: x
REAL(SP) :: bessy1_s
    Returns the Bessel function Y1(x) for positive x.
REAL(SP) :: xx,z
REAL(DP) :: y                    We'll accumulate polynomials in double precision.
REAL(DP), DIMENSION(5) :: p = (/1.0_dp,0.183105e-2_dp,&
    -0.3516396496e-4_dp,0.2457520174e-5_dp,-0.240337019e-6_dp/)
REAL(DP), DIMENSION(5) :: q = (/0.04687499995_dp,&
    -0.2002690873e-3_dp,0.8449199096e-5_dp,-0.88228987e-6_dp,&
    0.105787412e-6_dp/)
REAL(DP), DIMENSION(6) :: r = (/-0.4900604943e13_dp,&
    0.1275274390e13_dp,-0.5153438139e11_dp,0.7349264551e9_dp,&
    -0.4237922726e7_dp,0.8511937935e4_dp/)
REAL(DP), DIMENSION(7) :: s = (/0.2499580570e14_dp,&
    0.4244419664e12_dp,0.3733650367e10_dp,0.2245904002e8_dp,&
    0.1020426050e6_dp,0.3549632885e3_dp,1.0_dp/)
call assert(x > 0.0, 'bessy1_s arg')
if (abs(x) < 8.0) then           Rational function approximation of (6.5.8).
    y=x**2
    bessy1_s=x*(poly(y,r)/poly(y,s))+&
        0.636619772_sp*(bessj1(x)*log(x)-1.0_sp/x)
else                             Fitting function (6.5.10).
```

```
        z=8.0_sp/x
        y=z**2
        xx=x-2.356194491_sp
        bessy1_s=sqrt(0.636619772_sp/x)*(sin(xx)*&
            poly(y,p)+z*cos(xx)*poly(y,q))
    end if
    END FUNCTION bessy1_s


    FUNCTION bessy1_v(x)
    USE nrtype; USE nrutil, ONLY : assert,poly
    USE nr, ONLY : bessj1
    IMPLICIT NONE
    REAL(SP), DIMENSION(:), INTENT(IN) :: x
    REAL(SP), DIMENSION(size(x)) :: bessy1_v
    REAL(SP), DIMENSION(size(x)) :: xx,z
    REAL(DP), DIMENSION(size(x)) :: y
    LOGICAL(LGT), DIMENSION(size(x)) :: mask
    REAL(DP), DIMENSION(5) :: p = (/1.0_dp,0.183105e-2_dp,&
        -0.3516396496e-4_dp,0.2457520174e-5_dp,-0.240337019e-6_dp/)
    REAL(DP), DIMENSION(5) :: q = (/0.04687499995_dp,&
        -0.2002690873e-3_dp,0.8449199096e-5_dp,-0.88228987e-6_dp,&
        0.105787412e-6_dp/)
    REAL(DP), DIMENSION(6) :: r = (/-0.4900604943e13_dp,&
        0.1275274390e13_dp,-0.5153438139e11_dp,0.7349264551e9_dp,&
        -0.4237922726e7_dp,0.8511937935e4_dp/)
    REAL(DP), DIMENSION(7) :: s = (/0.2499580570e14_dp,&
        0.4244419664e12_dp,0.3733650367e10_dp,0.2245904002e8_dp,&
        0.1020426050e6_dp,0.3549632885e3_dp,1.0_dp/)
    call assert(all(x > 0.0), 'bessy1_v arg')
    mask = (abs(x) < 8.0)
    where (mask)
        y=x**2
        bessy1_v=x*(poly(y,r,mask)/poly(y,s,mask))+&
            0.636619772_sp*(bessj1(x)*log(x)-1.0_sp/x)
    elsewhere
        z=8.0_sp/x
        y=z**2
        xx=x-2.356194491_sp
        bessy1_v=sqrt(0.636619772_sp/x)*(sin(xx)*&
            poly(y,p,.not. mask)+z*cos(xx)*poly(y,q,.not. mask))
    end where
    END FUNCTION bessy1_v
```

⋆ ⋆ ⋆

```
    FUNCTION bessy_s(n,x)
    USE nrtype; USE nrutil, ONLY : assert
    USE nr, ONLY : bessy0,bessy1
    IMPLICIT NONE
    INTEGER(I4B), INTENT(IN) :: n
    REAL(SP), INTENT(IN) :: x
    REAL(SP) :: bessy_s
```

Returns the Bessel function $Y_n(x)$ for positive x and $n \geq 2$.

```
    INTEGER(I4B) :: j
    REAL(SP) :: by,bym,byp,tox
    call assert(n >= 2, x > 0.0, 'bessy_s args')
    tox=2.0_sp/x
    by=bessy1(x)                    Starting values for the recurrence.
    bym=bessy0(x)
    do j=1,n-1                      Recurrence (6.5.7).
```

```
        byp=j*tox*by-bym
        bym=by
        by=byp
    end do
    bessy_s=by
    END FUNCTION bessy_s
```

```
    FUNCTION bessy_v(n,x)
    USE nrtype; USE nrutil, ONLY : assert
    USE nr, ONLY : bessy0,bessy1
    IMPLICIT NONE
    INTEGER(I4B), INTENT(IN) :: n
    REAL(SP), DIMENSION(:), INTENT(IN) :: x
    REAL(SP), DIMENSION(size(x)) :: bessy_v
    INTEGER(I4B) :: j
    REAL(SP), DIMENSION(size(x)) :: by,bym,byp,tox
    call assert(n >= 2, all(x > 0.0), 'bessy_v args')
    tox=2.0_sp/x
    by=bessy1(x)
    bym=bessy0(x)
    do j=1,n-1
        byp=j*tox*by-bym
        bym=by
        by=byp
    end do
    bessy_v=by
    END FUNCTION bessy_v
```

f90 Notice that the vector routine is *exactly* the same as the scalar routine, but operates only on vectors, and that nothing in the routine is specific to any level of precision or kind type of real variable. Cases like this make us wish that Fortran 90 provided for "template" types that could automatically take the type and shape of the actual arguments. (Such facilities are available in other, more object-oriented languages such as C++.)

⋆ ⋆ ⋆

```
    FUNCTION bessj_s(n,x)
    USE nrtype; USE nrutil, ONLY : assert
    USE nr, ONLY : bessj0,bessj1
    IMPLICIT NONE
    INTEGER(I4B), INTENT(IN) :: n
    REAL(SP), INTENT(IN) :: x
    REAL(SP) :: bessj_s
    INTEGER(I4B), PARAMETER :: IACC=40,IEXP=maxexponent(x)/2
        Returns the Bessel function Jn(x) for any real x and n >= 2. Make the parameter IACC
        larger to increase accuracy.
    INTEGER(I4B) :: j,jsum,m
    REAL(SP) :: ax,bj,bjm,bjp,summ,tox
    call assert(n >= 2, 'bessj_s args')
    ax=abs(x)
    if (ax*ax <= 8.0_sp*tiny(x)) then                  Underflow limit.
        bessj_s=0.0
    else if (ax > real(n,sp)) then                     Upwards recurrence from J0 and J1.
        tox=2.0_sp/ax
        bjm=bessj0(ax)
        bj=bessj1(ax)
        do j=1,n-1
```

```
        bjp=j*tox*bj-bjm
        bjm=bj
        bj=bjp
    end do
    bessj_s=bj
else                                        Downwards recurrence from an even m
    tox=2.0_sp/ax                              here computed.
    m=2*((n+int(sqrt(real(IACC*n,sp))))/2)
    bessj_s=0.0
    jsum=0                                  jsum will alternate between 0 and 1; when
    summ=0.0                                   it is 1, we accumulate in sum the
    bjp=0.0                                    even terms in (5.5.16).
    bj=1.0
    do j=m,1,-1                             The downward recurrence.
        bjm=j*tox*bj-bjp
        bjp=bj
        bj=bjm
        if (exponent(bj) > IEXP) then       Renormalize to prevent overflows.
            bj=scale(bj,-IEXP)
            bjp=scale(bjp,-IEXP)
            bessj_s=scale(bessj_s,-IEXP)
            summ=scale(summ,-IEXP)
        end if
        if (jsum /= 0) summ=summ+bj         Accumulate the sum.
        jsum=1-jsum                         Change 0 to 1 or vice versa.
        if (j == n) bessj_s=bjp             Save the unnormalized answer.
    end do
    summ=2.0_sp*summ-bj                     Compute (5.5.16)
    bessj_s=bessj_s/summ                    and use it to normalize the answer.
end if
if (x < 0.0 .and. mod(n,2) == 1) bessj_s=-bessj_s
END FUNCTION bessj_s
```

The `bessj` routine does not conveniently parallelize with Fortran 90's language constructions, but Bessel functions are of sufficient importance that we feel the need for a parallel version nevertheless. The basic method adopted below is to encapsulate as contained vector functions two separate algorithms, one for the case $x \leq n$, the other for $x > n$. Both of these have masks as input arguments; within each routine, however, they immediately revert to the pack-unpack method. The choice to pack in the subsidiary routines, rather than in the main routine, is arbitrary; the main routine is supposed to be a little clearer this way.

f90 `if (exponent(bj) > IEXP) then...` In the Fortran 77 version of this routine, we scaled the variables by 10^{-10} whenever `bj` was bigger than 10^{10}. On a machine with a large exponent range, we could improve efficiency by scaling less often. In order to remain portable, however, we used the conservative value of 10^{10}. An elegant way of handling renormalization is provided by the Fortran 90 intrinsic functions that manipulate real numbers. We test with `if (exponent(bj) > IEXP)` and then if necessary renormalize with `bj=scale(bj,-IEXP)` and similarly for the other variables. Our conservative choice is to set `IEXP=maxexponent(x)/2`. Note that an added benefit of scaling this way is that only the exponent of each variable is modified; no roundoff error is introduced as it can be if we do a floating-point division instead.

```
FUNCTION bessj_v(n,xx)
USE nrtype; USE nrutil, ONLY : assert
USE nr, ONLY : bessj0,bessj1
IMPLICIT NONE
INTEGER(I4B), INTENT(IN) :: n
REAL(SP), DIMENSION(:), INTENT(IN) :: xx
REAL(SP), DIMENSION(size(xx)) :: bessj_v
INTEGER(I4B), PARAMETER :: IACC=40,IEXP=maxexponent(xx)/2
REAL(SP), DIMENSION(size(xx)) :: ax
LOGICAL(LGT), DIMENSION(size(xx)) :: mask,mask0
REAL(SP), DIMENSION(:), ALLOCATABLE :: x,bj,bjm,bjp,summ,tox,bessjle
LOGICAL(LGT), DIMENSION(:), ALLOCATABLE :: renorm
INTEGER(I4B) :: j,jsum,m,npak
call assert(n >= 2, 'bessj_v args')
ax=abs(xx)
mask = (ax <= real(n,sp))
mask0 = (ax*ax <= 8.0_sp*tiny(xx))
bessj_v=bessjle_v(n,ax,logical(mask .and. .not.mask0, kind=lgt))
bessj_v=merge(bessjgt_v(n,ax,.not. mask),bessj_v,.not. mask)
where (mask0) bessj_v=0.0
where (xx < 0.0 .and. mod(n,2) == 1) bessj_v=-bessj_v
CONTAINS

FUNCTION bessjgt_v(n,xx,mask)
IMPLICIT NONE
INTEGER(I4B), INTENT(IN) :: n
REAL(SP), DIMENSION(:), INTENT(IN) :: xx
LOGICAL(LGT), DIMENSION(size(xx)), INTENT(IN) :: mask
REAL(SP), DIMENSION(size(xx)) :: bessjgt_v
npak=count(mask)
if (npak == 0) RETURN
allocate(x(npak),bj(npak),bjm(npak),bjp(npak),tox(npak))
x=pack(xx,mask)
tox=2.0_sp/x
bjm=bessj0(x)
bj=bessj1(x)
do j=1,n-1
    bjp=j*tox*bj-bjm
    bjm=bj
    bj=bjp
end do
bessjgt_v=unpack(bj,mask,0.0_sp)
deallocate(x,bj,bjm,bjp,tox)
END FUNCTION bessjgt_v

FUNCTION bessjle_v(n,xx,mask)
IMPLICIT NONE
INTEGER(I4B), INTENT(IN) :: n
REAL(SP), DIMENSION(:), INTENT(IN) :: xx
LOGICAL(LGT), DIMENSION(size(xx)), INTENT(IN) :: mask
REAL(SP), DIMENSION(size(xx)) :: bessjle_v
npak=count(mask)
if (npak == 0) RETURN
allocate(x(npak),bj(npak),bjm(npak),bjp(npak),summ(npak), &
    bessjle(npak),tox(npak),renorm(npak))
x=pack(xx,mask)
tox=2.0_sp/x
m=2*((n+int(sqrt(real(IACC*n,sp))))/2)
bessjle=0.0
jsum=0
summ=0.0
bjp=0.0
bj=1.0
do j=m,1,-1
    bjm=j*tox*bj-bjp
```

```
        bjp=bj
        bj=bjm
        renorm = (exponent(bj)>IEXP)
        bj=merge(scale(bj,-IEXP),bj,renorm)
        bjp=merge(scale(bjp,-IEXP),bjp,renorm)
        bessjle=merge(scale(bessjle,-IEXP),bessjle,renorm)
        summ=merge(scale(summ,-IEXP),summ,renorm)
        if (jsum /= 0) summ=summ+bj
        jsum=1-jsum
        if (j == n) bessjle=bjp
    end do
    summ=2.0_sp*summ-bj
    bessjle=bessjle/summ
    bessjle_v=unpack(bessjle,mask,0.0_sp)
    deallocate(x,bj,bjm,bjp,summ,bessjle,tox,renorm)
    END FUNCTION bessjle_v
    END FUNCTION bessj_v
```

f90 `bessj_v=... bessj_v=merge(bessjgt_v(...),bessj_v,...)` The vector `bessj_v` is set once (with a mask) and then merged with *itself*, along with the vector result of the `bessjgt_v` call. Thus are the two evaluation methods combined. (A third case, where an argument is zero, is then handled by an immediately following `where`.)

* * *

```
FUNCTION bessi0_s(x)
USE nrtype; USE nrutil, ONLY : poly
IMPLICIT NONE
REAL(SP), INTENT(IN) :: x
REAL(SP) :: bessi0_s
    Returns the modified Bessel function I0(x) for any real x.
REAL(SP) :: ax
REAL(DP), DIMENSION(7) :: p = (/1.0_dp,3.5156229_dp,&
    3.0899424_dp,1.2067492_dp,0.2659732_dp,0.360768e-1_dp,&
    0.45813e-2_dp/)             Accumulate polynomials in double precision.
REAL(DP), DIMENSION(9) :: q = (/0.39894228_dp,0.1328592e-1_dp,&
    0.225319e-2_dp,-0.157565e-2_dp,0.916281e-2_dp,&
    -0.2057706e-1_dp,0.2635537e-1_dp,-0.1647633e-1_dp,&
    0.392377e-2_dp/)
ax=abs(x)
if (ax < 3.75) then             Polynomial fit.
    bessi0_s=poly(real((x/3.75_sp)**2,dp),p)
else
    bessi0_s=(exp(ax)/sqrt(ax))*poly(real(3.75_sp/ax,dp),q)
end if
END FUNCTION bessi0_s
```

```
FUNCTION bessi0_v(x)
USE nrtype; USE nrutil, ONLY : poly
IMPLICIT NONE
REAL(SP), DIMENSION(:), INTENT(IN) :: x
REAL(SP), DIMENSION(size(x)) :: bessi0_v
REAL(SP), DIMENSION(size(x)) :: ax
REAL(DP), DIMENSION(size(x)) :: y
LOGICAL(LGT), DIMENSION(size(x)) :: mask
REAL(DP), DIMENSION(7) :: p = (/1.0_dp,3.5156229_dp,&
    3.0899424_dp,1.2067492_dp,0.2659732_dp,0.360768e-1_dp,&
```

```
    0.45813e-2_dp/)
REAL(DP), DIMENSION(9) :: q = (/0.39894228_dp,0.1328592e-1_dp,&
    0.225319e-2_dp,-0.157565e-2_dp,0.916281e-2_dp,&
    -0.2057706e-1_dp,0.2635537e-1_dp,-0.1647633e-1_dp,&
    0.392377e-2_dp/)
ax=abs(x)
mask = (ax < 3.75)
where (mask)
    bessi0_v=poly(real((x/3.75_sp)**2,dp),p,mask)
elsewhere
    y=3.75_sp/ax
    bessi0_v=(exp(ax)/sqrt(ax))*poly(real(y,dp),q,.not. mask)
end where
END FUNCTION bessi0_v
```

⋆ ⋆ ⋆

```
FUNCTION bessk0_s(x)
USE nrtype; USE nrutil, ONLY : assert,poly
USE nr, ONLY : bessi0
IMPLICIT NONE
REAL(SP), INTENT(IN) :: x
REAL(SP) :: bessk0_s
```

Returns the modified Bessel function $K_0(x)$ for positive real x.

```
REAL(DP) :: y
```

Accumulate polynomials in double precision.

```
REAL(DP), DIMENSION(7) :: p = (/-0.57721566_dp,0.42278420_dp,&
    0.23069756_dp,0.3488590e-1_dp,0.262698e-2_dp,0.10750e-3_dp,&
    0.74e-5_dp/)
REAL(DP), DIMENSION(7) :: q = (/1.25331414_dp,-0.7832358e-1_dp,&
    0.2189568e-1_dp,-0.1062446e-1_dp,0.587872e-2_dp,&
    -0.251540e-2_dp,0.53208e-3_dp/)
call assert(x > 0.0, 'bessk0_s arg')
if (x <= 2.0) then
```

Polynomial fit.

```
    y=x*x/4.0_sp
    bessk0_s=(-log(x/2.0_sp)*bessi0(x))+poly(y,p)
else
    y=(2.0_sp/x)
    bessk0_s=(exp(-x)/sqrt(x))*poly(y,q)
end if
END FUNCTION bessk0_s
```

```
FUNCTION bessk0_v(x)
USE nrtype; USE nrutil, ONLY : assert,poly
USE nr, ONLY : bessi0
IMPLICIT NONE
REAL(SP), DIMENSION(:), INTENT(IN) :: x
REAL(SP), DIMENSION(size(x)) :: bessk0_v
REAL(DP), DIMENSION(size(x)) :: y
LOGICAL(LGT), DIMENSION(size(x)) :: mask
REAL(DP), DIMENSION(7) :: p = (/-0.57721566_dp,0.42278420_dp,&
    0.23069756_dp,0.3488590e-1_dp,0.262698e-2_dp,0.10750e-3_dp,&
    0.74e-5_dp/)
REAL(DP), DIMENSION(7) :: q = (/1.25331414_dp,-0.7832358e-1_dp,&
    0.2189568e-1_dp,-0.1062446e-1_dp,0.587872e-2_dp,&
    -0.251540e-2_dp,0.53208e-3_dp/)
call assert(all(x > 0.0), 'bessk0_v arg')
mask = (x <= 2.0)
where (mask)
    y=x*x/4.0_sp
    bessk0_v=(-log(x/2.0_sp)*bessi0(x))+poly(y,p,mask)
```

```
elsewhere
    y=(2.0_sp/x)
    bessk0_v=(exp(-x)/sqrt(x))*poly(y,q,.not. mask)
end where
END FUNCTION bessk0_v
```

⋆ ⋆ ⋆

```
FUNCTION bessi1_s(x)
USE nrtype; USE nrutil, ONLY : poly
IMPLICIT NONE
REAL(SP), INTENT(IN) :: x
REAL(SP) :: bessi1_s
    Returns the modified Bessel function I1(x) for any real x.
REAL(SP) :: ax
REAL(DP), DIMENSION(7) :: p = (/0.5_dp,0.87890594_dp,&
    0.51498869_dp,0.15084934_dp,0.2658733e-1_dp,&
    0.301532e-2_dp,0.32411e-3_dp/)
      Accumulate polynomials in double precision.
REAL(DP), DIMENSION(9) :: q = (/0.39894228_dp,-0.3988024e-1_dp,&
    -0.362018e-2_dp,0.163801e-2_dp,-0.1031555e-1_dp,&
    0.2282967e-1_dp,-0.2895312e-1_dp,0.1787654e-1_dp,&
    -0.420059e-2_dp/)
ax=abs(x)
if (ax < 3.75) then            Polynomial fit.
    bessi1_s=ax*poly(real((x/3.75_sp)**2,dp),p)
else
    bessi1_s=(exp(ax)/sqrt(ax))*poly(real(3.75_sp/ax,dp),q)
end if
if (x < 0.0) bessi1_s=-bessi1_s
END FUNCTION bessi1_s
```

```
FUNCTION bessi1_v(x)
USE nrtype; USE nrutil, ONLY : poly
IMPLICIT NONE
REAL(SP), DIMENSION(:), INTENT(IN) :: x
REAL(SP), DIMENSION(size(x)) :: bessi1_v
REAL(SP), DIMENSION(size(x)) :: ax
REAL(DP), DIMENSION(size(x)) :: y
LOGICAL(LGT), DIMENSION(size(x)) :: mask
REAL(DP), DIMENSION(7) :: p = (/0.5_dp,0.87890594_dp,&
    0.51498869_dp,0.15084934_dp,0.2658733e-1_dp,&
    0.301532e-2_dp,0.32411e-3_dp/)
REAL(DP), DIMENSION(9) :: q = (/0.39894228_dp,-0.3988024e-1_dp,&
    -0.362018e-2_dp,0.163801e-2_dp,-0.1031555e-1_dp,&
    0.2282967e-1_dp,-0.2895312e-1_dp,0.1787654e-1_dp,&
    -0.420059e-2_dp/)
ax=abs(x)
mask = (ax < 3.75)
where (mask)
    bessi1_v=ax*poly(real((x/3.75_sp)**2,dp),p,mask)
elsewhere
    y=3.75_sp/ax
    bessi1_v=(exp(ax)/sqrt(ax))*poly(real(y,dp),q,.not. mask)
end where
where (x < 0.0) bessi1_v=-bessi1_v
END FUNCTION bessi1_v
```

* * *

```
FUNCTION bessk1_s(x)
USE nrtype; USE nrutil, ONLY : assert,poly
USE nr, ONLY : bessi1
IMPLICIT NONE
REAL(SP), INTENT(IN) :: x
REAL(SP) :: bessk1_s
   Returns the modified Bessel function K_1(x) for positive real x.
REAL(DP) :: y                     Accumulate polynomials in double precision.
REAL(DP), DIMENSION(7) :: p = (/1.0_dp,0.15443144_dp,&
    -0.67278579_dp,-0.18156897_dp,-0.1919402e-1_dp,&
    -0.110404e-2_dp,-0.4686e-4_dp/)
REAL(DP), DIMENSION(7) :: q = (/1.25331414_dp,0.23498619_dp,&
    -0.3655620e-1_dp,0.1504268e-1_dp,-0.780353e-2_dp,&
    0.325614e-2_dp,-0.68245e-3_dp/)
call assert(x > 0.0, 'bessk1_s arg')
if (x <= 2.0) then               Polynomial fit.
    y=x*x/4.0_sp
    bessk1_s=(log(x/2.0_sp)*bessi1(x))+(1.0_sp/x)*poly(y,p)
else
    y=2.0_sp/x
    bessk1_s=(exp(-x)/sqrt(x))*poly(y,q)
end if
END FUNCTION bessk1_s
```

```
FUNCTION bessk1_v(x)
USE nrtype; USE nrutil, ONLY : assert,poly
USE nr, ONLY : bessi1
IMPLICIT NONE
REAL(SP), DIMENSION(:), INTENT(IN) :: x
REAL(SP), DIMENSION(size(x)) :: bessk1_v
REAL(DP), DIMENSION(size(x)) :: y
LOGICAL(LGT), DIMENSION(size(x)) :: mask
REAL(DP), DIMENSION(7) :: p = (/1.0_dp,0.15443144_dp,&
    -0.67278579_dp,-0.18156897_dp,-0.1919402e-1_dp,&
    -0.110404e-2_dp,-0.4686e-4_dp/)
REAL(DP), DIMENSION(7) :: q = (/1.25331414_dp,0.23498619_dp,&
    -0.3655620e-1_dp,0.1504268e-1_dp,-0.780353e-2_dp,&
    0.325614e-2_dp,-0.68245e-3_dp/)
call assert(all(x > 0.0), 'bessk1_v arg')
mask = (x <= 2.0)
where (mask)
    y=x*x/4.0_sp
    bessk1_v=(log(x/2.0_sp)*bessi1(x))+(1.0_sp/x)*poly(y,p,mask)
elsewhere
    y=2.0_sp/x
    bessk1_v=(exp(-x)/sqrt(x))*poly(y,q,.not. mask)
end where
END FUNCTION bessk1_v
```

* * *

```
FUNCTION bessk_s(n,x)
USE nrtype; USE nrutil, ONLY : assert
USE nr, ONLY : bessk0,bessk1
IMPLICIT NONE
INTEGER(I4B), INTENT(IN) :: n
REAL(SP), INTENT(IN) :: x
REAL(SP) :: bessk_s
    Returns the modified Bessel function Kn(x) for positive x and n >= 2.
INTEGER(I4B) :: j
REAL(SP) :: bk,bkm,bkp,tox
call assert(n >= 2, x > 0.0, 'bessk_s args')
tox=2.0_sp/x
bkm=bessk0(x)                    Upward recurrence for all x...
bk=bessk1(x)
do j=1,n-1                       ...and here it is.
    bkp=bkm+j*tox*bk
    bkm=bk
    bk=bkp
end do
bessk_s=bk
END FUNCTION bessk_s
```

```
FUNCTION bessk_v(n,x)
USE nrtype; USE nrutil, ONLY : assert
USE nr, ONLY : bessk0,bessk1
IMPLICIT NONE
INTEGER(I4B), INTENT(IN) :: n
REAL(SP), DIMENSION(:), INTENT(IN) :: x
REAL(SP), DIMENSION(size(x)) :: bessk_v
INTEGER(I4B) :: j
REAL(SP), DIMENSION(size(x)) :: bk,bkm,bkp,tox
call assert(n >= 2, all(x > 0.0), 'bessk_v args')
tox=2.0_sp/x
bkm=bessk0(x)
bk=bessk1(x)
do j=1,n-1
    bkp=bkm+j*tox*bk
    bkm=bk
    bk=bkp
end do
bessk_v=bk
END FUNCTION bessk_v
```

The scalar and vector versions of bessk are identical, and have no precision-specific constants, another example of where we would like to define a generic "template" function if the language had this facility.

⋆ ⋆ ⋆

```
FUNCTION bessi_s(n,x)
USE nrtype; USE nrutil, ONLY : assert
USE nr, ONLY : bessi0
IMPLICIT NONE
INTEGER(I4B), INTENT(IN) :: n
REAL(SP), INTENT(IN) :: x
REAL(SP) :: bessi_s
INTEGER(I4B), PARAMETER :: IACC=40,IEXP=maxexponent(x)/2
   Returns the modified Bessel function In(x) for any real x and n >= 2. Make the parameter
   IACC larger to increase accuracy.
INTEGER(I4B) :: j,m
REAL(SP) :: bi,bim,bip,tox
call assert(n >= 2, 'bessi_s args')
bessi_s=0.0
if (x*x <= 8.0_sp*tiny(x)) RETURN              Underflow limit.
tox=2.0_sp/abs(x)
bip=0.0
bi=1.0
m=2*((n+int(sqrt(real(IACC*n,sp)))))           Downward recurrence from even m.
do j=m,1,-1
    bim=bip+j*tox*bi                           The downward recurrence.
    bip=bi
    bi=bim
    if (exponent(bi) > IEXP) then              Renormalize to prevent overflows.
        bessi_s=scale(bessi_s,-IEXP)
        bi=scale(bi,-IEXP)
        bip=scale(bip,-IEXP)
    end if
    if (j == n) bessi_s=bip
end do
bessi_s=bessi_s*bessi0(x)/bi                   Normalize with bessi0.
if (x < 0.0 .and. mod(n,2) == 1) bessi_s=-bessi_s
END FUNCTION bessi_s
```

f90 `if (exponent(bi) > IEXP) then` See discussion of scaling for `bessj` on p. 1107.

```
FUNCTION bessi_v(n,x)
USE nrtype; USE nrutil, ONLY : assert
USE nr, ONLY : bessi0
IMPLICIT NONE
INTEGER(I4B), INTENT(IN) :: n
REAL(SP), DIMENSION(:), INTENT(IN) :: x
REAL(SP), DIMENSION(size(x)) :: bessi_v
INTEGER(I4B), PARAMETER :: IACC=40,IEXP=maxexponent(x)/2
INTEGER(I4B) :: j,m
REAL(SP), DIMENSION(size(x)) :: bi,bim,bip,tox
LOGICAL(LGT), DIMENSION(size(x)) :: mask
call assert(n >= 2, 'bessi_v args')
bessi_v=0.0
mask = (x <= 8.0_sp*tiny(x))
tox=2.0_sp/merge(2.0_sp,abs(x),mask)
bip=0.0
bi=1.0_sp
m=2*((n+int(sqrt(real(IACC*n,sp)))))
do j=m,1,-1
    bim=bip+j*tox*bi
    bip=bi
    bi=bim
    where (exponent(bi) > IEXP)
        bessi_v=scale(bessi_v,-IEXP)
```

```
        bi=scale(bi,-IEXP)
        bip=scale(bip,-IEXP)
    end where
    if (j == n) bessi_v=bip
end do
bessi_v=bessi_v*bessi0(x)/bi
where (mask) bessi_v=0.0_sp
where (x < 0.0 .and. mod(n,2) == 1) bessi_v=-bessi_v
END FUNCTION bessi_v
```

```
mask = (x == 0.0)
tox=2.0_sp/merge(2.0_sp,abs(x),mask)
```

For the special case $x = 0$, the value of the returned function should be zero; however, the evaluation of `tox` will give a divide check. We substitute an innocuous value for the zero cases, then fix up their answers at the end.

⋆ ⋆ ⋆

```
SUBROUTINE bessjy_s(x,xnu,rj,ry,rjp,ryp)
USE nrtype; USE nrutil, ONLY : assert,nrerror
USE nr, ONLY : beschb
IMPLICIT NONE
REAL(SP), INTENT(IN) :: x,xnu
REAL(SP), INTENT(OUT) :: rj,ry,rjp,ryp
INTEGER(I4B), PARAMETER :: MAXIT=10000
REAL(DP), PARAMETER :: XMIN=2.0_dp,EPS=1.0e-10_dp,FPMIN=1.0e-30_dp
```

Returns the Bessel functions `rj` $= J_\nu$, `ry` $= Y_\nu$ and their derivatives `rjp` $= J'_\nu$, `ryp` $= Y'_\nu$, for positive `x` and for `xnu` $= \nu \geq 0$. The relative accuracy is within one or two significant digits of EPS, except near a zero of one of the functions, where EPS controls its absolute accuracy. FPMIN is a number close to the machine's smallest floating-point number. All internal arithmetic is in double precision. To convert the entire routine to double precision, change the SP declaration above and decrease EPS to 10^{-16}. Also convert the subroutine `beschb`.

```
INTEGER(I4B) :: i,isign,l,nl
REAL(DP) :: a,b,c,d,del,del1,e,f,fact,fact2,fact3,ff,gam,gam1,gam2,&
    gammi,gampl,h,p,pimu,pimu2,q,r,rjl,rjl1,rjmu,rjp1,rjpl,rjtemp,&
    ry1,rymu,rymup,rytemp,sum,sum1,w,x2,xi,xi2,xmu,xmu2
COMPLEX(DPC) :: aa,bb,cc,dd,dl,pq
call assert(x > 0.0, xnu >= 0.0, 'bessjy args')
nl=merge(int(xnu+0.5_dp), max(0,int(xnu-x+1.5_dp)), x < XMIN)
```

`nl` is the number of downward recurrences of the J's and upward recurrences of Y's. `xmu` lies between $-1/2$ and $1/2$ for `x` < XMIN, while it is chosen so that `x` is greater than the turning point for `x` $\geq$ XMIN.

```
xmu=xnu-nl
xmu2=xmu*xmu
xi=1.0_dp/x
xi2=2.0_dp*xi
w=xi2/PI_D                          The Wronskian.
isign=1                             Evaluate CF1 by modified Lentz's method
h=xnu*xi                               (§5.2). isign keeps track of sign changes
if (h < FPMIN) h=FPMIN                 in the denominator.
b=xi2*xnu
d=0.0
c=h
do i=1,MAXIT
    b=b+xi2
    d=b-d
    if (abs(d) < FPMIN) d=FPMIN
    c=b-1.0_dp/c
```

```
        if (abs(c) < FPMIN) c=FPMIN
        d=1.0_dp/d
        del=c*d
        h=del*h
        if (d < 0.0) isign=-isign
        if (abs(del-1.0_dp) < EPS) exit
    end do
    if (i > MAXIT) call nrerror('x too large in bessjy; try asymptotic expansion')
    rjl=isign*FPMIN                       Initialize J_ν and J'_ν for downward recurrence.
    rjpl=h*rjl
    rjl1=rjl                              Store values for later rescaling.
    rjp1=rjpl
    fact=xnu*xi
    do l=nl,1,-1
        rjtemp=fact*rjl+rjpl
        fact=fact-xi
        rjpl=fact*rjtemp-rjl
        rjl=rjtemp
    end do
    if (rjl == 0.0) rjl=EPS
    f=rjpl/rjl                            Now have unnormalized J_μ and J'_μ.
    if (x < XMIN) then                    Use series.
        x2=0.5_dp*x
        pimu=PI_D*xmu
        if (abs(pimu) < EPS) then
            fact=1.0
        else
            fact=pimu/sin(pimu)
        end if
        d=-log(x2)
        e=xmu*d
        if (abs(e) < EPS) then
            fact2=1.0
        else
            fact2=sinh(e)/e
        end if
        call beschb(xmu,gam1,gam2,gampl,gammi)     Chebyshev evaluation of Γ1 and Γ2.
        ff=2.0_dp/PI_D*fact*(gam1*cosh(e)+gam2*fact2*d)     f0.
        e=exp(e)
        p=e/(gampl*PI_D)                                    p0.
        q=1.0_dp/(e*PI_D*gammi)                             q0.
        pimu2=0.5_dp*pimu
        if (abs(pimu2) < EPS) then
            fact3=1.0
        else
            fact3=sin(pimu2)/pimu2
        end if
        r=PI_D*pimu2*fact3*fact3
        c=1.0
        d=-x2*x2
        sum=ff+r*q
        sum1=p
        do i=1,MAXIT
            ff=(i*ff+p+q)/(i*i-xmu2)
            c=c*d/i
            p=p/(i-xmu)
            q=q/(i+xmu)
            del=c*(ff+r*q)
            sum=sum+del
            del1=c*p-i*del
            sum1=sum1+del1
            if (abs(del) < (1.0_dp+abs(sum))*EPS) exit
        end do
        if (i > MAXIT) call nrerror('bessy series failed to converge')
```

```
        rymu=-sum
        ry1=-sum1*xi2
        rymup=xmu*xi*rymu-ry1
        rjmu=w/(rymup-f*rymu)                    Equation (6.7.13).
    else                                         Evaluate CF2 by modified Lentz's method
        a=0.25_dp-xmu2                              (§5.2).
        pq=cmplx(-0.5_dp*xi,1.0_dp,kind=dpc)
        aa=cmplx(0.0_dp,xi*a,kind=dpc)
        bb=cmplx(2.0_dp*x,2.0_dp,kind=dpc)
        cc=bb+aa/pq
        dd=1.0_dp/bb
        pq=cc*dd*pq
        do i=2,MAXIT
            a=a+2*(i-1)
            bb=bb+cmplx(0.0_dp,2.0_dp,kind=dpc)
            dd=a*dd+bb
            if (absc(dd) < FPMIN) dd=FPMIN
            cc=bb+a/cc
            if (absc(cc) < FPMIN) cc=FPMIN
            dd=1.0_dp/dd
            dl=cc*dd
            pq=pq*dl
            if (absc(dl-1.0_dp) < EPS) exit
        end do
        if (i > MAXIT) call nrerror('cf2 failed in bessjy')
        p=real(pq)
        q=aimag(pq)
        gam=(p-f)/q                              Equations (6.7.6) – (6.7.10).
        rjmu=sqrt(w/((p-f)*gam+q))
        rjmu=sign(rjmu,rjl)
        rymu=rjmu*gam
        rymup=rymu*(p+q/gam)
        ry1=xmu*xi*rymu-rymup
    end if
    fact=rjmu/rjl
    rj=rjl1*fact                                 Scale original J_ν and J'_ν.
    rjp=rjp1*fact
    do i=1,nl                                    Upward recurrence of Y_ν.
        rytemp=(xmu+i)*xi2*ry1-rymu
        rymu=ry1
        ry1=rytemp
    end do
    ry=rymu
    ryp=xnu*xi*rymu-ry1
    CONTAINS

    FUNCTION absc(z)
    IMPLICIT NONE
    COMPLEX(DPC), INTENT(IN) :: z
    REAL(DP) :: absc
    absc=abs(real(z))+abs(aimag(z))
    END FUNCTION absc
    END SUBROUTINE bessjy_s
```

Yes there is a vector version bessjy_v. Its general scheme is to have a bunch of contained functions for various cases, and then combine their outputs (somewhat like bessj_v, above, but much more complicated). A listing runs to about four printed pages, and we judge it to be of not much interest, so we will not include it here. (It is included on the machine-readable media.)

⋆ ⋆ ⋆

```
SUBROUTINE beschb_s(x,gam1,gam2,gampl,gammi)
USE nrtype
USE nr, ONLY : chebev
IMPLICIT NONE
REAL(DP), INTENT(IN) :: x
REAL(DP), INTENT(OUT) :: gam1,gam2,gampl,gammi
INTEGER(I4B), PARAMETER :: NUSE1=5,NUSE2=5
```

Evaluates Γ_1 and Γ_2 by Chebyshev expansion for $|x| \leq 1/2$. Also returns $1/\Gamma(1+x)$ and $1/\Gamma(1-x)$. If converting to double precision, set NUSE1 $= 7$, NUSE2 $= 8$.

```
REAL(SP) :: xx
REAL(SP), DIMENSION(7) :: c1=(/-1.142022680371168_sp,&
    6.5165112670737e-3_sp,3.087090173086e-4_sp,-3.4706269649e-6_sp,&
    6.9437664e-9_sp,3.67795e-11_sp,-1.356e-13_sp/)
REAL(SP), DIMENSION(8) :: c2=(/1.843740587300905_sp,&
    -7.68528408447867e-2_sp,1.2719271366546e-3_sp,&
    -4.9717367042e-6_sp, -3.31261198e-8_sp,2.423096e-10_sp,&
    -1.702e-13_sp,-1.49e-15_sp/)
xx=8.0_dp*x*x-1.0_dp
gam1=chebev(-1.0_sp,1.0_sp,c1(1:NUSE1),xx)
gam2=chebev(-1.0_sp,1.0_sp,c2(1:NUSE2),xx)
gampl=gam2-x*gam1
gammi=gam2+x*gam1
END SUBROUTINE beschb_s
```

Multiply x by 2 to make range be -1 to 1, and then apply transformation for evaluating even Chebyshev series.

```
SUBROUTINE beschb_v(x,gam1,gam2,gampl,gammi)
USE nrtype
USE nr, ONLY : chebev
IMPLICIT NONE
REAL(DP), DIMENSION(:), INTENT(IN) :: x
REAL(DP), DIMENSION(:), INTENT(OUT) :: gam1,gam2,gampl,gammi
INTEGER(I4B), PARAMETER :: NUSE1=5,NUSE2=5
REAL(SP), DIMENSION(size(x)) :: xx
REAL(SP), DIMENSION(7) :: c1=(/-1.142022680371168_sp,&
    6.5165112670737e-3_sp,3.087090173086e-4_sp,-3.4706269649e-6_sp,&
    6.9437664e-9_sp,3.67795e-11_sp,-1.356e-13_sp/)
REAL(SP), DIMENSION(8) :: c2=(/1.843740587300905_sp,&
    -7.68528408447867e-2_sp,1.2719271366546e-3_sp,&
    -4.9717367042e-6_sp, -3.31261198e-8_sp,2.423096e-10_sp,&
    -1.702e-13_sp,-1.49e-15_sp/)
xx=8.0_dp*x*x-1.0_dp
gam1=chebev(-1.0_sp,1.0_sp,c1(1:NUSE1),xx)
gam2=chebev(-1.0_sp,1.0_sp,c2(1:NUSE2),xx)
gampl=gam2-x*gam1
gammi=gam2+x*gam1
END SUBROUTINE beschb_v
```

⋆ ⋆ ⋆

```
SUBROUTINE bessik(x,xnu,ri,rk,rip,rkp)
USE nrtype; USE nrutil, ONLY : assert,nrerror
USE nr, ONLY : beschb
IMPLICIT NONE
REAL(SP), INTENT(IN) :: x,xnu
REAL(SP), INTENT(OUT) :: ri,rk,rip,rkp
INTEGER(I4B), PARAMETER :: MAXIT=10000
REAL(SP), PARAMETER :: XMIN=2.0
REAL(DP), PARAMETER :: EPS=1.0e-10_dp,FPMIN=1.0e-30_dp
```

Returns the modified Bessel functions ri $= I_\nu$, rk $= K_\nu$ and their derivatives rip $= I'_\nu$, rkp $= K'_\nu$, for positive x and for xnu $= \nu \geq 0$. The relative accuracy is within one or two significant digits of EPS. FPMIN is a number close to the machine's smallest floating-point number. All internal arithmetic is in double precision. To convert the entire routine to double precision, change the REAL declaration above and decrease EPS to 10^{-16}. Also convert the subroutine beschb.

```
INTEGER(I4B) :: i,l,nl
REAL(DP) :: a,a1,b,c,d,del,del1,delh,dels,e,f,fact,fact2,ff,&
    gam1,gam2,gammi,gampl,h,p,pimu,q,q1,q2,qnew,&
    ril,ril1,rimu,rip1,ripl,ritemp,rk1,rkmu,rkmup,rktemp,&
    s,sum,sum1,x2,xi,xi2,xmu,xmu2
call assert(x > 0.0, xnu >= 0.0, 'bessik args')
nl=int(xnu+0.5_dp)                    nl is the number of downward recurrences
xmu=xnu-nl                                of the I's and upward recurrences
xmu2=xmu*xmu                              of K's. xmu lies between -1/2 and
xi=1.0_dp/x                               1/2.
xi2=2.0_dp*xi
h=xnu*xi                              Evaluate CF1 by modified Lentz's method
if (h < FPMIN) h=FPMIN                    (§5.2).
b=xi2*xnu
d=0.0
c=h
do i=1,MAXIT
    b=b+xi2
    d=1.0_dp/(b+d)                    Denominators cannot be zero here, so no
    c=b+1.0_dp/c                          need for special precautions.
    del=c*d
    h=del*h
    if (abs(del-1.0_dp) < EPS) exit
end do
if (i > MAXIT) call nrerror('x too large in bessik; try asymptotic expansion')
ril=FPMIN                             Initialize I_ν and I'_ν for downward recur-
ripl=h*ril                                rence.
ril1=ril                              Store values for later rescaling.
rip1=ripl
fact=xnu*xi
do l=nl,1,-1
    ritemp=fact*ril+ripl
    fact=fact-xi
    ripl=fact*ritemp+ril
    ril=ritemp
end do
f=ripl/ril                            Now have unnormalized I_μ and I'_μ.
if (x < XMIN) then                    Use series.
    x2=0.5_dp*x
    pimu=PI_D*xmu
    if (abs(pimu) < EPS) then
        fact=1.0
    else
        fact=pimu/sin(pimu)
    end if
    d=-log(x2)
    e=xmu*d
    if (abs(e) < EPS) then
        fact2=1.0
    else
        fact2=sinh(e)/e
    end if
    call beschb(xmu,gam1,gam2,gampl,gammi)   Chebyshev evaluation of Γ1 and Γ2.
    ff=fact*(gam1*cosh(e)+gam2*fact2*d)      f0.
    sum=ff
    e=exp(e)
    p=0.5_dp*e/gampl                         p0.
    q=0.5_dp/(e*gammi)                       q0.
    c=1.0
    d=x2*x2
```

```
    sum1=p
    do i=1,MAXIT
        ff=(i*ff+p+q)/(i*i-xmu2)
        c=c*d/i
        p=p/(i-xmu)
        q=q/(i+xmu)
        del=c*ff
        sum=sum+del
        del1=c*(p-i*ff)
        sum1=sum1+del1
        if (abs(del) < abs(sum)*EPS) exit
    end do
    if (i > MAXIT) call nrerror('bessk series failed to converge')
    rkmu=sum
    rk1=sum1*xi2
else                                    Evaluate CF2 by Steed's algorithm (§5.2),
    b=2.0_dp*(1.0_dp+x)                     which is OK because there can be no
    d=1.0_dp/b                              zero denominators.
    delh=d
    h=delh
    q1=0.0                              Initializations for recurrence (6.7.35).
    q2=1.0
    a1=0.25_dp-xmu2
    c=a1
    q=c                                 First term in equation (6.7.34).
    a=-a1
    s=1.0_dp+q*delh
    do i=2,MAXIT
        a=a-2*(i-1)
        c=-a*c/i
        qnew=(q1-b*q2)/a
        q1=q2
        q2=qnew
        q=q+c*qnew
        b=b+2.0_dp
        d=1.0_dp/(b+a*d)
        delh=(b*d-1.0_dp)*delh
        h=h+delh
        dels=q*delh
        s=s+dels
        if (abs(dels/s) < EPS) exit     Need only test convergence of sum, since
    end do                                  CF2 itself converges more quickly.
    if (i > MAXIT) call nrerror('bessik: failure to converge in cf2')
    h=a1*h
    rkmu=sqrt(PI_D/(2.0_dp*x))*exp(-x)/s    Omit the factor exp(-x) to scale all the
    rk1=rkmu*(xmu+x+0.5_dp-h)*xi            returned functions by exp(x) for x >=
end if                                      XMIN.
rkmup=xmu*xi*rkmu-rk1
rimu=xi/(f*rkmu-rkmup)                  Get I_mu from Wronskian.
ri=(rimu*ril1)/ril                      Scale original I_nu and I'_nu.
rip=(rimu*rip1)/ril
do i=1,nl                               Upward recurrence of K_nu.
    rktemp=(xmu+i)*xi2*rk1+rkmu
    rkmu=rk1
    rk1=rktemp
end do
rk=rkmu
rkp=xnu*xi*rkmu-rk1
END SUBROUTINE bessik
```

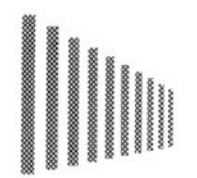

bessik does not readily parallelize, and we thus don't provide a vector version. Since airy, immediately following, requires bessik, we don't have a vector version of it, either.

⋆ ⋆ ⋆

```
SUBROUTINE airy(x,ai,bi,aip,bip)
USE nrtype
USE nr, ONLY : bessik,bessjy
IMPLICIT NONE
REAL(SP), INTENT(IN) :: x
REAL(SP), INTENT(OUT) :: ai,bi,aip,bip
   Returns Airy functions Ai(x), Bi(x), and their derivatives Ai'(x), Bi'(x).
REAL(SP) :: absx,ri,rip,rj,rjp,rk,rkp,rootx,ry,ryp,z
REAL(SP), PARAMETER :: THIRD=1.0_sp/3.0_sp,TWOTHR=2.0_sp/3.0_sp, &
    ONOVRT=0.5773502691896258_sp
absx=abs(x)
rootx=sqrt(absx)
z=TWOTHR*absx*rootx
if (x > 0.0) then
    call bessik(z,THIRD,ri,rk,rip,rkp)
    ai=rootx*ONOVRT*rk/PI
    bi=rootx*(rk/PI+2.0_sp*ONOVRT*ri)
    call bessik(z,TWOTHR,ri,rk,rip,rkp)
    aip=-x*ONOVRT*rk/PI
    bip=x*(rk/PI+2.0_sp*ONOVRT*ri)
else if (x < 0.0) then
    call bessjy(z,THIRD,rj,ry,rjp,ryp)
    ai=0.5_sp*rootx*(rj-ONOVRT*ry)
    bi=-0.5_sp*rootx*(ry+ONOVRT*rj)
    call bessjy(z,TWOTHR,rj,ry,rjp,ryp)
    aip=0.5_sp*absx*(ONOVRT*ry+rj)
    bip=0.5_sp*absx*(ONOVRT*rj-ry)
else                                Case x = 0.
    ai=0.3550280538878172_sp
    bi=ai/ONOVRT
    aip=-0.2588194037928068_sp
    bip=-aip/ONOVRT
end if
END SUBROUTINE airy
```

⋆ ⋆ ⋆

```
SUBROUTINE sphbes_s(n,x,sj,sy,sjp,syp)
USE nrtype; USE nrutil, ONLY : assert
USE nr, ONLY : bessjy
IMPLICIT NONE
INTEGER(I4B), INTENT(IN) :: n
REAL(SP), INTENT(IN) :: x
REAL(SP), INTENT(OUT) :: sj,sy,sjp,syp
   Returns spherical Bessel functions j_n(x), y_n(x), and their derivatives j'_n(x), y'_n(x) for
   integer n >= 0 and x > 0.
REAL(SP), PARAMETER :: RTPIO2=1.253314137315500_sp
REAL(SP) :: factor,order,rj,rjp,ry,ryp
call assert(n >= 0, x > 0.0, 'sphbes_s args')
order=n+0.5_sp
call bessjy(x,order,rj,ry,rjp,ryp)
factor=RTPIO2/sqrt(x)
sj=factor*rj
sy=factor*ry
```

```
sjp=factor*rjp-sj/(2.0_sp*x)
syp=factor*ryp-sy/(2.0_sp*x)
END SUBROUTINE sphbes_s
```

```
SUBROUTINE sphbes_v(n,x,sj,sy,sjp,syp)
USE nrtype; USE nrutil, ONLY : assert
USE nr, ONLY : bessjy
IMPLICIT NONE
INTEGER(I4B), INTENT(IN) :: n
REAL(SP), DIMENSION(:), INTENT(IN) :: x
REAL(SP), DIMENSION(:), INTENT(OUT) :: sj,sy,sjp,syp
REAL(SP), PARAMETER :: RTPIO2=1.253314137315500_sp
REAL(SP) :: order
REAL(SP), DIMENSION(size(x)) :: factor,rj,rjp,ry,ryp
call assert(n >= 0, all(x > 0.0), 'sphbes_v args')
order=n+0.5_sp
call bessjy(x,order,rj,ry,rjp,ryp)
factor=RTPIO2/sqrt(x)
sj=factor*rj
sy=factor*ry
sjp=factor*rjp-sj/(2.0_sp*x)
syp=factor*ryp-sy/(2.0_sp*x)
END SUBROUTINE sphbes_v
```

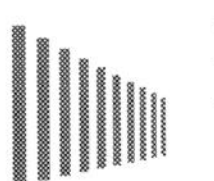

Note that sphbes_v uses (through overloading) bessjy_v. The listing of that routine was omitted above, but it is on the machine-readable media.

⋆ ⋆ ⋆

```
FUNCTION plgndr_s(l,m,x)
USE nrtype; USE nrutil, ONLY : arth,assert
IMPLICIT NONE
INTEGER(I4B), INTENT(IN) :: l,m
REAL(SP), INTENT(IN) :: x
REAL(SP) :: plgndr_s
    Computes the associated Legendre polynomial P_l^m(x). Here m and l are integers satisfying
    0 <= m <= l, while x lies in the range -1 <= x <= 1.
INTEGER(I4B) :: ll
REAL(SP) :: pll,pmm,pmmp1,somx2
call assert(m >= 0, m <= l, abs(x) <= 1.0, 'plgndr_s args')
pmm=1.0                               Compute P_m^m.
if (m > 0) then
    somx2=sqrt((1.0_sp-x)*(1.0_sp+x))
    pmm=product(arth(1.0_sp,2.0_sp,m))*somx2**m
    if (mod(m,2) == 1) pmm=-pmm
end if
if (l == m) then
    plgndr_s=pmm
else
    pmmp1=x*(2*m+1)*pmm               Compute P_{m+1}^m.
    if (l == m+1) then
        plgndr_s=pmmp1
    else                              Compute P_l^m, l > m+1.
        do ll=m+2,l
            pll=(x*(2*ll-1)*pmmp1-(ll+m-1)*pmm)/(ll-m)
            pmm=pmmp1
            pmmp1=pll
        end do
        plgndr_s=pll
```

```
    end if
end if
END FUNCTION plgndr_s
```

product(arth(1.0_sp,2.0_sp,m))
That is, $(2m-1)!!$

```
FUNCTION plgndr_v(l,m,x)
USE nrtype; USE nrutil, ONLY : arth,assert
IMPLICIT NONE
INTEGER(I4B), INTENT(IN) :: l,m
REAL(SP), DIMENSION(:), INTENT(IN) :: x
REAL(SP), DIMENSION(size(x)) :: plgndr_v
INTEGER(I4B) :: ll
REAL(SP), DIMENSION(size(x)) :: pll,pmm,pmmp1,somx2
call assert(m >= 0, m <= l, all(abs(x) <= 1.0), 'plgndr_v args')
pmm=1.0
if (m > 0) then
    somx2=sqrt((1.0_sp-x)*(1.0_sp+x))
    pmm=product(arth(1.0_sp,2.0_sp,m))*somx2**m
    if (mod(m,2) == 1) pmm=-pmm
end if
if (l == m) then
    plgndr_v=pmm
else
    pmmp1=x*(2*m+1)*pmm
    if (l == m+1) then
        plgndr_v=pmmp1
    else
        do ll=m+2,l
            pll=(x*(2*ll-1)*pmmp1-(ll+m-1)*pmm)/(ll-m)
            pmm=pmmp1
            pmmp1=pll
        end do
        plgndr_v=pll
    end if
end if
END FUNCTION plgndr_v
```

All those if's (not where's) may strike you as odd in a vector routine, but it is vectorized only on x, the dependent variable, not on the scalar indices l and m. Much harder to write a routine that is parallel for a vector of arbitrary triplets (l, m, x). Try it!

⋆ ⋆ ⋆

```
SUBROUTINE frenel(x,s,c)
USE nrtype; USE nrutil, ONLY : nrerror
IMPLICIT NONE
REAL(SP), INTENT(IN) :: x
REAL(SP), INTENT(OUT) :: s,c
INTEGER(I4B), PARAMETER :: MAXIT=100
REAL(SP), PARAMETER :: EPS=epsilon(x),FPMIN=tiny(x),BIG=huge(x)*EPS,&
    XMIN=1.5
```

Computes the Fresnel integrals $S(x)$ and $C(x)$ for all real x.
Parameters: MAXIT is the maximum number of iterations allowed; EPS is the relative error; FPMIN is a number near the smallest representable floating-point number; BIG is a number

```
    near the machine overflow limit; XMIN is the dividing line between using the series and
    continued fraction.
INTEGER(I4B) :: k,n
REAL(SP) :: a,ax,fact,pix2,sign,sum,sumc,sums,term,test
COMPLEX(SPC) :: b,cc,d,h,del,cs
LOGICAL(LGT) :: odd
ax=abs(x)
if (ax < sqrt(FPMIN)) then              Special case: avoid failure of convergence test be-
    s=0.0                                   cause of underflow.
    c=ax
else if (ax <= XMIN) then               Evaluate both series simultaneously.
    sum=0.0
    sums=0.0
    sumc=ax
    sign=1.0
    fact=PIO2*ax*ax
    odd=.true.
    term=ax
    n=3
    do k=1,MAXIT
        term=term*fact/k
        sum=sum+sign*term/n
        test=abs(sum)*EPS
        if (odd) then
            sign=-sign
            sums=sum
            sum=sumc
        else
            sumc=sum
            sum=sums
        end if
        if (term < test) exit
        odd=.not. odd
        n=n+2
    end do
    if (k > MAXIT) call nrerror('frenel: series failed')
    s=sums
    c=sumc
else                                    Evaluate continued fraction by modified Lentz's method
    pix2=PI*ax*ax                           (§5.2).
    b=cmplx(1.0_sp,-pix2,kind=spc)
    cc=BIG
    d=1.0_sp/b
    h=d
    n=-1
    do k=2,MAXIT
        n=n+2
        a=-n*(n+1)
        b=b+4.0_sp
        d=1.0_sp/(a*d+b)                Denominators cannot be zero.
        cc=b+a/cc
        del=cc*d
        h=h*del
        if (absc(del-1.0_sp) <= EPS) exit
    end do
    if (k > MAXIT) call nrerror('cf failed in frenel')
    h=h*cmplx(ax,-ax,kind=spc)
    cs=cmplx(0.5_sp,0.5_sp,kind=spc)*(1.0_sp-&
        cmplx(cos(0.5_sp*pix2),sin(0.5_sp*pix2),kind=spc)*h)
    c=real(cs)
    s=aimag(cs)
end if
if (x < 0.0) then                       Use antisymmetry.
    c=-c
```

```
    s=-s
end if
CONTAINS

FUNCTION absc(z)
IMPLICIT NONE
COMPLEX(SPC), INTENT(IN) :: z
REAL(SP) :: absc
absc=abs(real(z))+abs(aimag(z))
END FUNCTION absc
END SUBROUTINE frenel
```

f90 `b=cmplx(1.0_sp,-pix2,kind=spc)` It's a good idea *always* to include the `kind=` parameter when you use the `cmplx` intrinsic. The reason is that, perhaps counterintuitively, the result of `cmplx` is not determined by the kind of its arguments, but is rather the "default complex kind." Since that default may not be what you think it is (or what `spc` is defined to be), the desired kind should be specified explicitly.

`c=real(cs)` And why not specify a `kind=` parameter here, where it is also optionally allowed? Our answer is that the `real` intrinsic actually merges two different usages. When its argument is complex, it is the counterpart of `aimag` and returns a value whose kind is determined by the kind of its argument. In fact `aimag` doesn't even allow an optional kind parameter, so we never put one in the corresponding use of `real`. The other usage of `real` is for "casting," that is, converting one real type to another (e.g., double precision to single precision, or vice versa). Here we *always* include a kind parameter, since otherwise the result is the default real kind, with the same dangers mentioned in the previous paragraph.

* * *

```
SUBROUTINE cisi(x,ci,si)
USE nrtype; USE nrutil, ONLY : nrerror
IMPLICIT NONE
REAL(SP), INTENT(IN) :: x
REAL(SP), INTENT(OUT) :: ci,si
INTEGER(I4B), PARAMETER :: MAXIT=100
REAL(SP), PARAMETER :: EPS=epsilon(x),FPMIN=4.0_sp*tiny(x),&
    BIG=huge(x)*EPS,TMIN=2.0
    Computes the cosine and sine integrals Ci(x) and Si(x). Ci(0) is returned as a large negative
    number and no error message is generated. For x < 0 the routine returns Ci(-x) and you
    must supply the -iπ yourself.
    Parameters: MAXIT is the maximum number of iterations allowed; EPS is the relative error,
    or absolute error near a zero of Ci(x); FPMIN is a number near the smallest representable
    floating-point number; BIG is a number near the machine overflow limit; TMIN is the dividing
    line between using the series and continued fraction; EULER = γ (in nrtype).
INTEGER(I4B) :: i,k
REAL(SP) :: a,err,fact,sign,sum,sumc,sums,t,term
COMPLEX(SPC) :: h,b,c,d,del
LOGICAL(LGT) :: odd
t=abs(x)
if (t == 0.0) then                           Special case.
    si=0.0
    ci=-BIG
    RETURN
end if                                       Evaluate continued fraction by modified Lentz's
if (t > TMIN) then                               method (§5.2).
    b=cmplx(1.0_sp,t,kind=spc)
```

```
    c=BIG
    d=1.0_sp/b
    h=d
    do i=2,MAXIT
        a=-(i-1)**2
        b=b+2.0_sp
        d=1.0_sp/(a*d+b)                Denominators cannot be zero.
        c=b+a/c
        del=c*d
        h=h*del
        if (absc(del-1.0_sp) <= EPS) exit
    end do
    if (i > MAXIT) call nrerror('continued fraction failed in cisi')
    h=cmplx(cos(t),-sin(t),kind=spc)*h
    ci=-real(h)
    si=PIO2+aimag(h)
else                                    Evaluate both series simultaneously.
    if (t < sqrt(FPMIN)) then           Special case: avoid failure of convergence test
        sumc=0.0                            because of underflow.
        sums=t
    else
        sum=0.0
        sums=0.0
        sumc=0.0
        sign=1.0
        fact=1.0
        odd=.true.
        do k=1,MAXIT
            fact=fact*t/k
            term=fact/k
            sum=sum+sign*term
            err=term/abs(sum)
            if (odd) then
                sign=-sign
                sums=sum
                sum=sumc
            else
                sumc=sum
                sum=sums
            end if
            if (err < EPS) exit
            odd=.not. odd
        end do
        if (k > MAXIT) call nrerror('MAXIT exceeded in cisi')
    end if
    si=sums
    ci=sumc+log(t)+EULER
end if
if (x < 0.0) si=-si
CONTAINS

FUNCTION absc(z)
IMPLICIT NONE
COMPLEX(SPC), INTENT(IN) :: z
REAL(SP) :: absc
absc=abs(real(z))+abs(aimag(z))
END FUNCTION absc
END SUBROUTINE cisi
```

⋆ ⋆ ⋆

```
FUNCTION dawson_s(x)
USE nrtype; USE nrutil, ONLY : arth,geop
IMPLICIT NONE
REAL(SP), INTENT(IN) :: x
REAL(SP) :: dawson_s
    Returns Dawson's integral F(x) = exp(-x^2) \int_0^x exp(t^2)dt for any real x.
INTEGER(I4B), PARAMETER :: NMAX=6
REAL(SP), PARAMETER :: H=0.4_sp,A1=2.0_sp/3.0_sp,A2=0.4_sp,&
    A3=2.0_sp/7.0_sp
INTEGER(I4B) :: i,n0
REAL(SP) :: ec,x2,xp,xx
REAL(SP), DIMENSION(NMAX) :: d1,d2,e1
REAL(SP), DIMENSION(NMAX), SAVE :: c=(/ (0.0_sp,i=1,NMAX) /)
if (c(1) == 0.0) c(1:NMAX)=exp(-(arth(1,2,NMAX)*H)**2)
  Initialize c on first call.
if (abs(x) < 0.2_sp) then                    Use series expansion.
    x2=x**2
    dawson_s=x*(1.0_sp-A1*x2*(1.0_sp-A2*x2*(1.0_sp-A3*x2)))
else                                         Use sampling theorem representation.
    xx=abs(x)
    n0=2*nint(0.5_sp*xx/H)
    xp=xx-real(n0,sp)*H
    ec=exp(2.0_sp*xp*H)
    d1=arth(n0+1,2,NMAX)
    d2=arth(n0-1,-2,NMAX)
    e1=geop(ec,ec**2,NMAX)
    dawson_s=0.5641895835477563_sp*sign(exp(-xp**2),x)*&    Constant is 1/sqrt(pi).
        sum(c*(e1/d1+1.0_sp/(d2*e1)))
end if
END FUNCTION dawson_s
```

`REAL(SP), DIMENSION(NMAX), SAVE :: c=(/ (0.0_sp,i=1,NMAX) /)` This is one way to give initial values to an array. Actually, we're somewhat nervous about using the "implied do-loop" form of the array constructor, as above, because our parallel compilers might not always be smart enough to execute the constructor in parallel. In this case, with `NMAX=6`, the damage potential is quite minimal. An alternative way to initialize the array would be with a data statement, "`DATA c /NMAX*0.0_sp/`"; however, this is not considered good Fortran 90 style, and there is no reason to think that it would be faster.

`c(1:NMAX)=exp(-(arth(1,2,NMAX)*H)**2)` Another example where the `arth` function of `nrutil` comes in handy. Otherwise, this would be

```
do i=1,NMAX
   c(i)=exp(-((2.0_sp*i-1.0_sp)*H)**2)
end do
```

`arth(n0+1,2,NMAX)...arth(n0-1,-2,NMAX)...geop(ec,ec**2,NMAX)` These are not just notationally convenient for generating the sequences $(n_0+1, n_0+3, n_0+5, \ldots)$, $(n_0-1, n_0-3, n_0-5, \ldots)$, and $(\mathtt{ec}, \mathtt{ec}^3, \mathtt{ec}^5, \ldots)$. They also may allow parallelization with parallel versions of `arth` and `geop`, such as those in `nrutil`.

```
FUNCTION dawson_v(x)
USE nrtype; USE nrutil, ONLY : arth
IMPLICIT NONE
REAL(SP), DIMENSION(:), INTENT(IN) :: x
REAL(SP), DIMENSION(size(x)) :: dawson_v
INTEGER(I4B), PARAMETER :: NMAX=6
```

```
REAL(SP), PARAMETER :: H=0.4_sp,A1=2.0_sp/3.0_sp,A2=0.4_sp,&
   A3=2.0_sp/7.0_sp
INTEGER(I4B) :: i,n
REAL(SP), DIMENSION(size(x)) :: x2
REAL(SP), DIMENSION(NMAX), SAVE :: c=(/ (0.0_sp,i=1,NMAX) /)
LOGICAL(LGT), DIMENSION(size(x)) :: mask
if (c(1) == 0.0) c(1:NMAX)=exp(-(arth(1,2,NMAX)*H)**2)
mask = (abs(x) >= 0.2_sp)
dawson_v=dawsonseries_v(x,mask)
where (.not. mask)
    x2=x**2
    dawson_v=x*(1.0_sp-A1*x2*(1.0_sp-A2*x2*(1.0_sp-A3*x2)))
end where
CONTAINS

FUNCTION dawsonseries_v(xin,mask)
IMPLICIT NONE
REAL(SP), DIMENSION(:), INTENT(IN) :: xin
LOGICAL(LGT), DIMENSION(size(xin)), INTENT(IN) :: mask
REAL(SP), DIMENSION(size(xin)) :: dawsonseries_v
INTEGER(I4B), DIMENSION(:), ALLOCATABLE :: n0
REAL(SP), DIMENSION(:), ALLOCATABLE :: d1,d2,e1,e2,sm,xp,xx,x
n=count(mask)
if (n == 0) RETURN
allocate(n0(n),d1(n),d2(n),e1(n),e2(n),sm(n),xp(n),xx(n),x(n))
x=pack(xin,mask)
xx=abs(x)
n0=2*nint(0.5_sp*xx/H)
xp=xx-real(n0,sp)*H
e1=exp(2.0_sp*xp*H)
e2=e1**2
d1=n0+1.0_sp
d2=d1-2.0_sp
sm=0.0
do i=1,NMAX
    sm=sm+c(i)*(e1/d1+1.0_sp/(d2*e1))
    d1=d1+2.0_sp
    d2=d2-2.0_sp
    e1=e2*e1
end do
sm=0.5641895835477563_sp*sign(exp(-xp**2),x)*sm
dawsonseries_v=unpack(sm,mask,0.0_sp)
deallocate(n0,d1,d2,e1,e2,sm,xp,xx)
END FUNCTION dawsonseries_v
END FUNCTION dawson_v
```

`dawson_v=dawsonseries_v(x,mask)` Pass-the-buck method for getting masked values, see note to `bessj0_v` above, p. 1102. Within the contained `dawsonseries`, we use the pack-unpack method. Note that, unlike in `dawson_s`, the sums are done by do-loops, because the parallelization is already over the components of the vector argument.

⋆ ⋆ ⋆

```
FUNCTION rf_s(x,y,z)
USE nrtype; USE nrutil, ONLY : assert
IMPLICIT NONE
REAL(SP), INTENT(IN) :: x,y,z
REAL(SP) :: rf_s
REAL(SP), PARAMETER :: ERRTOL=0.08_sp,TINY=1.5e-38_sp,BIG=3.0e37_sp,&
    THIRD=1.0_sp/3.0_sp,&
```

```
    C1=1.0_sp/24.0_sp,C2=0.1_sp,C3=3.0_sp/44.0_sp,C4=1.0_sp/14.0_sp
```

Computes Carlson's elliptic integral of the first kind, $R_F(x, y, z)$. x, y, and z must be nonnegative, and at most one can be zero. TINY must be at least 5 times the machine underflow limit, BIG at most one-fifth the machine overflow limit.

```
REAL(SP) :: alamb,ave,delx,dely,delz,e2,e3,sqrtx,sqrty,sqrtz,xt,yt,zt
call assert(min(x,y,z) >= 0.0, min(x+y,x+z,y+z) >= TINY, &
    max(x,y,z) <= BIG, 'rf_s args')
xt=x
yt=y
zt=z
do
    sqrtx=sqrt(xt)
    sqrty=sqrt(yt)
    sqrtz=sqrt(zt)
    alamb=sqrtx*(sqrty+sqrtz)+sqrty*sqrtz
    xt=0.25_sp*(xt+alamb)
    yt=0.25_sp*(yt+alamb)
    zt=0.25_sp*(zt+alamb)
    ave=THIRD*(xt+yt+zt)
    delx=(ave-xt)/ave
    dely=(ave-yt)/ave
    delz=(ave-zt)/ave
    if (max(abs(delx),abs(dely),abs(delz)) <= ERRTOL) exit
end do
e2=delx*dely-delz**2
e3=delx*dely*delz
rf_s=(1.0_sp+(C1*e2-C2-C3*e3)*e2+C4*e3)/sqrt(ave)
END FUNCTION rf_s
```

```
FUNCTION rf_v(x,y,z)
USE nrtype; USE nrutil, ONLY : assert,assert_eq
IMPLICIT NONE
REAL(SP), DIMENSION(:), INTENT(IN) :: x,y,z
REAL(SP), DIMENSION(size(x)) :: rf_v
REAL(SP), PARAMETER :: ERRTOL=0.08_sp,TINY=1.5e-38_sp,BIG=3.0e37_sp,&
    THIRD=1.0_sp/3.0_sp,&
    C1=1.0_sp/24.0_sp,C2=0.1_sp,C3=3.0_sp/44.0_sp,C4=1.0_sp/14.0_sp
REAL(SP), DIMENSION(size(x)) :: alamb,ave,delx,dely,delz,e2,e3,&
    sqrtx,sqrty,sqrtz,xt,yt,zt
LOGICAL(LGT), DIMENSION(size(x)) :: converged
INTEGER(I4B) :: ndum
ndum=assert_eq(size(x),size(y),size(z),'rf_v')
call assert(all(min(x,y,z) >= 0.0), all(min(x+y,x+z,y+z) >= TINY), &
    all(max(x,y,z) <= BIG), 'rf_v args')
xt=x
yt=y
zt=z
converged=.false.
do
    where (.not. converged)
        sqrtx=sqrt(xt)
        sqrty=sqrt(yt)
        sqrtz=sqrt(zt)
        alamb=sqrtx*(sqrty+sqrtz)+sqrty*sqrtz
        xt=0.25_sp*(xt+alamb)
        yt=0.25_sp*(yt+alamb)
        zt=0.25_sp*(zt+alamb)
        ave=THIRD*(xt+yt+zt)
        delx=(ave-xt)/ave
        dely=(ave-yt)/ave
        delz=(ave-zt)/ave
        converged = (max(abs(delx),abs(dely),abs(delz)) <= ERRTOL)
```

```
      end where
      if (all(converged)) exit
   end do
   e2=delx*dely-delz**2
   e3=delx*dely*delz
   rf_v=(1.0_sp+(C1*e2-C2-C3*e3)*e2+C4*e3)/sqrt(ave)
   END FUNCTION rf_v
```

```
   FUNCTION rd_s(x,y,z)
   USE nrtype; USE nrutil, ONLY : assert
   IMPLICIT NONE
   REAL(SP), INTENT(IN) :: x,y,z
   REAL(SP) :: rd_s
   REAL(SP), PARAMETER :: ERRTOL=0.05_sp,TINY=1.0e-25_sp,BIG=4.5e21_sp,&
       C1=3.0_sp/14.0_sp,C2=1.0_sp/6.0_sp,C3=9.0_sp/22.0_sp,&
       C4=3.0_sp/26.0_sp,C5=0.25_sp*C3,C6=1.5_sp*C4
```

Computes Carlson's elliptic integral of the second kind, $R_D(x, y, z)$. x and y must be nonnegative, and at most one can be zero. z must be positive. TINY must be at least twice the negative 2/3 power of the machine overflow limit. BIG must be at most $0.1 \times$ ERRTOL times the negative 2/3 power of the machine underflow limit.

```
   REAL(SP) :: alamb,ave,delx,dely,delz,ea,eb,ec,ed,&
       ee,fac,sqrtx,sqrty,sqrtz,sum,xt,yt,zt
   call assert(min(x,y) >= 0.0, min(x+y,z) >= TINY, max(x,y,z) <= BIG, &
       'rd_s args')
   xt=x
   yt=y
   zt=z
   sum=0.0
   fac=1.0
   do
       sqrtx=sqrt(xt)
       sqrty=sqrt(yt)
       sqrtz=sqrt(zt)
       alamb=sqrtx*(sqrty+sqrtz)+sqrty*sqrtz
       sum=sum+fac/(sqrtz*(zt+alamb))
       fac=0.25_sp*fac
       xt=0.25_sp*(xt+alamb)
       yt=0.25_sp*(yt+alamb)
       zt=0.25_sp*(zt+alamb)
       ave=0.2_sp*(xt+yt+3.0_sp*zt)
       delx=(ave-xt)/ave
       dely=(ave-yt)/ave
       delz=(ave-zt)/ave
       if (max(abs(delx),abs(dely),abs(delz)) <= ERRTOL) exit
   end do
   ea=delx*dely
   eb=delz*delz
   ec=ea-eb
   ed=ea-6.0_sp*eb
   ee=ed+ec+ec
   rd_s=3.0_sp*sum+fac*(1.0_sp+ed*(-C1+C5*ed-C6*delz*ee)&
       +delz*(C2*ee+delz*(-C3*ec+delz*C4*ea)))/(ave*sqrt(ave))
   END FUNCTION rd_s
```

```
FUNCTION rd_v(x,y,z)
USE nrtype; USE nrutil, ONLY : assert,assert_eq
IMPLICIT NONE
REAL(SP), DIMENSION(:), INTENT(IN) :: x,y,z
REAL(SP), DIMENSION(size(x)) :: rd_v
REAL(SP), PARAMETER :: ERRTOL=0.05_sp,TINY=1.0e-25_sp,BIG=4.5e21_sp,&
    C1=3.0_sp/14.0_sp,C2=1.0_sp/6.0_sp,C3=9.0_sp/22.0_sp,&
    C4=3.0_sp/26.0_sp,C5=0.25_sp*C3,C6=1.5_sp*C4
REAL(SP), DIMENSION(size(x)) :: alamb,ave,delx,dely,delz,ea,eb,ec,ed,&
    ee,fac,sqrtx,sqrty,sqrtz,sum,xt,yt,zt
LOGICAL(LGT), DIMENSION(size(x)) :: converged
INTEGER(I4B) :: ndum
ndum=assert_eq(size(x),size(y),size(z),'rd_v')
call assert(all(min(x,y) >= 0.0), all(min(x+y,z) >= TINY), &
    all(max(x,y,z) <= BIG), 'rd_v args')
xt=x
yt=y
zt=z
sum=0.0
fac=1.0
converged=.false.
do
    where (.not. converged)
        sqrtx=sqrt(xt)
        sqrty=sqrt(yt)
        sqrtz=sqrt(zt)
        alamb=sqrtx*(sqrty+sqrtz)+sqrty*sqrtz
        sum=sum+fac/(sqrtz*(zt+alamb))
        fac=0.25_sp*fac
        xt=0.25_sp*(xt+alamb)
        yt=0.25_sp*(yt+alamb)
        zt=0.25_sp*(zt+alamb)
        ave=0.2_sp*(xt+yt+3.0_sp*zt)
        delx=(ave-xt)/ave
        dely=(ave-yt)/ave
        delz=(ave-zt)/ave
        converged = (all(max(abs(delx),abs(dely),abs(delz)) <= ERRTOL))
    end where
    if (all(converged)) exit
end do
ea=delx*dely
eb=delz*delz
ec=ea-eb
ed=ea-6.0_sp*eb
ee=ed+ec+ec
rd_v=3.0_sp*sum+fac*(1.0_sp+ed*(-C1+C5*ed-C6*delz*ee)&
    +delz*(C2*ee+delz*(-C3*ec+delz*C4*ea)))/(ave*sqrt(ave))
END FUNCTION rd_v
```

```
FUNCTION rj_s(x,y,z,p)
USE nrtype; USE nrutil, ONLY : assert
USE nr, ONLY : rc,rf
IMPLICIT NONE
REAL(SP), INTENT(IN) :: x,y,z,p
REAL(SP) :: rj_s
REAL(SP), PARAMETER :: ERRTOL=0.05_sp,TINY=2.5e-13_sp,BIG=9.0e11_sp,&
    C1=3.0_sp/14.0_sp,C2=1.0_sp/3.0_sp,C3=3.0_sp/22.0_sp,&
    C4=3.0_sp/26.0_sp,C5=0.75_sp*C3,C6=1.5_sp*C4,C7=0.5_sp*C2,&
    C8=C3+C3
```

Computes Carlson's elliptic integral of the third kind, $R_J(x, y, z, p)$. x, y, and z must be nonnegative, and at most one can be zero. p must be nonzero. If $p < 0$, the Cauchy principal value is returned. TINY must be at least twice the cube root of the machine underflow limit, BIG at most one-fifth the cube root of the machine overflow limit.

```
REAL(SP) :: a,alamb,alpha,ave,b,bet,delp,delx,&
    dely,delz,ea,eb,ec,ed,ee,fac,pt,rho,sqrtx,sqrty,sqrtz,&
    sm,tau,xt,yt,zt
call assert(min(x,y,z) >= 0.0, min(x+y,x+z,y+z,abs(p)) >= TINY, &
    max(x,y,z,abs(p)) <= BIG, 'rj_s args')
sm=0.0
fac=1.0
if (p > 0.0) then
    xt=x
    yt=y
    zt=z
    pt=p
else
    xt=min(x,y,z)
    zt=max(x,y,z)
    yt=x+y+z-xt-zt
    a=1.0_sp/(yt-p)
    b=a*(zt-yt)*(yt-xt)
    pt=yt+b
    rho=xt*zt/yt
    tau=p*pt/yt
end if
do
    sqrtx=sqrt(xt)
    sqrty=sqrt(yt)
    sqrtz=sqrt(zt)
    alamb=sqrtx*(sqrty+sqrtz)+sqrty*sqrtz
    alpha=(pt*(sqrtx+sqrty+sqrtz)+sqrtx*sqrty*sqrtz)**2
    bet=pt*(pt+alamb)**2
    sm=sm+fac*rc(alpha,bet)
    fac=0.25_sp*fac
    xt=0.25_sp*(xt+alamb)
    yt=0.25_sp*(yt+alamb)
    zt=0.25_sp*(zt+alamb)
    pt=0.25_sp*(pt+alamb)
    ave=0.2_sp*(xt+yt+zt+pt+pt)
    delx=(ave-xt)/ave
    dely=(ave-yt)/ave
    delz=(ave-zt)/ave
    delp=(ave-pt)/ave
    if (max(abs(delx),abs(dely),abs(delz),abs(delp)) <= ERRTOL) exit
end do
ea=delx*(dely+delz)+dely*delz
eb=delx*dely*delz
ec=delp**2
ed=ea-3.0_sp*ec
ee=eb+2.0_sp*delp*(ea-ec)
rj_s=3.0_sp*sm+fac*(1.0_sp+ed*(-C1+C5*ed-C6*ee)+eb*(C7+delp*(-C8&
    +delp*C4))+delp*ea*(C2-delp*C3)-C2*delp*ec)/(ave*sqrt(ave))
if (p <= 0.0) rj_s=a*(b*rj_s+3.0_sp*(rc(rho,tau)-rf(xt,yt,zt)))
END FUNCTION rj_s
```

```
FUNCTION rj_v(x,y,z,p)
USE nrtype; USE nrutil, ONLY : assert,assert_eq
USE nr, ONLY : rc,rf
IMPLICIT NONE
REAL(SP), DIMENSION(:), INTENT(IN) :: x,y,z,p
REAL(SP), DIMENSION(size(x)) :: rj_v
REAL(SP), PARAMETER :: ERRTOL=0.05_sp,TINY=2.5e-13_sp,BIG=9.0e11_sp,&
    C1=3.0_sp/14.0_sp,C2=1.0_sp/3.0_sp,C3=3.0_sp/22.0_sp,&
```

```
    C4=3.0_sp/26.0_sp,C5=0.75_sp*C3,C6=1.5_sp*C4,C7=0.5_sp*C2,&
    C8=C3+C3
REAL(SP), DIMENSION(size(x)) :: a,alamb,alpha,ave,b,bet,delp,delx,&
    dely,delz,ea,eb,ec,ed,ee,fac,pt,rho,sqrtx,sqrty,sqrtz,&
    sm,tau,xt,yt,zt
LOGICAL(LGT), DIMENSION(size(x)) :: mask
INTEGER(I4B) :: ndum
ndum=assert_eq(size(x),size(y),size(z),size(p),'rj_v')
call assert(all(min(x,y,z) >= 0.0), all(min(x+y,x+z,y+z,abs(p)) >= TINY), &
    all(max(x,y,z,abs(p)) <= BIG), 'rj_v args')
sm=0.0
fac=1.0
where (p > 0.0)
    xt=x
    yt=y
    zt=z
    pt=p
elsewhere
    xt=min(x,y,z)
    zt=max(x,y,z)
    yt=x+y+z-xt-zt
    a=1.0_sp/(yt-p)
    b=a*(zt-yt)*(yt-xt)
    pt=yt+b
    rho=xt*zt/yt
    tau=p*pt/yt
end where
mask=.false.
do
    where (.not. mask)
        sqrtx=sqrt(xt)
        sqrty=sqrt(yt)
        sqrtz=sqrt(zt)
        alamb=sqrtx*(sqrty+sqrtz)+sqrty*sqrtz
        alpha=(pt*(sqrtx+sqrty+sqrtz)+sqrtx*sqrty*sqrtz)**2
        bet=pt*(pt+alamb)**2
        sm=sm+fac*rc(alpha,bet)
        fac=0.25_sp*fac
        xt=0.25_sp*(xt+alamb)
        yt=0.25_sp*(yt+alamb)
        zt=0.25_sp*(zt+alamb)
        pt=0.25_sp*(pt+alamb)
        ave=0.2_sp*(xt+yt+zt+pt+pt)
        delx=(ave-xt)/ave
        dely=(ave-yt)/ave
        delz=(ave-zt)/ave
        delp=(ave-pt)/ave
        mask = (max(abs(delx),abs(dely),abs(delz),abs(delp)) <= ERRTOL)
    end where
    if (all(mask)) exit
end do
ea=delx*(dely+delz)+dely*delz
eb=delx*dely*delz
ec=delp**2
ed=ea-3.0_sp*ec
ee=eb+2.0_sp*delp*(ea-ec)
rj_v=3.0_sp*sm+fac*(1.0_sp+ed*(-C1+C5*ed-C6*ee)+eb*(C7+delp*(-C8&
    +delp*C4))+delp*ea*(C2-delp*C3)-C2*delp*ec)/(ave*sqrt(ave))
mask = (p <= 0.0)
where (mask) rj_v=a*(b*rj_v+&
    unpack(3.0_sp*(rc(pack(rho,mask),pack(tau,mask))-&
    rf(pack(xt,mask),pack(yt,mask),pack(zt,mask))),mask,0.0_sp))
END FUNCTION rj_v
```

f90 `unpack(3.0_sp*(rc(pack(rho,mask),pack(tau,mask))...),mask,0.0_sp)`
If you're willing to put up with fairly unreadable code, you can use the pack-unpack trick (for getting a masked subset of components out of a vector function) right in-line, as here. Of course the "outer level" that is seen by the enclosing `where` construction has to contain only objects that have the same shape as the `mask` that goes with the `where`. Because it is so hard to read, we don't like to do this very often. An alternative would be to use `CONTAINS` to incorporate short, masked "wrapper functions" for the functions used in this way.

```
FUNCTION rc_s(x,y)
USE nrtype; USE nrutil, ONLY : assert
IMPLICIT NONE
REAL(SP), INTENT(IN) :: x,y
REAL(SP) :: rc_s
REAL(SP), PARAMETER :: ERRTOL=0.04_sp,TINY=1.69e-38_sp,&
    SQRTNY=1.3e-19_sp,BIG=3.0e37_sp,TNBG=TINY*BIG,&
    COMP1=2.236_sp/SQRTNY,COMP2=TNBG*TNBG/25.0_sp,&
    THIRD=1.0_sp/3.0_sp,&
    C1=0.3_sp,C2=1.0_sp/7.0_sp,C3=0.375_sp,C4=9.0_sp/22.0_sp
```

Computes Carlson's degenerate elliptic integral, $R_C(x,y)$. x must be nonnegative and y must be nonzero. If $y < 0$, the Cauchy principal value is returned. TINY must be at least 5 times the machine underflow limit, BIG at most one-fifth the machine maximum overflow limit.

```
REAL(SP) :: alamb,ave,s,w,xt,yt
call assert( (/x >= 0.0,y /= 0.0,x+abs(y) >= TINY,x+abs(y) <= BIG, &
    y >= -COMP1 .or. x <= 0.0 .or. x >= COMP2/),'rc_s')
if (y > 0.0) then
    xt=x
    yt=y
    w=1.0
else
    xt=x-y
    yt=-y
    w=sqrt(x)/sqrt(xt)
end if
do
    alamb=2.0_sp*sqrt(xt)*sqrt(yt)+yt
    xt=0.25_sp*(xt+alamb)
    yt=0.25_sp*(yt+alamb)
    ave=THIRD*(xt+yt+yt)
    s=(yt-ave)/ave
    if (abs(s) <= ERRTOL) exit
end do
rc_s=w*(1.0_sp+s*s*(C1+s*(C2+s*(C3+s*C4))))/sqrt(ave)
END FUNCTION rc_s
```

```
FUNCTION rc_v(x,y)
USE nrtype; USE nrutil, ONLY : assert,assert_eq
IMPLICIT NONE
REAL(SP), DIMENSION(:), INTENT(IN) :: x,y
REAL(SP), DIMENSION(size(x)) :: rc_v
REAL(SP), PARAMETER :: ERRTOL=0.04_sp,TINY=1.69e-38_sp,&
    SQRTNY=1.3e-19_sp,BIG=3.0e37_sp,TNBG=TINY*BIG,&
    COMP1=2.236_sp/SQRTNY,COMP2=TNBG*TNBG/25.0_sp,&
    THIRD=1.0_sp/3.0_sp,&
    C1=0.3_sp,C2=1.0_sp/7.0_sp,C3=0.375_sp,C4=9.0_sp/22.0_sp
REAL(SP), DIMENSION(size(x)) :: alamb,ave,s,w,xt,yt
LOGICAL(LGT), DIMENSION(size(x)) :: converged
INTEGER(I4B) :: ndum
```

```
ndum=assert_eq(size(x),size(y),'rc_v')
call assert( (/all(x >= 0.0),all(y /= 0.0),all(x+abs(y) >= TINY), &
    all(x+abs(y) <= BIG),all(y >= -COMP1 .or. x <= 0.0 &
    .or. x >= COMP2) /),'rc_v')
where (y > 0.0)
    xt=x
    yt=y
    w=1.0
elsewhere
    xt=x-y
    yt=-y
    w=sqrt(x)/sqrt(xt)
end where
converged=.false.
do
    where (.not. converged)
        alamb=2.0_sp*sqrt(xt)*sqrt(yt)+yt
        xt=0.25_sp*(xt+alamb)
        yt=0.25_sp*(yt+alamb)
        ave=THIRD*(xt+yt+yt)
        s=(yt-ave)/ave
        converged = (abs(s) <= ERRTOL)
    end where
    if (all(converged)) exit
end do
rc_v=w*(1.0_sp+s*s*(C1+s*(C2+s*(C3+s*C4))))/sqrt(ave)
END FUNCTION rc_v
```

⋆ ⋆ ⋆

```
FUNCTION ellf_s(phi,ak)
USE nrtype
USE nr, ONLY : rf
IMPLICIT NONE
REAL(SP), INTENT(IN) :: phi,ak
REAL(SP) :: ellf_s
```

Legendre elliptic integral of the 1st kind $F(\phi, k)$, evaluated using Carlson's function R_F. The argument ranges are $0 \leq \phi \leq \pi/2$, $0 \leq k \sin\phi \leq 1$.

```
REAL(SP) :: s
s=sin(phi)
ellf_s=s*rf(cos(phi)**2,(1.0_sp-s*ak)*(1.0_sp+s*ak),1.0_sp)
END FUNCTION ellf_s
```

```
FUNCTION ellf_v(phi,ak)
USE nrtype; USE nrutil, ONLY : assert_eq
USE nr, ONLY : rf
IMPLICIT NONE
REAL(SP), DIMENSION(:), INTENT(IN) :: phi,ak
REAL(SP), DIMENSION(size(phi)) :: ellf_v
REAL(SP), DIMENSION(size(phi)) :: s
INTEGER(I4B) :: ndum
ndum=assert_eq(size(phi),size(ak),'ellf_v')
s=sin(phi)
ellf_v=s*rf(cos(phi)**2,(1.0_sp-s*ak)*(1.0_sp+s*ak),&
    spread(1.0_sp,1,size(phi)))
END FUNCTION ellf_v
```

```
FUNCTION elle_s(phi,ak)
USE nrtype
USE nr, ONLY : rd,rf
IMPLICIT NONE
REAL(SP), INTENT(IN) :: phi,ak
REAL(SP) :: elle_s
```

Legendre elliptic integral of the 2nd kind $E(\phi, k)$, evaluated using Carlson's functions R_D and R_F. The argument ranges are $0 \le \phi \le \pi/2$, $0 \le k \sin\phi \le 1$.

```
REAL(SP) :: cc,q,s
s=sin(phi)
cc=cos(phi)**2
q=(1.0_sp-s*ak)*(1.0_sp+s*ak)
elle_s=s*(rf(cc,q,1.0_sp)-((s*ak)**2)*rd(cc,q,1.0_sp)/3.0_sp)
END FUNCTION elle_s
```

```
FUNCTION elle_v(phi,ak)
USE nrtype; USE nrutil, ONLY : assert_eq
USE nr, ONLY : rd,rf
IMPLICIT NONE
REAL(SP), DIMENSION(:), INTENT(IN) :: phi,ak
REAL(SP), DIMENSION(size(phi)) :: elle_v
REAL(SP), DIMENSION(size(phi)) :: cc,q,s
INTEGER(I4B) :: ndum
ndum=assert_eq(size(phi),size(ak),'elle_v')
s=sin(phi)
cc=cos(phi)**2
q=(1.0_sp-s*ak)*(1.0_sp+s*ak)
elle_v=s*(rf(cc,q,spread(1.0_sp,1,size(phi)))-((s*ak)**2)*&
    rd(cc,q,spread(1.0_sp,1,size(phi)))/3.0_sp)
END FUNCTION elle_v
```

f90 `rd(cc,q,spread(1.0_sp,1,size(phi)))` See note to `erf_v`, p. 1094 above.

```
FUNCTION ellpi_s(phi,en,ak)
USE nrtype
USE nr, ONLY : rf,rj
IMPLICIT NONE
REAL(SP), INTENT(IN) :: phi,en,ak
REAL(SP) :: ellpi_s
```

Legendre elliptic integral of the 3rd kind $\Pi(\phi, n, k)$, evaluated using Carlson's functions R_J and R_F. (Note that the sign convention on n is opposite that of Abramowitz and Stegun.) The ranges of ϕ and k are $0 \le \phi \le \pi/2$, $0 \le k \sin\phi \le 1$.

```
REAL(SP) :: cc,enss,q,s
s=sin(phi)
enss=en*s*s
cc=cos(phi)**2
q=(1.0_sp-s*ak)*(1.0_sp+s*ak)
ellpi_s=s*(rf(cc,q,1.0_sp)-enss*rj(cc,q,1.0_sp,1.0_sp+enss)/3.0_sp)
END FUNCTION ellpi_s
```

```
FUNCTION ellpi_v(phi,en,ak)
USE nrtype
USE nr, ONLY : rf,rj
IMPLICIT NONE
REAL(SP), DIMENSION(:), INTENT(IN) :: phi,en,ak
REAL(SP), DIMENSION(size(phi)) :: ellpi_v
REAL(SP), DIMENSION(size(phi)) :: cc,enss,q,s
s=sin(phi)
enss=en*s*s
cc=cos(phi)**2
q=(1.0_sp-s*ak)*(1.0_sp+s*ak)
ellpi_v=s*(rf(cc,q,spread(1.0_sp,1,size(phi)))-enss*&
    rj(cc,q,spread(1.0_sp,1,size(phi)),1.0_sp+enss)/3.0_sp)
END FUNCTION ellpi_v
```

⋆ ⋆ ⋆

```
SUBROUTINE sncndn(uu,emmc,sn,cn,dn)
USE nrtype; USE nrutil, ONLY : nrerror
IMPLICIT NONE
REAL(SP), INTENT(IN) :: uu,emmc
REAL(SP), INTENT(OUT) :: sn,cn,dn
```

Returns the Jacobian elliptic functions sn(u, k_c), cn(u, k_c), and dn(u, k_c). Here uu $= u$, while emmc $= k_c^2$.

```
REAL(SP), PARAMETER :: CA=0.0003_sp        The accuracy is the square of CA.
INTEGER(I4B), PARAMETER :: MAXIT=13
INTEGER(I4B) :: i,ii,l
REAL(SP) :: a,b,c,d,emc,u
REAL(SP), DIMENSION(MAXIT) :: em,en
LOGICAL(LGT) :: bo
emc=emmc
u=uu
if (emc /= 0.0) then
    bo=(emc < 0.0)
    if (bo) then
        d=1.0_sp-emc
        emc=-emc/d
        d=sqrt(d)
        u=d*u
    end if
    a=1.0
    dn=1.0
    do i=1,MAXIT
        l=i
        em(i)=a
        emc=sqrt(emc)
        en(i)=emc
        c=0.5_sp*(a+emc)
        if (abs(a-emc) <= CA*a) exit
        emc=a*emc
        a=c
    end do
    if (i > MAXIT) call nrerror('sncndn: convergence failed')
    u=c*u
    sn=sin(u)
    cn=cos(u)
    if (sn /= 0.0) then
        a=cn/sn
        c=a*c
        do ii=l,1,-1
            b=em(ii)
```

```
            a=c*a
            c=dn*c
            dn=(en(ii)+a)/(b+a)
            a=c/b
        end do
        a=1.0_sp/sqrt(c**2+1.0_sp)
        sn=sign(a,sn)
        cn=c*sn
    end if
    if (bo) then
        a=dn
        dn=cn
        cn=a
        sn=sn/d
    end if
else
    cn=1.0_sp/cosh(u)
    dn=cn
    sn=tanh(u)
end if
END SUBROUTINE sncndn
```

⋆ ⋆ ⋆

```
MODULE hypgeo_info
USE nrtype
COMPLEX(SPC) :: hypgeo_aa,hypgeo_bb,hypgeo_cc,hypgeo_dz,hypgeo_z0
END MODULE hypgeo_info
```

```
FUNCTION hypgeo(a,b,c,z)
USE nrtype
USE hypgeo_info
USE nr, ONLY : bsstep,hypdrv,hypser,odeint
IMPLICIT NONE
COMPLEX(SPC), INTENT(IN) :: a,b,c,z
COMPLEX(SPC) :: hypgeo
REAL(SP), PARAMETER :: EPS=1.0e-6_sp
```

Complex hypergeometric function $_2F_1$ for complex $a, b, c,$ and z, by direct integration of the hypergeometric equation in the complex plane. The branch cut is taken to lie along the real axis, Re $z > 1$.

Parameter: EPS is an accuracy parameter.

```
COMPLEX(SPC), DIMENSION(2) :: y
REAL(SP), DIMENSION(4) :: ry
if (real(z)**2+aimag(z)**2 <= 0.25) then        Use series...
    call hypser(a,b,c,z,hypgeo,y(2))
    RETURN
else if (real(z) < 0.0) then                    ...or pick a starting point for the path
    hypgeo_z0=cmplx(-0.5_sp,0.0_sp,kind=spc)        integration.
else if (real(z) <= 1.0) then
    hypgeo_z0=cmplx(0.5_sp,0.0_sp,kind=spc)
else
    hypgeo_z0=cmplx(0.0_sp,sign(0.5_sp,aimag(z)),kind=spc)
end if
hypgeo_aa=a                                     Load the module variables, used to pass
hypgeo_bb=b                                         parameters "over the head" of odeint
hypgeo_cc=c                                         to hypdrv.
hypgeo_dz=z-hypgeo_z0
call hypser(hypgeo_aa,hypgeo_bb,hypgeo_cc,hypgeo_z0,y(1),y(2))
  Get starting function and derivative.
ry(1:4:2)=real(y)
```

```
ry(2:4:2)=aimag(y)
call odeint(ry,0.0_sp,1.0_sp,EPS,0.1_sp,0.0001_sp,hypdrv,bsstep)
  The arguments to odeint are the vector of independent variables, the starting and ending
  values of the dependent variable, the accuracy parameter, an initial guess for stepsize, a
  minimum stepsize, and the names of the derivative routine and the (here Bulirsch-Stoer)
  stepping routine.
y=cmplx(ry(1:4:2),ry(2:4:2),kind=spc)
hypgeo=y(1)
END FUNCTION hypgeo
```

```
SUBROUTINE hypser(a,b,c,z,series,deriv)
USE nrtype; USE nrutil, ONLY : nrerror
IMPLICIT NONE
COMPLEX(SPC), INTENT(IN) :: a,b,c,z
COMPLEX(SPC), INTENT(OUT) :: series,deriv
   Returns the hypergeometric series 2F1 and its derivative, iterating to machine accuracy.
   For cabs(z) <= 1/2 convergence is quite rapid.
INTEGER(I4B) :: n
INTEGER(I4B), PARAMETER :: MAXIT=1000
COMPLEX(SPC) :: aa,bb,cc,fac,temp
deriv=cmplx(0.0_sp,0.0_sp,kind=spc)
fac=cmplx(1.0_sp,0.0_sp,kind=spc)
temp=fac
aa=a
bb=b
cc=c
do n=1,MAXIT
    fac=((aa*bb)/cc)*fac
    deriv=deriv+fac
    fac=fac*z/n
    series=temp+fac
    if (series == temp) RETURN
    temp=series
    aa=aa+1.0
    bb=bb+1.0
    cc=cc+1.0
end do
call nrerror('hypser: convergence failure')
END SUBROUTINE hypser
```

```
SUBROUTINE hypdrv(s,ry,rdyds)
USE nrtype
USE hypgeo_info
IMPLICIT NONE
REAL(SP), INTENT(IN) :: s
REAL(SP), DIMENSION(:), INTENT(IN) :: ry
REAL(SP), DIMENSION(:), INTENT(OUT) :: rdyds
   Derivative subroutine for the hypergeometric equation; see text equation (5.14.4).
COMPLEX(SPC), DIMENSION(2) :: y,dyds
COMPLEX(SPC) :: z
y=cmplx(ry(1:4:2),ry(2:4:2),kind=spc)
z=hypgeo_z0+s*hypgeo_dz
dyds(1)=y(2)*hypgeo_dz
dyds(2)=((hypgeo_aa*hypgeo_bb)*y(1)-(hypgeo_cc-&
    ((hypgeo_aa+hypgeo_bb)+1.0_sp)*z)*y(2))*hypgeo_dz/(z*(1.0_sp-z))
rdyds(1:4:2)=real(dyds)
rdyds(2:4:2)=aimag(dyds)
END SUBROUTINE hypdrv
```

f90 Notice that the real array (of length 4) `ry` is immediately mapped into a complex array of length 2, and that the process is reversed at the end of the routine with `rdyds`. In Fortran 77 no such mapping is necessary: the calling program sends real arguments, and the Fortran 77 `hypdrv` simply interprets what is sent as complex. Fortran 90's stronger typing does not encourage (and, practically, does not allow) this convenience; but it is a small price to pay for the vastly increased error-checking capabilities of a strongly typed language.

Chapter B7. Random Numbers

One might think that good random number generators, including those in Volume 1, should last forever. The world of computing changes very rapidly, however:

- When Volume 1 was published, it was unusual, except on the fastest supercomputers, to "exhaust" a 32-bit random number generator, that is, to call for all 2^{32} sequential random values in its periodic sequence. Now, this is feasible, and not uncommon, on fast desktop workstations. A useful generator today must have a minimum of 64 bits of state space, and generally somewhat more.
- Before Fortran 90, the Fortran language had no standardized calling sequence for random numbers. Now, although there is still no standard *algorithm* defined by the language (rightly, we think), there is at least a standard calling sequence, exemplified in the intrinsics `random_number` and `random_seed`.
- The rise of parallel computing places new algorithmic demands on random generators. The classic algorithms, which compute each random value from the previous one, evidently need generalization to a parallel environment.
- New algorithms and techniques have been discovered, in some cases significantly faster than their predecessors.

These are the reasons that we have decided to implement, in Fortran 90, different uniform random number generators from those in Volume 1's Fortran 77 implementations. We hasten to add that there is nothing wrong with any of the generators in Volume 1. That volume's `ran0` and `ran1` routines are, to our knowledge, completely adequate as 32-bit generators; `ran2` has a 64-bit state space, and our previous offer of $1000 for *any* demonstrated failure in the algorithm has never yet been claimed (see [1]).

Before we launch into the discussion of parallelizable generators with Fortran 90 calling conventions, we want to attend to the continuing needs of longtime "`x=ran(idum)`" users with purely serial machines. If you are a satisfied user of Volume 1's `ran0`, `ran1`, or `ran2` Fortran 77 versions, you are in this group. The following routine, `ran`, preserves those routines' calling conventions, is considerably faster than `ran2`, and does not suffer from the old `ran0` or `ran1`'s 32-bit period exhaustion limitation. It is completely portable to all Fortran 90 environments. We recommend `ran` as the plug-compatible replacement for the old `ran0`, `ran1`, and `ran2`, and we happily offer exactly the same $1000 reward terms as were (and are still) offered on the old `ran2`.

```
FUNCTION ran(idum)
IMPLICIT NONE
INTEGER, PARAMETER :: K4B=selected_int_kind(9)
INTEGER(K4B), INTENT(INOUT) :: idum
REAL :: ran
    "Minimal" random number generator of Park and Miller combined with a Marsaglia shift
    sequence. Returns a uniform random deviate between 0.0 and 1.0 (exclusive of the endpoint
    values). This fully portable, scalar generator has the "traditional" (not Fortran 90) calling
    sequence with a random deviate as the returned function value: call with idum a negative
    integer to initialize; thereafter, do not alter idum except to reinitialize. The period of this
    generator is about 3.1 × 10^18.
INTEGER(K4B), PARAMETER :: IA=16807,IM=2147483647,IQ=127773,IR=2836
REAL, SAVE :: am
INTEGER(K4B), SAVE :: ix=-1,iy=-1,k
if (idum <= 0 .or. iy < 0) then                 Initialize.
    am=nearest(1.0,-1.0)/IM
    iy=ior(ieor(888889999,abs(idum)),1)
    ix=ieor(777755555,abs(idum))
    idum=abs(idum)+1                            Set idum positive.
end if
ix=ieor(ix,ishft(ix,13))                        Marsaglia shift sequence with period 2^32 - 1.
ix=ieor(ix,ishft(ix,-17))
ix=ieor(ix,ishft(ix,5))
k=iy/IQ                                         Park-Miller sequence by Schrage's method,
iy=IA*(iy-k*IQ)-IR*k                                period 2^31 - 2.
if (iy < 0) iy=iy+IM
ran=am*ior(iand(IM,ieor(ix,iy)),1)              Combine the two generators with masking to
END FUNCTION ran                                    ensure nonzero value.
```

This is a good place to discuss a new bit of algorithmics that has crept into ran, above, and even more strongly affects all of our new random number generators, below. Consider:

```
ix=ieor(ix,ishft(ix,13))
ix=ieor(ix,ishft(ix,-17))
ix=ieor(ix,ishft(ix,5))
```

These lines update a 32-bit integer ix, which cycles pseudo-randomly through a full period of $2^{32} - 1$ values (excluding zero) before repeating. Generators of this type have been extensively explored by Marsaglia (see [2]), who has kindly communicated some additional results to us in advance of publication. For convenience, we will refer to generators of this sort as "Marsaglia shift registers."

Useful properties of Marsaglia shift registers are (i) they are very fast on most machines, since they use only fast logical operations, and (ii) the bit-mixing that they induce is quite different in character from that induced by arithmetic operations such as are used in linear congruential generators (see Volume 1) or lagged Fibonacci generators (see below). Thus, the combination of a Marsaglia shift register with another, algorithmically quite different generator is a powerful way to suppress any residual correlations or other weaknesses in the other generator. Indeed, Marsaglia finds (and we concur) that the above generator (with constants $13, -17, 5$, as shown) is *by itself* about as good as any 32-bit random generator.

Here is a very brief outline of the theory behind these generators: Consider the 32 bits of the integer as components in a vector of length 32, in a linear space where addition and multiplication are done modulo 2. Noting that exclusive-or (ieor) is the same as addition, each of the three lines in the updating can be written as the action of a 32×32 matrix on a vector, where the matrix is all zeros except for ones on the diagonal, and on exactly one super- or subdiagonal (corresponding to positive or negative second arguments in `ishft`). Denote this matrix as $\mathbf{S}_k$, where k is the shift argument. Then, one full step of updating (three lines of code, above) corresponds to multiplication by the matrix $\mathbf{T} \equiv \mathbf{S}_{k_3}\mathbf{S}_{k_2}\mathbf{S}_{k_1}$.

One next needs to find triples of integers (k_1, k_2, k_3), for example $(13, -17, 5)$, that give the full $M \equiv 2^{32} - 1$ period. Necessary and sufficient conditions are that $\mathbf{T}^M = \mathbf{1}$ (the identity matrix), and that $\mathbf{T}^N \neq \mathbf{1}$ for these five values of N: $N = 3 \times 5 \times 17 \times 257$, $N = 3 \times 5 \times 17 \times 65537$, $N = 3 \times 5 \times 257 \times 65537$, $N = 3 \times 17 \times 257 \times 65537$, $N = 5 \times 17 \times 257 \times 65537$. (Note that each of the five prime factors of M is omitted one at a time to get the five values of N.) The required large powers of $\mathbf{T}$ are readily computed by successive squarings, requiring only on the order of $32^3 \log M$ operations. With this machinery, one can find full-period triples (k_1, k_2, k_3) by exhaustive search, at reasonable cost.

Not all such triples are equally good as generators of random integers, however. Marsaglia subjects candidate values to a battery of tests for randomness, and we have ourselves applied various tests. This stage of winnowing is as much art as science, because all 32-bit generators can be made to exhibit signs of failure due to period exhaustion (if for no other reason). "Good" triples, in order of our preference, are $(13, -17, 5)$, $(5, -13, 6)$, $(5, -9, 7)$, $(13, -17, 15)$, $(16, -7, 11)$. When a full-period triple is good, its reverse is also full-period, and also generally good. A good *quadruple* due to Marsaglia (generalizing the above in the obvious way) is $(-4, 8, -1, 5)$. We would not recommend relying on any single Marsaglia shift generator (nor on any other simple generator) *by itself*. Two or more generators, of quite different types, should be combined [1].

⋆ ⋆ ⋆

Let us now discuss explicitly the needs of *parallel* random number generators. The general scheme, from the user's perspective, is that of Fortran 90's intrinsic `random_number`: A statement like `call ran1(harvest)` (where `ran1` will be one of our portable replacements for the compiler-dependent `random_number`) should fill the real array `harvest` with pseudo-random real values in the range $(0, 1)$. Of course, we want the underlying machinery to be completely parallel, that is, no do-loops of order $N \equiv$ `size(harvest)`.

A first design decision is whether to replicate the state-space across the parallel dimension N, i.e., whether to reserve storage for essentially N scalar generators. Although there are various schemes that avoid doing this (e.g., mapping a single, smaller, state space into N different output values on each call), we think that it is a memory cost well worth paying in return for achieving a less exotic (and thus better tested) algorithm. However, this choice dictates that we must keep the state space *per component* quite small. We have settled on five or fewer 32-bit words of state space per component as a reasonable limit. Some otherwise interesting and well tested methods (such as Knuth's subtractive generator, implemented in Volume 1 as `ran3`) are ruled out by this constraint.

A second design decision is how to initialize the parallel state space, so that different parallel components produce different sequences, and so that there is an acceptable degree of randomness *across* the parallel dimension, as well as *between successive calls* of the generator. Each component starts its life with one and only one unique identifier, its component index n in the range $1 \ldots N$. One is tempted simply to hash the values n into the corresponding components of initial state space. "Random" hashing is a bad idea, however, because different n's will produce identical 32-bit hash results by chance when N is no larger than $\sim 2^{16}$. We therefore prefer to use a kind of reversible pseudo-encryption (similar to the routine psdes in Volume 1 and below) which guarantees causally that different n's produce different state space initializations.

f90 The machinery for allocating, deallocating, and initializing the state space, including provision of a user interface for getting or putting the contents of the state space (as in the intrinsic random_seed) is fairly complicated. Rather than duplicate it in each different random generator that we provide, we have consolidated it in a single module, ran_state, whose contents we will now discuss. Such a discussion is necessarily technical, if not arcane; on first reading, you may wish to skip ahead to the actual new routines ran0, ran1, and ran2. If you do so, you will need to know only that ran_state provides each vector random routine with five 32-bit vectors of state information, denoted iran, jran, kran, mran, nran. (The overloaded scalar generators have five corresponding 32-bit scalars, denoted iran0, etc.)

```
MODULE ran_state
    This module supports the random number routines ran0, ran1, ran2, and ran3. It pro-
    vides each generator with five integers (for vector versions, five vectors of integers), for
    use as internal state space. The first three integers (iran, jran, kran) are maintained
    as nonnegative values, while the last two (mran, nran) have 32-bit nonzero values. Also
    provided by this module is support for initializing or reinitializing the state space to a desired
    standard sequence number, hashing the initial values to random values, and allocating and
    deallocating the internal workspace.
USE nrtype
IMPLICIT NONE
INTEGER, PARAMETER :: K4B=selected_int_kind(9)
   Independent of the usual integer kind I4B, we need a kind value for (ideally) 32-bit integers.
INTEGER(K4B), PARAMETER :: hg=huge(1_K4B), hgm=-hg, hgng=hgm-1
INTEGER(K4B), SAVE :: lenran=0, seq=0
INTEGER(K4B), SAVE :: iran0,jran0,kran0,nran0,mran0,rans
INTEGER(K4B), DIMENSION(:,:), POINTER, SAVE :: ranseeds
INTEGER(K4B), DIMENSION(:), POINTER, SAVE :: iran,jran,kran, &
    nran,mran,ranv
REAL(SP), SAVE :: amm
INTERFACE ran_hash                    Scalar and vector versions of the hashing procedure.
    MODULE PROCEDURE ran_hash_s, ran_hash_v
END INTERFACE
CONTAINS
```

(We here intersperse discussion with the listing of the module.) The module defines K4B as an integer KIND that is intended to be 32 bits. If your machine doesn't have 32-bit integers (hard to believe!) this will be caught later, and an error message generated. The definition of the parameters hg, hgm, and hgng makes an assumption about 32-bit integers that goes beyond the strict Fortran 90 integer model, that the magnitude of the most negative representable integer is greater by one than that of the most positive representable integer. This is a property of the *two's complement arithmetic* that is used on virtually all modern machines (see, e.g., [3]).

The global variables rans (for scalar) and ranv (for vector) are used by all of our routines to store the *integer* value associated with the most recently returned call. You can access these (with a "USE ran_state" statement) if you want integer, rather than real, random deviates.

The first routine, ran_init, is called by routines later in the chapter to initialize their state space. It is *not* intended to be called from a user's program.

```
SUBROUTINE ran_init(length)
USE nrtype; USE nrutil, ONLY : arth,nrerror,reallocate
IMPLICIT NONE
INTEGER(K4B), INTENT(IN) :: length
    Initialize or reinitialize the random generator state space to vectors of size length. The
    saved variable seq is hashed (via calls to the module routine ran_hash) to create unique
    starting seeds, different for each vector component.
INTEGER(K4B) :: new,j,hgt
if (length < lenran) RETURN                     Simply return if enough space is already al-
hgt=hg                                              located.
  The following lines check that kind value K4B is in fact a 32-bit integer with the usual properties
  that we expect it to have (under negation and wrap-around addition). If all of these tests are
  satisfied, then the routines that use this module are portable, even though they go beyond
  Fortran 90's integer model.
if (hg /= 2147483647) call nrerror('ran_init: arith assump 1 fails')
if (hgng >= 0) call nrerror('ran_init: arith assump 2 fails')
if (hgt+1 /= hgng) call nrerror('ran_init: arith assump 3 fails')
if (not(hg) >= 0) call nrerror('ran_init: arith assump 4 fails')
if (not(hgng) < 0) call nrerror('ran_init: arith assump 5 fails')
if (hg+hgng >= 0) call nrerror('ran_init: arith assump 6 fails')
if (not(-1_k4b) < 0) call nrerror('ran_init: arith assump 7 fails')
if (not(0_k4b) >= 0) call nrerror('ran_init: arith assump 8 fails')
if (not(1_k4b) >= 0) call nrerror('ran_init: arith assump 9 fails')
if (lenran > 0) then                            Reallocate space, or ...
    ranseeds=>reallocate(ranseeds,length,5)
    ranv=>reallocate(ranv,length-1)
    new=lenran+1
else                                            allocate space.
    allocate(ranseeds(length,5))
    allocate(ranv(length-1))
    new=1                                       Index of first location not yet initialized.
    amm=nearest(1.0_sp,-1.0_sp)/hgng
      Use of nearest is to ensure that returned random deviates are strictly less than 1.0.
    if (amm*hgng >= 1.0 .or. amm*hgng <= 0.0) &
        call nrerror('ran_init: arth assump 10 fails')
end if
  Set starting values, unique by seq and vector component.
ranseeds(new:,1)=seq
ranseeds(new:,2:5)=spread(arth(new,1,size(ranseeds(new:,1))),2,4)
do j=1,4                                        Hash them.
    call ran_hash(ranseeds(new:,j),ranseeds(new:,j+1))
end do
where (ranseeds(new:,1:3) < 0) &                Enforce nonnegativity.
    ranseeds(new:,1:3)=not(ranseeds(new:,1:3))
where (ranseeds(new:,4:5) == 0) ranseeds(new:,4:5)=1    Enforce nonzero.
if (new == 1) then                              Set scalar seeds.
    iran0=ranseeds(1,1)
    jran0=ranseeds(1,2)
    kran0=ranseeds(1,3)
    mran0=ranseeds(1,4)
    nran0=ranseeds(1,5)
    rans=nran0
end if
if (length > 1) then                            Point to vector seeds.
    iran => ranseeds(2:,1)
    jran => ranseeds(2:,2)
    kran => ranseeds(2:,3)
    mran => ranseeds(2:,4)
    nran => ranseeds(2:,5)
    ranv = nran
```

```
end if
lenran=length
END SUBROUTINE ran_init
```

f90 `hgt=hg ... if (hgt+1 /= hgng)` Bit of dirty laundry here! We are testing whether the most positive integer hg wraps around to the most negative integer hgng when 1 is added to it. We can't just write hg+1, since some compilers will evaluate this at compile time and return an overflow error message. If your compiler sees through the charade of the temporary variable hgt, you'll have to find another way to trick it.

`amm=nearest(1.0_sp,-1.0_sp)/hgng...` Logically, amm should be a parameter; but the nearest intrinsic is trouble-prone in the initialization expression for a parameter (named constant), so we compute this at run time. We then check that amm, when multiplied by the largest possible negative integer, does not equal or exceed unity. (Our random deviates are guaranteed never to equal zero or unity exactly.)

You might wonder why amm is negative, and why we multiply it by negative integers to get positive random deviates. The answer, which will become manifest in the random generators given below, is that we want to use the fast not operation on integers to convert them to nonzero values of all one sign. This is possible if the conversion is to negative values, since not(i) is negative for all nonnegative i. If the conversion were to positive values, we would have problems both with zero (its sign bit is already positive) and hgng (since not(hgng) is generally zero).

```
iran0=ranseeds(1,1) ...
iran => ranseeds(2:,1)...
```

The initial state information is stored in ranseeds, a two-dimensional array whose column (second) index ranges from 1 to 5 over the state variables. ranseeds(1,:) is reserved for scalar random generators, while ranseeds(2:,:) is for vector-parallel generators. The ranseeds array is made available to vector generators through the pointers iran, jran, kran, mran, and nran. The corresponding scalar values, iran0,..., nran0 are simply global variables, not pointers, because the overhead of addressing a scalar through a pointer is often too great. (We will have to copy these scalar values back into ranseeds when it, rarely, needs to be addressed as an array.)

`call ran_hash(...)` Unique, and random, initial state information is obtained by putting a user-settable "sequence number" into iran, a component number into jran, and hashing this pair. Then jran and kran are hashed, kran and mran are hashed, and so forth.

```
SUBROUTINE ran_deallocate
    User interface to release the workspace used by the random number routines.
if (lenran > 0) then
    deallocate(ranseeds,ranv)
    nullify(ranseeds,ranv,iran,jran,kran,mran,nran)
    lenran = 0
end if
END SUBROUTINE ran_deallocate
```

The above routine is supplied as a user interface for deallocating all the state space storage.

```
SUBROUTINE ran_seed(sequence,size,put,get)
IMPLICIT NONE
INTEGER, OPTIONAL, INTENT(IN) :: sequence
INTEGER, OPTIONAL, INTENT(OUT) :: size
INTEGER, DIMENSION(:), OPTIONAL, INTENT(IN) :: put
INTEGER, DIMENSION(:), OPTIONAL, INTENT(OUT) :: get
   User interface for seeding the random number routines. Syntax is exactly like Fortran 90's
   random_seed routine, with one additional argument keyword: sequence, set to any inte-
   ger value, causes an immediate new initialization, seeded by that integer.
if (present(size)) then
    size=5*lenran
else if (present(put)) then
    if (lenran == 0) RETURN
    ranseeds=reshape(put,shape(ranseeds))
    where (ranseeds(:,1:3) < 0) ranseeds(:,1:3)=not(ranseeds(:,1:3))
      Enforce nonnegativity and nonzero conditions on any user-supplied seeds.
    where (ranseeds(:,4:5) == 0) ranseeds(:,4:5)=1
    iran0=ranseeds(1,1)
    jran0=ranseeds(1,2)
    kran0=ranseeds(1,3)
    mran0=ranseeds(1,4)
    nran0=ranseeds(1,5)
else if (present(get)) then
    if (lenran == 0) RETURN
    ranseeds(1,1:5)=(/ iran0,jran0,kran0,mran0,nran0 /)
    get=reshape(ranseeds,shape(get))
else if (present(sequence)) then
    call ran_deallocate
    seq=sequence
end if
END SUBROUTINE ran_seed
```

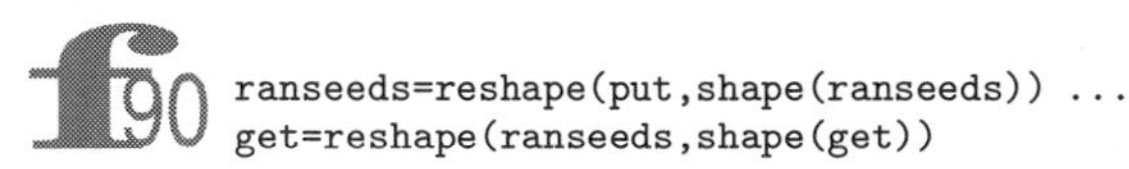

```
ranseeds=reshape(put,shape(ranseeds)) ...
get=reshape(ranseeds,shape(get))
```

Fortran 90's convention is that random state space is a one-dimensional array, so we map to this on both the get and put keywords.

```
iran0=...jran0=...kran0=...
ranseeds(1,1:5)=(/ iran0,jran0,kran0,mran0,nran0 /)
```

It's much more convenient to set a vector from a bunch of scalars then the other way around.

```
SUBROUTINE ran_hash_s(il,ir)
IMPLICIT NONE
INTEGER(K4B), INTENT(INOUT) :: il,ir
   DES-like hashing of two 32-bit integers, using shifts, xor's, and adds to make the internal
   nonlinear function.
INTEGER(K4B) :: is,j
do j=1,4
    is=ir
    ir=ieor(ir,ishft(ir,5))+1422217823           The various constants are chosen to give
    ir=ieor(ir,ishft(ir,-16))+1842055030            good bit mixing and should not be
    ir=ieor(ir,ishft(ir,9))+80567781                changed.
    ir=ieor(il,ir)
    il=is
end do
END SUBROUTINE ran_hash_s
```

```
SUBROUTINE ran_hash_v(il,ir)
IMPLICIT NONE
INTEGER(K4B), DIMENSION(:), INTENT(INOUT) :: il,ir
   Vector version of ran_hash_s.
INTEGER(K4B), DIMENSION(size(il)) :: is
INTEGER(K4B) :: j
do j=1,4
    is=ir
    ir=ieor(ir,ishft(ir,5))+1422217823
    ir=ieor(ir,ishft(ir,-16))+1842055030
    ir=ieor(ir,ishft(ir,9))+80567781
    ir=ieor(il,ir)
    il=is
end do
END SUBROUTINE ran_hash_v

END MODULE ran_state
```

The lines

```
ir=ieor(ir,ishft(ir,5))+1422217823
ir=ieor(ir,ishft(ir,-16))+1842055030
ir=ieor(ir,ishft(ir,9))+80567781
```

are *not* a Marsaglia shift sequence, though they resemble one. Instead, they implement a fast, nonlinear function on `ir` that we use as the "S-box" in a DES-like hashing algorithm. (See Volume 1, §7.5.) The triplet $(5, -16, 9)$ is *not* chosen to give a full period Marsaglia sequence — it doesn't. Instead it is chosen as being particularly good at separating in Hamming distance (i.e., number of nonidentical bits) two initially close values of `ir` (e.g., differing by only one bit). The large integer constants are chosen by a similar criterion. Note that the wrap-around of addition without generating an overflow error condition, which was tested in `ran_init`, is relied upon here.

⋆ ⋆ ⋆

```
SUBROUTINE ran0_s(harvest)
USE nrtype
USE ran_state, ONLY: K4B,amm,lenran,ran_init,iran0,jran0,kran0,nran0,rans
IMPLICIT NONE
REAL(SP), INTENT(OUT) :: harvest
   Lagged Fibonacci generator combined with a Marsaglia shift sequence. Returns as harvest
   a uniform random deviate between 0.0 and 1.0 (exclusive of the endpoint values). This gen-
   erator has the same calling and initialization conventions as Fortran 90's random_number
   routine. Use ran_seed to initialize or reinitialize to a particular sequence. The period of
   this generator is about 2.0 × 10^28, and it fully vectorizes. Validity of the integer model
   assumed by this generator is tested at initialization.
if (lenran < 1) call ran_init(1)                   Initialization routine in ran_state.
rans=iran0-kran0                                   Update Fibonacci generator, which
if (rans < 0) rans=rans+2147483579_k4b                has period p^2+p+1, p = 2^31 -
iran0=jran0                                           69.
jran0=kran0
kran0=rans
nran0=ieor(nran0,ishft(nran0,13))                  Update Marsaglia shift sequence with
nran0=ieor(nran0,ishft(nran0,-17))                    period 2^32 - 1.
nran0=ieor(nran0,ishft(nran0,5))
rans=ieor(nran0,rans)                              Combine the generators.
harvest=amm*merge(rans,not(rans), rans<0 )         Make the result positive definite (note
END SUBROUTINE ran0_s                                 that amm is negative).
```

```
SUBROUTINE ran0_v(harvest)
USE nrtype
USE ran_state, ONLY: K4B,amm,lenran,ran_init,iran,jran,kran,nran,ranv
IMPLICIT NONE
REAL(SP), DIMENSION(:), INTENT(OUT) :: harvest
INTEGER(K4B) :: n
n=size(harvest)
if (lenran < n+1) call ran_init(n+1)
ranv(1:n)=iran(1:n)-kran(1:n)
where (ranv(1:n) < 0) ranv(1:n)=ranv(1:n)+2147483579_k4b
iran(1:n)=jran(1:n)
jran(1:n)=kran(1:n)
kran(1:n)=ranv(1:n)
nran(1:n)=ieor(nran(1:n),ishft(nran(1:n),13))
nran(1:n)=ieor(nran(1:n),ishft(nran(1:n),-17))
nran(1:n)=ieor(nran(1:n),ishft(nran(1:n),5))
ranv(1:n)=ieor(nran(1:n),ranv(1:n))
harvest=amm*merge(ranv(1:n),not(ranv(1:n)), ranv(1:n)<0 )
END SUBROUTINE ran0_v
```

This is the simplest, and fastest, of the generators provided. It combines a subtractive Fibonacci generator (Number 6 in ref. [1], and one of the generators in Marsaglia and Zaman's mzran) with a Marsaglia shift sequence. On typical machines it is only 20% or so faster than ran1, however; so we recommend the latter preferentially. While we know of no weakness in ran0, we are not offering a prize for finding a weakness. ran0 does have the feature, useful if you have a machine with nonstandard arithmetic, that it does not go beyond Fortran 90's assumed integer model.

Note that ran0_s and ran0_v are overloaded by the module nr onto the single name ran0 (and similarly for the routines below).

⋆ ⋆ ⋆

```
SUBROUTINE ran1_s(harvest)
USE nrtype
USE ran_state, ONLY: K4B,amm,lenran,ran_init, &
    iran0,jran0,kran0,nran0,mran0,rans
IMPLICIT NONE
REAL(SP), INTENT(OUT) :: harvest
```

Lagged Fibonacci generator combined with two Marsaglia shift sequences. On output, returns as harvest a uniform random deviate between 0.0 and 1.0 (exclusive of the endpoint values). This generator has the same calling and initialization conventions as Fortran 90's random_number routine. Use ran_seed to initialize or reinitialize to a particular sequence. The period of this generator is about 8.5×10^{37}, and it fully vectorizes. Validity of the integer model assumed by this generator is tested at initialization.

```
if (lenran < 1) call ran_init(1)             Initialization routine in ran_state.
rans=iran0-kran0                             Update Fibonacci generator, which
if (rans < 0) rans=rans+2147483579_k4b          has period p^2+p+1, p = 2^31 - 69.
iran0=jran0
jran0=kran0
kran0=rans
nran0=ieor(nran0,ishft(nran0,13))            Update Marsaglia shift sequence.
nran0=ieor(nran0,ishft(nran0,-17))
nran0=ieor(nran0,ishft(nran0,5))
```

Once only per cycle, advance sequence by 1, shortening its period to $2^{32}-2$.

```
if (nran0 == 1) nran0=270369_k4b
mran0=ieor(mran0,ishft(mran0,5))             Update Marsaglia shift sequence with
mran0=ieor(mran0,ishft(mran0,-13))              period 2^32 - 1.
mran0=ieor(mran0,ishft(mran0,6))
```

```
rans=ieor(nran0,rans)+mran0
```

Combine the generators. The above statement has wrap-around addition.

```
harvest=amm*merge(rans,not(rans), rans<0 )
END SUBROUTINE ran1_s
```

Make the result positive definite (note that amm is negative).

```
SUBROUTINE ran1_v(harvest)
USE nrtype
USE ran_state, ONLY: K4B,amm,lenran,ran_init, &
    iran,jran,kran,nran,mran,ranv
IMPLICIT NONE
REAL(SP), DIMENSION(:), INTENT(OUT) :: harvest
INTEGER(K4B) :: n
n=size(harvest)
if (lenran < n+1) call ran_init(n+1)
ranv(1:n)=iran(1:n)-kran(1:n)
where (ranv(1:n) < 0) ranv(1:n)=ranv(1:n)+2147483579_k4b
iran(1:n)=jran(1:n)
jran(1:n)=kran(1:n)
kran(1:n)=ranv(1:n)
nran(1:n)=ieor(nran(1:n),ishft(nran(1:n),13))
nran(1:n)=ieor(nran(1:n),ishft(nran(1:n),-17))
nran(1:n)=ieor(nran(1:n),ishft(nran(1:n),5))
where (nran(1:n) == 1) nran(1:n)=270369_k4b
mran(1:n)=ieor(mran(1:n),ishft(mran(1:n),5))
mran(1:n)=ieor(mran(1:n),ishft(mran(1:n),-13))
mran(1:n)=ieor(mran(1:n),ishft(mran(1:n),6))
ranv(1:n)=ieor(nran(1:n),ranv(1:n))+mran(1:n)
harvest=amm*merge(ranv(1:n),not(ranv(1:n)), ranv(1:n)<0 )
END SUBROUTINE ran1_v
```

The routine ran1 combines *three* fast generators: the two used in ran0, plus an additional (different) Marsaglia shift sequence. The last generator is combined via an addition that can wrap-around.

We think that, within the limits of its floating-point precision, ran1 provides perfect random numbers. We will pay \$1000 to the first reader who convinces us otherwise (by exhibiting a statistical test that ran1 fails in a nontrivial way, excluding the ordinary limitations of a floating-point representation).

⋆ ⋆ ⋆

```
SUBROUTINE ran2_s(harvest)
USE nrtype
USE ran_state, ONLY: K4B,amm,lenran,ran_init, &
    iran0,jran0,kran0,nran0,mran0,rans
IMPLICIT NONE
REAL(SP), INTENT(OUT) :: harvest
```

Lagged Fibonacci generator combined with a Marsaglia shift sequence and a linear congruential generator. Returns as harvest a uniform random deviate between 0.0 and 1.0 (exclusive of the endpoint values). This generator has the same calling and initialization conventions as Fortran 90's random_number routine. Use ran_seed to initialize or reinitialize to a particular sequence. The period of this generator is about 8.5×10^{37}, and it fully vectorizes. Validity of the integer model assumed by this generator is tested at initialization.

```
if (lenran < 1) call ran_init(1)
```

Initialization routine in ran_state.

```
rans=iran0-kran0
if (rans < 0) rans=rans+2147483579_k4b
iran0=jran0
jran0=kran0
kran0=rans
```

Update Fibonacci generator, which has period p^2+p+1, $p = 2^{31} - 69$.

```
nran0=ieor(nran0,ishft(nran0,13))                    Update Marsaglia shift sequence with
nran0=ieor(nran0,ishft(nran0,-17))                       period 2^32 - 1.
nran0=ieor(nran0,ishft(nran0,5))
rans=iand(mran0,65535)
  Update the sequence m <- 69069m + 820265819 mod 2^32 using shifts instead of multiplies.
  Wrap-around addition (tested at initialization) is used.
mran0=ishft(3533*ishft(mran0,-16)+rans,16)+ &
    3533*rans+820265819_k4b
rans=ieor(nran0,kran0)+mran0                         Combine the generators.
harvest=amm*merge(rans,not(rans), rans<0 )           Make the result positive definite (note
END SUBROUTINE ran2_s                                    that amm is negative).
```

```
SUBROUTINE ran2_v(harvest)
USE nrtype
USE ran_state, ONLY: K4B,amm,lenran,ran_init, &
    iran,jran,kran,nran,mran,ranv
IMPLICIT NONE
REAL(SP), DIMENSION(:), INTENT(OUT) :: harvest
INTEGER(K4B) :: n
n=size(harvest)
if (lenran < n+1) call ran_init(n+1)
ranv(1:n)=iran(1:n)-kran(1:n)
where (ranv(1:n) < 0) ranv(1:n)=ranv(1:n)+2147483579_k4b
iran(1:n)=jran(1:n)
jran(1:n)=kran(1:n)
kran(1:n)=ranv(1:n)
nran(1:n)=ieor(nran(1:n),ishft(nran(1:n),13))
nran(1:n)=ieor(nran(1:n),ishft(nran(1:n),-17))
nran(1:n)=ieor(nran(1:n),ishft(nran(1:n),5))
ranv(1:n)=iand(mran(1:n),65535)
mran(1:n)=ishft(3533*ishft(mran(1:n),-16)+ranv(1:n),16)+ &
    3533*ranv(1:n)+820265819_k4b
ranv(1:n)=ieor(nran(1:n),kran(1:n))+mran(1:n)
harvest=amm*merge(ranv(1:n),not(ranv(1:n)), ranv(1:n)<0 )
END SUBROUTINE ran2_v
```

ran2, for use by readers whose caution is extreme, also combines three generators. The difference from ran1 is that each generator is based on a completely different method from the other two. The third generator, in this case, is a linear congruential generator, modulo 2^{32}. This generator relies extensively on wrap-around addition (which is automatically tested at initialization). On machines with fast arithmetic, ran2 is on the order of only 20% slower than ran1. We offer a $1000 bounty on ran2, with the same terms as for ran1, above.

⋆ ⋆ ⋆

```
SUBROUTINE expdev_s(harvest)
USE nrtype
USE nr, ONLY : ran1
IMPLICIT NONE
REAL(SP), INTENT(OUT) :: harvest
  Returns in harvest an exponentially distributed, positive, random deviate of unit mean,
  using ran1 as the source of uniform deviates.
REAL(SP) :: dum
call ran1(dum)
harvest=-log(dum)              We use the fact that ran1 never returns exactly 0 or 1.
END SUBROUTINE expdev_s
```

```
SUBROUTINE expdev_v(harvest)
USE nrtype
USE nr, ONLY : ran1
IMPLICIT NONE
REAL(SP), DIMENSION(:), INTENT(OUT) :: harvest
REAL(SP), DIMENSION(size(harvest)) :: dum
call ran1(dum)
harvest=-log(dum)
END SUBROUTINE expdev_v
```

f90 `call ran1(dum)` The only noteworthy thing about this line is its simplicity: Once all the machinery is in place, the random number generators are self-initializing (to the sequence defined by `seq` $= 0$), and (via overloading) usable with both scalar and vector arguments.

⋆ ⋆ ⋆

```
SUBROUTINE gasdev_s(harvest)
USE nrtype
USE nr, ONLY : ran1
IMPLICIT NONE
REAL(SP), INTENT(OUT) :: harvest
   Returns in harvest a normally distributed deviate with zero mean and unit variance, using
   ran1 as the source of uniform deviates.
REAL(SP) :: rsq,v1,v2
REAL(SP), SAVE :: g
LOGICAL, SAVE :: gaus_stored=.false.
if (gaus_stored) then                      We have an extra deviate handy,
    harvest=g                              so return it,
    gaus_stored=.false.                    and unset the flag.
else                                       We don't have an extra deviate handy, so
    do
        call ran1(v1)                      pick two uniform numbers in the square ex-
        call ran1(v2)                          tending from -1 to +1 in each direction,
        v1=2.0_sp*v1-1.0_sp
        v2=2.0_sp*v2-1.0_sp
        rsq=v1**2+v2**2                    see if they are in the unit circle,
        if (rsq > 0.0 .and. rsq < 1.0) exit
    end do                                 otherwise try again.
    rsq=sqrt(-2.0_sp*log(rsq)/rsq)         Now make the Box-Muller transformation to
    harvest=v1*rsq                             get two normal deviates. Return one and
    g=v2*rsq                                   save the other for next time.
    gaus_stored=.true.                     Set flag.
end if
END SUBROUTINE gasdev_s
```

```
SUBROUTINE gasdev_v(harvest)
USE nrtype; USE nrutil, ONLY : array_copy
USE nr, ONLY : ran1
IMPLICIT NONE
REAL(SP), DIMENSION(:), INTENT(OUT) :: harvest
REAL(SP), DIMENSION(size(harvest)) :: rsq,v1,v2
REAL(SP), ALLOCATABLE, DIMENSION(:), SAVE :: g
INTEGER(I4B) :: n,ng,nn,m
INTEGER(I4B), SAVE :: last_allocated=0
LOGICAL, SAVE :: gaus_stored=.false.
LOGICAL, DIMENSION(size(harvest)) :: mask
n=size(harvest)
if (n /= last_allocated) then
```

```
    if (last_allocated /= 0) deallocate(g)
    allocate(g(n))
    last_allocated=n
    gaus_stored=.false.
end if
if (gaus_stored) then
    harvest=g
    gaus_stored=.false.
else
    ng=1
    do
        if (ng > n) exit
        call ran1(v1(ng:n))
        call ran1(v2(ng:n))
        v1(ng:n)=2.0_sp*v1(ng:n)-1.0_sp
        v2(ng:n)=2.0_sp*v2(ng:n)-1.0_sp
        rsq(ng:n)=v1(ng:n)**2+v2(ng:n)**2
        mask(ng:n)=(rsq(ng:n)>0.0 .and. rsq(ng:n)<1.0)
        call array_copy(pack(v1(ng:n),mask(ng:n)),v1(ng:),nn,m)
        v2(ng:ng+nn-1)=pack(v2(ng:n),mask(ng:n))
        rsq(ng:ng+nn-1)=pack(rsq(ng:n),mask(ng:n))
        ng=ng+nn
    end do
    rsq=sqrt(-2.0_sp*log(rsq)/rsq)
    harvest=v1*rsq
    g=v2*rsq
    gaus_stored=.true.
end if
END SUBROUTINE gasdev_v
```

`if (n /= last_allocated) ...` We make the assumption that, in most cases, the size of `harvest` will not change between successive calls. Therefore, if it *does* change, we don't try to save the previously generated deviates that, half the time, will be around. If your use has rapidly varying sizes (or, even worse, calls alternating between two different sizes), you should remedy this inefficiency in the obvious way.

`call array_copy(pack(v1(ng:n),mask(ng:n)),v1(ng:),nn,m)` This is a variant of the pack-unpack method (see note to `factrl`, p. 1087). Different here is that we don't care which random deviates end up in which component. Thus, we can simply keep packing successful returns into `v1` and `v2` until they are full.

f90 Note also the use of `array_copy`, since we don't know in advance the length of the array returned by `pack`.

⋆ ⋆ ⋆

```
FUNCTION gamdev(ia)
USE nrtype; USE nrutil, ONLY : assert
USE nr, ONLY : ran1
IMPLICIT NONE
INTEGER(I4B), INTENT(IN) :: ia
REAL(SP) :: gamdev
    Returns a deviate distributed as a gamma distribution of integer order ia, i.e., a waiting
    time to the iath event in a Poisson process of unit mean, using ran1 as the source of
    uniform deviates.
REAL(SP) :: am,e,h,s,x,y,v(2),arr(5)
call assert(ia >= 1, 'gamdev arg')
if (ia < 6) then                                Use direct method, adding waiting times.
```

```
    call ran1(arr(1:ia))
    x=-log(product(arr(1:ia)))
else                                          Use rejection method.
    do
        call ran1(v)
        v(2)=2.0_sp*v(2)-1.0_sp               These three lines generate the tangent of a
        if (dot_product(v,v) > 1.0) cycle         random angle, i.e., are equivalent to
        y=v(2)/v(1)                               y = tan(πran(idum)).
        am=ia-1
        s=sqrt(2.0_sp*am+1.0_sp)
        x=s*y+am                              We decide whether to reject x:
        if (x <= 0.0) cycle                   Reject in region of zero probability.
        e=(1.0_sp+y**2)*exp(am*log(x/am)-s*y)     Ratio of probability function to
        call ran1(h)                                  comparison function.
        if (h <= e) exit                      Reject on basis of a second uniform deviate.
    end do
end if
gamdev=x
END FUNCTION gamdev
```

`x=-log(product(arr(1:ia)))` Why take the `log` of the product instead of the sum of the `log`s? Because `log` is assumed to be slower than multiply.

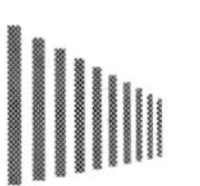

We don't have vector versions of the less commonly used deviate generators, `gamdev`, `poidev`, and `bnldev`.

⋆ ⋆ ⋆

```
FUNCTION poidev(xm)
USE nrtype
USE nr, ONLY : gammln,ran1
IMPLICIT NONE
REAL(SP), INTENT(IN) :: xm
REAL(SP) :: poidev
    Returns as a floating-point number an integer value that is a random deviate drawn from a
    Poisson distribution of mean xm, using ran1 as a source of uniform random deviates.
REAL(SP) :: em,harvest,t,y
REAL(SP), SAVE :: alxm,g,oldm=-1.0_sp,sq
  oldm is a flag for whether xm has changed since last call.
if (xm < 12.0) then                           Use direct method.
    if (xm /= oldm) then
        oldm=xm
        g=exp(-xm)                            If xm is new, compute the exponential.
    end if
    em=-1
    t=1.0
    do
        em=em+1.0_sp                          Instead of adding exponential deviates it is
        call ran1(harvest)                        equivalent to multiply uniform deviates.
        t=t*harvest                               We never actually have to take the log;
        if (t <= g) exit                          merely compare to the pre-computed ex-
    end do                                        ponential.
else                                          Use rejection method.
    if (xm /= oldm) then                      If xm has changed since the last call, then pre-
        oldm=xm                                   compute some functions that occur be-
        sq=sqrt(2.0_sp*xm)                        low.
        alxm=log(xm)
        g=xm*alxm-gammln(xm+1.0_sp)           The function gammln is the natural log of the
    end if                                        gamma function, as given in §6.1.
    do
```

```
        do
            call ran1(harvest)              y is a deviate from a Lorentzian comparison
            y=tan(PI*harvest)                   function.
            em=sq*y+xm                      em is y, shifted and scaled.
            if (em >= 0.0) exit             Reject if in regime of zero probability.
        end do
        em=int(em)                          The trick for integer-valued distributions.
        t=0.9_sp*(1.0_sp+y**2)*exp(em*alxm-gammln(em+1.0_sp)-g)
          The ratio of the desired distribution to the comparison function; we accept or reject
          by comparing it to another uniform deviate. The factor 0.9 is chosen so that t never
          exceeds 1.
        call ran1(harvest)
        if (harvest <= t) exit
    end do
end if
poidev=em
END FUNCTION poidev
```

⋆ ⋆ ⋆

```
FUNCTION bnldev(pp,n)
USE nrtype
USE nr, ONLY : gammln,ran1
IMPLICIT NONE
REAL(SP), INTENT(IN) :: pp
INTEGER(I4B), INTENT(IN) :: n
REAL(SP) :: bnldev
    Returns as a floating-point number an integer value that is a random deviate drawn from a
    binomial distribution of n trials each of probability pp, using ran1 as a source of uniform
    random deviates.
INTEGER(I4B) :: j
INTEGER(I4B), SAVE :: nold=-1
REAL(SP) :: am,em,g,h,p,sq,t,y,arr(24)
REAL(SP), SAVE :: pc,plog,pclog,en,oldg,pold=-1.0      Arguments from previous calls.
p=merge(pp,1.0_sp-pp, pp <= 0.5_sp )
  The binomial distribution is invariant under changing pp to 1.-pp, if we also change the
  answer to n minus itself; we'll remember to do this below.
am=n*p                                  This is the mean of the deviate to be produced.
if (n < 25) then                        Use the direct method while n is not too large.
    call ran1(arr(1:n))                     This can require up to 25 calls to ran1.
    bnldev=count(arr(1:n)<p)
else if (am < 1.0) then                 If fewer than one event is expected out of 25
    g=exp(-am)                              or more trials, then the distribution is quite
    t=1.0                                   accurately Poisson. Use direct Poisson method.
    do j=0,n
        call ran1(h)
        t=t*h
        if (t < g) exit
    end do
    bnldev=merge(j,n, j <= n)
else                                    Use the rejection method.
    if (n /= nold) then                 If n has changed, then compute useful quanti-
        en=n                                ties.
        oldg=gammln(en+1.0_sp)
        nold=n
    end if
    if (p /= pold) then                 If p has changed, then compute useful quanti-
        pc=1.0_sp-p                         ties.
        plog=log(p)
        pclog=log(pc)
        pold=p
```

```
    end if
    sq=sqrt(2.0_sp*am*pc)               The following code should by now seem familiar:
    do                                      rejection method with a Lorentzian compar-
        call ran1(h)                        ison function.
        y=tan(PI*h)
        em=sq*y+am
        if (em < 0.0 .or. em >= en+1.0_sp) cycle     Reject.
        em=int(em)                      Trick for integer-valued distribution.
        t=1.2_sp*sq*(1.0_sp+y**2)*exp(oldg-gammln(em+1.0_sp)-&
            gammln(en-em+1.0_sp)+em*plog+(en-em)*pclog)
        call ran1(h)
        if (h <= t) exit                Reject. This happens about 1.5 times per devi-
    end do                                  ate, on average.
    bnldev=em
end if
if (p /= pp) bnldev=n-bnldev            Remember to undo the symmetry transforma-
END FUNCTION bnldev                         tion.
```

⋆ ⋆ ⋆

f90 The routines psdes and psdes_safe both perform *exactly* the same hashing as was done by the Fortran 77 routine psdes. The difference is that psdes makes assumptions about arithmetic that go beyond the strict Fortran 90 model, while psdes_safe makes no such assumptions. The disadvantage of psdes_safe is that it is significantly slower, performing most of its arithmetic in double-precision reals that are then converted to integers with Fortran 90's modulo intrinsic.

In fact the nonsafe version, psdes, works fine on almost all machines and compilers that we have tried. There is a reason for this: Our assumed integer model is the same as the C language unsigned int, and virtually all modern computers and compilers have a lot of C hidden inside. If psdes and psdes_safe produce identical output on your system for any hundred or so different input values, you can be quite confident about using the faster version exclusively.

At the other end of things, note that in the very unlikely case that your system fails on the ran_hash routine in the ran_state module (you will have learned this from error messages generated by ran_init), you can substitute psdes_safe for ran_hash: They are plug-compatible.

```
SUBROUTINE psdes_s(lword,rword)
USE nrtype
IMPLICIT NONE
INTEGER(I4B), INTENT(INOUT) :: lword,rword
INTEGER(I4B), PARAMETER :: NITER=4
    "Pseudo-DES" hashing of the 64-bit word (lword,irword). Both 32-bit arguments are
    returned hashed on all bits. Note that this version of the routine assumes properties of
    integer arithmetic that go beyond the Fortran 90 model, though they are compatible with
    unsigned integers in C.
INTEGER(I4B), DIMENSION(4), SAVE :: C1,C2
DATA C1 /Z'BAA96887',Z'1E17D32C',Z'03BCDC3C',Z'0F33D1B2'/
DATA C2 /Z'4B0F3B58',Z'E874F0C3',Z'6955C5A6',Z'55A7CA46'/
INTEGER(I4B) :: i,ia,ib,iswap,itmph,itmpl
do i=1,NITER                        Perform niter iterations of DES logic, using a simpler
    iswap=rword                         (noncryptographic) nonlinear function instead of DES's.
    ia=ieor(rword,C1(i))            The bit-rich constants C1 and (below) C2 guarantee lots
    itmpl=iand(ia,65535)                of nonlinear mixing.
    itmph=iand(ishft(ia,-16),65535)
```

```
    ib=itmpl**2+not(itmph**2)
    ia=ior(ishft(ib,16),iand(ishft(ib,-16),65535))
    rword=ieor(lword,ieor(C2(i),ia)+itmpl*itmph)
    lword=iswap
end do
END SUBROUTINE psdes_s


SUBROUTINE psdes_v(lword,rword)
USE nrtype; USE nrutil, ONLY : assert_eq
IMPLICIT NONE
INTEGER(I4B), DIMENSION(:), INTENT(INOUT) :: lword,rword
INTEGER(I4B), PARAMETER :: NITER=4
INTEGER(I4B), DIMENSION(4), SAVE :: C1,C2
DATA C1 /Z'BAA96887',Z'1E17D32C',Z'03BCDC3C',Z'0F33D1B2'/
DATA C2 /Z'4B0F3B58',Z'E874F0C3',Z'6955C5A6',Z'55A7CA46'/
INTEGER(I4B), DIMENSION(size(lword)) :: ia,ib,iswap,itmph,itmpl
INTEGER(I4B) :: i
i=assert_eq(size(lword),size(rword),'psdes_v')
do i=1,NITER
    iswap=rword
    ia=ieor(rword,C1(i))
    itmpl=iand(ia,65535)
    itmph=iand(ishft(ia,-16),65535)
    ib=itmpl**2+not(itmph**2)
    ia=ior(ishft(ib,16),iand(ishft(ib,-16),65535))
    rword=ieor(lword,ieor(C2(i),ia)+itmpl*itmph)
    lword=iswap
end do
END SUBROUTINE psdes_v


SUBROUTINE psdes_safe_s(lword,rword)
USE nrtype
IMPLICIT NONE
INTEGER(I4B), INTENT(INOUT) :: lword,rword
INTEGER(I4B), PARAMETER :: NITER=4
```

"Pseudo-DES" hashing of the 64-bit word (`lword,irword`). Both 32-bit arguments are returned hashed on all bits. This is a slower version of the routine that makes no assumptions outside of the Fortran 90 integer model.

```
INTEGER(I4B), DIMENSION(4), SAVE :: C1,C2
DATA C1 /Z'BAA96887',Z'1E17D32C',Z'03BCDC3C',Z'0F33D1B2'/
DATA C2 /Z'4B0F3B58',Z'E874F0C3',Z'6955C5A6',Z'55A7CA46'/
INTEGER(I4B) :: i,ia,ib,iswap
REAL(DP) :: alo,ahi
do i=1,NITER
    iswap=rword
    ia=ieor(rword,C1(i))
    alo=real(iand(ia,65535),dp)
    ahi=real(iand(ishft(ia,-16),65535),dp)
    ib=modint(alo*alo+real(not(modint(ahi*ahi)),dp))
    ia=ior(ishft(ib,16),iand(ishft(ib,-16),65535))
    rword=ieor(lword,modint(real(ieor(C2(i),ia),dp)+alo*ahi))
    lword=iswap
end do
CONTAINS

FUNCTION modint(x)
REAL(DP), INTENT(IN) :: x
INTEGER(I4B) :: modint
REAL(DP) :: a
REAL(DP), PARAMETER :: big=huge(modint), base=big+big+2.0_dp
a=modulo(x,base)
```

```
if (a > big) a=a-base
modint=nint(a,kind=i4b)
END FUNCTION modint
END SUBROUTINE psdes_safe_s


SUBROUTINE psdes_safe_v(lword,rword)
USE nrtype; USE nrutil, ONLY : assert_eq
IMPLICIT NONE
INTEGER(I4B), DIMENSION(:), INTENT(INOUT) :: lword,rword
INTEGER(I4B), PARAMETER :: NITER=4
INTEGER(I4B), SAVE :: C1(4),C2(4)
DATA C1 /Z'BAA96887',Z'1E17D32C',Z'03BCDC3C',Z'0F33D1B2'/
DATA C2 /Z'4B0F3B58',Z'E874F0C3',Z'6955C5A6',Z'55A7CA46'/
INTEGER(I4B), DIMENSION(size(lword)) :: ia,ib,iswap
REAL(DP), DIMENSION(size(lword)) :: alo,ahi
INTEGER(I4B) :: i
i=assert_eq(size(lword),size(rword),'psdes_safe_v')
do i=1,NITER
    iswap=rword
    ia=ieor(rword,C1(i))
    alo=real(iand(ia,65535),dp)
    ahi=real(iand(ishft(ia,-16),65535),dp)
    ib=modint(alo*alo+real(not(modint(ahi*ahi)),dp))
    ia=ior(ishft(ib,16),iand(ishft(ib,-16),65535))
    rword=ieor(lword,modint(real(ieor(C2(i),ia),dp)+alo*ahi))
    lword=iswap
end do
CONTAINS
FUNCTION modint(x)
REAL(DP), DIMENSION(:), INTENT(IN) :: x
INTEGER(I4B), DIMENSION(size(x)) :: modint
REAL(DP), DIMENSION(size(x)) :: a
REAL(DP), PARAMETER :: big=huge(modint), base=big+big+2.0_dp
a=modulo(x,base)
where (a > big) a=a-base
modint=nint(a,kind=i4b)
END FUNCTION modint
END SUBROUTINE psdes_safe_v
```

`FUNCTION modint(x)` This embedded routine takes a double-precision real argument, and returns it as an integer mod 2^{32} (correctly wrapping it to negative to take into account that Fortran 90 has no unsigned integers).

⋆ ⋆ ⋆

```
SUBROUTINE ran3_s(harvest)
USE nrtype
USE ran_state, ONLY: K4B,amm,lenran,ran_init,ran_hash,mran0,nran0,rans
IMPLICIT NONE
REAL(SP), INTENT(OUT) :: harvest
    Random number generation by DES-like hashing of two 32-bit words, using the algorithm
    ran_hash. Returns as harvest a uniform random deviate between 0.0 and 1.0 (exclusive
    of the endpoint values).
INTEGER(K4B) :: temp
if (lenran < 1) call ran_init(1)                Initialize.
nran0=ieor(nran0,ishft(nran0,13))               Two Marsaglia shift sequences are
nran0=ieor(nran0,ishft(nran0,-17))                  maintained as input to the hash-
nran0=ieor(nran0,ishft(nran0,5))                    ing. The period of the combined
if (nran0 == 1) nran0=270369_k4b                    generator is about 1.8 × 10^19.
```

```
rans=nran0
mran0=ieor(mran0,ishft(mran0,5))
mran0=ieor(mran0,ishft(mran0,-13))
mran0=ieor(mran0,ishft(mran0,6))
temp=mran0
call ran_hash(temp,rans)                          Hash.
harvest=amm*merge(rans,not(rans), rans<0 )        Make the result positive definite (note
END SUBROUTINE ran3_s                                 that amm is negative).
```

```
SUBROUTINE ran3_v(harvest)
USE nrtype
USE ran_state, ONLY: K4B,amm,lenran,ran_init,ran_hash,mran,nran,ranv
IMPLICIT NONE
REAL(SP), DIMENSION(:), INTENT(OUT) :: harvest
INTEGER(K4B), DIMENSION(size(harvest)) :: temp
INTEGER(K4B) :: n
n=size(harvest)
if (lenran < n+1) call ran_init(n+1)
nran(1:n)=ieor(nran(1:n),ishft(nran(1:n),13))
nran(1:n)=ieor(nran(1:n),ishft(nran(1:n),-17))
nran(1:n)=ieor(nran(1:n),ishft(nran(1:n),5))
where (nran(1:n) == 1) nran(1:n)=270369_k4b
ranv(1:n)=nran(1:n)
mran(1:n)=ieor(mran(1:n),ishft(mran(1:n),5))
mran(1:n)=ieor(mran(1:n),ishft(mran(1:n),-13))
mran(1:n)=ieor(mran(1:n),ishft(mran(1:n),6))
temp=mran(1:n)
call ran_hash(temp,ranv(1:n))
harvest=amm*merge(ranv(1:n),not(ranv(1:n)), ranv(1:n)<0 )
END SUBROUTINE ran3_v
```

As given, ran3 uses the ran_hash function in the module ran_state as its DES surrogate. That function is sufficiently fast to make ran3 only about a factor of 2 slower than our baseline recommended generator ran1. The slower routine psdes and (even slower) psdes_safe are plug-compatible with ran_hash, and could be substituted for it in this routine.

⋆ ⋆ ⋆

```
FUNCTION irbit1(iseed)
USE nrtype
IMPLICIT NONE
INTEGER(I4B), INTENT(INOUT) :: iseed
INTEGER(I4B) :: irbit1
   Returns as an integer a random bit, based on the 18 low-significance bits in iseed (which
   is modified for the next call).
if (btest(iseed,17) .neqv. btest(iseed,4) .neqv. btest(iseed,1) &
    .neqv. btest(iseed,0)) then
    iseed=ibset(ishft(iseed,1),0)           Leftshift the seed and put a 1 in its bit 1.
    irbit1=1
else                                        But if the XOR calculation gave a 0,
    iseed=ishft(iseed,1)                    then put that in bit 1 instead.
    irbit1=0
end if
END FUNCTION irbit1
```

```
FUNCTION irbit2(iseed)
USE nrtype
IMPLICIT NONE
INTEGER(I4B), INTENT(INOUT) :: iseed
INTEGER(I4B) :: irbit2
   Returns as an integer a random bit, based on the 18 low-significance bits in iseed (which
   is modified for the next call).
INTEGER(I4B), PARAMETER :: IB1=1,IB2=2,IB5=16,MASK=IB1+IB2+IB5
if (btest(iseed,17)) then            Change all masked bits, shift, and put 1 into bit 1.
   iseed=ibset(ishft(ieor(iseed,MASK),1),0)
   irbit2=1
else                                 Shift and put 0 into bit 1.
   iseed=ibclr(ishft(iseed,1),0)
   irbit2=0
end if
END FUNCTION irbit2
```

⋆ ⋆ ⋆

```
SUBROUTINE sobseq(x,init)
USE nrtype; USE nrutil, ONLY : nrerror
IMPLICIT NONE
REAL(SP), DIMENSION(:), INTENT(OUT) :: x
INTEGER(I4B), OPTIONAL, INTENT(IN) :: init
INTEGER(I4B), PARAMETER :: MAXBIT=30,MAXDIM=6
   When the optional integer init is present, internally initializes a set of MAXBIT direction
   numbers for each of MAXDIM different Sobol' sequences. Otherwise returns as the vector x
   of length N the next values from N of these sequences. (N must not be changed between
   initializations.)
REAL(SP), SAVE :: fac
INTEGER(I4B) :: i,im,ipp,j,k,l
INTEGER(I4B), DIMENSION(:,:), ALLOCATABLE:: iu
INTEGER(I4B), SAVE :: in
INTEGER(I4B), DIMENSION(MAXDIM), SAVE :: ip,ix,mdeg
INTEGER(I4B), DIMENSION(MAXDIM*MAXBIT), SAVE :: iv
DATA ip /0,1,1,2,1,4/, mdeg /1,2,3,3,4,4/, ix /6*0/
DATA iv /6*1,3,1,3,3,1,1,5,7,7,3,3,5,15,11,5,15,13,9,156*0/
if (present(init)) then                  Initialize, don't return a vector.
   ix=0
   in=0
   if (iv(1) /= 1) RETURN
   fac=1.0_sp/2.0_sp**MAXBIT
   allocate(iu(MAXDIM,MAXBIT))
   iu=reshape(iv,shape(iu))              To allow both 1D and 2D addressing.
   do k=1,MAXDIM
      do j=1,mdeg(k)                     Stored values require only normalization.
         iu(k,j)=iu(k,j)*2**(MAXBIT-j)
      end do
      do j=mdeg(k)+1,MAXBIT              Use the recurrence to get other values.
         ipp=ip(k)
         i=iu(k,j-mdeg(k))
         i=ieor(i,i/2**mdeg(k))
         do l=mdeg(k)-1,1,-1
            if (btest(ipp,0)) i=ieor(i,iu(k,j-l))
            ipp=ipp/2
         end do
         iu(k,j)=i
      end do
   end do
   iv=reshape(iu,shape(iv))
   deallocate(iu)
```

```
else                                   Calculate the next vector in the sequence.
    im=in
    do j=1,MAXBIT                      Find the rightmost zero bit.
        if (.not. btest(im,0)) exit
        im=im/2
    end do
    if (j > MAXBIT) call nrerror('MAXBIT too small in sobseq')
    im=(j-1)*MAXDIM
    j=min(size(x),MAXDIM)
    ix(1:j)=ieor(ix(1:j),iv(1+im:j+im))
      XOR the appropriate direction number into each component of the vector and convert
      to a floating number.
    x(1:j)=ix(1:j)*fac
    in=in+1                            Increment the counter.
end if
END SUBROUTINE sobseq
```

f90 `if (present(init)) then ... allocate(iu(...)) ... iu=reshape(...)` Wanting to avoid the deprecated `EQUIVALENCE` statement, we must reshape `iv` into a two-dimensional array, then un-reshape it after we are done. This is done only once, at initialization time, so there is no serious inefficiency introduced.

⋆ ⋆ ⋆

```
SUBROUTINE vegas(region,func,init,ncall,itmx,nprn,tgral,sd,chi2a)
USE nrtype
USE nr, ONLY : ran1
IMPLICIT NONE
REAL(SP), DIMENSION(:), INTENT(IN) :: region
INTEGER(I4B), INTENT(IN) :: init,ncall,itmx,nprn
REAL(SP), INTENT(OUT) :: tgral,sd,chi2a
INTERFACE
    FUNCTION func(pt,wgt)
    USE nrtype
    IMPLICIT NONE
    REAL(SP), DIMENSION(:), INTENT(IN) :: pt
    REAL(SP), INTENT(IN) :: wgt
    REAL(SP) :: func
    END FUNCTION func
END INTERFACE
REAL(SP), PARAMETER :: ALPH=1.5_sp,TINY=1.0e-30_sp
INTEGER(I4B), PARAMETER :: MXDIM=10,NDMX=50
    Performs Monte Carlo integration of a user-supplied d-dimensional function func over a
    rectangular volume specified by region, a vector of length 2d consisting of d "lower left"
    coordinates of the region followed by d "upper right" coordinates. The integration consists of
    itmx iterations, each with approximately ncall calls to the function. After each iteration
    the grid is refined; more than 5 or 10 iterations are rarely useful. The input flag init
    signals whether this call is a new start, or a subsequent call for additional iterations (see
    comments below). The input flag nprn (normally 0) controls the amount of diagnostic
    output. Returned answers are tgral (the best estimate of the integral), sd (its standard
    deviation), and chi2a (χ^2 per degree of freedom, an indicator of whether consistent results
    are being obtained). See text for further details.
INTEGER(I4B), SAVE :: i,it,j,k,mds,nd,ndim,ndo,ng,npg      Best make everything static,
INTEGER(I4B), DIMENSION(MXDIM), SAVE :: ia,kg                  allowing restarts.
REAL(SP), SAVE :: calls,dv2g,dxg,f,f2,f2b,fb,rc,ti,tsi,wgt,xjac,xn,xnd,xo,harvest
REAL(SP), DIMENSION(NDMX,MXDIM), SAVE :: d,di,xi
REAL(SP), DIMENSION(MXDIM), SAVE :: dt,dx,x
REAL(SP), DIMENSION(NDMX), SAVE :: r,xin
REAL(DP), SAVE :: schi,si,swgt
```

```
ndim=size(region)/2
if (init <= 0) then                          Normal entry. Enter here on a cold start.
    mds=1                                    Change to mds=0 to disable stratified sam-
    ndo=1                                        pling, i.e., use importance sampling only.
    xi(1,:)=1.0
end if
if (init <= 1) then                          Enter here to inherit the grid from a previous
    si=0.0                                       call, but not its answers.
    swgt=0.0
    schi=0.0
end if
if (init <= 2) then                          Enter here to inherit the previous grid and its
    nd=NDMX                                      answers.
    ng=1
    if (mds /= 0) then                       Set up for stratification.
        ng=(ncall/2.0_sp+0.25_sp)**(1.0_sp/ndim)
        mds=1
        if ((2*ng-NDMX) >= 0) then
            mds=-1
            npg=ng/NDMX+1
            nd=ng/npg
            ng=npg*nd
        end if
    end if
    k=ng**ndim
    npg=max(ncall/k,2)
    calls=real(npg,sp)*real(k,sp)
    dxg=1.0_sp/ng
    dv2g=(calls*dxg**ndim)**2/npg/npg/(npg-1.0_sp)
    xnd=nd
    dxg=dxg*xnd
    dx(1:ndim)=region(1+ndim:2*ndim)-region(1:ndim)
    xjac=1.0_sp/calls*product(dx(1:ndim))
    if (nd /= ndo) then                      Do binning if necessary.
        r(1:max(nd,ndo))=1.0
        do j=1,ndim
            call rebin(ndo/xnd,nd,r,xin,xi(:,j))
        end do
        ndo=nd
    end if
    if (nprn >= 0) write(*,200) ndim,calls,it,itmx,nprn,&
        ALPH,mds,nd,(j,region(j),j,region(j+ndim),j=1,ndim)
end if
do it=1,itmx                                 Main iteration loop. Can enter here (init ≥
    ti=0.0                                       3) to do an additional itmx iterations
    tsi=0.0                                      with all other parameters unchanged.
    kg(:)=1
    d(1:nd,:)=0.0
    di(1:nd,:)=0.0
    iterate: do
        fb=0.0
        f2b=0.0
        do k=1,npg
            wgt=xjac
            do j=1,ndim
                call ran1(harvest)
                xn=(kg(j)-harvest)*dxg+1.0_sp
                ia(j)=max(min(int(xn),NDMX),1)
                if (ia(j) > 1) then
                    xo=xi(ia(j),j)-xi(ia(j)-1,j)
                    rc=xi(ia(j)-1,j)+(xn-ia(j))*xo
                else
                    xo=xi(ia(j),j)
                    rc=(xn-ia(j))*xo
```

```
                end if
                x(j)=region(j)+rc*dx(j)
                wgt=wgt*xo*xnd
            end do
            f=wgt*func(x(1:ndim),wgt)
            f2=f*f
            fb=fb+f
            f2b=f2b+f2
            do j=1,ndim
                di(ia(j),j)=di(ia(j),j)+f
                if (mds >= 0) d(ia(j),j)=d(ia(j),j)+f2
            end do
        end do
        f2b=sqrt(f2b*npg)
        f2b=(f2b-fb)*(f2b+fb)
        if (f2b <= 0.0) f2b=TINY
        ti=ti+fb
        tsi=tsi+f2b
        if (mds < 0) then                      Use stratified sampling.
            do j=1,ndim
                d(ia(j),j)=d(ia(j),j)+f2b
            end do
        end if
        do k=ndim,1,-1
            kg(k)=mod(kg(k),ng)+1
            if (kg(k) /= 1) cycle iterate
        end do
        exit iterate
    end do iterate
    tsi=tsi*dv2g                               Compute final results for this iteration.
    wgt=1.0_sp/tsi
    si=si+real(wgt,dp)*real(ti,dp)
    schi=schi+real(wgt,dp)*real(ti,dp)**2
    swgt=swgt+real(wgt,dp)
    tgral=si/swgt
    chi2a=max((schi-si*tgral)/(it-0.99_dp),0.0_dp)
    sd=sqrt(1.0_sp/swgt)
    tsi=sqrt(tsi)
    if (nprn >= 0) then
        write(*,201) it,ti,tsi,tgral,sd,chi2a
        if (nprn /= 0) then
            do j=1,ndim
                write(*,202) j,(xi(i,j),di(i,j),&
                    i=1+nprn/2,nd,nprn)
            end do
        end if
    end if
    do j=1,ndim                                Refine the grid. Consult references to under-
        xo=d(1,j)                                  stand the subtlety of this procedure. The
        xn=d(2,j)                                  refinement is damped, to avoid rapid,
        d(1,j)=(xo+xn)/2.0_sp                      destabilizing changes, and also compressed
        dt(j)=d(1,j)                               in range by the exponent ALPH.
        do i=2,nd-1
            rc=xo+xn
            xo=xn
            xn=d(i+1,j)
            d(i,j)=(rc+xn)/3.0_sp
            dt(j)=dt(j)+d(i,j)
        end do
        d(nd,j)=(xo+xn)/2.0_sp
        dt(j)=dt(j)+d(nd,j)
    end do
    where (d(1:nd,:) < TINY) d(1:nd,:)=TINY
    do j=1,ndim
```

```
            r(1:nd)=((1.0_sp-d(1:nd,j)/dt(j))/(log(dt(j))-log(d(1:nd,j))))**ALPH
            rc=sum(r(1:nd))
            call rebin(rc/xnd,nd,r,xin,xi(:,j))
        end do
    end do
200 format(/' input parameters for vegas: ndim=',i3,' ncall=',f8.0&
        /28x,' it=',i5,' itmx=',i5&
        /28x,' nprn=',i3,' alph=',f5.2/28x,' mds=',i3,' nd=',i4&
        /(30x,'xl(',i2,')= ',g11.4,' xu(',i2,')= ',g11.4))
201 format(/' iteration no.',I3,': ','integral =',g14.7,' +/- ',g9.2,&
        /' all iterations: integral =',g14.7,' +/- ',g9.2,&
        ' chi**2/it''n =',g9.2)
202 format(/' data for axis ',I2/' X delta i ',&
        ' x delta i ',' x delta i ',&
        /(1x,f7.5,1x,g11.4,5x,f7.5,1x,g11.4,5x,f7.5,1x,g11.4))
    CONTAINS

    SUBROUTINE rebin(rc,nd,r,xin,xi)
    IMPLICIT NONE
    REAL(SP), INTENT(IN) :: rc
    INTEGER(I4B), INTENT(IN) :: nd
    REAL(SP), DIMENSION(:), INTENT(IN) :: r
    REAL(SP), DIMENSION(:), INTENT(OUT) :: xin
    REAL(SP), DIMENSION(:), INTENT(INOUT) :: xi
```

Utility routine used by vegas, to rebin a vector of densities xi into new bins defined by a vector r.

```
    INTEGER(I4B) :: i,k
    REAL(SP) :: dr,xn,xo
    k=0
    xo=0.0
    dr=0.0
    do i=1,nd-1
        do
            if (rc <= dr) exit
            k=k+1
            dr=dr+r(k)
        end do
        if (k > 1) xo=xi(k-1)
        xn=xi(k)
        dr=dr-rc
        xin(i)=xn-(xn-xo)*dr/r(k)
    end do
    xi(1:nd-1)=xin(1:nd-1)
    xi(nd)=1.0
    END SUBROUTINE rebin
    END SUBROUTINE vegas
```

⋆ ⋆ ⋆

```
    RECURSIVE SUBROUTINE miser(func,regn,ndim,npts,dith,ave,var)
    USE nrtype; USE nrutil, ONLY : assert_eq
    IMPLICIT NONE
    INTERFACE
        FUNCTION func(x)
        USE nrtype
        IMPLICIT NONE
        REAL(SP) :: func
        REAL(SP), DIMENSION(:), INTENT(IN) :: x
        END FUNCTION func
    END INTERFACE
    REAL(SP), DIMENSION(:), INTENT(IN) :: regn
    INTEGER(I4B), INTENT(IN) :: ndim,npts
```

```
REAL(SP), INTENT(IN) :: dith
REAL(SP), INTENT(OUT) :: ave,var
REAL(SP), PARAMETER :: PFAC=0.1_sp,TINY=1.0e-30_sp,BIG=1.0e30_sp
INTEGER(I4B), PARAMETER :: MNPT=15,MNBS=60
```

Monte Carlo samples a user-supplied ndim-dimensional function func in a rectangular volume specified by region, a 2×ndim vector consisting of ndim "lower-left" coordinates of the region followed by ndim "upper-right" coordinates. The function is sampled a total of npts times, at locations determined by the method of recursive stratified sampling. The mean value of the function in the region is returned as ave; an estimate of the statistical uncertainty of ave (square of standard deviation) is returned as var. The input parameter dith should normally be set to zero, but can be set to (e.g.) 0.1 if func's active region falls on the boundary of a power-of-2 subdivision of region.

Parameters: PFAC is the fraction of remaining function evaluations used *at each stage* to explore the variance of func. At least MNPT function evaluations are performed in any terminal subregion; a subregion is further bisected only if at least MNBS function evaluations are available.

```
REAL(SP), DIMENSION(:), ALLOCATABLE :: regn_temp
INTEGER(I4B) :: j,jb,n,ndum,npre,nptl,nptr
INTEGER(I4B), SAVE :: iran=0
REAL(SP) :: avel,varl,fracl,fval,rgl,rgm,rgr,&
    s,sigl,siglb,sigr,sigrb,sm,sm2,sumb,sumr
REAL(SP), DIMENSION(:), ALLOCATABLE :: fmaxl,fmaxr,fminl,fminr,pt,rmid
ndum=assert_eq(size(regn),2*ndim,'miser')
allocate(pt(ndim))
if (npts < MNBS) then                    Too few points to bisect; do straight Monte
    sm=0.0                                   Carlo.
    sm2=0.0
    do n=1,npts
        call ranpt(pt,regn)
        fval=func(pt)
        sm=sm+fval
        sm2=sm2+fval**2
    end do
    ave=sm/npts
    var=max(TINY,(sm2-sm**2/npts)/npts**2)
else                                     Do the preliminary (uniform) sampling.
    npre=max(int(npts*PFAC),MNPT)
    allocate(rmid(ndim),fmaxl(ndim),fmaxr(ndim),fminl(ndim),fminr(ndim))
    fminl(:)=BIG                         Initialize the left and right bounds for each
    fminr(:)=BIG                             dimension.
    fmaxl(:)=-BIG
    fmaxr(:)=-BIG
    do j=1,ndim
        iran=mod(iran*2661+36979,175000)
        s=sign(dith,real(iran-87500,sp))
        rmid(j)=(0.5_sp+s)*regn(j)+(0.5_sp-s)*regn(ndim+j)
    end do
    do n=1,npre                          Loop over the points in the sample.
        call ranpt(pt,regn)
        fval=func(pt)
        where (pt <= rmid)               Find the left and right bounds for each di-
            fminl=min(fminl,fval)            mension.
            fmaxl=max(fmaxl,fval)
        elsewhere
            fminr=min(fminr,fval)
            fmaxr=max(fmaxr,fval)
        end where
    end do
    sumb=BIG                             Choose which dimension jb to bisect.
    jb=0
    siglb=1.0
    sigrb=1.0
    do j=1,ndim
        if (fmaxl(j) > fminl(j) .and. fmaxr(j) > fminr(j)) then
```

```
            sigl=max(TINY,(fmaxl(j)-fminl(j))**(2.0_sp/3.0_sp))
            sigr=max(TINY,(fmaxr(j)-fminr(j))**(2.0_sp/3.0_sp))
            sumr=sigl+sigr                    Equation (7.8.24); see text.
            if (sumr <= sumb) then
                sumb=sumr
                jb=j
                siglb=sigl
                sigrb=sigr
            end if
        end if
    end do
    deallocate(fminr,fminl,fmaxr,fmaxl)
    if (jb == 0) jb=1+(ndim*iran)/175000      MNPT may be too small.
    rgl=regn(jb)                              Apportion the remaining points between left
    rgm=rmid(jb)                                  and right.
    rgr=regn(ndim+jb)
    fracl=abs((rgm-rgl)/(rgr-rgl))
    nptl=(MNPT+(npts-npre-2*MNPT)*fracl*siglb/ &     Equation (7.8.23).
        (fracl*siglb+(1.0_sp-fracl)*sigrb))
    nptr=npts-npre-nptl
    allocate(regn_temp(2*ndim))
    regn_temp(:)=regn(:)
    regn_temp(ndim+jb)=rmid(jb)               Set region to left.
    call miser(func,regn_temp,ndim,nptl,dith,avel,varl)
      Dispatch recursive call; will return back here eventually.
    regn_temp(jb)=rmid(jb)
    regn_temp(ndim+jb)=regn(ndim+jb)          Set region to right.
    call miser(func,regn_temp,ndim,nptr,dith,ave,var)
      Dispatch recursive call; will return back here eventually.
    deallocate(regn_temp)
    ave=fracl*avel+(1-fracl)*ave              Combine left and right regions by equation
    var=fracl*fracl*varl+(1-fracl)*(1-fracl)*var     (7.8.11) (1st line).
    deallocate(rmid)
end if
deallocate(pt)
CONTAINS

SUBROUTINE ranpt(pt,region)
USE nr, ONLY : ran1
IMPLICIT NONE
REAL(SP), DIMENSION(:), INTENT(OUT) :: pt
REAL(SP), DIMENSION(:), INTENT(IN) :: region
    Returns a uniformly random point pt in a rectangular region of dimension d. Used by
    miser; calls ran1 for uniform deviates.
INTEGER(I4B) :: n
call ran1(pt)
n=size(pt)
pt(1:n)=region(1:n)+(region(n+1:2*n)-region(1:n))*pt(1:n)
END SUBROUTINE ranpt
END SUBROUTINE miser
```

f90 The Fortran 90 version of this routine is much more straightforward than the Fortran 77 version, because Fortran 90 allows recursion. (In fact, this routine is modeled on the C version of miser, which was recursive from the start.)

CITED REFERENCES AND FURTHER READING:

Marsaglia, G., and Zaman, A. 1994, *Computers in Physics*, vol. 8, pp. 117–121. [1]

Marsaglia, G. 1985, *Linear Algebra and Its Applications*, vol. 67, pp. 147-156. [2]

Harbison, S.P., and Steele, G.L. 1991, *C: A Reference Manual*, Third Edition, §5.1.1. [3]

Chapter B8. Sorting

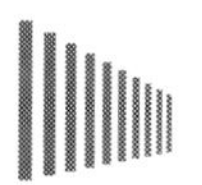

Caution! If you are expecting to sort efficiently on a parallel machine, whether its parallelism is small-scale or massive, you almost certainly want to use library routines that are specific to your hardware.

We include in this chapter translations into Fortran 90 of the general purpose *serial* sorting routines that are in Volume 1, augmented by several new routines that give pedagogical demonstrations of how parallel sorts can be achieved with Fortran 90 parallel constructions and intrinsics. However, we intend the above word "pedagogical" to be taken seriously: these new, supposedly parallel, routines are *not* likely to be competitive with machine-specific library routines. Neither do they compete successfully on serial machines with the all-serial routines provided (namely `sort`, `sort2`, `sort3`, `indexx`, and `select`).

⋆ ⋆ ⋆

```
SUBROUTINE sort_pick(arr)
USE nrtype
IMPLICIT NONE
REAL(SP), DIMENSION(:), INTENT(INOUT) :: arr
   Sorts an array arr into ascending numerical order, by straight insertion. arr is replaced
   on output by its sorted rearrangement.
INTEGER(I4B) :: i,j,n
REAL(SP) :: a
n=size(arr)
do j=2,n                          Pick out each element in turn.
   a=arr(j)
   do i=j-1,1,-1                  Look for the place to insert it.
      if (arr(i) <= a) exit
      arr(i+1)=arr(i)
   end do
   arr(i+1)=a                     Insert it.
end do
END SUBROUTINE sort_pick
```

Not only is `sort_pick` (renamed from Volume 1's `piksrt`) *not parallelizable*, but also, even worse, it is an N^2 routine. It is meant to be invoked only for the most trivial sorting jobs, say, $N < 20$.

⋆ ⋆ ⋆

```
SUBROUTINE sort_shell(arr)
USE nrtype
IMPLICIT NONE
REAL(SP), DIMENSION(:), INTENT(INOUT) :: arr
    Sorts an array arr into ascending numerical order by Shell's method (diminishing increment
    sort). arr is replaced on output by its sorted rearrangement.
INTEGER(I4B) :: i,j,inc,n
REAL(SP) :: v
n=size(arr)
inc=1
do                              Determine the starting increment.
    inc=3*inc+1
    if (inc > n) exit
end do
do                              Loop over the partial sorts.
    inc=inc/3
    do i=inc+1,n                Outer loop of straight insertion.
        v=arr(i)
        j=i
        do                      Inner loop of straight insertion.
            if (arr(j-inc) <= v) exit
            arr(j)=arr(j-inc)
            j=j-inc
            if (j <= inc) exit
        end do
        arr(j)=v
    end do
    if (inc <= 1) exit
end do
END SUBROUTINE sort_shell
```

The routine sort_shell is renamed from Volume 1's shell. Shell's Method, a diminishing increment sort, is not directly parallelizable. However, one can write a fully parallel routine (though not an especially fast one — see remarks at beginning of this chapter) in much the same spirit:

```
SUBROUTINE sort_byreshape(arr)
USE nrtype; USE nrutil, ONLY : swap
IMPLICIT NONE
REAL(SP), DIMENSION(:), INTENT(INOUT) :: arr
    Sort an array arr by bubble sorting a succession of reshapings into array slices. The method
    is similar to Shell sort, but allows parallelization within the vectorized masked swap calls.
REAL(SP), DIMENSION(:,:), ALLOCATABLE :: tab
REAL(SP), PARAMETER :: big=huge(arr)
INTEGER(I4B) :: inc,n,m
n=size(arr)
inc=1
do                                          Find the largest increment that fits.
    inc=2*inc+1
    if (inc > n) exit
end do
do                                          Loop over the different shapes for the reshaped
    inc=inc/2                                   array.
    m=(n+inc-1)/inc
    allocate(tab(inc,m))                    Allocate space and reshape the array. big en-
    tab=reshape(arr, (/inc,m/) , (/big/) )      sures that fill elements stay at the
    do                                          end.
            Bubble sort all the rows in parallel.
        call swap(tab(:,1:m-1:2),tab(:,2:m:2), &
            tab(:,1:m-1:2)>tab(:,2:m:2))
        call swap(tab(:,2:m-1:2),tab(:,3:m:2), &
            tab(:,2:m-1:2)>tab(:,3:m:2))
```

```
        if (all(tab(:,1:m-1) <= tab(:,2:m))) exit
    end do
    arr=reshape(tab,shape(arr))        Put the array back together for the next shape.
    deallocate(tab)
    if (inc <= 1) exit
end do
END SUBROUTINE sort_byreshape
```

The basic idea is to reshape the given one-dimensional array into a succession of two-dimensional arrays, starting with "tall and narrow" (many rows, few columns), and ending up with "short and wide" (many columns, few rows). At each stage we sort all the rows in parallel by a bubble sort, giving something close to Shell's diminishing increments.

⋆ ⋆ ⋆

We now arrive at those routines, based on the Quicksort algorithm, that we actually intend for use with general N on serial machines:

```
SUBROUTINE sort(arr)
USE nrtype; USE nrutil, ONLY : swap,nrerror
IMPLICIT NONE
REAL(SP), DIMENSION(:), INTENT(INOUT) :: arr
INTEGER(I4B), PARAMETER :: NN=15, NSTACK=50
   Sorts an array arr into ascending numerical order using the Quicksort algorithm. arr is
   replaced on output by its sorted rearrangement.
   Parameters: NN is the size of subarrays sorted by straight insertion and NSTACK is the
   required auxiliary storage.
REAL(SP) :: a
INTEGER(I4B) :: n,k,i,j,jstack,l,r
INTEGER(I4B), DIMENSION(NSTACK) :: istack
n=size(arr)
jstack=0
l=1
r=n
do
    if (r-l < NN) then                 Insertion sort when subarray small enough.
        do j=l+1,r
            a=arr(j)
            do i=j-1,l,-1
                if (arr(i) <= a) exit
                arr(i+1)=arr(i)
            end do
            arr(i+1)=a
        end do
        if (jstack == 0) RETURN
        r=istack(jstack)               Pop stack and begin a new round of partition-
        l=istack(jstack-1)                 ing.
        jstack=jstack-2
    else                               Choose median of left, center, and right elements
        k=(l+r)/2                          as partitioning element a. Also rearrange so
        call swap(arr(k),arr(l+1))         that a(l) ≤ a(l+1) ≤ a(r).
        call swap(arr(l),arr(r),arr(l)>arr(r))
        call swap(arr(l+1),arr(r),arr(l+1)>arr(r))
        call swap(arr(l),arr(l+1),arr(l)>arr(l+1))
        i=l+1                          Initialize pointers for partitioning.
        j=r
        a=arr(l+1)                     Partitioning element.
        do                             Here is the meat.
            do                         Scan up to find element >= a.
                i=i+1
```

```
                if (arr(i) >= a) exit
            end do
            do                                  Scan down to find element <= a.
                j=j-1
                if (arr(j) <= a) exit
            end do
            if (j < i) exit                     Pointers crossed. Exit with partitioning complete.
            call swap(arr(i),arr(j))                Exchange elements.
        end do
        arr(l+1)=arr(j)                         Insert partitioning element.
        arr(j)=a
        jstack=jstack+2
          Push pointers to larger subarray on stack; process smaller subarray immediately.
        if (jstack > NSTACK) call nrerror('sort: NSTACK too small')
        if (r-i+1 >= j-l) then
            istack(jstack)=r
            istack(jstack-1)=i
            r=j-1
        else
            istack(jstack)=j-1
            istack(jstack-1)=l
            l=i
        end if
    end if
end do
END SUBROUTINE sort
```

f90 `call swap(...) ... call swap(...)` One might think twice about putting all these external function calls (to `nrutil` routines) in the inner loop of something as streamlined as a sort routine, but here they are executed only once for each partitioning.

`call swap(arr(i),arr(j))` This call *is* in a loop, but not the innermost loop. Most modern machines are very fast at the "context changes" implied by subroutine calls and returns; but in a time-critical context you might code this swap in-line and see if there is any timing difference.

```
SUBROUTINE sort2(arr,slave)
USE nrtype; USE nrutil, ONLY : assert_eq
USE nr, ONLY : indexx
IMPLICIT NONE
REAL(SP), DIMENSION(:), INTENT(INOUT) :: arr,slave
    Sorts an array arr into ascending order using Quicksort, while making the corresponding
    rearrangement of the same-size array slave. The sorting and rearrangement are performed
    by means of an index array.
INTEGER(I4B) :: ndum
INTEGER(I4B), DIMENSION(size(arr)) :: index
ndum=assert_eq(size(arr),size(slave),'sort2')
call indexx(arr,index)                  Make the index array.
arr=arr(index)                          Sort arr.
slave=slave(index)                      Rearrange slave.
END SUBROUTINE sort2
```

* * *

A close surrogate for the Quicksort partition-exchange algorithm can be coded, parallelizable, by using Fortran 90's `pack` intrinsic. On real compilers, unfortunately, the resulting code is not very efficient as compared with (on serial machines) the tightness of `sort`'s inner loop, above, or (on parallel machines) supplied library sort routines. We illustrate the principle nevertheless in the following routine.

```
RECURSIVE SUBROUTINE sort_bypack(arr)
USE nrtype; USE nrutil, ONLY : array_copy,swap
IMPLICIT NONE
REAL(SP), DIMENSION(:), INTENT(INOUT) :: arr
    Sort an array arr by recursively applying the Fortran 90 pack intrinsic. The method is
    similar to Quicksort, but this variant allows parallelization by the Fortran 90 compiler.
REAL(SP) :: a
INTEGER(I4B) :: n,k,nl,nerr
INTEGER(I4B), SAVE :: level=0
LOGICAL, DIMENSION(:), ALLOCATABLE, SAVE :: mask
REAL(SP), DIMENSION(:), ALLOCATABLE, SAVE :: temp
n=size(arr)
if (n <= 1) RETURN
k=(1+n)/2
call swap(arr(1),arr(k),arr(1)>arr(k))       Pivot element is median of first, middle,
call swap(arr(k),arr(n),arr(k)>arr(n))           and last.
call swap(arr(1),arr(k),arr(1)>arr(k))
if (n <= 3) RETURN
level=level+1                                Keep track of recursion level to avoid al-
if (level == 1) allocate(mask(n),temp(n))        location overhead.
a=arr(k)
mask(1:n) = (arr <= a)                       Which elements move to left?
mask(k) = .false.
call array_copy(pack(arr,mask(1:n)),temp,nl,nerr)      Move them.
mask(k) = .true.
temp(nl+2:n)=pack(arr,.not. mask(1:n))       Move others to right.
temp(nl+1)=a
arr=temp(1:n)
call sort_bypack(arr(1:nl))                  And recurse.
call sort_bypack(arr(nl+2:n))
if (level == 1) deallocate(mask,temp)
level=level-1
END SUBROUTINE sort_bypack
```

⋆ ⋆ ⋆

The following routine, sort_heap, is renamed from Volume 1's hpsort.

```
SUBROUTINE sort_heap(arr)
USE nrtype
USE nrutil, ONLY : swap
IMPLICIT NONE
REAL(SP), DIMENSION(:), INTENT(INOUT) :: arr
    Sorts an array arr into ascending numerical order using the Heapsort algorithm. arr is
    replaced on output by its sorted rearrangement.
INTEGER(I4B) :: i,n
n=size(arr)
do i=n/2,1,-1
        The index i, which here determines the "left" range of the sift-down, i.e., the element to
        be sifted down, is decremented from n/2 down to 1 during the "hiring" (heap creation)
        phase.
    call sift_down(i,n)
end do
do i=n,2,-1
        Here the "right" range of the sift-down is decremented from n-1 down to 1 during the
        "retirement-and-promotion" (heap selection) phase.
    call swap(arr(1),arr(i))                 Clear a space at the end of the array, and
    call sift_down(1,i-1)                        retire the top of the heap into it.
end do
CONTAINS

SUBROUTINE sift_down(l,r)
INTEGER(I4B), INTENT(IN) :: l,r
```

Carry out the sift-down on element arr(l) to maintain the heap structure.

```
INTEGER(I4B) :: j,jold
REAL(SP) :: a
a=arr(l)
jold=l
j=l+l
do                                          "Do while j <= r:"
    if (j > r) exit
    if (j < r) then
        if (arr(j) < arr(j+1)) j=j+1        Compare to the better underling.
    end if
    if (a >= arr(j)) exit                   Found a's level. Terminate the sift-down. Oth-
    arr(jold)=arr(j)                            erwise, demote a and continue.
    jold=j
    j=j+j
end do
arr(jold)=a                                 Put a into its slot.
END SUBROUTINE sift_down
END SUBROUTINE sort_heap
```

⋆ ⋆ ⋆

Another opportunity provided by Fortran 90 for a fully parallelizable sort, at least pedagogically, is to use the language's allowed access to the actual floating-point representation and to code a radix sort [1] on its bits. This is *not* efficient, but it illustrates some Fortran 90 language features perhaps worthy of study for other applications.

```
SUBROUTINE sort_radix(arr)
USE nrtype; USE nrutil, ONLY : array_copy,nrerror
IMPLICIT NONE
REAL(SP), DIMENSION(:), INTENT(INOUT) :: arr
    Sort an array arr by radix sort on its bits.
INTEGER(I4B), DIMENSION(size(arr)) :: narr,temp
LOGICAL, DIMENSION(size(arr)) :: msk
INTEGER(I4B) :: k,negm,ib,ia,n,nl,nerr
    Because we are going to transfer reals to integers, we must check that the number of bits
    is the same in each:
ib=bit_size(narr)
ia=ceiling(log(real(maxexponent(arr)-minexponent(arr),sp))/log(2.0_sp)) &
    + digits(arr)
if (ib /= ia) call nrerror('sort_radix: bit sizes not compatible')
negm=not(ishftc(1,-1))                      Mask for all bits except sign bit.
n=size(arr)
narr=transfer(arr,narr,n)
where (btest(narr,ib-1)) narr=ieor(narr,negm)     Flip all bits on neg. numbers.
do k=0,ib-2
        Work from low- to high-order bits, and partition the array according to the value of the
        bit.
    msk=btest(narr,k)
    call array_copy(pack(narr,.not. msk),temp,nl,nerr)
    temp(nl+1:n)=pack(narr,msk)
    narr=temp
end do
msk=btest(narr,ib-1)                        The sign bit gets separate treat-
call array_copy(pack(narr,msk),temp,nl,nerr)    ment, since here 1 comes be-
temp(nl+1:n)=pack(narr,.not. msk)               fore 0.
narr=temp
where (btest(narr,ib-1)) narr=ieor(narr,negm)     Unflip all bits on neg. numbers.
arr=transfer(narr,arr,n)
END SUBROUTINE sort_radix
```

⋆ ⋆ ⋆

We overload the generic name `indexx` with two specific implementations, one for SP floating values, the other for I4B integers. (You can of course add more overloadings if you need them.)

```
SUBROUTINE indexx_sp(arr,index)
USE nrtype; USE nrutil, ONLY : arth,assert_eq,nrerror,swap
IMPLICIT NONE
REAL(SP), DIMENSION(:), INTENT(IN) :: arr
INTEGER(I4B), DIMENSION(:), INTENT(OUT) :: index
INTEGER(I4B), PARAMETER :: NN=15, NSTACK=50
```

Indexes an array `arr`, i.e., outputs the array `index` of length N such that `arr(index(`j`))` is in ascending order for $j = 1, 2, \ldots, N$. The input quantity `arr` is not changed.

```
REAL(SP) :: a
INTEGER(I4B) :: n,k,i,j,indext,jstack,l,r
INTEGER(I4B), DIMENSION(NSTACK) :: istack
n=assert_eq(size(index),size(arr),'indexx_sp')
index=arth(1,1,n)
jstack=0
l=1
r=n
do
    if (r-l < NN) then
        do j=l+1,r
            indext=index(j)
            a=arr(indext)
            do i=j-1,l,-1
                if (arr(index(i)) <= a) exit
                index(i+1)=index(i)
            end do
            index(i+1)=indext
        end do
        if (jstack == 0) RETURN
        r=istack(jstack)
        l=istack(jstack-1)
        jstack=jstack-2
    else
        k=(l+r)/2
        call swap(index(k),index(l+1))
        call icomp_xchg(index(l),index(r))
        call icomp_xchg(index(l+1),index(r))
        call icomp_xchg(index(l),index(l+1))
        i=l+1
        j=r
        indext=index(l+1)
        a=arr(indext)
        do
            do
                i=i+1
                if (arr(index(i)) >= a) exit
            end do
            do
                j=j-1
                if (arr(index(j)) <= a) exit
            end do
            if (j < i) exit
            call swap(index(i),index(j))
        end do
        index(l+1)=index(j)
        index(j)=indext
        jstack=jstack+2
        if (jstack > NSTACK) call nrerror('indexx: NSTACK too small')
        if (r-i+1 >= j-l) then
            istack(jstack)=r
```

```
            istack(jstack-1)=i
            r=j-1
        else
            istack(jstack)=j-1
            istack(jstack-1)=l
            l=i
        end if
    end if
end do
CONTAINS

SUBROUTINE icomp_xchg(i,j)
INTEGER(I4B), INTENT(INOUT) :: i,j
INTEGER(I4B) :: swp
if (arr(j) < arr(i)) then
    swp=i
    i=j
    j=swp
end if
END SUBROUTINE icomp_xchg
END SUBROUTINE indexx_sp
```

```
SUBROUTINE indexx_i4b(iarr,index)
USE nrtype; USE nrutil, ONLY : arth,assert_eq,nrerror,swap
IMPLICIT NONE
INTEGER(I4B), DIMENSION(:), INTENT(IN) :: iarr
INTEGER(I4B), DIMENSION(:), INTENT(OUT) :: index
INTEGER(I4B), PARAMETER :: NN=15, NSTACK=50
INTEGER(I4B) :: a
INTEGER(I4B) :: n,k,i,j,indext,jstack,l,r
INTEGER(I4B), DIMENSION(NSTACK) :: istack
n=assert_eq(size(index),size(iarr),'indexx_sp')
index=arth(1,1,n)
jstack=0
l=1
r=n
do
    if (r-l < NN) then
        do j=l+1,r
            indext=index(j)
            a=iarr(indext)
            do i=j-1,1,-1
                if (iarr(index(i)) <= a) exit
                index(i+1)=index(i)
            end do
            index(i+1)=indext
        end do
        if (jstack == 0) RETURN
        r=istack(jstack)
        l=istack(jstack-1)
        jstack=jstack-2
    else
        k=(l+r)/2
        call swap(index(k),index(l+1))
        call icomp_xchg(index(l),index(r))
        call icomp_xchg(index(l+1),index(r))
        call icomp_xchg(index(l),index(l+1))
        i=l+1
        j=r
        indext=index(l+1)
        a=iarr(indext)
        do
            do
```

```
                i=i+1
                if (iarr(index(i)) >= a) exit
            end do
            do
                j=j-1
                if (iarr(index(j)) <= a) exit
            end do
            if (j < i) exit
            call swap(index(i),index(j))
        end do
        index(l+1)=index(j)
        index(j)=indext
        jstack=jstack+2
        if (jstack > NSTACK) call nrerror('indexx: NSTACK too small')
        if (r-i+1 >= j-l) then
            istack(jstack)=r
            istack(jstack-1)=i
            r=j-1
        else
            istack(jstack)=j-1
            istack(jstack-1)=l
            l=i
        end if
    end if
end do
CONTAINS

SUBROUTINE icomp_xchg(i,j)
INTEGER(I4B), INTENT(INOUT) :: i,j
INTEGER(I4B) :: swp
if (iarr(j) < iarr(i)) then
    swp=i
    i=j
    j=swp
end if
END SUBROUTINE icomp_xchg
END SUBROUTINE indexx_i4b
```

* * *

```
SUBROUTINE sort3(arr,slave1,slave2)
USE nrtype; USE nrutil, ONLY : assert_eq
USE nr, ONLY : indexx
IMPLICIT NONE
REAL(SP), DIMENSION(:), INTENT(INOUT) :: arr,slave1,slave2
```

Sorts an array `arr` into ascending order using Quicksort, while making the corresponding rearrangement of the same-size arrays `slave1` and `slave2`. The sorting and rearrangement are performed by means of an index array.

```
INTEGER(I4B) :: ndum
INTEGER(I4B), DIMENSION(size(arr)) :: index
ndum=assert_eq(size(arr),size(slave1),size(slave2),'sort3')
call indexx(arr,index)              Make the index array.
arr=arr(index)                      Sort arr.
slave1=slave1(index)                Rearrange slave1,
slave2=slave2(index)                and slave2.
END SUBROUTINE sort3
```

* * *

```
FUNCTION rank(index)
USE nrtype; USE nrutil, ONLY : arth
IMPLICIT NONE
INTEGER(I4B), DIMENSION(:), INTENT(IN) :: index
INTEGER(I4B), DIMENSION(size(index)) :: rank
   Given index as output from the routine indexx, this routine returns a same-size array
   rank, the corresponding table of ranks.
rank(index(:))=arth(1,1,size(index))
END FUNCTION rank
```

⋆ ⋆ ⋆

Just as in the case of sort, where an approximation of the underlying Quicksort partition-exchange algorithm can be captured with the Fortran 90 pack intrinsic, the same can be done with indexx. As before, although it is in principle parallelizable by the compiler, it is likely not competitive with library routines.

```
RECURSIVE SUBROUTINE index_bypack(arr,index,partial)
USE nrtype; USE nrutil, ONLY : array_copy,arth,assert_eq
IMPLICIT NONE
REAL(SP), DIMENSION(:), INTENT(IN) :: arr
INTEGER(I4B), DIMENSION(:), INTENT(INOUT) :: index
INTEGER, OPTIONAL, INTENT(IN) :: partial
   Indexes an array arr, i.e., outputs the array index of length N such that arr(index(j))
   is in ascending order for j = 1,2,...,N. The method is to apply recursively the Fortran
   90 pack intrinsic. This is similar to Quicksort, but allows parallelization by the Fortran 90
   compiler. partial is an optional argument that is used only internally on the recursive calls.
REAL(SP) :: a
INTEGER(I4B) :: n,k,nl,indext,nerr
INTEGER(I4B), SAVE :: level=0
LOGICAL, DIMENSION(:), ALLOCATABLE, SAVE :: mask
INTEGER(I4B), DIMENSION(:), ALLOCATABLE, SAVE :: temp
if (present(partial)) then
    n=size(index)
else
    n=assert_eq(size(index),size(arr),'indexx_bypack')
    index=arth(1,1,n)
end if
if (n <= 1) RETURN
k=(1+n)/2
call icomp_xchg(index(1),index(k))                  Pivot element is median of first, mid-
call icomp_xchg(index(k),index(n))                     dle, and last.
call icomp_xchg(index(1),index(k))
if (n <= 3) RETURN
level=level+1                                       Keep track of recursion level to avoid
if (level == 1) allocate(mask(n),temp(n))              allocation overhead.
indext=index(k)
a=arr(indext)
mask(1:n) = (arr(index) <= a)                       Which elements move to left?
mask(k) = .false.
call array_copy(pack(index,mask(1:n)),temp,nl,nerr)    Move them.
mask(k) = .true.
temp(nl+2:n)=pack(index,.not. mask(1:n))            Move others to right.
temp(nl+1)=indext
index=temp(1:n)
call index_bypack(arr,index(1:nl),partial=1)        And recurse.
call index_bypack(arr,index(nl+2:n),partial=1)
if (level == 1) deallocate(mask,temp)
level=level-1
```

```
CONTAINS

SUBROUTINE icomp_xchg(i,j)
IMPLICIT NONE
INTEGER(I4B), INTENT(INOUT) :: i,j
```
Swap or don't swap integer arguments, depending on the ordering of their corresponding elements in an array arr.
```
INTEGER(I4B) :: swp
if (arr(j) < arr(i)) then
    swp=i
    i=j
    j=swp
end if
END SUBROUTINE icomp_xchg
END SUBROUTINE index_bypack
```

⋆ ⋆ ⋆

```
FUNCTION select(k,arr)
USE nrtype; USE nrutil, ONLY : assert,swap
IMPLICIT NONE
INTEGER(I4B), INTENT(IN) :: k
REAL(SP), DIMENSION(:), INTENT(INOUT) :: arr
REAL(SP) :: select
```
Returns the kth smallest value in the array arr. The input array will be rearranged to have this value in location arr(k), with all smaller elements moved to arr(1:k-1) (in arbitrary order) and all larger elements in arr(k+1:) (also in arbitrary order).
```
INTEGER(I4B) :: i,r,j,l,n
REAL(SP) :: a
n=size(arr)
call assert(k >= 1, k <= n, 'select args')
l=1
r=n
do
    if (r-l <= 1) then                      Active partition contains 1 or 2 elements.
        if (r-l == 1) call swap(arr(l),arr(r),arr(l)>arr(r))   Active partition con-
        select=arr(k)                                            tains 2 elements.
        RETURN
    else                                    Choose median of left, center, and right elements
        i=(l+r)/2                               as partitioning element a. Also rearrange so
        call swap(arr(i),arr(l+1))              that arr(l) ≤ arr(l+1) ≤ arr(r).
        call swap(arr(l),arr(r),arr(l)>arr(r))
        call swap(arr(l+1),arr(r),arr(l+1)>arr(r))
        call swap(arr(l),arr(l+1),arr(l)>arr(l+1))
        i=l+1                               Initialize pointers for partitioning.
        j=r
        a=arr(l+1)                          Partitioning element.
        do                                  Here is the meat.
            do                              Scan up to find element > a.
                i=i+1
                if (arr(i) >= a) exit
            end do
            do                              Scan down to find element < a.
                j=j-1
                if (arr(j) <= a) exit
            end do
            if (j < i) exit                 Pointers crossed. Exit with partitioning complete.
            call swap(arr(i),arr(j))            Exchange elements.
        end do
        arr(l+1)=arr(j)                     Insert partitioning element.
        arr(j)=a
        if (j >= k) r=j-1                   Keep active the partition that contains the kth
                                                element.
```

```
        if (j <= k) l=i
    end if
end do
END FUNCTION select
```

⋆ ⋆ ⋆

The following routine, select_inplace, is renamed from Volume 1's selip.

```
FUNCTION select_inplace(k,arr)
USE nrtype
USE nr, ONLY : select
IMPLICIT NONE
INTEGER(I4B), INTENT(IN) :: k
REAL(SP), DIMENSION(:), INTENT(IN) :: arr
REAL(SP) :: select_inplace
    Returns the kth smallest value in the array arr, without altering the input array. In Fortran
    90's assumed memory-rich environment, we just call select in scratch space.
REAL(SP), DIMENSION(size(arr)) :: tarr
tarr=arr
select_inplace=select(k,tarr)
END FUNCTION select_inplace
```

f90 Volume 1's selip routine uses an entirely different algorithm, for the purpose of avoiding any additional memory allocation beyond that of the input array. Fortran 90 presumes a richer memory environment, so select_inplace simply does the obvious (destructive) selection in scratch space. You can of course use the old selip if your in-core or in-cache memory is at a premium.

```
FUNCTION select_bypack(k,arr)
USE nrtype; USE nrutil, ONLY : array_copy,assert,swap
IMPLICIT NONE
INTEGER(I4B), INTENT(IN) :: k
REAL(SP), DIMENSION(:), INTENT(INOUT) :: arr
REAL(SP) :: select_bypack
    Returns the kth smallest value in the array arr. The input array will be rearranged to have
    this value in location arr(k), with all smaller elements moved to arr(1:k-1) (in arbitrary
    order) and all larger elements in arr(k+1:) (also in arbitrary order). This implementation
    allows parallelization in the Fortran 90 pack intrinsic.
LOGICAL, DIMENSION(size(arr)) :: mask
REAL(SP), DIMENSION(size(arr)) :: temp
INTEGER(I4B) :: i,r,j,l,n,nl,nerr
REAL(SP) :: a
n=size(arr)
call assert(k >= 1, k <= n, 'select_bypack args')
l=1                                                  Initial left and right bounds.
r=n
do                                                   Keep partitioning until desired el-
    if (r-l <= 1) exit                                   ement is found.
    i=(l+r)/2
    call swap(arr(l),arr(i),arr(l)>arr(i))           Pivot element is median of first,
    call swap(arr(i),arr(r),arr(i)>arr(r))               middle, and last.
    call swap(arr(l),arr(i),arr(l)>arr(i))
    a=arr(i)
    mask(l:r) = (arr(l:r) <= a)                      Which elements move to left?
    mask(i) = .false.
    call array_copy(pack(arr(l:r),mask(l:r)),temp(l:),nl,nerr)     Move them.
    j=l+nl
```

```
    mask(i) = .true.
    temp(j+1:r)=pack(arr(l:r),.not. mask(l:r))      Move others to right.
    temp(j)=a
    arr(l:r)=temp(l:r)
    if (k > j) then                                  Reset bounds to whichever side
        l=j+1                                            has the desired element.
    else if (k < j) then
        r=j-1
    else
        l=j
        r=j
    end if
end do
if (r-l == 1) call swap(arr(l),arr(r),arr(l)>arr(r))    Case of only two left.
select_bypack=arr(k)
END FUNCTION select_bypack
```

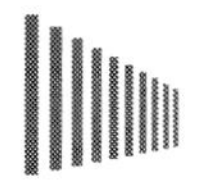

The above routine select_bypack is parallelizable, but as discussed above (sort_bypack, index_bypack) it is generally not very efficient.

⋆ ⋆ ⋆

The following routine, select_heap, is renamed from Volume 1's hpsel.

```
SUBROUTINE select_heap(arr,heap)
USE nrtype; USE nrutil, ONLY : nrerror,swap
USE nr, ONLY : sort
IMPLICIT NONE
REAL(SP), DIMENSION(:), INTENT(IN) :: arr
REAL(SP), DIMENSION(:), INTENT(OUT) :: heap
    Returns in heap, an array of length M, the largest M elements of the array arr of length
    N, with heap(1) guaranteed to be the the Mth largest element. The array arr is not
    altered. For efficiency, this routine should be used only when M << N.
INTEGER(I4B) :: i,j,k,m,n
m=size(heap)
n=size(arr)
if (m > n/2 .or. m < 1) call nrerror('probable misuse of select_heap')
heap=arr(1:m)
call sort(heap)                     Create initial heap by overkill! We assume m << n.
do i=m+1,n                          For each remaining element...
    if (arr(i) > heap(1)) then          Put it on the heap?
        heap(1)=arr(i)
        j=1
        do                          Sift down.
            k=2*j
            if (k > m) exit
            if (k /= m) then
                if (heap(k) > heap(k+1)) k=k+1
            end if
            if (heap(j) <= heap(k)) exit
            call swap(heap(k),heap(j))
            j=k
        end do
    end if
end do
END SUBROUTINE select_heap
```

⋆ ⋆ ⋆

```
FUNCTION eclass(lista,listb,n)
USE nrtype; USE nrutil, ONLY : arth,assert_eq
IMPLICIT NONE
INTEGER(I4B), DIMENSION(:), INTENT(IN) :: lista,listb
INTEGER(I4B), INTENT(IN) :: n
INTEGER(I4B), DIMENSION(n) :: eclass
```

Given M equivalences between pairs of `n` individual elements in the form of the input arrays `lista` and `listb` of length M, this routine returns in an array of length `n` the number of the equivalence class of each of the `n` elements, integers between 1 and `n` (not all such integers used).

```
INTEGER :: j,k,l,m
m=assert_eq(size(lista),size(listb),'eclass')
eclass(1:n)=arth(1,1,n)              Initialize each element its own class.
do l=1,m                             For each piece of input information...
    j=lista(l)
    do                               Track first element up to its ancestor.
        if (eclass(j) == j) exit
        j=eclass(j)
    end do
    k=listb(l)
    do                               Track second element up to its ancestor.
        if (eclass(k) == k) exit
        k=eclass(k)
    end do
    if (j /= k) eclass(j)=k          If they are not already related, make them so.
end do
do j=1,n                             Final sweep up to highest ancestors.
    do
        if (eclass(j) == eclass(eclass(j))) exit
        eclass(j)=eclass(eclass(j))
    end do
end do
END FUNCTION eclass
```

```
FUNCTION eclazz(equiv,n)
USE nrtype; USE nrutil, ONLY : arth
IMPLICIT NONE
INTERFACE
    FUNCTION equiv(i,j)
    USE nrtype
    IMPLICIT NONE
    LOGICAL(LGT) :: equiv
    INTEGER(I4B), INTENT(IN) :: i,j
    END FUNCTION equiv
END INTERFACE
INTEGER(I4B), INTENT(IN) :: n
INTEGER(I4B), DIMENSION(n) :: eclazz
```

Given a user-supplied logical function `equiv` that tells whether a pair of elements, each in the range `1...n`, are related, return in an array of length `n` equivalence class numbers for each element.

```
INTEGER :: i,j
eclazz(1:n)=arth(1,1,n)
do i=2,n                                  Loop over first element of all pairs.
    do j=1,i-1                            Loop over second element of all pairs.
        eclazz(j)=eclazz(eclazz(j))       Sweep it up this much.
        if (equiv(i,j)) eclazz(eclazz(eclazz(j)))=i
          Good exercise for the reader to figure out why this much ancestry is necessary!
    end do
end do
do i=1,n                                  Only this much sweeping is needed finally.
    eclazz(i)=eclazz(eclazz(i))
end do
END FUNCTION eclazz
```

CITED REFERENCES AND FURTHER READING:

Knuth, D.E. 1973, *Sorting and Searching*, vol. 3 of *The Art of Computer Programming* (Reading, MA: Addison-Wesley), §5.2.5. [1]

Chapter B9. Root Finding and Nonlinear Sets of Equations

```
SUBROUTINE scrsho(func)
USE nrtype
IMPLICIT NONE
INTERFACE
    FUNCTION func(x)
    USE nrtype
    IMPLICIT NONE
    REAL(SP), INTENT(IN) :: x
    REAL(SP) :: func
    END FUNCTION func
END INTERFACE
INTEGER(I4B), PARAMETER :: ISCR=60,JSCR=21
```

For interactive "dumb terminal" use. Produce a crude graph of the function func over the prompted-for interval x1,x2. Query for another plot until the user signals satisfaction.
Parameters: Number of horizontal and vertical positions in display.

```
INTEGER(I4B) :: i,j,jz
REAL(SP) :: dx,dyj,x,x1,x2,ybig,ysml
REAL(SP), DIMENSION(ISCR) :: y
CHARACTER(1), DIMENSION(ISCR,JSCR) :: scr
CHARACTER(1) :: blank=' ',zero='-',yy='l',xx='-',ff='x'
do
    write (*,*) ' Enter x1,x2 (= to stop)'          Query for another plot; quit if x1=x2.
    read (*,*) x1,x2
    if (x1 == x2) RETURN
    scr(1,1:JSCR)=yy                             Fill vertical sides with character 'l'.
    scr(ISCR,1:JSCR)=yy
    scr(2:ISCR-1,1)=xx                           Fill top, bottom with character '-'.
    scr(2:ISCR-1,JSCR)=xx
    scr(2:ISCR-1,2:JSCR-1)=blank                 Fill interior with blanks.
    dx=(x2-x1)/(ISCR-1)
    x=x1
    do i=1,ISCR                                  Evaluate the function at equal intervals.
        y(i)=func(x)
        x=x+dx
    end do
    ysml=min(minval(y(:)),0.0_sp)                Limits will include 0.
    ybig=max(maxval(y(:)),0.0_sp)
    if (ybig == ysml) ybig=ysml+1.0              Be sure to separate top and bottom.
    dyj=(JSCR-1)/(ybig-ysml)
    jz=1-ysml*dyj                                Note which row corresponds to 0.
    scr(1:ISCR,jz)=zero
    do i=1,ISCR                                  Place an indicator at function height and 0.
        j=1+(y(i)-ysml)*dyj
        scr(i,j)=ff
    end do
    write (*,'(1x,1p,e10.3,1x,80a1)') ybig,(scr(i,JSCR),i=1,ISCR)
    do j=JSCR-1,2,-1                             Display.
        write (*,'(12x,80a1)') (scr(i,j),i=1,ISCR)
    end do
    write (*,'(1x,1p,e10.3,1x,80a1)') ysml,(scr(i,1),i=1,ISCR)
```

```
    write (*,'(12x,1p,e10.3,40x,e10.3)') x1,x2
end do
END SUBROUTINE scrsho
```

f90 `CHARACTER(1), DIMENSION(ISCR,JSCR) :: scr` In Fortran 90, the length of variables of type character should be declared as `CHARACTER(1)` or `CHARACTER(len=1)` (for a variable of length 1), rather than the older form `CHARACTER*1`. While the older form is still legal syntax, the newer one is more consistent with the syntax of other type declarations. (For variables of length 1, you can actually omit the length specifier entirely, and just say `CHARACTER`.)

⋆ ⋆ ⋆

```
SUBROUTINE zbrac(func,x1,x2,succes)
USE nrtype; USE nrutil, ONLY : nrerror
IMPLICIT NONE
REAL(SP), INTENT(INOUT) :: x1,x2
LOGICAL(LGT), INTENT(OUT) :: succes
INTERFACE
    FUNCTION func(x)
    USE nrtype
    IMPLICIT NONE
    REAL(SP), INTENT(IN) :: x
    REAL(SP) :: func
    END FUNCTION func
END INTERFACE
INTEGER(I4B), PARAMETER :: NTRY=50
REAL(SP), PARAMETER :: FACTOR=1.6_sp
    Given a function func and an initial guessed range x1 to x2, the routine expands the range
    geometrically until a root is bracketed by the returned values x1 and x2 (in which case
    succes returns as .true.) or until the range becomes unacceptably large (in which case
    succes returns as .false.).
INTEGER(I4B) :: j
REAL(SP) :: f1,f2
if (x1 == x2) call nrerror('zbrac: you have to guess an initial range')
f1=func(x1)
f2=func(x2)
succes=.true.
do j=1,NTRY
    if ((f1 > 0.0 .and. f2 < 0.0) .or. &
        (f1 < 0.0 .and. f2 > 0.0)) RETURN
    if (abs(f1) < abs(f2)) then
        x1=x1+FACTOR*(x1-x2)
        f1=func(x1)
    else
        x2=x2+FACTOR*(x2-x1)
        f2=func(x2)
    end if
end do
succes=.false.
END SUBROUTINE zbrac
```

⋆ ⋆ ⋆

```
SUBROUTINE zbrak(func,x1,x2,n,xb1,xb2,nb)
USE nrtype; USE nrutil, ONLY : arth
IMPLICIT NONE
INTEGER(I4B), INTENT(IN) :: n
INTEGER(I4B), INTENT(OUT) :: nb
REAL(SP), INTENT(IN) :: x1,x2
REAL(SP), DIMENSION(:), POINTER :: xb1,xb2
INTERFACE
    FUNCTION func(x)
    USE nrtype
    IMPLICIT NONE
    REAL(SP), INTENT(IN) :: x
    REAL(SP) :: func
    END FUNCTION func
END INTERFACE
```

Given a function func defined on the interval from x1-x2 subdivide the interval into n equally spaced segments, and search for zero crossings of the function. nb is returned as the number of bracketing pairs xb1(1:nb), xb2(1:nb) that are found. xb1 and xb2 are pointers to arrays of length nb that are dynamically allocated by the routine.

```
INTEGER(I4B) :: i
REAL(SP) :: dx
REAL(SP), DIMENSION(0:n) :: f,x
LOGICAL(LGT), DIMENSION(1:n) :: mask
LOGICAL(LGT), SAVE :: init=.true.
if (init) then
    init=.false.
    nullify(xb1,xb2)
end if
if (associated(xb1)) deallocate(xb1)
if (associated(xb2)) deallocate(xb2)
dx=(x2-x1)/n                          Determine the spacing appropriate to the mesh.
x=x1+dx*arth(0,1,n+1)
do i=0,n                              Evaluate the function at the mesh points.
    f(i)=func(x(i))
end do
mask=f(1:n)*f(0:n-1) <= 0.0           Record where the sign changes occur.
nb=count(mask)                        Number of sign changes.
allocate(xb1(nb),xb2(nb))
xb1(1:nb)=pack(x(0:n-1),mask)         Store the bounds of each bracket.
xb2(1:nb)=pack(x(1:n),mask)
END SUBROUTINE zbrak
```

This routine shows how to return arrays xb1 and xb2 whose size is not known in advance. The coding is explained in the subsection on pointers in §21.5.

⋆ ⋆ ⋆

```
FUNCTION rtbis(func,x1,x2,xacc)
USE nrtype; USE nrutil, ONLY : nrerror
IMPLICIT NONE
REAL(SP), INTENT(IN) :: x1,x2,xacc
REAL(SP) :: rtbis
INTERFACE
    FUNCTION func(x)
    USE nrtype
    IMPLICIT NONE
    REAL(SP), INTENT(IN) :: x
    REAL(SP) :: func
    END FUNCTION func
END INTERFACE
```

```
INTEGER(I4B), PARAMETER :: MAXIT=40
```

Using bisection, find the root of a function func known to lie between x1 and x2. The root, returned as rtbis, will be refined until its accuracy is ±xacc.

Parameter: MAXIT is the maximum allowed number of bisections.

```
INTEGER(I4B) :: j
REAL(SP) :: dx,f,fmid,xmid
fmid=func(x2)
f=func(x1)
if (f*fmid >= 0.0) call nrerror('rtbis: root must be bracketed')
if (f < 0.0) then                 Orient the search so that f>0 lies at x+dx.
    rtbis=x1
    dx=x2-x1
else
    rtbis=x2
    dx=x1-x2
end if
do j=1,MAXIT                      Bisection loop.
    dx=dx*0.5_sp
    xmid=rtbis+dx
    fmid=func(xmid)
    if (fmid <= 0.0) rtbis=xmid
    if (abs(dx) < xacc .or. fmid == 0.0) RETURN
end do
call nrerror('rtbis: too many bisections')
END FUNCTION rtbis
```

* * *

```
FUNCTION rtflsp(func,x1,x2,xacc)
USE nrtype; USE nrutil, ONLY : nrerror,swap
IMPLICIT NONE
REAL(SP), INTENT(IN) :: x1,x2,xacc
REAL(SP) :: rtflsp
INTERFACE
    FUNCTION func(x)
    USE nrtype
    IMPLICIT NONE
    REAL(SP), INTENT(IN) :: x
    REAL(SP) :: func
    END FUNCTION func
END INTERFACE
INTEGER(I4B), PARAMETER :: MAXIT=30
```

Using the false position method, find the root of a function func known to lie between x1 and x2. The root, returned as rtflsp, is refined until its accuracy is ±xacc.

Parameter: MAXIT is the maximum allowed number of iterations.

```
INTEGER(I4B) :: j
REAL(SP) :: del,dx,f,fh,fl,xh,xl
fl=func(x1)
fh=func(x2)                        Be sure the interval brackets a root.
if ((fl > 0.0 .and. fh > 0.0) .or. &
    (fl < 0.0 .and. fh < 0.0)) call &
    nrerror('rtflsp: root must be bracketed between arguments')
if (fl < 0.0) then                 Identify the limits so that xl corresponds to
    xl=x1                              the low side.
    xh=x2
else
    xl=x2
    xh=x1
    call swap(fl,fh)
end if
dx=xh-xl
```

```
do j=1,MAXIT                                    False position loop.
    rtflsp=xl+dx*fl/(fl-fh)                     Increment with respect to latest value.
    f=func(rtflsp)
    if (f < 0.0) then                           Replace appropriate limit.
        del=xl-rtflsp
        xl=rtflsp
        fl=f
    else
        del=xh-rtflsp
        xh=rtflsp
        fh=f
    end if
    dx=xh-xl
    if (abs(del) < xacc .or. f == 0.0) RETURN   Convergence.
end do
call nrerror('rtflsp exceed maximum iterations')
END FUNCTION rtflsp
```

⋆ ⋆ ⋆

```
FUNCTION rtsec(func,x1,x2,xacc)
USE nrtype; USE nrutil, ONLY : nrerror,swap
IMPLICIT NONE
REAL(SP), INTENT(IN) :: x1,x2,xacc
REAL(SP) :: rtsec
INTERFACE
    FUNCTION func(x)
    USE nrtype
    IMPLICIT NONE
    REAL(SP), INTENT(IN) :: x
    REAL(SP) :: func
    END FUNCTION func
END INTERFACE
INTEGER(I4B), PARAMETER :: MAXIT=30
```

Using the secant method, find the root of a function `func` thought to lie between `x1` and `x2`. The root, returned as `rtsec`, is refined until its accuracy is ±`xacc`.
Parameter: `MAXIT` is the maximum allowed number of iterations.

```
INTEGER(I4B) :: j
REAL(SP) :: dx,f,fl,xl
fl=func(x1)
f=func(x2)
if (abs(fl) < abs(f)) then                      Pick the bound with the smaller function value
    rtsec=x1                                        as the most recent guess.
    xl=x2
    call swap(fl,f)
else
    xl=x1
    rtsec=x2
end if
do j=1,MAXIT                                    Secant loop.
    dx=(xl-rtsec)*f/(f-fl)                      Increment with respect to latest value.
    xl=rtsec
    fl=f
    rtsec=rtsec+dx
    f=func(rtsec)
    if (abs(dx) < xacc .or. f == 0.0) RETURN    Convergence.
end do
call nrerror('rtsec: exceed maximum iterations')
END FUNCTION rtsec
```

⋆ ⋆ ⋆

```
FUNCTION zriddr(func,x1,x2,xacc)
USE nrtype; USE nrutil, ONLY : nrerror
IMPLICIT NONE
REAL(SP), INTENT(IN) :: x1,x2,xacc
REAL(SP) :: zriddr
INTERFACE
   FUNCTION func(x)
   USE nrtype
   IMPLICIT NONE
   REAL(SP), INTENT(IN) :: x
   REAL(SP) :: func
   END FUNCTION func
END INTERFACE
INTEGER(I4B), PARAMETER :: MAXIT=60
```

Using Ridders' method, return the root of a function `func` known to lie between `x1` and `x2`. The root, returned as `zriddr`, will be refined to an approximate accuracy `xacc`.

```
REAL(SP), PARAMETER :: UNUSED=-1.11e30_sp
INTEGER(I4B) :: j
REAL(SP) :: fh,fl,fm,fnew,s,xh,xl,xm,xnew
fl=func(x1)
fh=func(x2)
if ((fl > 0.0 .and. fh < 0.0) .or. (fl < 0.0 .and. fh > 0.0)) then
   xl=x1
   xh=x2
   zriddr=UNUSED                       Any highly unlikely value, to simplify logic
   do j=1,MAXIT                           below.
      xm=0.5_sp*(xl+xh)
      fm=func(xm)                      First of two function evaluations per it-
      s=sqrt(fm**2-fl*fh)                 eration.
      if (s == 0.0) RETURN
      xnew=xm+(xm-xl)*(sign(1.0_sp,fl-fh)*fm/s)      Updating formula.
      if (abs(xnew-zriddr) <= xacc) RETURN
      zriddr=xnew
      fnew=func(zriddr)                Second of two function evaluations per
      if (fnew == 0.0) RETURN             iteration.
      if (sign(fm,fnew) /= fm) then    Bookkeeping to keep the root bracketed
         xl=xm                            on next iteration.
         fl=fm
         xh=zriddr
         fh=fnew
      else if (sign(fl,fnew) /= fl) then
         xh=zriddr
         fh=fnew
      else if (sign(fh,fnew) /= fh) then
         xl=zriddr
         fl=fnew
      else
         call nrerror('zriddr: never get here')
      end if
      if (abs(xh-xl) <= xacc) RETURN
   end do
   call nrerror('zriddr: exceeded maximum iterations')
else if (fl == 0.0) then
   zriddr=x1
else if (fh == 0.0) then
   zriddr=x2
else
   call nrerror('zriddr: root must be bracketed')
end if
END FUNCTION zriddr
```

⋆ ⋆ ⋆

```
FUNCTION zbrent(func,x1,x2,tol)
USE nrtype; USE nrutil, ONLY : nrerror
IMPLICIT NONE
REAL(SP), INTENT(IN) :: x1,x2,tol
REAL(SP) :: zbrent
INTERFACE
    FUNCTION func(x)
    USE nrtype
    IMPLICIT NONE
    REAL(SP), INTENT(IN) :: x
    REAL(SP) :: func
    END FUNCTION func
END INTERFACE
INTEGER(I4B), PARAMETER :: ITMAX=100
REAL(SP), PARAMETER :: EPS=epsilon(x1)
    Using Brent's method, find the root of a function func known to lie between x1 and x2.
    The root, returned as zbrent, will be refined until its accuracy is tol.
    Parameters: Maximum allowed number of iterations, and machine floating-point precision.
INTEGER(I4B) :: iter
REAL(SP) :: a,b,c,d,e,fa,fb,fc,p,q,r,s,tol1,xm
a=x1
b=x2
fa=func(a)
fb=func(b)
if ((fa > 0.0 .and. fb > 0.0) .or. (fa < 0.0 .and. fb < 0.0)) &
    call nrerror('root must be bracketed for zbrent')
c=b
fc=fb
do iter=1,ITMAX
    if ((fb > 0.0 .and. fc > 0.0) .or. (fb < 0.0 .and. fc < 0.0)) then
        c=a                                  Rename a, b, c and adjust bounding in-
        fc=fa                                    terval d.
        d=b-a
        e=d
    end if
    if (abs(fc) < abs(fb)) then
        a=b
        b=c
        c=a
        fa=fb
        fb=fc
        fc=fa
    end if
    tol1=2.0_sp*EPS*abs(b)+0.5_sp*tol        Convergence check.
    xm=0.5_sp*(c-b)
    if (abs(xm) <= tol1 .or. fb == 0.0) then
        zbrent=b
        RETURN
    end if
    if (abs(e) >= tol1 .and. abs(fa) > abs(fb)) then
        s=fb/fa                              Attempt inverse quadratic interpolation.
        if (a == c) then
            p=2.0_sp*xm*s
            q=1.0_sp-s
        else
            q=fa/fc
            r=fb/fc
            p=s*(2.0_sp*xm*q*(q-r)-(b-a)*(r-1.0_sp))
            q=(q-1.0_sp)*(r-1.0_sp)*(s-1.0_sp)
        end if
        if (p > 0.0) q=-q                    Check whether in bounds.
        p=abs(p)
        if (2.0_sp*p < min(3.0_sp*xm*q-abs(tol1*q),abs(e*q))) then
            e=d                              Accept interpolation.
```

```
                d=p/q
            else
                d=xm                                Interpolation failed; use bisection.
                e=d
            end if
        else                                        Bounds decreasing too slowly; use bisec-
            d=xm                                        tion.
            e=d
        end if
        a=b                                         Move last best guess to a.
        fa=fb
        b=b+merge(d,sign(tol1,xm), abs(d) > tol1 )      Evaluate new trial root.
        fb=func(b)
    end do
    call nrerror('zbrent: exceeded maximum iterations')
    zbrent=b
    END FUNCTION zbrent
```

REAL(SP), PARAMETER :: EPS=epsilon(x1) The routine zbrent works best when EPS is *exactly* the machine precision. The Fortran 90 intrinsic function epsilon allows us to code this in a portable fashion.

```
FUNCTION rtnewt(funcd,x1,x2,xacc)
USE nrtype; USE nrutil, ONLY : nrerror
IMPLICIT NONE
REAL(SP), INTENT(IN) :: x1,x2,xacc
REAL(SP) :: rtnewt
INTERFACE
    SUBROUTINE funcd(x,fval,fderiv)
    USE nrtype
    IMPLICIT NONE
    REAL(SP), INTENT(IN) :: x
    REAL(SP), INTENT(OUT) :: fval,fderiv
    END SUBROUTINE funcd
END INTERFACE
INTEGER(I4B), PARAMETER :: MAXIT=20
    Using the Newton-Raphson method, find the root of a function known to lie in the interval
    [x1, x2]. The root rtnewt will be refined until its accuracy is known within ±xacc. funcd
    is a user-supplied subroutine that returns both the function value and the first derivative of
    the function.
    Parameter: MAXIT is the maximum number of iterations.
INTEGER(I4B) :: j
REAL(SP) :: df,dx,f
rtnewt=0.5_sp*(x1+x2)                       Initial guess.
do j=1,MAXIT
    call funcd(rtnewt,f,df)
    dx=f/df
    rtnewt=rtnewt-dx
    if ((x1-rtnewt)*(rtnewt-x2) < 0.0)&
        call nrerror('rtnewt: values jumped out of brackets')
    if (abs(dx) < xacc) RETURN              Convergence.
end do
call nrerror('rtnewt exceeded maximum iterations')
END FUNCTION rtnewt
```

⋆ ⋆ ⋆

```
FUNCTION rtsafe(funcd,x1,x2,xacc)
USE nrtype; USE nrutil, ONLY : nrerror
IMPLICIT NONE
REAL(SP), INTENT(IN) :: x1,x2,xacc
REAL(SP) :: rtsafe
INTERFACE
    SUBROUTINE funcd(x,fval,fderiv)
    USE nrtype
    IMPLICIT NONE
    REAL(SP), INTENT(IN) :: x
    REAL(SP), INTENT(OUT) :: fval,fderiv
    END SUBROUTINE funcd
END INTERFACE
INTEGER(I4B), PARAMETER :: MAXIT=100
```

Using a combination of Newton-Raphson and bisection, find the root of a function bracketed between `x1` and `x2`. The root, returned as the function value `rtsafe`, will be refined until its accuracy is known within $\pm$`xacc`. `funcd` is a user-supplied subroutine that returns both the function value and the first derivative of the function.

Parameter: `MAXIT` is the maximum allowed number of iterations.

```
INTEGER(I4B) :: j
REAL(SP) :: df,dx,dxold,f,fh,fl,temp,xh,xl
call funcd(x1,fl,df)
call funcd(x2,fh,df)
if ((fl > 0.0 .and. fh > 0.0) .or. &
    (fl < 0.0 .and. fh < 0.0)) &
    call nrerror('root must be bracketed in rtsafe')
if (fl == 0.0) then
    rtsafe=x1
    RETURN
else if (fh == 0.0) then
    rtsafe=x2
    RETURN
else if (fl < 0.0) then                   Orient the search so that f(xl) < 0.
    xl=x1
    xh=x2
else
    xh=x1
    xl=x2
end if
rtsafe=0.5_sp*(x1+x2)                     Initialize the guess for root,
dxold=abs(x2-x1)                          the "stepsize before last,"
dx=dxold                                  and the last step.
call funcd(rtsafe,f,df)
do j=1,MAXIT                              Loop over allowed iterations.
    if (((rtsafe-xh)*df-f)*((rtsafe-xl)*df-f) > 0.0 .or. &
        abs(2.0_sp*f) > abs(dxold*df) ) then
          Bisect if Newton out of range, or not decreasing fast enough.
        dxold=dx
        dx=0.5_sp*(xh-xl)
        rtsafe=xl+dx
        if (xl == rtsafe) RETURN          Change in root is negligible.
    else                                  Newton step acceptable. Take it.
        dxold=dx
        dx=f/df
        temp=rtsafe
        rtsafe=rtsafe-dx
        if (temp == rtsafe) RETURN
    end if
    if (abs(dx) < xacc) RETURN            Convergence criterion.
    call funcd(rtsafe,f,df)               One new function evaluation per iteration.
    if (f < 0.0) then                     Maintain the bracket on the root.
        xl=rtsafe
    else
        xh=rtsafe
```

```
    end if
end do
call nrerror('rtsafe: exceeded maximum iterations')
END FUNCTION rtsafe
```

⋆ ⋆ ⋆

```
SUBROUTINE laguer(a,x,its)
USE nrtype; USE nrutil, ONLY : nrerror,poly,poly_term
IMPLICIT NONE
INTEGER(I4B), INTENT(OUT) :: its
COMPLEX(SPC), INTENT(INOUT) :: x
COMPLEX(SPC), DIMENSION(:), INTENT(IN) :: a
REAL(SP), PARAMETER :: EPS=epsilon(1.0_sp)
INTEGER(I4B), PARAMETER :: MR=8,MT=10,MAXIT=MT*MR
```

Given an array of $M+1$ complex coefficients a of the polynomial $\sum_{i=1}^{M+1} \mathrm{a}(i)x^{i-1}$, and given a complex value x, this routine improves x by Laguerre's method until it converges, within the achievable roundoff limit, to a root of the given polynomial. The number of iterations taken is returned as its.

Parameters: EPS is the estimated fractional roundoff error. We try to break (rare) limit cycles with MR different fractional values, once every MT steps, for MAXIT total allowed iterations.

```
INTEGER(I4B) :: iter,m
REAL(SP) :: abx,abp,abm,err
COMPLEX(SPC) :: dx,x1,f,g,h,sq,gp,gm,g2
COMPLEX(SPC), DIMENSION(size(a)) :: b,d
REAL(SP), DIMENSION(MR) :: frac = &
    (/ 0.5_sp,0.25_sp,0.75_sp,0.13_sp,0.38_sp,0.62_sp,0.88_sp,1.0_sp /)
      Fractions used to break a limit cycle.
m=size(a)-1
do iter=1,MAXIT                             Loop over iterations up to allowed maximum.
    its=iter
    abx=abs(x)
    b(m+1:1:-1)=poly_term(a(m+1:1:-1),x)       Efficient computation of the polynomial
    d(m:1:-1)=poly_term(b(m+1:2:-1),x)             and its first two derivatives.
    f=poly(x,d(2:m))
    err=EPS*poly(abx,abs(b(1:m+1)))          Esimate of roundoff in evaluating polynomial.
    if (abs(b(1)) <= err) RETURN             We are on the root.
    g=d(1)/b(1)                              The generic case: Use Laguerre's formula.
    g2=g*g
    h=g2-2.0_sp*f/b(1)
    sq=sqrt((m-1)*(m*h-g2))
    gp=g+sq
    gm=g-sq
    abp=abs(gp)
    abm=abs(gm)
    if (abp < abm) gp=gm
    if (max(abp,abm) > 0.0) then
        dx=m/gp
    else
        dx=exp(cmplx(log(1.0_sp+abx),iter,kind=spc))
    end if
    x1=x-dx
    if (x == x1) RETURN                      Converged.
    if (mod(iter,MT) /= 0) then
        x=x1
    else                                     Every so often we take a fractional step, to
        x=x-dx*frac(iter/MT)                     break any limit cycle (itself a rare occur-
    end if                                       rence).
end do
call nrerror('laguer: too many iterations')
```

Very unusual — can occur only for complex roots. Try a different starting guess for the root.

```
END SUBROUTINE laguer
```

f90 `b(m+1:1:-1)=poly_term...f=poly(x,d(2:m))` The `poly_term` function in `nrutil` tabulates the partial sums of a polynomial, while `poly` evaluates the polynomial at `x`. In this example, we use `poly_term` on the coefficient array in reverse order, so that the value of the polynomial ends up in `b(1)` and the value of its first derivative in `d(1)`.

`dx=exp(cmplx(log(1.0_sp+abx),iter,kind=spc))` The intrinsic function `cmplx` returns a quantity of type default complex unless the `kind` argument is present. To facilitate converting our routines from single to double precision, we always include the `kind` argument explicitly so that when you redefine `spc` in `nrtype` to be double-precision complex the conversions are carried out correctly.

* * *

```
SUBROUTINE zroots(a,roots,polish)
USE nrtype; USE nrutil, ONLY : assert_eq,poly_term
USE nr, ONLY : laguer,indexx
IMPLICIT NONE
COMPLEX(SPC), DIMENSION(:), INTENT(IN) :: a
COMPLEX(SPC), DIMENSION(:), INTENT(OUT) :: roots
LOGICAL(LGT), INTENT(IN) :: polish
REAL(SP), PARAMETER :: EPS=1.0e-6_sp
```

Given the array of $M+1$ complex coefficients `a` of the polynomial $\sum_{i=1}^{M+1} \mathrm{a}(i)x^{i-1}$, this routine successively calls `laguer` and finds all M complex `roots`. The logical variable `polish` should be input as `.true.` if polishing (also by Laguerre's method) is desired, `.false.` if the roots will be subsequently polished by other means.

Parameter: EPS is a small number.

```
INTEGER(I4B) :: j,its,m
INTEGER(I4B), DIMENSION(size(roots)) :: indx
COMPLEX(SPC) :: x
COMPLEX(SPC), DIMENSION(size(a)) :: ad
m=assert_eq(size(roots),size(a)-1,'zroots')
ad(:)=a(:)                                  Copy of coefficients for successive deflation.
do j=m,1,-1                                 Loop over each root to be found.
    x=cmplx(0.0_sp,kind=spc)
      Start at zero to favor convergence to smallest remaining root.
    call laguer(ad(1:j+1),x,its)            Find the root.
    if (abs(aimag(x)) <= 2.0_sp*EPS**2*abs(real(x))) &
        x=cmplx(real(x),kind=spc)
    roots(j)=x
    ad(j:1:-1)=poly_term(ad(j+1:2:-1),x)        Forward deflation.
end do
if (polish) then
    do j=1,m                                Polish the roots using the undeflated coeffi-
        call laguer(a(:),roots(j),its)          cients.
    end do
end if
call indexx(real(roots),indx)               Sort roots by their real parts.
roots=roots(indx)
END SUBROUTINE zroots
```

f90 `x=cmplx(0.0_sp,kind=spc)...x=cmplx(real(x),kind=spc)` See the discussion of why we include `kind=spc` just above. Note that while `real(x)` returns type default real if `x` is integer or real, it returns single or double precision correctly if `x` is complex.

* * *

```
SUBROUTINE zrhqr(a,rtr,rti)
USE nrtype; USE nrutil, ONLY : assert_eq,nrerror
USE nr, ONLY : balanc,hqr,indexx
IMPLICIT NONE
REAL(SP), DIMENSION(:), INTENT(IN) :: a
REAL(SP), DIMENSION(:), INTENT(OUT) :: rtr,rti
```

Find all the roots of a polynomial with real coefficients, $\sum_{i=1}^{M+1} \mathtt{a}(i)x^{i-1}$, given the array of $M+1$ coefficients a. The method is to construct an upper Hessenberg matrix whose eigenvalues are the desired roots, and then use the routines balanc and hqr. The real and imaginary parts of the M roots are returned in rtr and rti, respectively.

```
INTEGER(I4B) :: k,m
INTEGER(I4B), DIMENSION(size(rtr)) :: indx
REAL(SP), DIMENSION(size(a)-1,size(a)-1) :: hess
m=assert_eq(size(rtr),size(rti),size(a)-1,'zrhqr')
if (a(m+1) == 0.0) call &
    nrerror('zrhqr: Last value of array a must not be 0')
hess(1,:)=-a(m:1:-1)/a(m+1)          Construct the matrix.
hess(2:m,:)=0.0
do k=1,m-1
    hess(k+1,k)=1.0
end do
call balanc(hess)                    Find its eigenvalues.
call hqr(hess,rtr,rti)
call indexx(rtr,indx)                Sort roots by their real parts.
rtr=rtr(indx)
rti=rti(indx)
END SUBROUTINE zrhqr
```

⋆ ⋆ ⋆

```
SUBROUTINE qroot(p,b,c,eps)
USE nrtype; USE nrutil, ONLY : nrerror
USE nr, ONLY : poldiv
IMPLICIT NONE
REAL(SP), DIMENSION(:), INTENT(IN) :: p
REAL(SP), INTENT(INOUT) :: b,c
REAL(SP), INTENT(IN) :: eps
INTEGER(I4B), PARAMETER :: ITMAX=20
REAL(SP), PARAMETER :: TINY=1.0e-6_sp
```

Given an array of N coefficients p of a polynomial of degree $N-1$, and trial values for the coefficients of a quadratic factor $x^2 + \mathtt{b}x + \mathtt{c}$, improve the solution until the coefficients b, c change by less than eps. The routine poldiv of §5.3 is used.

Parameters: ITMAX is the maximum number of iterations, TINY is a small number.

```
INTEGER(I4B) :: iter,n
REAL(SP) :: delb,delc,div,r,rb,rc,s,sb,sc
REAL(SP), DIMENSION(3) :: d
REAL(SP), DIMENSION(size(p)) :: q,qq,rem
n=size(p)
d(3)=1.0
do iter=1,ITMAX
    d(2)=b
    d(1)=c
    call poldiv(p,d,q,rem)
    s=rem(1)                              First division gives r,s.
    r=rem(2)
    call poldiv(q(1:n-1),d(:),qq(1:n-1),rem(1:n-1))
    sc=-rem(1)                            Second division gives partial r,s with respect
    rc=-rem(2)                                to c.
    sb=-c*rc
    rb=sc-b*rc
    div=1.0_sp/(sb*rc-sc*rb)              Solve 2x2 equation.
```

```
    delb=(r*sc-s*rc)*div
    delc=(-r*sb+s*rb)*div
    b=b+delb
    c=c+delc
    if ((abs(delb) <= eps*abs(b) .or. abs(b) < TINY) .and. &
        (abs(delc) <= eps*abs(c) .or. abs(c) < TINY)) RETURN    Coefficients converged.
end do
call nrerror('qroot: too many iterations')
END SUBROUTINE qroot
```

⋆ ⋆ ⋆

```
SUBROUTINE mnewt(ntrial,x,tolx,tolf,usrfun)
USE nrtype
USE nr, ONLY : lubksb,ludcmp
IMPLICIT NONE
INTEGER(I4B), INTENT(IN) :: ntrial
REAL(SP), INTENT(IN) :: tolx,tolf
REAL(SP), DIMENSION(:), INTENT(INOUT) :: x
INTERFACE
    SUBROUTINE usrfun(x,fvec,fjac)
    USE nrtype
    IMPLICIT NONE
    REAL(SP), DIMENSION(:), INTENT(IN) :: x
    REAL(SP), DIMENSION(:), INTENT(OUT) :: fvec
    REAL(SP), DIMENSION(:,:), INTENT(OUT) :: fjac
    END SUBROUTINE usrfun
END INTERFACE
```

Given an initial guess x for a root in N dimensions, take ntrial Newton-Raphson steps to improve the root. Stop if the root converges in either summed absolute variable increments tolx or summed absolute function values tolf.

```
INTEGER(I4B) :: i
INTEGER(I4B), DIMENSION(size(x)) :: indx
REAL(SP) :: d
REAL(SP), DIMENSION(size(x)) :: fvec,p
REAL(SP), DIMENSION(size(x),size(x)) :: fjac
do i=1,ntrial
    call usrfun(x,fvec,fjac)
      User subroutine supplies function values at x in fvec and Jacobian matrix in fjac.
    if (sum(abs(fvec)) <= tolf) RETURN      Check function convergence.
    p=-fvec                                 Right-hand side of linear equations.
    call ludcmp(fjac,indx,d)                Solve linear equations using LU decom-
    call lubksb(fjac,indx,p)                    position.
    x=x+p                                   Update solution.
    if (sum(abs(p)) <= tolx) RETURN         Check root convergence.
end do
END SUBROUTINE mnewt
```

⋆ ⋆ ⋆

```
SUBROUTINE lnsrch(xold,fold,g,p,x,f,stpmax,check,func)
USE nrtype; USE nrutil, ONLY : assert_eq,nrerror,vabs
IMPLICIT NONE
REAL(SP), DIMENSION(:), INTENT(IN) :: xold,g
REAL(SP), DIMENSION(:), INTENT(INOUT) :: p
REAL(SP), INTENT(IN) :: fold,stpmax
REAL(SP), DIMENSION(:), INTENT(OUT) :: x
REAL(SP), INTENT(OUT) :: f
LOGICAL(LGT), INTENT(OUT) :: check
INTERFACE
    FUNCTION func(x)
    USE nrtype
    IMPLICIT NONE
    REAL(SP) :: func
    REAL(SP), DIMENSION(:), INTENT(IN) :: x
    END FUNCTION func
END INTERFACE
REAL(SP), PARAMETER :: ALF=1.0e-4_sp,TOLX=epsilon(x)
```

Given an N-dimensional point xold, the value of the function and gradient there, fold and g, and a direction p, finds a new point x along the direction p from xold where the function func has decreased "sufficiently." xold, g, p, and x are all arrays of length N. The new function value is returned in f. stpmax is an input quantity that limits the length of the steps so that you do not try to evaluate the function in regions where it is undefined or subject to overflow. p is usually the Newton direction. The output quantity check is false on a normal exit. It is true when x is too close to xold. In a minimization algorithm, this usually signals convergence and can be ignored. However, in a zero-finding algorithm the calling program should check whether the convergence is spurious.

Parameters: ALF ensures sufficient decrease in function value; TOLX is the convergence criterion on Δx.

```
INTEGER(I4B) :: ndum
REAL(SP) :: a,alam,alam2,alamin,b,disc,f2,pabs,rhs1,rhs2,slope,tmplam
ndum=assert_eq(size(g),size(p),size(x),size(xold),'lnsrch')
check=.false.
pabs=vabs(p(:))
if (pabs > stpmax) p(:)=p(:)*stpmax/pabs          Scale if attempted step is too big.
slope=dot_product(g,p)
if (slope >= 0.0) call nrerror('roundoff problem in lnsrch')
alamin=TOLX/maxval(abs(p(:))/max(abs(xold(:)),1.0_sp))    Compute λmin.
alam=1.0                                          Always try full Newton step first.
do                                                Start of iteration loop.
    x(:)=xold(:)+alam*p(:)
    f=func(x)
    if (alam < alamin) then                       Convergence on Δx. For zero find-
        x(:)=xold(:)                                  ing, the calling program should
        check=.true.                                  verify the convergence.
        RETURN
    else if (f <= fold+ALF*alam*slope) then       Sufficient function decrease.
        RETURN
    else                                          Backtrack.
        if (alam == 1.0) then                     First time.
            tmplam=-slope/(2.0_sp*(f-fold-slope))
        else                                      Subsequent backtracks.
            rhs1=f-fold-alam*slope
            rhs2=f2-fold-alam2*slope
            a=(rhs1/alam**2-rhs2/alam2**2)/(alam-alam2)
            b=(-alam2*rhs1/alam**2+alam*rhs2/alam2**2)/&
                (alam-alam2)
            if (a == 0.0) then
                tmplam=-slope/(2.0_sp*b)
            else
                disc=b*b-3.0_sp*a*slope
                if (disc < 0.0) then
                    tmplam=0.5_sp*alam
                else if (b <= 0.0) then
```

```
                    tmplam=(-b+sqrt(disc))/(3.0_sp*a)
                else
                    tmplam=-slope/(b+sqrt(disc))
                end if
            end if
            if (tmplam > 0.5_sp*alam) tmplam=0.5_sp*alam     λ ≤ 0.5λ1.
        end if
    end if
    alam2=alam
    f2=f
    alam=max(tmplam,0.1_sp*alam)                   λ ≥ 0.1λ1.
end do                                             Try again.
END SUBROUTINE lnsrch
```

```
SUBROUTINE newt(x,check)
USE nrtype; USE nrutil, ONLY : nrerror,vabs
USE nr, ONLY : fdjac,lnsrch,lubksb,ludcmp
USE fminln                                      Communicates with fmin.
IMPLICIT NONE
REAL(SP), DIMENSION(:), INTENT(INOUT) :: x
LOGICAL(LGT), INTENT(OUT) :: check
INTEGER(I4B), PARAMETER :: MAXITS=200
REAL(SP), PARAMETER :: TOLF=1.0e-4_sp,TOLMIN=1.0e-6_sp,TOLX=epsilon(x),&
    STPMX=100.0
```

Given an initial guess x for a root in N dimensions, find the root by a globally convergent Newton's method. The length N vector of functions to be zeroed, called fvec in the routine below, is returned by a user-supplied routine that *must* be called funcv and have the declaration FUNCTION funcv(x). The output quantity check is false on a normal return and true if the routine has converged to a local minimum of the function fmin defined below. In this case try restarting from a different initial guess.
Parameters: MAXITS is the maximum number of iterations; TOLF sets the convergence criterion on function values; TOLMIN sets the criterion for deciding whether spurious convergence to a minimum of fmin has occurred; TOLX is the convergence criterion on $\delta\mathbf{x}$; STPMX is the scaled maximum step length allowed in line searches.

```
INTEGER(I4B) :: its
INTEGER(I4B), DIMENSION(size(x)) :: indx
REAL(SP) :: d,f,fold,stpmax
REAL(SP), DIMENSION(size(x)) :: g,p,xold
REAL(SP), DIMENSION(size(x)), TARGET :: fvec
REAL(SP), DIMENSION(size(x),size(x)) :: fjac
fmin_fvecp=>fvec
f=fmin(x)                                      fvec is also computed by this call.
if (maxval(abs(fvec(:))) < 0.01_sp*TOLF) then      Test for initial guess being a root.
    check=.false.                                     Use more stringent test than
    RETURN                                            simply TOLF.
end if
stpmax=STPMX*max(vabs(x(:)),real(size(x),sp))      Calculate stpmax for line searches.
do its=1,MAXITS                                Start of iteration loop.
    call fdjac(x,fvec,fjac)
      If analytic Jacobian is available, you can replace the routine fdjac below with your own
      routine.
    g(:)=matmul(fvec(:),fjac(:,:))             Compute ∇f for the line search.
    xold(:)=x(:)                               Store x,
    fold=f                                     and f.
    p(:)=-fvec(:)                              Right-hand side for linear equations.
    call ludcmp(fjac,indx,d)                   Solve linear equations by LU decomposition.
    call lubksb(fjac,indx,p)
    call lnsrch(xold,fold,g,p,x,f,stpmax,check,fmin)
      lnsrch returns new x and f. It also calculates fvec at the new x when it calls fmin.
    if (maxval(abs(fvec(:))) < TOLF) then          Test for convergence on function val-
        check=.false.                                 ues.
        RETURN
```

```
    end if
    if (check) then                          Check for gradient of f zero, i.e., spurious
        check=(maxval(abs(g(:))*max(abs(x(:)),1.0_sp) / &       convergence.
            max(f,0.5_sp*size(x))) < TOLMIN)
        RETURN                               Test for convergence on δx.
    end if
    if (maxval(abs(x(:)-xold(:))/max(abs(x(:)),1.0_sp)) < TOLX) &
        RETURN
end do
call nrerror('MAXITS exceeded in newt')
END SUBROUTINE newt
```

f90 `USE fminln` Here we have an example of how to pass an array `fvec` to a function `fmin` without making it an argument of `fmin`. In the language of §21.5, we are using Method 2: We define a pointer `fmin_fvecp` in the module `fminln`:

```
REAL(SP), DIMENSION(:), POINTER :: fmin_fvecp
```

`fvec` itself is declared as an automatic array of the appropriate size in `newt`:

```
REAL(SP), DIMENSION(size(x)), TARGET :: fvec
```

On entry into `newt`, the pointer is associated:

```
fmin_fvecp=>fvec
```

The pointer is then used in `fmin` as a synonym for `fvec`. If you are sufficiently paranoid, you can test whether `fmin_fvecp` has in fact been associated on entry into `fmin`. Heeding our admonition always to deallocate memory when it no longer is needed, you may ask where the deallocation takes place in this example. Answer: On exit from `newt`, the automatic array `fvec` is automatically freed.

The Method 1 way of setting up this task is to declare an allocatable array in the module:

```
REAL(SP), DIMENSION(:), ALLOCATABLE :: fvec
```

On entry into `newt` we allocate it appropriately:

```
allocate(fvec,size(x))
```

and it can now be used in both `newt` and `fmin`. Of course, we must remember to deallocate explicitly `fvec` on exit from `newt`. If we forget, all kinds of bad things would happen on a second call to `newt`. The status of `fvec` on the first return from `newt` becomes undefined. The status cannot be tested with `if(allocated(...))`, and `fvec` may not be referenced in any way. If we tried to guard against this by adding the `SAVE` attribute to the declaration of `fvec`, then we would generate an error from trying to allocate an already-allocated array.

```
SUBROUTINE fdjac(x,fvec,df)
USE nrtype; USE nrutil, ONLY : assert_eq
IMPLICIT NONE
REAL(SP), DIMENSION(:), INTENT(IN) :: fvec
REAL(SP), DIMENSION(:), INTENT(INOUT) :: x
REAL(SP), DIMENSION(:,:), INTENT(OUT) :: df
INTERFACE
```

```
    FUNCTION funcv(x)
    USE nrtype
    IMPLICIT NONE
    REAL(SP), DIMENSION(:), INTENT(IN) :: x
    REAL(SP), DIMENSION(size(x)) :: funcv
    END FUNCTION funcv
END INTERFACE
REAL(SP), PARAMETER :: EPS=1.0e-4_sp
```

Computes forward-difference approximation to Jacobian. On input, x is the point at which the Jacobian is to be evaluated, and fvec is the vector of function values at the point, both arrays of length N. df is the $N \times N$ output Jacobian. FUNCTION funcv(x) is a fixed-name, user-supplied routine that returns the vector of functions at x.

Parameter: EPS is the approximate square root of the machine precision.

```
INTEGER(I4B) :: j,n
REAL(SP), DIMENSION(size(x)) :: xsav,xph,h
n=assert_eq(size(x),size(fvec),size(df,1),size(df,2),'fdjac')
xsav=x
h=EPS*abs(xsav)
where (h == 0.0) h=EPS
xph=xsav+h                                  Trick to reduce finite precision error.
h=xph-xsav
do j=1,n
    x(j)=xph(j)
    df(:,j)=(funcv(x)-fvec(:))/h(j)         Forward difference formula.
    x(j)=xsav(j)
end do
END SUBROUTINE fdjac
```

```
MODULE fminln
USE nrtype; USE nrutil, ONLY : nrerror
REAL(SP), DIMENSION(:), POINTER :: fmin_fvecp
CONTAINS
FUNCTION fmin(x)
IMPLICIT NONE
REAL(SP), DIMENSION(:), INTENT(IN) :: x
REAL(SP) :: fmin
```

Returns $f = \frac{1}{2}\mathbf{F}\cdot\mathbf{F}$ at x. FUNCTION funcv(x) is a fixed-name, user-supplied routine that returns the vector of functions at x. The pointer fmin_vecp communicates the function values back to newt.

```
INTERFACE
    FUNCTION funcv(x)
    USE nrtype
    IMPLICIT NONE
    REAL(SP), DIMENSION(:), INTENT(IN) :: x
    REAL(SP), DIMENSION(size(x)) :: funcv
    END FUNCTION funcv
END INTERFACE
if (.not. associated(fmin_fvecp)) call &
    nrerror('fmin: problem with pointer for returned values')
fmin_fvecp=funcv(x)
fmin=0.5_sp*dot_product(fmin_fvecp,fmin_fvecp)
END FUNCTION fmin
END MODULE fminln
```

⋆ ⋆ ⋆

```
SUBROUTINE broydn(x,check)
USE nrtype; USE nrutil, ONLY : get_diag,lower_triangle,nrerror,&
    outerprod,put_diag,unit_matrix,vabs
USE nr, ONLY : fdjac,lnsrch,qrdcmp,qrupdt,rsolv
USE fminln                                  Communicates with fmin.
IMPLICIT NONE
REAL(SP), DIMENSION(:), INTENT(INOUT) :: x
LOGICAL(LGT), INTENT(OUT) :: check
INTEGER(I4B), PARAMETER :: MAXITS=200
REAL(SP), PARAMETER :: EPS=epsilon(x),TOLF=1.0e-4_sp,TOLMIN=1.0e-6_sp,&
    TOLX=EPS,STPMX=100.0
```

Given an initial guess x for a root in N dimensions, find the root by Broyden's method embedded in a globally convergent strategy. The length N vector of functions to be zeroed, called fvec in the routine below, is returned by a user-supplied routine that *must* be called funcv and have the declaration FUNCTION funcv(x). The subroutine fdjac and the function fmin from newt are used. The output quantity check is false on a normal return and true if the routine has converged to a local minimum of the function fmin or if Broyden's method can make no further progress. In this case try restarting from a different initial guess.

Parameters: MAXITS is the maximum number of iterations; EPS is the machine precision; TOLF sets the convergence criterion on function values; TOLMIN sets the criterion for deciding whether spurious convergence to a minimum of fmin has occurred; TOLX is the convergence criterion on $\delta\mathbf{x}$; STPMX is the scaled maximum step length allowed in line searches.

```
INTEGER(I4B) :: i,its,k,n
REAL(SP) :: f,fold,stpmax
REAL(SP), DIMENSION(size(x)), TARGET :: fvec
REAL(SP), DIMENSION(size(x)) :: c,d,fvcold,g,p,s,t,w,xold
REAL(SP), DIMENSION(size(x),size(x)) :: qt,r
LOGICAL :: restrt,sing
fmin_fvecp=>fvec
n=size(x)
f=fmin(x)                                   fvec is also computed by this call.
if (maxval(abs(fvec(:))) < 0.01_sp*TOLF) then       Test for initial guess being a root.
    check=.false.                                   Use more stringent test than
    RETURN                                          simply TOLF.
end if
stpmax=STPMX*max(vabs(x(:)),real(n,sp))     Calculate stpmax for line searches.
restrt=.true.                               Ensure initial Jacobian gets computed.
do its=1,MAXITS                             Start of iteration loop.
    if (restrt) then
        call fdjac(x,fvec,r)                Initialize or reinitialize Jacobian in r.
        call qrdcmp(r,c,d,sing)             QR decomposition of Jacobian.
        if (sing) call nrerror('singular Jacobian in broydn')
        call unit_matrix(qt)                Form Q^T explicitly.
        do k=1,n-1
            if (c(k) /= 0.0) then
                qt(k:n,:)=qt(k:n,:)-outerprod(r(k:n,k),&
                    matmul(r(k:n,k),qt(k:n,:)))/c(k)
            end if
        end do
        where (lower_triangle(n,n)) r(:,:)=0.0
        call put_diag(d(:),r(:,:))          Form R explicitly.
    else                                    Carry out Broyden update.
        s(:)=x(:)-xold(:)                   s = δx.
        do i=1,n                            t = R · s.
            t(i)=dot_product(r(i,i:n),s(i:n))
        end do
        w(:)=fvec(:)-fvcold(:)-matmul(t(:),qt(:,:))     w = δF − B · s.
        where (abs(w(:)) < EPS*(abs(fvec(:))+abs(fvcold(:)))) &
            w(:)=0.0                        Don't update with noisy components of
        if (any(w(:) /= 0.0)) then              w.
            t(:)=matmul(qt(:,:),w(:))       t = Q^T · w.
            s(:)=s(:)/dot_product(s,s)      Store s/(s · s) in s.
```

```
            call qrupdt(r,qt,t,s)                Update R and Q^T.
            d(:)=get_diag(r(:,:))                Diagonal of R stored in d.
            if (any(d(:) == 0.0)) &
                call nrerror('r singular in broydn')
        end if
    end if
    p(:)=-matmul(qt(:,:),fvec(:))                r.h.s. for linear equations is -Q^T · F.
    do i=1,n                                     Compute ∇f ≈ (Q · R)^T · F for the line
        g(i)=-dot_product(r(1:i,i),p(1:i))           search.
    end do
    xold(:)=x(:)                                 Store x, F, and f.
    fvcold(:)=fvec(:)
    fold=f
    call rsolv(r,d,p)                            Solve linear equations.
    call lnsrch(xold,fold,g,p,x,f,stpmax,check,fmin)
      lnsrch returns new x and f. It also calculates fvec at the new x when it calls fmin.
    if (maxval(abs(fvec(:))) < TOLF) then           Test for convergence on function val-
        check=.false.                                   ues.
        RETURN
    end if
    if (check) then                              True if line search failed to find a new
        if (restrt .or. maxval(abs(g(:))*max(abs(x(:)), &          x.
            1.0_sp)/max(f,0.5_sp*n)) < TOLMIN) RETURN
              If restrt is true we have failure: We have already tried reinitializing the Jaco-
              bian. The other test is for gradient of f zero, i.e., spurious convergence.
        restrt=.true.                            Try reinitializing the Jacobian.
    else                                         Successful step; will use Broyden update
        restrt=.false.                               for next step.
        if (maxval((abs(x(:)-xold(:)))/max(abs(x(:)), &
            1.0_sp)) < TOLX) RETURN              Test for convergence on δx.
    end if
end do
call nrerror('MAXITS exceeded in broydn')
END SUBROUTINE broydn
```

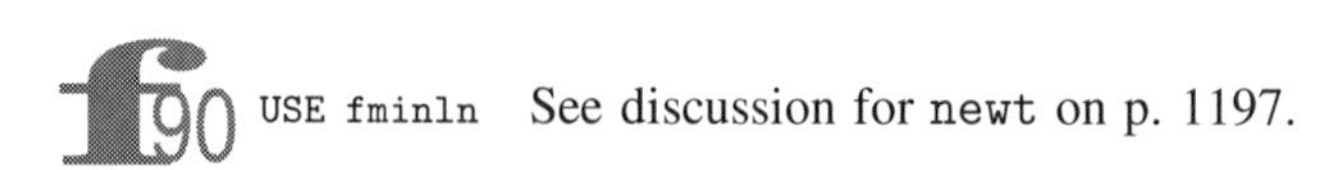

`USE fminln` See discussion for `newt` on p. 1197.

`qt(k:n,:)=...outerprod...matmul` Another example of the coding of equation (22.1.6).

`where (lower_triangle(n,n))...` The `lower_triangle` function in `nrutil` returns a lower triangular logical mask. As used here, the mask is true everywhere in the lower triangle of an $n \times n$ matrix, excluding the diagonal. An optional integer argument `extra` allows additional diagonals to be set to true. With `extra=1` the lower triangle including the diagonal would be true.

`call put_diag(d(:),r(:,:))` This subroutine in `nrutil` sets the diagonal values of the matrix `r` to the values of the vector `d`. It is overloaded so that `d` could be a scalar, in which case the scalar value would be broadcast onto the diagonal of `r`.

Chapter B10. Minimization or Maximization of Functions

```
SUBROUTINE mnbrak(ax,bx,cx,fa,fb,fc,func)
USE nrtype; USE nrutil, ONLY : swap
IMPLICIT NONE
REAL(SP), INTENT(INOUT) :: ax,bx
REAL(SP), INTENT(OUT) :: cx,fa,fb,fc
INTERFACE
    FUNCTION func(x)
    USE nrtype
    IMPLICIT NONE
    REAL(SP), INTENT(IN) :: x
    REAL(SP) :: func
    END FUNCTION func
END INTERFACE
REAL(SP), PARAMETER :: GOLD=1.618034_sp,GLIMIT=100.0_sp,TINY=1.0e-20_sp
```

Given a function func, and given distinct initial points ax and bx, this routine searches in the downhill direction (defined by the function as evaluated at the initial points) and returns new points ax, bx, cx that bracket a minimum of the function. Also returned are the function values at the three points, fa, fb, and fc.

Parameters: GOLD is the default ratio by which successive intervals are magnified; GLIMIT is the maximum magnification allowed for a parabolic-fit step.

```
REAL(SP) :: fu,q,r,u,ulim
fa=func(ax)
fb=func(bx)
if (fb > fa) then                      Switch roles of a and b so that we
    call swap(ax,bx)                       can go downhill in the direction
    call swap(fa,fb)                       from a to b.
end if
cx=bx+GOLD*(bx-ax)                     First guess for c.
fc=func(cx)
do                                     Do-while-loop: Keep returning here
    if (fb < fc) RETURN                    until we bracket.
```

Compute u by parabolic extrapolation from a, b, c. TINY is used to prevent any possible division by zero.

```
    r=(bx-ax)*(fb-fc)
    q=(bx-cx)*(fb-fa)
    u=bx-((bx-cx)*q-(bx-ax)*r)/(2.0_sp*sign(max(abs(q-r),TINY),q-r))
    ulim=bx+GLIMIT*(cx-bx)
```

We won't go farther than this. Test various possibilities:

```
    if ((bx-u)*(u-cx) > 0.0) then      Parabolic u is between b and c: try
        fu=func(u)                         it.
        if (fu < fc) then              Got a minimum between b and c.
            ax=bx
            fa=fb
            bx=u
            fb=fu
            RETURN
        else if (fu > fb) then         Got a minimum between a and u.
            cx=u
            fc=fu
            RETURN
```

```
        end if
        u=cx+GOLD*(cx-bx)                      Parabolic fit was no use. Use default
        fu=func(u)                                 magnification.
    else if ((cx-u)*(u-ulim) > 0.0) then       Parabolic fit is between c and its al-
        fu=func(u)                                 lowed limit.
        if (fu < fc) then
            bx=cx
            cx=u
            u=cx+GOLD*(cx-bx)
            call shft(fb,fc,fu,func(u))
        end if
    else if ((u-ulim)*(ulim-cx) >= 0.0) then   Limit parabolic u to maximum al-
        u=ulim                                     lowed value.
        fu=func(u)
    else                                       Reject parabolic u, use default mag-
        u=cx+GOLD*(cx-bx)                          nification.
        fu=func(u)
    end if
    call shft(ax,bx,cx,u)
    call shft(fa,fb,fc,fu)                     Eliminate oldest point and continue.
end do
CONTAINS

SUBROUTINE shft(a,b,c,d)
REAL(SP), INTENT(OUT) :: a
REAL(SP), INTENT(INOUT) :: b,c
REAL(SP), INTENT(IN) :: d
a=b
b=c
c=d
END SUBROUTINE shft
END SUBROUTINE mnbrak
```

f90 `call shft...` There are three places in `mnbrak` where we need to shift four variables around. Rather than repeat code, we make `shft` an internal subroutine, coming after a `CONTAINS` statement. It is invisible to all procedures except `mnbrak`.

⋆ ⋆ ⋆

```
FUNCTION golden(ax,bx,cx,func,tol,xmin)
USE nrtype
IMPLICIT NONE
REAL(SP), INTENT(IN) :: ax,bx,cx,tol
REAL(SP), INTENT(OUT) :: xmin
REAL(SP) :: golden
INTERFACE
    FUNCTION func(x)
    USE nrtype
    IMPLICIT NONE
    REAL(SP), INTENT(IN) :: x
    REAL(SP) :: func
    END FUNCTION func
END INTERFACE
REAL(SP), PARAMETER :: R=0.61803399_sp,C=1.0_sp-R
```

Given a function `func`, and given a bracketing triplet of abscissas `ax`, `bx`, `cx` (such that `bx` is between `ax` and `cx`, and `func(bx)` is less than both `func(ax)` and `func(cx)`), this routine performs a golden section search for the minimum, isolating it to a fractional precision of about `tol`. The abscissa of the minimum is returned as `xmin`, and the minimum

```
    function value is returned as golden, the returned function value.
    Parameters: The golden ratios.
REAL(SP) :: f1,f2,x0,x1,x2,x3
x0=ax                                           At any given time we will keep track of
x3=cx                                               four points, x0,x1,x2,x3.
if (abs(cx-bx) > abs(bx-ax)) then               Make x0 to x1 the smaller segment,
    x1=bx
    x2=bx+C*(cx-bx)                             and fill in the new point to be tried.
else
    x2=bx
    x1=bx-C*(bx-ax)
end if
f1=func(x1)
f2=func(x2)
  The initial function evaluations. Note that we never need to evaluate the function at the
  original endpoints.
do                                              Do-while-loop: We keep returning here.
    if (abs(x3-x0) <= tol*(abs(x1)+abs(x2))) exit
    if (f2 < f1) then                           One possible outcome,
        call shft3(x0,x1,x2,R*x2+C*x3)          its housekeeping,
        call shft2(f1,f2,func(x2))              and a new function evaluation.
    else                                        The other outcome,
        call shft3(x3,x2,x1,R*x1+C*x0)
        call shft2(f2,f1,func(x1))              and its new function evaluation.
    end if
end do                                          Back to see if we are done.
if (f1 < f2) then                               We are done. Output the best of the two
    golden=f1                                       current values.
    xmin=x1
else
    golden=f2
    xmin=x2
end if
CONTAINS

SUBROUTINE shft2(a,b,c)
REAL(SP), INTENT(OUT) :: a
REAL(SP), INTENT(INOUT) :: b
REAL(SP), INTENT(IN) :: c
a=b
b=c
END SUBROUTINE shft2

SUBROUTINE shft3(a,b,c,d)
REAL(SP), INTENT(OUT) :: a
REAL(SP), INTENT(INOUT) :: b,c
REAL(SP), INTENT(IN) :: d
a=b
b=c
c=d
END SUBROUTINE shft3
END FUNCTION golden
```

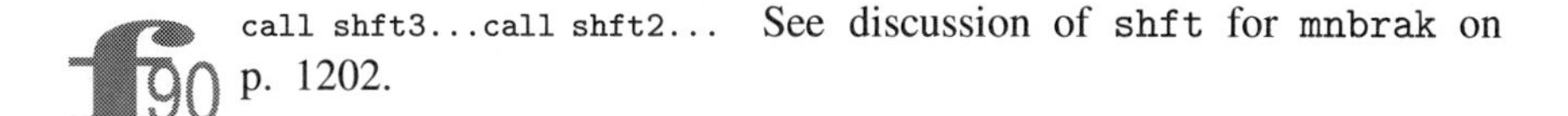

call shft3...call shft2... See discussion of shft for mnbrak on p. 1202.

* * *

```
FUNCTION brent(ax,bx,cx,func,tol,xmin)
USE nrtype; USE nrutil, ONLY : nrerror
IMPLICIT NONE
REAL(SP), INTENT(IN) :: ax,bx,cx,tol
REAL(SP), INTENT(OUT) :: xmin
REAL(SP) :: brent
INTERFACE
    FUNCTION func(x)
    USE nrtype
    IMPLICIT NONE
    REAL(SP), INTENT(IN) :: x
    REAL(SP) :: func
    END FUNCTION func
END INTERFACE
INTEGER(I4B), PARAMETER :: ITMAX=100
REAL(SP), PARAMETER :: CGOLD=0.3819660_sp,ZEPS=1.0e-3_sp*epsilon(ax)
```

Given a function func, and given a bracketing triplet of abscissas ax, bx, cx (such that bx is between ax and cx, and func(bx) is less than both func(ax) and func(cx)), this routine isolates the minimum to a fractional precision of about tol using Brent's method. The abscissa of the minimum is returned as xmin, and the minimum function value is returned as brent, the returned function value.

Parameters: Maximum allowed number of iterations; golden ratio; and a small number that protects against trying to achieve fractional accuracy for a minimum that happens to be exactly zero.

```
INTEGER(I4B) :: iter
REAL(SP) :: a,b,d,e,etemp,fu,fv,fw,fx,p,q,r,tol1,tol2,u,v,w,x,xm
a=min(ax,cx)                                  a and b must be in ascending order, though
b=max(ax,cx)                                      the input abscissas need not be.
v=bx                                          Initializations...
w=v
x=v
e=0.0                                         This will be the distance moved on the step
fx=func(x)                                        before last.
fv=fx
fw=fx
do iter=1,ITMAX                               Main program loop.
    xm=0.5_sp*(a+b)
    tol1=tol*abs(x)+ZEPS
    tol2=2.0_sp*tol1
    if (abs(x-xm) <= (tol2-0.5_sp*(b-a))) then          Test for done here.
        xmin=x                                Arrive here ready to exit with best values.
        brent=fx
        RETURN
    end if
    if (abs(e) > tol1) then                   Construct a trial parabolic fit.
        r=(x-w)*(fx-fv)
        q=(x-v)*(fx-fw)
        p=(x-v)*q-(x-w)*r
        q=2.0_sp*(q-r)
        if (q > 0.0) p=-p
        q=abs(q)
        etemp=e
        e=d
        if (abs(p) >= abs(0.5_sp*q*etemp) .or. &
            p <= q*(a-x) .or. p >= q*(b-x)) then
              The above conditions determine the acceptability of the parabolic fit. Here it is
              not o.k., so we take the golden section step into the larger of the two segments.
            e=merge(a-x,b-x, x >= xm )
            d=CGOLD*e
        else                                  Take the parabolic step.
            d=p/q
            u=x+d
            if (u-a < tol2 .or. b-u < tol2) d=sign(tol1,xm-x)
        end if
```

```
    else                                        Take the golden section step into the larger
        e=merge(a-x,b-x, x >= xm )                  of the two segments.
        d=CGOLD*e
    end if
    u=merge(x+d,x+sign(tol1,d), abs(d) >= tol1 )
      Arrive here with d computed either from parabolic fit, or else from golden section.
    fu=func(u)
      This is the one function evaluation per iteration.
    if (fu <= fx) then                          Now we have to decide what to do with our
        if (u >= x) then                            function evaluation. Housekeeping follows:
            a=x
        else
            b=x
        end if
        call shft(v,w,x,u)
        call shft(fv,fw,fx,fu)
    else
        if (u < x) then
            a=u
        else
            b=u
        end if
        if (fu <= fw .or. w == x) then
            v=w
            fv=fw
            w=u
            fw=fu
        else if (fu <= fv .or. v == x .or. v == w) then
            v=u
            fv=fu
        end if
    end if
end do                                          Done with housekeeping. Back for another
call nrerror('brent: exceed maximum iterations')    iteration.
CONTAINS

SUBROUTINE shft(a,b,c,d)
REAL(SP), INTENT(OUT) :: a
REAL(SP), INTENT(INOUT) :: b,c
REAL(SP), INTENT(IN) :: d
a=b
b=c
c=d
END SUBROUTINE shft
END FUNCTION brent
```

⋆ ⋆ ⋆

```
FUNCTION dbrent(ax,bx,cx,func,dfunc,tol,xmin)
USE nrtype; USE nrutil, ONLY : nrerror
IMPLICIT NONE
REAL(SP), INTENT(IN) :: ax,bx,cx,tol
REAL(SP), INTENT(OUT) :: xmin
REAL(SP) :: dbrent
INTERFACE
    FUNCTION func(x)
    USE nrtype
    IMPLICIT NONE
    REAL(SP), INTENT(IN) :: x
    REAL(SP) :: func
    END FUNCTION func

    FUNCTION dfunc(x)
```

```
    USE nrtype
    IMPLICIT NONE
    REAL(SP), INTENT(IN) :: x
    REAL(SP) :: dfunc
    END FUNCTION dfunc
END INTERFACE
INTEGER(I4B), PARAMETER :: ITMAX=100
REAL(SP), PARAMETER :: ZEPS=1.0e-3_sp*epsilon(ax)
```

Given a function `func` and its derivative function `dfunc`, and given a bracketing triplet of abscissas `ax`, `bx`, `cx` [such that `bx` is between `ax` and `cx`, and `func(bx)` is less than both `func(ax)` and `func(cx)`], this routine isolates the minimum to a fractional precision of about `tol` using a modification of Brent's method that uses derivatives. The abscissa of the minimum is returned as `xmin`, and the minimum function value is returned as `dbrent`, the returned function value.

Parameters: Maximum allowed number of iterations, and a small number that protects against trying to achieve fractional accuracy for a minimum that happens to be exactly zero.

```
INTEGER(I4B) :: iter
REAL(SP) :: a,b,d,d1,d2,du,dv,dw,dx,e,fu,fv,fw,fx,olde,tol1,tol2,&
    u,u1,u2,v,w,x,xm
```

Comments following will point out only differences from the routine `brent`. Read that routine first.

```
LOGICAL :: ok1,ok2                 Will be used as flags for whether pro-
a=min(ax,cx)                          posed steps are acceptable or not.
b=max(ax,cx)
v=bx
w=v
x=v
e=0.0
fx=func(x)
fv=fx
fw=fx
dx=dfunc(x)                        All our housekeeping chores are dou-
dv=dx                                 bled by the necessity of moving
dw=dx                                 derivative values around as well
do iter=1,ITMAX                       as function values.
    xm=0.5_sp*(a+b)
    tol1=tol*abs(x)+ZEPS
    tol2=2.0_sp*tol1
    if (abs(x-xm) <= (tol2-0.5_sp*(b-a))) exit
    if (abs(e) > tol1) then
        d1=2.0_sp*(b-a)            Initialize these d's to an out-of-bracket
        d2=d1                         value.
        if (dw /= dx) d1=(w-x)*dx/(dx-dw)      Secant method with each point.
        if (dv /= dx) d2=(v-x)*dx/(dx-dv)
```

Which of these two estimates of `d` shall we take? We will insist that they be within the bracket, and on the side pointed to by the derivative at `x`:

```
        u1=x+d1
        u2=x+d2
        ok1=((a-u1)*(u1-b) > 0.0) .and. (dx*d1 <= 0.0)
        ok2=((a-u2)*(u2-b) > 0.0) .and. (dx*d2 <= 0.0)
        olde=e                     Movement on the step before last.
        e=d
        if (ok1 .or. ok2) then     Take only an acceptable d, and if
            if (ok1 .and. ok2) then    both are acceptable, then take
                d=merge(d1,d2, abs(d1) < abs(d2))    the smallest one.
            else
                d=merge(d1,d2,ok1)
            end if
            if (abs(d) <= abs(0.5_sp*olde)) then
                u=x+d
                if (u-a < tol2 .or. b-u < tol2) &
                    d=sign(tol1,xm-x)
            else
```

```
                e=merge(a,b, dx >= 0.0)-x
                  Decide which segment by the sign of the derivative.
                d=0.5_sp*e                         Bisect, not golden section.
            end if
        else
            e=merge(a,b, dx >= 0.0)-x
            d=0.5_sp*e                             Bisect, not golden section.
        end if
    else
        e=merge(a,b, dx >= 0.0)-x
        d=0.5_sp*e                                 Bisect, not golden section.
    end if
    if (abs(d) >= tol1) then
        u=x+d
        fu=func(u)
    else
        u=x+sign(tol1,d)
        fu=func(u)                                 If the minimum step in the downhill
        if (fu > fx) exit                              direction takes us uphill, then we
    end if                                             are done.
    du=dfunc(u)                                    Now all the housekeeping, sigh.
    if (fu <= fx) then
        if (u >= x) then
            a=x
        else
            b=x
        end if
        call mov3(v,fv,dv,w,fw,dw)
        call mov3(w,fw,dw,x,fx,dx)
        call mov3(x,fx,dx,u,fu,du)
    else
        if (u < x) then
            a=u
        else
            b=u
        end if
        if (fu <= fw .or. w == x) then
            call mov3(v,fv,dv,w,fw,dw)
            call mov3(w,fw,dw,u,fu,du)
        else if (fu <= fv .or. v == x .or. v == w) then
            call mov3(v,fv,dv,u,fu,du)
        end if
    end if
end do
if (iter > ITMAX) call nrerror('dbrent: exceeded maximum iterations')
xmin=x
dbrent=fx
CONTAINS

SUBROUTINE mov3(a,b,c,d,e,f)
REAL(SP), INTENT(IN) :: d,e,f
REAL(SP), INTENT(OUT) :: a,b,c
a=d
b=e
c=f
END SUBROUTINE mov3
END FUNCTION dbrent
```

⋆ ⋆ ⋆

```
SUBROUTINE amoeba(p,y,ftol,func,iter)
USE nrtype; USE nrutil, ONLY : assert_eq,imaxloc,iminloc,nrerror,swap
IMPLICIT NONE
INTEGER(I4B), INTENT(OUT) :: iter
REAL(SP), INTENT(IN) :: ftol
REAL(SP), DIMENSION(:), INTENT(INOUT) :: y
REAL(SP), DIMENSION(:,:), INTENT(INOUT) :: p
INTERFACE
    FUNCTION func(x)
    USE nrtype
    IMPLICIT NONE
    REAL(SP), DIMENSION(:), INTENT(IN) :: x
    REAL(SP) :: func
    END FUNCTION func
END INTERFACE
INTEGER(I4B), PARAMETER :: ITMAX=5000
REAL(SP), PARAMETER :: TINY=1.0e-10
```

Minimization of the function func in N dimensions by the downhill simplex method of Nelder and Mead. The $(N+1)\times N$ matrix p is input. Its $N+1$ rows are N-dimensional vectors that are the vertices of the starting simplex. Also input is the vector y of length $N+1$, whose components must be preinitialized to the values of func evaluated at the $N+1$ vertices (rows) of p; and ftol the fractional convergence tolerance to be achieved in the function value (n.b.!). On output, p and y will have been reset to $N+1$ new points all within ftol of a minimum function value, and iter gives the number of function evaluations taken.

Parameters: The maximum allowed number of function evaluations, and a small number.

```
INTEGER(I4B) :: ihi,ndim                    Global variables.
REAL(SP), DIMENSION(size(p,2)) :: psum
call amoeba_private
CONTAINS

SUBROUTINE amoeba_private
IMPLICIT NONE
INTEGER(I4B) :: i,ilo,inhi
REAL(SP) :: rtol,ysave,ytry,ytmp
ndim=assert_eq(size(p,2),size(p,1)-1,size(y)-1,'amoeba')
iter=0
psum(:)=sum(p(:,:),dim=1)
do                                          Iteration loop.
    ilo=iminloc(y(:))                       Determine which point is the highest (worst),
    ihi=imaxloc(y(:))                           next-highest, and lowest (best).
    ytmp=y(ihi)
    y(ihi)=y(ilo)
    inhi=imaxloc(y(:))
    y(ihi)=ytmp
    rtol=2.0_sp*abs(y(ihi)-y(ilo))/(abs(y(ihi))+abs(y(ilo))+TINY)
      Compute the fractional range from highest to lowest and return if satisfactory.
    if (rtol < ftol) then                   If returning, put best point and value in slot
        call swap(y(1),y(ilo))                  1.
        call swap(p(1,:),p(ilo,:))
        RETURN
    end if
    if (iter >= ITMAX) call nrerror('ITMAX exceeded in amoeba')
      Begin a new iteration. First extrapolate by a factor -1 through the face of the simplex
      across from the high point, i.e., reflect the simplex from the high point.
    ytry=amotry(-1.0_sp)
    iter=iter+1
    if (ytry <= y(ilo)) then                Gives a result better than the best point, so
        ytry=amotry(2.0_sp)                     try an additional extrapolation by a fac-
        iter=iter+1                             tor of 2.
    else if (ytry >= y(inhi)) then          The reflected point is worse than the sec-
        ysave=y(ihi)                            ond highest, so look for an intermediate
        ytry=amotry(0.5_sp)                     lower point, i.e., do a one-dimensional
        iter=iter+1                             contraction.
```

```
        if (ytry >= ysave) then
            Can't seem to get rid of that high point. Better contract around the lowest
            (best) point.
            p(:,:)=0.5_sp*(p(:,:)+spread(p(ilo,:),1,size(p,1)))
            do i=1,ndim+1
                if (i /= ilo) y(i)=func(p(i,:))
            end do
            iter=iter+ndim                      Keep track of function evaluations.
            psum(:)=sum(p(:,:),dim=1)
        end if
    end if
end do                                          Go back for the test of doneness and the next
END SUBROUTINE amoeba_private                       iteration.

FUNCTION amotry(fac)
IMPLICIT NONE
REAL(SP), INTENT(IN) :: fac
REAL(SP) :: amotry
    Extrapolates by a factor fac through the face of the simplex across from the high point,
    tries it, and replaces the high point if the new point is better.
REAL(SP) :: fac1,fac2,ytry
REAL(SP), DIMENSION(size(p,2)) :: ptry
fac1=(1.0_sp-fac)/ndim
fac2=fac1-fac
ptry(:)=psum(:)*fac1-p(ihi,:)*fac2
ytry=func(ptry)                                 Evaluate the function at the trial point.
if (ytry < y(ihi)) then                         If it's better than the highest, then replace
    y(ihi)=ytry                                     the highest.
    psum(:)=psum(:)-p(ihi,:)+ptry(:)
    p(ihi,:)=ptry(:)
end if
amotry=ytry
END FUNCTION amotry
END SUBROUTINE amoeba
```

f90 The only action taken by the subroutine `amoeba` is to call the internal subroutine `amoeba_private`. Why this structure? The reason has to do with meeting the twin goals of data hiding (especially for "safe" scope of variables) and program readability. The situation is this: Logically, `amoeba` does most of the calculating, but calls an internal subroutine `amotry` at several different points, with several values of the parameter `fac`. However, `fac` is not the only piece of data that must be shared with `amotry`; the latter also needs access to several shared variables (`ihi`, `ndim`, `psum`) and arguments of `amoeba` (`p`, `y`, `func`).

The obvious (but not best) way of coding this would be to put the computational guts in `amoeba`, with `amotry` as the sole internal subprogram. Assuming that `fac` is passed as an argument to `amotry` (it being the parameter that is being rapidly altered), one must decide whether to pass all the other quantities to `amotry` (i) as additional arguments (as is done in the Fortran 77 version), or (ii) "automatically," i.e., doing nothing except using the fact that an internal subprogram has automatic access to all of its host's entities. Each of these choices has strong disadvantages. Choice (i) is inefficient (all those arguments) and also obscures the fact that `fac` is the primary changing argument. Choice (ii) makes the program extremely difficult to read, because it wouldn't be obvious without careful cross-comparison of the routines *which* variables in `amoeba` are actually global variables that are used by `amotry`.

Choice (ii) is also "unsafe scoping" because it gives a nontrivially complicated internal subprogram, `amotry`, access to all the variables in its host. A common and difficult-to-find bug is the accidental alteration of a variable that one "thought" was local, but is actually shared. (Simple variables like i, j, and n are the most common culprits.)

We are therefore led to reject both choice (i) and choice (ii) in favor of a structure previously described in the subsection on Scope, Visibility, and Data Hiding in §21.5. The guts of amoeba are put in amoeba_private, a *sister routine* to amotry. These two siblings have mutually private name spaces. However, any variables that they need to share (including the top-level arguments of amoeba) are declared as variables in the enclosing amoeba routine. The presence of these "global variables" serves as a warning flag to the reader that data are shared between routines.

An alternative attractive way of coding the above situation would be to use a module containing amoeba and amotry. Everything would be declared private except the name amoeba. The global variables would be at the top level, and the arguments of amoeba that need to be passed to amotry would be handled by pointers among the global variables. Unfortunately, Fortran 90 does not support pointers to functions. Sigh!

ilo=iminloc...ihi=imaxloc... See discussion of these functions on p. 1017.

call swap(y(1)...call swap(p(1,:)... Here the swap routine in nrutil is called once with a scalar argument and once with a vector argument. Inside nrutil scalar and vector versions have been overloaded onto the single name swap, hiding all the implementation details from the calling routine.

⋆ ⋆ ⋆

```
SUBROUTINE powell(p,xi,ftol,iter,fret)
USE nrtype; USE nrutil, ONLY : assert_eq,nrerror
USE nr, ONLY : linmin
IMPLICIT NONE
REAL(SP), DIMENSION(:), INTENT(INOUT) :: p
REAL(SP), DIMENSION(:,:), INTENT(INOUT) :: xi
INTEGER(I4B), INTENT(OUT) :: iter
REAL(SP), INTENT(IN) :: ftol
REAL(SP), INTENT(OUT) :: fret
INTERFACE
    FUNCTION func(p)
    USE nrtype
    IMPLICIT NONE
    REAL(SP), DIMENSION(:), INTENT(IN) :: p
    REAL(SP) :: func
    END FUNCTION func
END INTERFACE
INTEGER(I4B), PARAMETER :: ITMAX=200
REAL(SP), PARAMETER :: TINY=1.0e-25_sp
```

Minimization of a function func of N variables. (func is not an argument, it is a fixed function name.) Input consists of an initial starting point p, a vector of length N; an initial $N \times N$ matrix xi whose columns contain the initial set of directions (usually the N unit vectors); and ftol, the fractional tolerance in the function value such that failure to decrease by more than this amount on one iteration signals doneness. On output, p is set to the best point found, xi is the then-current direction set, fret is the returned function value at p, and iter is the number of iterations taken. The routine linmin is used.

Parameters: Maximum allowed iterations, and a small number.

```
INTEGER(I4B) :: i,ibig,n
REAL(SP) :: del,fp,fptt,t
REAL(SP), DIMENSION(size(p)) :: pt,ptt,xit
n=assert_eq(size(p),size(xi,1),size(xi,2),'powell')
fret=func(p)
```

```
pt(:)=p(:)                              Save the initial point.
iter=0
do
    iter=iter+1
    fp=fret
    ibig=0
    del=0.0                             Will be the biggest function decrease.
    do i=1,n                            Loop over all directions in the set.
        xit(:)=xi(:,i)                  Copy the direction,
        fptt=fret
        call linmin(p,xit,fret)         minimize along it,
        if (fptt-fret > del) then       and record it if it is the largest decrease so
            del=fptt-fret                   far.
            ibig=i
        end if
    end do
    if (2.0_sp*(fp-fret) <= ftol*(abs(fp)+abs(fret))+TINY) RETURN
      Termination criterion.
    if (iter == ITMAX) call &
        nrerror('powell exceeding maximum iterations')
    ptt(:)=2.0_sp*p(:)-pt(:)            Construct the extrapolated point and the av-
    xit(:)=p(:)-pt(:)                       erage direction moved. Save the old start-
    pt(:)=p(:)                              ing point.
    fptt=func(ptt)                      Function value at extrapolated point.
    if (fptt >= fp) cycle               One reason not to use new direction.
    t=2.0_sp*(fp-2.0_sp*fret+fptt)*(fp-fret-del)**2-del*(fp-fptt)**2
    if (t >= 0.0) cycle                 Other reason not to use new direction.
    call linmin(p,xit,fret)             Move to minimum of the new direction,
    xi(:,ibig)=xi(:,n)                  and save the new direction.
    xi(:,n)=xit(:)
end do                                  Back for another iteration.
END SUBROUTINE powell
```

⋆ ⋆ ⋆

```
MODULE f1dim_mod                Used for communication from linmin to f1dim.
USE nrtype
INTEGER(I4B) :: ncom
REAL(SP), DIMENSION(:), POINTER :: pcom,xicom
CONTAINS
FUNCTION f1dim(x)
IMPLICIT NONE
REAL(SP), INTENT(IN) :: x
REAL(SP) :: f1dim
  Used by linmin as the one-dimensional function passed to mnbrak and brent.
INTERFACE
    FUNCTION func(x)
    USE nrtype
    REAL(SP), DIMENSION(:), INTENT(IN) :: x
    REAL(SP) :: func
    END FUNCTION func
END INTERFACE
REAL(SP), DIMENSION(:), ALLOCATABLE :: xt
allocate(xt(ncom))
xt(:)=pcom(:)+x*xicom(:)
f1dim=func(xt)
deallocate(xt)
END FUNCTION f1dim
END MODULE f1dim_mod
```

```
SUBROUTINE linmin(p,xi,fret)
USE nrtype; USE nrutil, ONLY : assert_eq
USE nr, ONLY : mnbrak,brent
USE f1dim_mod
IMPLICIT NONE
REAL(SP), INTENT(OUT) :: fret
REAL(SP), DIMENSION(:), TARGET, INTENT(INOUT) :: p,xi
REAL(SP), PARAMETER :: TOL=1.0e-4_sp
```

Given an N-dimensional point p and an N-dimensional direction xi, both vectors of length N, moves and resets p to where the fixed-name function func takes on a minimum along the direction xi from p, and replaces xi by the actual vector displacement that p was moved. Also returns as fret the value of func at the returned location p. This is actually all accomplished by calling the routines mnbrak and brent.

Parameter: Tolerance passed to brent.

```
REAL(SP) :: ax,bx,fa,fb,fx,xmin,xx
ncom=assert_eq(size(p),size(xi),'linmin')
pcom=>p                         Communicate the global variables to f1dim.
xicom=>xi
ax=0.0                          Initial guess for brackets.
xx=1.0
call mnbrak(ax,xx,bx,fa,fx,fb,f1dim)
fret=brent(ax,xx,bx,f1dim,TOL,xmin)
xi=xmin*xi                      Construct the vector results to return.
p=p+xi
END SUBROUTINE linmin
```

f90 USE f1dim_mod At first sight this situation is like the one involving USE fminln in newt on p. 1197: We want to pass arrays p and xi from linmin to f1dim without having them be arguments of f1dim. If you recall the discussion in §21.5 and on p. 1197, there are two ways of effecting this: via pointers or via allocatable arrays. There is an important difference here, however. The arrays p and xi are themselves arguments of linmin, and so cannot be allocatable arrays in the module. If we did want to use allocatable arrays in the module, we would have to copy p and xi into them. The pointer implementation is much more elegant, since no unnecessary copying is required. The construction here is identical to the one in fminln and newt, except that p and xi are arguments instead of automatic arrays.

* * *

```
MODULE df1dim_mod               Used for communication from dlinmin to f1dim and df1dim.
USE nrtype
INTEGER(I4B) :: ncom
REAL(SP), DIMENSION(:), POINTER :: pcom,xicom
CONTAINS

FUNCTION f1dim(x)
IMPLICIT NONE
REAL(SP), INTENT(IN) :: x
REAL(SP) :: f1dim
```

Used by dlinmin as the one-dimensional function passed to mnbrak.

```
INTERFACE
    FUNCTION func(x)
    USE nrtype
    REAL(SP), DIMENSION(:), INTENT(IN) :: x
    REAL(SP) :: func
    END FUNCTION func
END INTERFACE
REAL(SP), DIMENSION(:), ALLOCATABLE :: xt
```

```
allocate(xt(ncom))
xt(:)=pcom(:)+x*xicom(:)
f1dim=func(xt)
deallocate(xt)
END FUNCTION f1dim

FUNCTION df1dim(x)
IMPLICIT NONE
REAL(SP), INTENT(IN) :: x
REAL(SP) :: df1dim
    Used by dlinmin as the one-dimensional function passed to dbrent.
INTERFACE
    FUNCTION dfunc(x)
    USE nrtype
    REAL(SP), DIMENSION(:), INTENT(IN) :: x
    REAL(SP), DIMENSION(size(x)) :: dfunc
    END FUNCTION dfunc
END INTERFACE
REAL(SP), DIMENSION(:), ALLOCATABLE :: xt,df
allocate(xt(ncom),df(ncom))
xt(:)=pcom(:)+x*xicom(:)
df(:)=dfunc(xt)
df1dim=dot_product(df,xicom)
deallocate(xt,df)
END FUNCTION df1dim
END MODULE df1dim_mod
```

```
SUBROUTINE dlinmin(p,xi,fret)
USE nrtype; USE nrutil, ONLY : assert_eq
USE nr, ONLY : mnbrak,dbrent
USE df1dim_mod
IMPLICIT NONE
REAL(SP), INTENT(OUT) :: fret
REAL(SP), DIMENSION(:), TARGET :: p,xi
REAL(SP), PARAMETER :: TOL=1.0e-4_sp
    Given an N-dimensional point p and an N-dimensional direction xi, both vectors of length
    N, moves and resets p to where the fixed-name function func takes on a minimum along
    the direction xi from p, and replaces xi by the actual vector displacement that p was
    moved. Also returns as fret the value of func at the returned location p. This is actually
    all accomplished by calling the routines mnbrak and dbrent. dfunc is a fixed-name user-
    supplied function that computes the gradient of func.
    Parameter: Tolerance passed to dbrent.
REAL(SP) :: ax,bx,fa,fb,fx,xmin,xx
ncom=assert_eq(size(p),size(xi),'dlinmin')
pcom=>p                         Communicate the global variables to f1dim.
xicom=>xi
ax=0.0                          Initial guess for brackets.
xx=1.0
call mnbrak(ax,xx,bx,fa,fx,fb,f1dim)
fret=dbrent(ax,xx,bx,f1dim,df1dim,TOL,xmin)
xi=xmin*xi                      Construct the vector results to return.
p=p+xi
END SUBROUTINE dlinmin
```

USE df1dim_mod See discussion of USE f1dim_mod on p. 1212.

★ ★ ★

```
SUBROUTINE frprmn(p,ftol,iter,fret)
USE nrtype; USE nrutil, ONLY : nrerror
USE nr, ONLY : linmin
IMPLICIT NONE
INTEGER(I4B), INTENT(OUT) :: iter
REAL(SP), INTENT(IN) :: ftol
REAL(SP), INTENT(OUT) :: fret
REAL(SP), DIMENSION(:), INTENT(INOUT) :: p
INTERFACE
    FUNCTION func(p)
    USE nrtype
    IMPLICIT NONE
    REAL(SP), DIMENSION(:), INTENT(IN) :: p
    REAL(SP) :: func
    END FUNCTION func

    FUNCTION dfunc(p)
    USE nrtype
    IMPLICIT NONE
    REAL(SP), DIMENSION(:), INTENT(IN) :: p
    REAL(SP), DIMENSION(size(p)) :: dfunc
    END FUNCTION dfunc
END INTERFACE
INTEGER(I4B), PARAMETER :: ITMAX=200
REAL(SP), PARAMETER :: EPS=1.0e-10_sp
```

Given a starting point `p` that is a vector of length N, Fletcher-Reeves-Polak-Ribiere minimization is performed on a function `func`, using its gradient as calculated by a routine `dfunc`. The convergence tolerance on the function value is input as `ftol`. Returned quantities are `p` (the location of the minimum), `iter` (the number of iterations that were performed), and `fret` (the minimum value of the function). The routine `linmin` is called to perform line minimizations.

Parameters: `ITMAX` is the maximum allowed number of iterations; EPS is a small number to rectify the special case of converging to exactly zero function value.

```
INTEGER(I4B) :: its
REAL(SP) :: dgg,fp,gam,gg
REAL(SP), DIMENSION(size(p)) :: g,h,xi
fp=func(p)                           Initializations.
xi=dfunc(p)
g=-xi
h=g
xi=h
do its=1,ITMAX                       Loop over iterations.
    iter=its
    call linmin(p,xi,fret)           Next statement is the normal return:
    if (2.0_sp*abs(fret-fp) <= ftol*(abs(fret)+abs(fp)+EPS)) RETURN
    fp=fret
    xi=dfunc(p)
    gg=dot_product(g,g)
!   dgg=dot_product(xi,xi)           This statement for Fletcher-Reeves.
    dgg=dot_product(xi+g,xi)         This statement for Polak-Ribiere.
    if (gg == 0.0) RETURN            Unlikely. If gradient is exactly zero then we are al-
    gam=dgg/gg                           ready done.
    g=-xi
    h=g+gam*h
    xi=h
end do
call nrerror('frprmn: maximum iterations exceeded')
END SUBROUTINE frprmn
```

⋆ ⋆ ⋆

```
SUBROUTINE dfpmin(p,gtol,iter,fret,func,dfunc)
USE nrtype; USE nrutil, ONLY : nrerror,outerprod,unit_matrix,vabs
USE nr, ONLY : lnsrch
IMPLICIT NONE
INTEGER(I4B), INTENT(OUT) :: iter
REAL(SP), INTENT(IN) :: gtol
REAL(SP), INTENT(OUT) :: fret
REAL(SP), DIMENSION(:), INTENT(INOUT) :: p
INTERFACE
    FUNCTION func(p)
    USE nrtype
    IMPLICIT NONE
    REAL(SP), DIMENSION(:), INTENT(IN) :: p
    REAL(SP) :: func
    END FUNCTION func

    FUNCTION dfunc(p)
    USE nrtype
    IMPLICIT NONE
    REAL(SP), DIMENSION(:), INTENT(IN) :: p
    REAL(SP), DIMENSION(size(p)) :: dfunc
    END FUNCTION dfunc
END INTERFACE
INTEGER(I4B), PARAMETER :: ITMAX=200
REAL(SP), PARAMETER :: STPMX=100.0_sp,EPS=epsilon(p),TOLX=4.0_sp*EPS
```

Given a starting point p that is a vector of length N, the Broyden-Fletcher-Goldfarb-Shanno variant of Davidon-Fletcher-Powell minimization is performed on a function func, using its gradient as calculated by a routine dfunc. The convergence requirement on zeroing the gradient is input as gtol. Returned quantities are p (the location of the minimum), iter (the number of iterations that were performed), and fret (the minimum value of the function). The routine lnsrch is called to perform approximate line minimizations.
Parameters: ITMAX is the maximum allowed number of iterations; STPMX is the scaled maximum step length allowed in line searches; EPS is the machine precision; TOLX is the convergence criterion on x values.

```
INTEGER(I4B) :: its
LOGICAL :: check
REAL(SP) :: den,fac,fad,fae,fp,stpmax,sumdg,sumxi
REAL(SP), DIMENSION(size(p)) :: dg,g,hdg,pnew,xi
REAL(SP), DIMENSION(size(p),size(p)) :: hessin
fp=func(p)                                  Calculate starting function value and gradi-
g=dfunc(p)                                      ent.
call unit_matrix(hessin)                    Initialize inverse Hessian to the unit matrix.
xi=-g                                       Initial line direction.
stpmax=STPMX*max(vabs(p),real(size(p),sp))
do its=1,ITMAX                              Main loop over the iterations.
    iter=its
    call lnsrch(p,fp,g,xi,pnew,fret,stpmax,check,func)
      The new function evaluation occurs in lnsrch; save the function value in fp for the next
      line search. It is usually safe to ignore the value of check.
    fp=fret
    xi=pnew-p                               Update the line direction,
    p=pnew                                  and the current point.
    if (maxval(abs(xi)/max(abs(p),1.0_sp)) < TOLX) RETURN
      Test for convergence on Δx.
    dg=g                                    Save the old gradient,
    g=dfunc(p)                              and get the new gradient.
    den=max(fret,1.0_sp)
    if (maxval(abs(g)*max(abs(p),1.0_sp)/den) < gtol) RETURN
      Test for convergence on zero gradient.
    dg=g-dg                                 Compute difference of gradients,
    hdg=matmul(hessin,dg)                   and difference times current matrix.
    fac=dot_product(dg,xi)                  Calculate dot products for the denominators.
    fae=dot_product(dg,hdg)
    sumdg=dot_product(dg,dg)
```

```
    sumxi=dot_product(xi,xi)
    if (fac > sqrt(EPS*sumdg*sumxi)) then        Skip update if fac not sufficiently
        fac=1.0_sp/fac                             positive.
        fad=1.0_sp/fae
        dg=fac*xi-fad*hdg                     Vector that makes BFGS different from DFP.
        hessin=hessin+fac*outerprod(xi,xi)-&      The BFGS updating formula.
            fad*outerprod(hdg,hdg)+fae*outerprod(dg,dg)
    end if
    xi=-matmul(hessin,g)                      Now calculate the next direction to go,
end do                                        and go back for another iteration.
call nrerror('dfpmin: too many iterations')
END SUBROUTINE dfpmin
```

f90 `call unit_matrix(hessin)` The `unit_matrix` routine in `nrutil` does exactly what its name suggests. The routine `dfpmin` makes use of `outerprod` from `nrutil`, as well as the matrix intrinsics `matmul` and `dot_product`, to simplify and parallelize the coding.

⋆ ⋆ ⋆

```
SUBROUTINE simplx(a,m1,m2,m3,icase,izrov,iposv)
USE nrtype; USE nrutil, ONLY : arth,assert_eq,ifirstloc,imaxloc,&
    nrerror,outerprod,swap
IMPLICIT NONE
REAL(SP), DIMENSION(:,:), INTENT(INOUT) :: a
INTEGER(I4B), INTENT(IN) :: m1,m2,m3
INTEGER(I4B), INTENT(OUT) :: icase
INTEGER(I4B), DIMENSION(:), INTENT(OUT) :: izrov,iposv
REAL(SP), PARAMETER :: EPS=1.0e-6_sp
```

Simplex method for linear programming. Input parameters a, `m1`, `m2`, and `m3`, and output parameters a, `icase`, `izrov`, and `iposv` are described above the routine in Vol. 1. Dimensions are $(M+2)\times(N+1)$ for a, M for `iposv`, N for `izrov`, with $\texttt{m1}+\texttt{m2}+\texttt{m3}=M$. Parameter: EPS is the absolute precision, which should be adjusted to the scale of your variables.

```
INTEGER(I4B) :: ip,k,kh,kp,nl1,m,n
INTEGER(I4B), DIMENSION(size(a,2)) :: l1
INTEGER(I4B), DIMENSION(m2) :: l3
REAL(SP) :: bmax
LOGICAL(LGT) :: init
m=assert_eq(size(a,1)-2,size(iposv),'simplx: m')
n=assert_eq(size(a,2)-1,size(izrov),'simplx: n')
if (m /= m1+m2+m3) call nrerror('simplx: bad input constraint counts')
if (any(a(2:m+1,1) < 0.0)) call nrerror('bad input tableau in simplx')
```

Constants b_i must be nonnegative.

```
nl1=n
l1(1:n)=arth(1,1,n)
```

Initialize index list of columns admissible for exchange.

```
izrov(:)=l1(1:n)                              Initially make all variables right-hand.
iposv(:)=n+arth(1,1,m)
```

Initial left-hand variables. `m1` type constraints are represented by having their slack variable initially left-hand, with no artificial variable. `m2` type constraints have their slack variable initially left-hand, with a minus sign, and their artificial variable handled implicitly during their first exchange. `m3` type constraints have their artificial variable initially left-hand.

```
init=.true.
phase1: do
    if (init) then                            Initial pass only.
        if (m2+m3 == 0) exit phase1           Origin is a feasible solution. Go to phase two.
        init=.false.
        l3(1:m2)=1
```

Initialize list of `m2` constraints whose slack variables have never been exchanged out of the initial basis.

```
        a(m+2,1:n+1)=-sum(a(m1+2:m+1,1:n+1),dim=1)      Compute the auxiliary objec-
    end if                                                 tive function.
```

```
    if (nl1 > 0) then
        kp=l1(imaxloc(a(m+2,l1(1:nl1)+1)))          Find the maximum coefficient of the
        bmax=a(m+2,kp+1)                                auxiliary objective function.
    else
        bmax=0.0
    end if
    phase1a: do
        if (bmax <= EPS .and. a(m+2,1) < -EPS) then
              Auxiliary objective function is still negative and can't be improved, hence no
              feasible solution exists.
            icase=-1
            RETURN
        else if (bmax <= EPS .and. a(m+2,1) <= EPS) then
              Auxiliary objective function is zero and can't be improved. This signals that we
              have a feasible starting vector. Clean out the artificial variables corresponding
              to any remaining equality constraints and then eventually exit phase one.
            do ip=m1+m2+1,m
                if (iposv(ip) == ip+n) then         Found an artificial variable for an equal-
                    if (nl1 > 0) then                   ity constraint.
                        kp=l1(imaxloc(abs(a(ip+1,l1(1:nl1)+1))))
                        bmax=a(ip+1,kp+1)
                    else
                        bmax=0.0
                    end if
                    if (bmax > EPS) exit phase1a     Exchange with column correspond-
                end if                                   ing to maximum pivot ele-
            end do                                       ment in row.
            where (spread(l3(1:m2),2,n+1) == 1) &
                a(m1+2:m1+m2+1,1:n+1)=-a(m1+2:m1+m2+1,1:n+1)
                  Change sign of row for any m2 constraints still present from the initial basis.
            exit phase1                         Go to phase two.
        end if
        call simp1                              Locate a pivot element (phase one).
        if (ip == 0) then                       Maximum of auxiliary objective function is
            icase=-1                                unbounded, so no feasible solution ex-
            RETURN                                  ists.
        end if
        exit phase1a
    end do phase1a
    call simp2(m+1,n)                           Exchange a left- and a right-hand variable.
    if (iposv(ip) >= n+m1+m2+1) then            Exchanged out an artificial variable for an
        k=ifirstloc(l1(1:nl1) == kp)                equality constraint. Make sure it stays
        nl1=nl1-1                                   out by removing it from the l1 list.
        l1(k:nl1)=l1(k+1:nl1+1)
    else
        kh=iposv(ip)-m1-n
        if (kh >= 1) then                       Exchanged out an m2 type constraint.
            if (l3(kh) /= 0) then               If it's the first time, correct the pivot col-
                l3(kh)=0                            umn for the minus sign and the implicit
                a(m+2,kp+1)=a(m+2,kp+1)+1.0_sp          artificial variable.
                a(1:m+2,kp+1)=-a(1:m+2,kp+1)
            end if
        end if
    end if
    call swap(izrov(kp),iposv(ip))              Update lists of left- and right-hand variables.
end do phase1                                   If still in phase one, go back again.
phase2: do
      We have an initial feasible solution. Now optimize it.
    if (nl1 > 0) then
        kp=l1(imaxloc(a(1,l1(1:nl1)+1)))        Test the z-row for doneness.
        bmax=a(1,kp+1)
    else
        bmax=0.0
    end if
```

```
    if (bmax <= EPS) then                    Done. Solution found. Return with the good
        icase=0                                  news.
        RETURN
    end if
    call simp1                               Locate a pivot element (phase two).
    if (ip == 0) then                        Objective function is unbounded. Report and
        icase=1                                  return.
        RETURN
    end if
    call simp2(m,n)                          Exchange a left- and a right-hand variable,
    call swap(izrov(kp),iposv(ip))           update lists of left- and right-hand variables,
end do phase2                                and return for another iteration.
CONTAINS

SUBROUTINE simp1
    Locate a pivot element, taking degeneracy into account.
IMPLICIT NONE
INTEGER(I4B) :: i,k
REAL(SP) :: q,q0,q1,qp
ip=0
i=ifirstloc(a(2:m+1,kp+1) < -EPS)
if (i > m) RETURN                            No possible pivots. Return with message.
q1=-a(i+1,1)/a(i+1,kp+1)
ip=i
do i=ip+1,m
    if (a(i+1,kp+1) < -EPS) then
        q=-a(i+1,1)/a(i+1,kp+1)
        if (q < q1) then
            ip=i
            q1=q
        else if (q == q1) then               We have a degeneracy.
            do k=1,n
                qp=-a(ip+1,k+1)/a(ip+1,kp+1)
                q0=-a(i+1,k+1)/a(i+1,kp+1)
                if (q0 /= qp) exit
            end do
            if (q0 < qp) ip=i
        end if
    end if
end do
END SUBROUTINE simp1

SUBROUTINE simp2(i1,k1)
IMPLICIT NONE
INTEGER(I4B), INTENT(IN) :: i1,k1
    Matrix operations to exchange a left-hand and right-hand variable (see text).
INTEGER(I4B) :: ip1,kp1
REAL(SP) :: piv
INTEGER(I4B), DIMENSION(k1) :: icol
INTEGER(I4B), DIMENSION(i1) :: irow
INTEGER(I4B), DIMENSION(max(i1,k1)+1) :: itmp
ip1=ip+1
kp1=kp+1
piv=1.0_sp/a(ip1,kp1)
itmp(1:k1+1)=arth(1,1,k1+1)
icol=pack(itmp(1:k1+1),itmp(1:k1+1) /= kp1)
itmp(1:i1+1)=arth(1,1,i1+1)
irow=pack(itmp(1:i1+1),itmp(1:i1+1) /= ip1)
a(irow,kp1)=a(irow,kp1)*piv
a(irow,icol)=a(irow,icol)-outerprod(a(irow,kp1),a(ip1,icol))
a(ip1,icol)=-a(ip1,icol)*piv
a(ip1,kp1)=piv
END SUBROUTINE simp2
END SUBROUTINE simplx
```

f90 `main_procedure: do` The routine `simplx` makes extensive use of named do-loops to control the program flow. The various `exit` statements have the names of the do-loops attached to them so we can easily tell where control is being transferred to. We believe that it is almost never necessary to use `goto` statements: Code will always be clearer with well-constructed block structures.

`phase1a: do...end do phase1a` This is not a real do-loop: It is executed only once, as you can see from the unconditional `exit` before the `end do`. We use this construction to define a block of code that is traversed once but that has several possible exit points.

```
where (spread(l3(1:m12-m1),2,n+1) == 1) &
    a(m1+2:m12+1,1:n+1)=-a(m1+2:m12+1,1:n+1)
```

These lines are equivalent to

```
do i=m1+1,m12
    if (l3(i-m1) == 1) a(i+1,1:n+1)=-a(i+1,1:n+1)
end do
```

⋆ ⋆ ⋆

```
SUBROUTINE anneal(x,y,iorder)
USE nrtype; USE nrutil, ONLY : arth,assert_eq,swap
USE nr, ONLY : ran1
IMPLICIT NONE
INTEGER(I4B), DIMENSION(:), INTENT(INOUT) :: iorder
REAL(SP), DIMENSION(:), INTENT(IN) :: x,y
    This algorithm finds the shortest round-trip path to N cities whose coordinates are in the
    length N arrays x, y. The length N array iorder specifies the order in which the cities are
    visited. On input, the elements of iorder may be set to any permutation of the numbers
    1...N. This routine will return the best alternative path it can find.
INTEGER(I4B), DIMENSION(6) :: n
INTEGER(I4B) :: i1,i2,j,k,nlimit,ncity,nn,nover,nsucc
REAL(SP) :: de,harvest,path,t,tfactr
LOGICAL(LGT) :: ans
ncity=assert_eq(size(x),size(y),size(iorder),'anneal')
nover=100*ncity                    Maximum number of paths tried at any temperature,
nlimit=10*ncity                    and of successful path changes before continuing.
tfactr=0.9_sp                      Annealing schedule: t is reduced by this factor on
t=0.5_sp                               each step.
path=sum(alen_v(x(iorder(1:ncity-1)),x(iorder(2:ncity)),&
    y(iorder(1:ncity-1)),y(iorder(2:ncity))))        Calculate initial path length.
i1=iorder(ncity)                             Close the loop by tying path ends to-
i2=iorder(1)                                     gether.
path=path+alen(x(i1),x(i2),y(i1),y(i2))
do j=1,100                                   Try up to 100 temperature steps.
    nsucc=0
    do k=1,nover
        do
            call ran1(harvest)
            n(1)=1+int(ncity*harvest)                Choose beginning of segment...
            call ran1(harvest)
            n(2)=1+int((ncity-1)*harvest)            ...and end of segment.
            if (n(2) >= n(1)) n(2)=n(2)+1
            nn=1+mod((n(1)-n(2)+ncity-1),ncity)      nn is the number of cities not on
            if (nn >= 3) exit                            the segment.
        end do
```

```
        call ran1(harvest)
          Decide whether to do a reversal or a transport.
        if (harvest < 0.5_sp) then                 Do a transport.
            call ran1(harvest)
            n(3)=n(2)+int(abs(nn-2)*harvest)+1
            n(3)=1+mod(n(3)-1,ncity)               Transport to a location not on the path.
            call trncst(x,y,iorder,n,de)           Calculate cost.
            call metrop(de,t,ans)                  Consult the oracle.
            if (ans) then
                nsucc=nsucc+1
                path=path+de
                call trnspt(iorder,n)              Carry out the transport.
            end if
        else                                       Do a path reversal.
            call revcst(x,y,iorder,n,de)           Calculate cost.
            call metrop(de,t,ans)                  Consult the oracle.
            if (ans) then
                nsucc=nsucc+1
                path=path+de
                call revers(iorder,n)              Carry out the reversal.
            end if
        end if
        if (nsucc >= nlimit) exit                  Finish early if we have enough successful
    end do                                             changes.
    write(*,*)
    write(*,*) 'T =',t,' Path Length =',path
    write(*,*) 'Successful Moves: ',nsucc
    t=t*tfactr                                     Annealing schedule.
    if (nsucc == 0) RETURN                         If no success, we are done.
end do
CONTAINS

FUNCTION alen(x1,x2,y1,y2)
IMPLICIT NONE
REAL(SP), INTENT(IN) :: x1,x2,y1,y2
REAL(SP) :: alen
   Computes distance between two cities.
alen=sqrt((x2-x1)**2+(y2-y1)**2)
END FUNCTION alen

FUNCTION alen_v(x1,x2,y1,y2)
IMPLICIT NONE
REAL(SP), DIMENSION(:), INTENT(IN) :: x1,x2,y1,y2
REAL(SP), DIMENSION(size(x1)) :: alen_v
   Computes distances between pairs of cities.
alen_v=sqrt((x2-x1)**2+(y2-y1)**2)
END FUNCTION alen_v

SUBROUTINE metrop(de,t,ans)
IMPLICIT NONE
REAL(SP), INTENT(IN) :: de,t
LOGICAL(LGT), INTENT(OUT) :: ans
   Metropolis algorithm. ans is a logical variable that issues a verdict on whether to accept a
   reconfiguration that leads to a change de in the objective function. If de<0, ans=.true.,
   while if de>0, ans is only .true. with probability exp(-de/t), where t is a temperature
   determined by the annealing schedule.
call ran1(harvest)
ans=(de < 0.0) .or. (harvest < exp(-de/t))
END SUBROUTINE metrop

SUBROUTINE revcst(x,y,iorder,n,de)
IMPLICIT NONE
REAL(SP), DIMENSION(:), INTENT(IN) :: x,y
INTEGER(I4B), DIMENSION(:), INTENT(IN) :: iorder
INTEGER(I4B), DIMENSION(:), INTENT(INOUT) :: n
REAL(SP), INTENT(OUT) :: de
```

This subroutine returns the value of the cost function for a proposed path reversal. The arrays x and y give the coordinates of these cities. iorder holds the present itinerary. The first two values n(1) and n(2) of array n give the starting and ending cities along the path segment which is to be reversed. On output, de is the cost of making the reversal. The actual reversal is not performed by this routine.

```
INTEGER(I4B) :: ncity
REAL(SP), DIMENSION(4) :: xx,yy
ncity=size(x)
n(3)=1+mod((n(1)+ncity-2),ncity)        Find the city before n(1) ...
n(4)=1+mod(n(2),ncity)                  ...and the city after n(2).
xx(1:4)=x(iorder(n(1:4)))               Find coordinates for the four cities involved.
yy(1:4)=y(iorder(n(1:4)))
de=-alen(xx(1),xx(3),yy(1),yy(3))&      Calculate cost of disconnecting the segment
    -alen(xx(2),xx(4),yy(2),yy(4))&        at both ends and reconnecting in the op-
    +alen(xx(1),xx(4),yy(1),yy(4))&        posite order.
    +alen(xx(2),xx(3),yy(2),yy(3))
END SUBROUTINE revcst

SUBROUTINE revers(iorder,n)
IMPLICIT NONE
INTEGER(I4B), DIMENSION(:), INTENT(INOUT) :: iorder
INTEGER(I4B), DIMENSION(:), INTENT(IN) :: n
```

This routine performs a path segment reversal. iorder is an input array giving the present itinerary. The vector n has as its first four elements the first and last cities n(1), n(2) of the path segment to be reversed, and the two cities n(3) and n(4) that immediately precede and follow this segment. n(3) and n(4) are found by subroutine revcst. On output, iorder contains the segment from n(1) to n(2) in reversed order.

```
INTEGER(I4B) :: j,k,l,nn,ncity
ncity=size(iorder)
nn=(1+mod(n(2)-n(1)+ncity,ncity))/2     This many cities must be swapped to effect
do j=1,nn                                  the reversal.
    k=1+mod((n(1)+j-2),ncity)           Start at the ends of the segment and swap
    l=1+mod((n(2)-j+ncity),ncity)          pairs of cities, moving toward the cen-
    call swap(iorder(k),iorder(l))         ter.
end do
END SUBROUTINE revers

SUBROUTINE trncst(x,y,iorder,n,de)
IMPLICIT NONE
REAL(SP), DIMENSION(:), INTENT(IN) :: x,y
INTEGER(I4B), DIMENSION(:), INTENT(IN) :: iorder
INTEGER(I4B), DIMENSION(:), INTENT(INOUT) :: n
REAL(SP), INTENT(OUT) :: de
```

This subroutine returns the value of the cost function for a proposed path segment transport. Arrays x and y give the city coordinates. iorder is an array giving the present itinerary. The first three elements of array n give the starting and ending cities of the path to be transported, and the point among the remaining cities after which it is to be inserted. On output, de is the cost of the change. The actual transport is not performed by this routine.

```
INTEGER(I4B) :: ncity
REAL(SP), DIMENSION(6) :: xx,yy
ncity=size(x)
n(4)=1+mod(n(3),ncity)                  Find the city following n(3) ...
n(5)=1+mod((n(1)+ncity-2),ncity)        ...and the one preceding n(1) ...
n(6)=1+mod(n(2),ncity)                  ...and the one following n(2).
xx(1:6)=x(iorder(n(1:6)))               Determine coordinates for the six cities in-
yy(1:6)=y(iorder(n(1:6)))                  volved.
de=-alen(xx(2),xx(6),yy(2),yy(6))&      Calculate the cost of disconnecting the path
    -alen(xx(1),xx(5),yy(1),yy(5))&        segment from n(1) to n(2), opening a
    -alen(xx(3),xx(4),yy(3),yy(4))&        space between n(3) and n(4), connect-
    +alen(xx(1),xx(3),yy(1),yy(3))&        ing the segment in the space, and con-
    +alen(xx(2),xx(4),yy(2),yy(4))&        necting n(5) to n(6).
    +alen(xx(5),xx(6),yy(5),yy(6))
END SUBROUTINE trncst

SUBROUTINE trnspt(iorder,n)
IMPLICIT NONE
```

```
INTEGER(I4B), DIMENSION(:), INTENT(INOUT) :: iorder
INTEGER(I4B), DIMENSION(:), INTENT(IN) :: n
```

This routine does the actual path transport, once metrop has approved. iorder is an input array giving the present itinerary. The array n has as its six elements the beginning n(1) and end n(2) of the path to be transported, the adjacent cities n(3) and n(4) between which the path is to be placed, and the cities n(5) and n(6) that precede and follow the path. n(4), n(5), and n(6) are calculated by subroutine trncst. On output, iorder is modified to reflect the movement of the path segment.

```
INTEGER(I4B) :: m1,m2,m3,nn,ncity
INTEGER(I4B), DIMENSION(size(iorder)) :: jorder
ncity=size(iorder)
m1=1+mod((n(2)-n(1)+ncity),ncity)          Find number of cities from n(1) to n(2) ...
m2=1+mod((n(5)-n(4)+ncity),ncity)          ...and the number from n(4) to n(5)
m3=1+mod((n(3)-n(6)+ncity),ncity)          ...and the number from n(6) to n(3).
jorder(1:m1)=iorder(1+mod((arth(1,1,m1)+n(1)-2),ncity))   Copy the chosen segment.
nn=m1
jorder(nn+1:nn+m2)=iorder(1+mod((arth(1,1,m2)+n(4)-2),ncity))
  Then copy the segment from n(4) to n(5).
nn=nn+m2
jorder(nn+1:nn+m3)=iorder(1+mod((arth(1,1,m3)+n(6)-2),ncity))
  Finally, the segment from n(6) to n(3).
iorder(1:ncity)=jorder(1:ncity)            Copy jorder back into iorder.
END SUBROUTINE trnspt
END SUBROUTINE anneal
```

⋆ ⋆ ⋆

```
SUBROUTINE amebsa(p,y,pb,yb,ftol,func,iter,temptr)
USE nrtype; USE nrutil, ONLY : assert_eq,imaxloc,iminloc,swap
USE nr, ONLY : ran1
IMPLICIT NONE
INTEGER(I4B), INTENT(INOUT) :: iter
REAL(SP), INTENT(INOUT) :: yb
REAL(SP), INTENT(IN) :: ftol,temptr
REAL(SP), DIMENSION(:), INTENT(INOUT) :: y,pb
REAL(SP), DIMENSION(:,:), INTENT(INOUT) :: p
INTERFACE
    FUNCTION func(x)
    USE nrtype
    IMPLICIT NONE
    REAL(SP), DIMENSION(:), INTENT(IN) :: x
    REAL(SP) :: func
    END FUNCTION func
END INTERFACE
INTEGER(I4B), PARAMETER :: NMAX=200
```

Minimization of the N-dimensional function func by simulated annealing combined with the downhill simplex method of Nelder and Mead. The $(N+1)\times N$ matrix p is input. Its $N+1$ rows are N-dimensional vectors that are the vertices of the starting simplex. Also input is the vector y of length $N+1$, whose components must be preinitialized to the values of func evaluated at the $N+1$ vertices (rows) of p; ftol, the fractional convergence tolerance to be achieved in the function value for an early return; iter, and temptr. The routine makes iter function evaluations at an annealing temperature temptr, then returns. You should then decrease temptr according to your annealing schedule, reset iter, and call the routine again (leaving other arguments unaltered between calls). If iter is returned with a positive value, then early convergence and return occurred. If you initialize yb to a very large value on the first call, then yb and pb (an array of length N) will subsequently return the best function value and point ever encountered (even if it is no longer a point in the simplex).

```
INTEGER(I4B) :: ihi,ndim                   Global variables.
REAL(SP) :: yhi
REAL(SP), DIMENSION(size(p,2)) :: psum
call amebsa_private
```

```
CONTAINS

SUBROUTINE amebsa_private
INTEGER(I4B) :: i,ilo,inhi
REAL(SP) :: rtol,ylo,ynhi,ysave,ytry
REAL(SP), DIMENSION(size(y)) :: yt,harvest
ndim=assert_eq(size(p,2),size(p,1)-1,size(y)-1,size(pb),'amebsa')
psum(:)=sum(p(:,:),dim=1)
do                                              Iteration loop.
    call ran1(harvest)
    yt(:)=y(:)-temptr*log(harvest)
      Whenever we "look at" a vertex, it gets a random thermal fluctuation.
    ilo=iminloc(yt(:))                          Determine which point is the highest (worst),
    ylo=yt(ilo)                                     next-highest, and lowest (best).
    ihi=imaxloc(yt(:))
    yhi=yt(ihi)
    yt(ihi)=ylo
    inhi=imaxloc(yt(:))
    ynhi=yt(inhi)
    rtol=2.0_sp*abs(yhi-ylo)/(abs(yhi)+abs(ylo))
      Compute the fractional range from highest to lowest and return if satisfactory.
    if (rtol < ftol .or. iter < 0) then         If returning, put best point and value in
        call swap(y(1),y(ilo))                      slot 1.
        call swap(p(1,:),p(ilo,:))
        RETURN
    end if
      Begin a new iteration. First extrapolate by a factor -1 through the face of the simplex
      across from the high point, i.e., reflect the simplex from the high point.
    ytry=amotsa(-1.0_sp)
    iter=iter-1
    if (ytry <= ylo) then                       Gives a result better than the best point, so
        ytry=amotsa(2.0_sp)                         try an additional extrapolation by a fac-
        iter=iter-1                                 tor of 2.
    else if (ytry >= ynhi) then                 The reflected point is worse than the second-
        ysave=yhi                                   highest, so look for an intermediate lower
        ytry=amotsa(0.5_sp)                         point, i.e., do a one-dimensional contrac-
        iter=iter-1                                 tion.
        if (ytry >= ysave) then
              Can't seem to get rid of that high point. Better contract around the lowest
              (best) point.
            p(:,:)=0.5_sp*(p(:,:)+spread(p(ilo,:),1,size(p,1)))
            do i=1,ndim+1
                if (i /= ilo) y(i)=func(p(i,:))
            end do
            iter=iter-ndim                      Keep track of function evaluations.
            psum(:)=sum(p(:,:),dim=1)
        end if
    end if
end do
END SUBROUTINE amebsa_private

FUNCTION amotsa(fac)
IMPLICIT NONE
REAL(SP), INTENT(IN) :: fac
REAL(SP) :: amotsa
    Extrapolates by a factor fac through the face of the simplex across from the high point,
    tries it, and replaces the high point if the new point is better.
REAL(SP) :: fac1,fac2,yflu,ytry,harv
REAL(SP), DIMENSION(size(p,2)) :: ptry
fac1=(1.0_sp-fac)/ndim
fac2=fac1-fac
ptry(:)=psum(:)*fac1-p(ihi,:)*fac2
ytry=func(ptry)
if (ytry <= yb) then                            Save the best-ever.
    pb(:)=ptry(:)
```

```
    yb=ytry
end if
call ran1(harv)
yflu=ytry+temptr*log(harv)
if (yflu < yhi) then
    y(ihi)=ytry
    yhi=yflu
    psum(:)=psum(:)-p(ihi,:)+ptry(:)
    p(ihi,:)=ptry(:)
end if
amotsa=yflu
END FUNCTION amotsa
END SUBROUTINE amebsa
```

We *added* a thermal fluctuation to all the current vertices, but we *subtract* it here, so as to give the simplex a thermal Brownian motion: It *likes* to accept any suggested change.

See the discussion of amoeba on p. 1209 for why the routine is coded this way.

Chapter B11. Eigensystems

```
SUBROUTINE jacobi(a,d,v,nrot)
USE nrtype; USE nrutil, ONLY : assert_eq,get_diag,nrerror,unit_matrix,&
    upper_triangle
IMPLICIT NONE
INTEGER(I4B), INTENT(OUT) :: nrot
REAL(SP), DIMENSION(:), INTENT(OUT) :: d
REAL(SP), DIMENSION(:,:), INTENT(INOUT) :: a
REAL(SP), DIMENSION(:,:), INTENT(OUT) :: v
```

Computes all eigenvalues and eigenvectors of a real symmetric $N \times N$ matrix a. On output, elements of a above the diagonal are destroyed. d is a vector of length N that returns the eigenvalues of a. v is an $N \times N$ matrix whose columns contain, on output, the normalized eigenvectors of a. nrot returns the number of Jacobi rotations that were required.

```
INTEGER(I4B) :: i,ip,iq,n
REAL(SP) :: c,g,h,s,sm,t,tau,theta,tresh
REAL(SP), DIMENSION(size(d)) :: b,z
n=assert_eq((/size(a,1),size(a,2),size(d),size(v,1),size(v,2)/),'jacobi')
call unit_matrix(v(:,:))                      Initialize v to the identity matrix.
b(:)=get_diag(a(:,:))                         Initialize b and d to the diagonal of
d(:)=b(:)                                         a.
z(:)=0.0                                      This vector will accumulate terms of
nrot=0                                            the form ta_pq as in eq. (11.1.14).
do i=1,50
    sm=sum(abs(a),mask=upper_triangle(n,n))   Sum off-diagonal elements.
    if (sm == 0.0) RETURN
```

The normal return, which relies on quadratic convergence to machine underflow.

```
    tresh=merge(0.2_sp*sm/n**2,0.0_sp, i < 4 )
```

On the first three sweeps, we will rotate only if tresh exceeded.

```
    do ip=1,n-1
        do iq=ip+1,n
            g=100.0_sp*abs(a(ip,iq))
```

After four sweeps, skip the rotation if the off-diagonal element is small.

```
            if ((i > 4) .and. (abs(d(ip))+g == abs(d(ip))) &
                .and. (abs(d(iq))+g == abs(d(iq)))) then
                a(ip,iq)=0.0
            else if (abs(a(ip,iq)) > tresh) then
                h=d(iq)-d(ip)
                if (abs(h)+g == abs(h)) then
                    t=a(ip,iq)/h                  t = 1/(2θ)
                else
                    theta=0.5_sp*h/a(ip,iq)       Equation (11.1.10).
                    t=1.0_sp/(abs(theta)+sqrt(1.0_sp+theta**2))
                    if (theta < 0.0) t=-t
                end if
                c=1.0_sp/sqrt(1+t**2)
                s=t*c
                tau=s/(1.0_sp+c)
                h=t*a(ip,iq)
                z(ip)=z(ip)-h
                z(iq)=z(iq)+h
                d(ip)=d(ip)-h
                d(iq)=d(iq)+h
                a(ip,iq)=0.0
```

```
                call jrotate(a(1:ip-1,ip),a(1:ip-1,iq))
                  Case of rotations 1 ≤ j < p.
                call jrotate(a(ip,ip+1:iq-1),a(ip+1:iq-1,iq))
                  Case of rotations p < j < q.
                call jrotate(a(ip,iq+1:n),a(iq,iq+1:n))
                  Case of rotations q < j ≤ n.
                call jrotate(v(:,ip),v(:,iq))
                nrot=nrot+1
            end if
        end do
    end do
    b(:)=b(:)+z(:)
    d(:)=b(:)                        Update d with the sum of ta_pq,
    z(:)=0.0                         and reinitialize z.
end do
call nrerror('too many iterations in jacobi')
CONTAINS

SUBROUTINE jrotate(a1,a2)
REAL(SP), DIMENSION(:), INTENT(INOUT) :: a1,a2
REAL(SP), DIMENSION(size(a1)) :: wk1
wk1(:)=a1(:)
a1(:)=a1(:)-s*(a2(:)+a1(:)*tau)
a2(:)=a2(:)+s*(wk1(:)-a2(:)*tau)
END SUBROUTINE jrotate
END SUBROUTINE jacobi
```

As discussed in Volume 1, `jacobi` is generally not competitive with `tqli` in terms of efficiency. However, `jacobi` can be parallelized whereas `tqli` uses an intrinsically serial algorithm. The version of `jacobi` implemented here is likely to be adequate for a small-scale parallel (SSP) machine, but is probably still not competitive with `tqli`. For a massively multiprocessor (MMP) machine, the order of the rotations needs to be chosen in a more complicated pattern than here so that the rotations can be executed in parallel. In this case the Jacobi algorithm may well turn out to be the method of choice. Parallel replacements for `tqli` based on a divide and conquer algorithm have also been proposed. See the discussion after `tqli` on p. 1229.

f90 `call unit_matrix...b(:)=get_diag...` These routines in `nrutil` both require access to the diagonal of a matrix, an operation that is not conveniently provided for in Fortran 90. We have split them off into `nrutil` in case your compiler provides parallel library routines so you can replace our standard versions.

`sm=sum(abs(a),mask=upper_triangle(n,n))` The `upper_triangle` function in `nrutil` returns an upper triangular logical mask. As used here, the mask is true everywhere in the upper triangle of an `n` × `n` matrix, excluding the diagonal. An optional integer argument `extra` allows additional diagonals to be set to true. With `extra=1` the upper triangle including the diagonal would be true. By using the mask, we can conveniently sum over the desired matrix elements in parallel.

`SUBROUTINE jrotate(a1,a2)` This internal subroutine also uses the values of `s` and `tau` from the calling subroutine `jacobi`. Variables in the calling routine are visible to an internal subprogram, but you should be circumspect in making use of this fact. It is easy to overwrite a value in the calling program inadvertently, and it is often difficult to figure out the logic of an internal routine if not all its variables are declared explicitly. However, jrotate is so simple that there is no danger here.

⋆ ⋆ ⋆

```
SUBROUTINE eigsrt(d,v)
USE nrtype; USE nrutil, ONLY : assert_eq,imaxloc,swap
IMPLICIT NONE
REAL(SP), DIMENSION(:), INTENT(INOUT) :: d
REAL(SP), DIMENSION(:,:), INTENT(INOUT) :: v
```

Given the eigenvalues d and eigenvectors v as output from jacobi (§11.1) or tqli (§11.3), this routine sorts the eigenvalues into descending order, and rearranges the columns of v correspondingly. The method is straight insertion.

```
INTEGER(I4B) :: i,j,n
n=assert_eq(size(d),size(v,1),size(v,2),'eigsrt')
do i=1,n-1
    j=imaxloc(d(i:n))+i-1
    if (j /= i) then
        call swap(d(i),d(j))
        call swap(v(:,i),v(:,j))
    end if
end do
END SUBROUTINE eigsrt
```

f90 j=imaxloc... See discussion of imaxloc on p. 1017.

call swap... See discussion of overloaded versions of swap after amoeba on p. 1210.

⋆ ⋆ ⋆

```
SUBROUTINE tred2(a,d,e,novectors)
USE nrtype; USE nrutil, ONLY : assert_eq,outerprod
IMPLICIT NONE
REAL(SP), DIMENSION(:,:), INTENT(INOUT) :: a
REAL(SP), DIMENSION(:), INTENT(OUT) :: d,e
LOGICAL(LGT), OPTIONAL, INTENT(IN) :: novectors
```

Householder reduction of a real, symmetric, $N \times N$ matrix a. On output, a is replaced by the orthogonal matrix $\mathbf{Q}$ effecting the transformation. d returns the diagonal elements of the tridiagonal matrix, and e the off-diagonal elements, with e(1)=0. If the optional argument novectors is present, only eigenvalues are to be found subsequently, in which case a contains no useful information on output.

```
INTEGER(I4B) :: i,j,l,n
REAL(SP) :: f,g,h,hh,scale
REAL(SP), DIMENSION(size(a,1)) :: gg
LOGICAL(LGT), SAVE :: yesvec=.true.
n=assert_eq(size(a,1),size(a,2),size(d),size(e),'tred2')
if (present(novectors)) yesvec=.not. novectors
do i=n,2,-1
    l=i-1
    h=0.0
    if (l > 1) then
        scale=sum(abs(a(i,1:l)))
        if (scale == 0.0) then              Skip transformation.
            e(i)=a(i,l)
        else
            a(i,1:l)=a(i,1:l)/scale         Use scaled a's for transformation.
            h=sum(a(i,1:l)**2)              Form σ in h.
```

```
            f=a(i,l)
            g=-sign(sqrt(h),f)
            e(i)=scale*g
            h=h-f*g                                  Now h is equation (11.2.4).
            a(i,l)=f-g                               Store u in the ith row of a.
            if (yesvec) a(1:l,i)=a(i,1:l)/h          Store u/H in ith column of a.
            do j=1,l                                 Store elements of p in temporarily
                e(j)=(dot_product(a(j,1:j),a(i,1:j)) &    unused elements of e.
                +dot_product(a(j+1:l,j),a(i,j+1:l)))/h
            end do
            f=dot_product(e(1:l),a(i,1:l))
            hh=f/(h+h)                               Form K, equation (11.2.11).
            e(1:l)=e(1:l)-hh*a(i,1:l)
              Form q and store in e overwriting p.
            do j=1,l                                 Reduce a, equation (11.2.13).
                a(j,1:j)=a(j,1:j)-a(i,j)*e(1:j)-e(j)*a(i,1:j)
            end do
        end if
    else
        e(i)=a(i,l)
    end if
    d(i)=h
end do
if (yesvec) d(1)=0.0
e(1)=0.0
do i=1,n                                         Begin accumulation of transforma-
    if (yesvec) then                                 tion matrices.
        l=i-1
        if (d(i) /= 0.0) then
              This block skipped when i=1. Use u and u/H stored in a to form P·Q.
            gg(1:l)=matmul(a(i,1:l),a(1:l,1:l))
            a(1:l,1:l)=a(1:l,1:l)-outerprod(a(1:l,i),gg(1:l))
        end if
        d(i)=a(i,i)
        a(i,i)=1.0                               Reset row and column of a to iden-
        a(i,1:l)=0.0                                 tity matrix for next iteration.
        a(1:l,i)=0.0
    else
        d(i)=a(i,i)
    end if
end do
END SUBROUTINE tred2
```

f90 This routine gives a nice example of the usefulness of optional arguments. The routine is written under the assumption that usually you will want to find both eigenvalues and eigenvectors. In this case you just supply the arguments a, d, and e. If, however, you want only eigenvalues, you supply the additional logical argument `novectors` with the value `.true.`. The routine then skips the unnecessary computations. Supplying `novectors` with the value `.false.` has the same effect as omitting it.

⋆ ⋆ ⋆

```
SUBROUTINE tqli(d,e,z)
USE nrtype; USE nrutil, ONLY : assert_eq,nrerror
USE nr, ONLY : pythag
IMPLICIT NONE
REAL(SP), DIMENSION(:), INTENT(INOUT) :: d,e
REAL(SP), DIMENSION(:,:), OPTIONAL, INTENT(INOUT) :: z
    QL algorithm with implicit shifts, to determine the eigenvalues and eigenvectors of a real,
    symmetric, tridiagonal matrix, or of a real, symmetric matrix previously reduced by tred2
```

§11.2. d is a vector of length N. On input, its elements are the diagonal elements of the tridiagonal matrix. On output, it returns the eigenvalues. The vector e inputs the subdiagonal elements of the tridiagonal matrix, with e(1) arbitrary. On output e is destroyed. When finding only the eigenvalues, the optional argument z is omitted. If the eigenvectors of a tridiagonal matrix are desired, the $N \times N$ matrix z is input as the identity matrix. If the eigenvectors of a matrix that has been reduced by tred2 are required, then z is input as the matrix output by tred2. In either case, the kth column of z returns the normalized eigenvector corresponding to d(k).

```
INTEGER(I4B) :: i,iter,l,m,n,ndum
REAL(SP) :: b,c,dd,f,g,p,r,s
REAL(SP), DIMENSION(size(e)) :: ff
n=assert_eq(size(d),size(e),'tqli: n')
if (present(z)) ndum=assert_eq(n,size(z,1),size(z,2),'tqli: ndum')
e(:)=eoshift(e(:),1)                       Convenient to renumber the elements of
do l=1,n                                       e.
    iter=0
    iterate: do
        do m=l,n-1                         Look for a single small subdiagonal ele-
            dd=abs(d(m))+abs(d(m+1))           ment to split the matrix.
            if (abs(e(m))+dd == dd) exit
        end do
        if (m == l) exit iterate
        if (iter == 30) call nrerror('too many iterations in tqli')
        iter=iter+1
        g=(d(l+1)-d(l))/(2.0_sp*e(l))      Form shift.
        r=pythag(g,1.0_sp)
        g=d(m)-d(l)+e(l)/(g+sign(r,g))     This is d_m - k_s.
        s=1.0
        c=1.0
        p=0.0
        do i=m-1,l,-1                      A plane rotation as in the original QL,
            f=s*e(i)                           followed by Givens rotations to re-
            b=c*e(i)                           store tridiagonal form.
            r=pythag(f,g)
            e(i+1)=r
            if (r == 0.0) then             Recover from underflow.
                d(i+1)=d(i+1)-p
                e(m)=0.0
                cycle iterate
            end if
            s=f/r
            c=g/r
            g=d(i+1)-p
            r=(d(i)-g)*s+2.0_sp*c*b
            p=s*r
            d(i+1)=g+p
            g=c*r-b
            if (present(z)) then           Form eigenvectors.
                ff(1:n)=z(1:n,i+1)
                z(1:n,i+1)=s*z(1:n,i)+c*ff(1:n)
                z(1:n,i)=c*z(1:n,i)-s*ff(1:n)
            end if
        end do
        d(l)=d(l)-p
        e(l)=g
        e(m)=0.0
    end do iterate
end do
END SUBROUTINE tqli
```

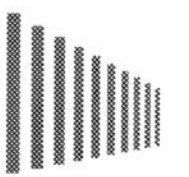

The routine tqli is intrinsically serial. A parallel replacement based on a divide and conquer algorithm has been proposed [1,2]. The idea is to split the tridiagonal matrix recursively into two tridiagonal matrices of half the size plus a correction. Given the eigensystems of the two smaller tridiagonal matrices, it is possible to join them together and add in the effect of the correction. When some small size of tridiagonal matrix is reached during the recursive splitting, its eigensystem is found directly with a routine like `tqli`. Each of these small problems is independent and can be assigned to an independent processor. The procedures for sewing together can also be done independently. For very large matrices, this algorithm can be an order of magnitude faster than `tqli` even on a serial machine, and no worse than a factor of 2 or 3 slower, depending on the matrix. Unfortunately the parallelism is not well expressed in Fortran 90. Also, the sewing together requires quite involved coding. For an implementation see the LAPACK routine SSTEDC. Another parallel strategy for eigensystems uses inverse iteration, where each eigenvalue and eigenvector can be found independently [3].

f90 This routine uses `z` as an optional argument that is required only if eigenvectors are being found as well as eigenvalues.

`iterate: do` See discussion of named do loops after `simplx` on p. 1219.

⋆ ⋆ ⋆

```
SUBROUTINE balanc(a)
USE nrtype; USE nrutil, ONLY : assert_eq
IMPLICIT NONE
REAL(SP), DIMENSION(:,:), INTENT(INOUT) :: a
REAL(SP), PARAMETER :: RADX=radix(a),SQRADX=RADX**2
   Given an N × N matrix a, this routine replaces it by a balanced matrix with identical
   eigenvalues. A symmetric matrix is already balanced and is unaffected by this procedure.
   The parameter RADX is the machine's floating-point radix.
INTEGER(I4B) :: i,last,ndum
REAL(SP) :: c,f,g,r,s
ndum=assert_eq(size(a,1),size(a,2),'balanc')
do
    last=1
    do i=1,size(a,1)                          Calculate row and column norms.
        c=sum(abs(a(:,i)))-a(i,i)
        r=sum(abs(a(i,:)))-a(i,i)
        if (c /= 0.0 .and. r /= 0.0) then     If both are nonzero,
            g=r/RADX
            f=1.0
            s=c+r
            do                                find the integer power of the ma-
                if (c >= g) exit                  chine radix that comes closest to
                f=f*RADX                          balancing the matrix.
                c=c*SQRADX
            end do
            g=r*RADX
            do
                if (c <= g) exit
                f=f/RADX
                c=c/SQRADX
            end do
            if ((c+r)/f < 0.95_sp*s) then
                last=0
                g=1.0_sp/f
                a(i,:)=a(i,:)*g               Apply similarity transformation.
                a(:,i)=a(:,i)*f
            end if
        end if
```

```
    end do
    if (last /= 0) exit
end do
END SUBROUTINE balanc
```

f90 `REAL(SP), PARAMETER :: RADX=radix(a)...` Fortran 90 provides a nice collection of numeric inquiry intrinsic functions. Here we find the machine's floating-point radix. Note that only the type of the argument a affects the returned function value.

⋆ ⋆ ⋆

```
SUBROUTINE elmhes(a)
USE nrtype; USE nrutil, ONLY : assert_eq,imaxloc,outerprod,swap
IMPLICIT NONE
REAL(SP), DIMENSION(:,:), INTENT(INOUT) :: a
   Reduction to Hessenberg form by the elimination method. The real, nonsymmetric, N × N
   matrix a is replaced by an upper Hessenberg matrix with identical eigenvalues. Recom-
   mended, but not required, is that this routine be preceded by balanc. On output, the
   Hessenberg matrix is in elements a(i,j) with i ≤ j + 1. Elements with i > j + 1 are to be
   thought of as zero, but are returned with random values.
INTEGER(I4B) :: i,m,n
REAL(SP) :: x
REAL(SP), DIMENSION(size(a,1)) :: y
n=assert_eq(size(a,1),size(a,2),'elmhes')
do m=2,n-1                                     m is called r + 1 in the text.
    i=imaxloc(abs(a(m:n,m-1)))+m-1             Find the pivot.
    x=a(i,m-1)
    if (i /= m) then                           Interchange rows and columns.
        call swap(a(i,m-1:n),a(m,m-1:n))
        call swap(a(:,i),a(:,m))
    end if
    if (x /= 0.0) then                         Carry out the elimination.
        y(m+1:n)=a(m+1:n,m-1)/x
        a(m+1:n,m-1)=y(m+1:n)
        a(m+1:n,m:n)=a(m+1:n,m:n)-outerprod(y(m+1:n),a(m,m:n))
        a(:,m)=a(:,m)+matmul(a(:,m+1:n),y(m+1:n))
    end if
end do
END SUBROUTINE elmhes
```

f90 `y(m+1:n)=...` If the four lines of code starting here were all coded for a serial machine in a single do-loop starting with `do i=m+1,n` (see Volume 1), it would pay to test whether y was zero because the next three lines could then be skipped for that value of i. There is no convenient way to do this here, even with a `where`, since the shape of the arrays on each of the three lines is different. For a parallel machine it is probably best just to do a few unnecessary multiplies and skip the test for zero values of y.

⋆ ⋆ ⋆

```
SUBROUTINE hqr(a,wr,wi)
USE nrtype; USE nrutil, ONLY : assert_eq,diagadd,nrerror,upper_triangle
IMPLICIT NONE
REAL(SP), DIMENSION(:), INTENT(OUT) :: wr,wi
REAL(SP), DIMENSION(:,:), INTENT(INOUT) :: a
```

Finds all eigenvalues of an $N \times N$ upper Hessenberg matrix a. On input a can be exactly as output from elmhes §11.5; on output it is destroyed. The real and imaginary parts of the N eigenvalues are returned in wr and wi, respectively.

```
INTEGER(I4B) :: i,its,k,l,m,n,nn,mnnk
REAL(SP) :: anorm,p,q,r,s,t,u,v,w,x,y,z
REAL(SP), DIMENSION(size(a,1)) :: pp
n=assert_eq(size(a,1),size(a,2),size(wr),size(wi),'hqr')
anorm=sum(abs(a),mask=upper_triangle(n,n,extra=2))
```

Compute matrix norm for possible use in locating single small subdiagonal element.

```
nn=n
t=0.0                                   Gets changed only by an exceptional shift.
do                                      Begin search for next eigenvalue: "Do while
    if (nn < 1) exit                        nn >= 1".
    its=0
    iterate: do                         Begin iteration.
        do l=nn,2,-1                    Look for single small subdiagonal element.
            s=abs(a(l-1,l-1))+abs(a(l,l))
            if (s == 0.0) s=anorm
            if (abs(a(l,l-1))+s == s) exit
        end do
        x=a(nn,nn)
        if (l == nn) then               One root found.
            wr(nn)=x+t
            wi(nn)=0.0
            nn=nn-1
            exit iterate                Go back for next eigenvalue.
        end if
        y=a(nn-1,nn-1)
        w=a(nn,nn-1)*a(nn-1,nn)
        if (l == nn-1) then             Two roots found ...
            p=0.5_sp*(y-x)
            q=p**2+w
            z=sqrt(abs(q))
            x=x+t
            if (q >= 0.0) then          ...a real pair ...
                z=p+sign(z,p)
                wr(nn)=x+z
                wr(nn-1)=wr(nn)
                if (z /= 0.0) wr(nn)=x-w/z
                wi(nn)=0.0
                wi(nn-1)=0.0
            else                        ...a complex pair.
                wr(nn)=x+p
                wr(nn-1)=wr(nn)
                wi(nn)=z
                wi(nn-1)=-z
            end if
            nn=nn-2
            exit iterate                Go back for next eigenvalue.
        end if
```

No roots found. Continue iteration.

```
        if (its == 30) call nrerror('too many iterations in hqr')
        if (its == 10 .or. its == 20) then      Form exceptional shift.
            t=t+x
            call diagadd(a(1:nn,1:nn),-x)
            s=abs(a(nn,nn-1))+abs(a(nn-1,nn-2))
            x=0.75_sp*s
            y=x
            w=-0.4375_sp*s**2
```

```
end if
its=its+1
do m=nn-2,1,-1                          Form shift and then look for 2 consecu-
    z=a(m,m)                               tive small subdiagonal elements.
    r=x-z
    s=y-z
    p=(r*s-w)/a(m+1,m)+a(m,m+1)         Equation (11.6.23).
    q=a(m+1,m+1)-z-r-s
    r=a(m+2,m+1)
    s=abs(p)+abs(q)+abs(r)              Scale to prevent overflow or underflow.
    p=p/s
    q=q/s
    r=r/s
    if (m == 1) exit
    u=abs(a(m,m-1))*(abs(q)+abs(r))
    v=abs(p)*(abs(a(m-1,m-1))+abs(z)+abs(a(m+1,m+1)))
    if (u+v == v) exit                  Equation (11.6.26).
end do
do i=m+2,nn
    a(i,i-2)=0.0
    if (i /= m+2) a(i,i-3)=0.0
end do
do k=m,nn-1                             Double QR step on rows 1 to nn and
    if (k /= m) then                       columns m to nn.
        p=a(k,k-1)                      Begin setup of Householder vector.
        q=a(k+1,k-1)
        r=0.0
        if (k /= nn-1) r=a(k+2,k-1)
        x=abs(p)+abs(q)+abs(r)
        if (x /= 0.0) then
            p=p/x                       Scale to prevent overflow or underflow.
            q=q/x
            r=r/x
        end if
    end if
    s=sign(sqrt(p**2+q**2+r**2),p)
    if (s /= 0.0) then
        if (k == m) then
            if (l /= m) a(k,k-1)=-a(k,k-1)
        else
            a(k,k-1)=-s*x
        end if
        p=p+s                           Equations (11.6.24).
        x=p/s
        y=q/s
        z=r/s
        q=q/p
        r=r/p                           Ready for row modification.
        pp(k:nn)=a(k,k:nn)+q*a(k+1,k:nn)
        if (k /= nn-1) then
            pp(k:nn)=pp(k:nn)+r*a(k+2,k:nn)
            a(k+2,k:nn)=a(k+2,k:nn)-pp(k:nn)*z
        end if
        a(k+1,k:nn)=a(k+1,k:nn)-pp(k:nn)*y
        a(k,k:nn)=a(k,k:nn)-pp(k:nn)*x
        mnnk=min(nn,k+3)                Column modification.
        pp(l:mnnk)=x*a(l:mnnk,k)+y*a(l:mnnk,k+1)
        if (k /= nn-1) then
            pp(l:mnnk)=pp(l:mnnk)+z*a(l:mnnk,k+2)
            a(l:mnnk,k+2)=a(l:mnnk,k+2)-pp(l:mnnk)*r
        end if
        a(l:mnnk,k+1)=a(l:mnnk,k+1)-pp(l:mnnk)*q
        a(l:mnnk,k)=a(l:mnnk,k)-pp(l:mnnk)
    end if
```

```
        end do
    end do iterate                          Go back for next iteration on current eigen-
end do                                          value.
END SUBROUTINE hqr
```

`anorm=sum(abs(a),mask=upper_triangle(n,n,extra=2)` See the discussion of `upper_triangle` after `jacobi` on p. 1226. Setting `extra=2` here picks out the upper Hessenberg part of the matrix.

`iterate: do` We use a named loop to improve the readability and structuring of the routine. The if-blocks that test for one or two roots end with `exit iterate`, transferring control back to the outermost loop and thus starting a search for the next root.

`call diagadd...` The routines that operate on the diagonal of a matrix are collected in `nrutil` partly so you can write clear code and partly in the hope that compiler writers will provide parallel library routines. Fortran 90 does not provide convenient parallel access to the diagonal of a matrix.

CITED REFERENCES AND FURTHER READING:

Golub, G.H., and Van Loan, C.F. 1989, *Matrix Computations*, 2nd ed. (Baltimore: Johns Hopkins University Press), §8.6 and references therein. [1]

Sorensen, D.C., and Tang, P.T.P. 1991, *SIAM Journal on Numerical Analysis*, vol. 28, pp. 1752–1775. [2]

Lo, S.-S., Philippe, B., and Sameh, A. 1987, *SIAM Journal on Scientific and Statistical Computing*, vol. 8, pp. s155–s165. [3]

Chapter B12. Fast Fourier Transform

The algorithms underlying the parallel routines in this chapter are described in §22.4. As described there, the basic building block is a routine for simultaneously taking the FFT of each row of a two-dimensional matrix:

```
SUBROUTINE fourrow_sp(data,isign)
USE nrtype; USE nrutil, ONLY : assert,swap
IMPLICIT NONE
COMPLEX(SPC), DIMENSION(:,:), INTENT(INOUT) :: data
INTEGER(I4B), INTENT(IN) :: isign
    Replaces each row (constant first index) of data(1:M,1:N) by its discrete Fourier trans-
    form (transform on second index), if isign is input as 1; or replaces each row of data
    by N times its inverse discrete Fourier transform, if isign is input as -1. N must be an
    integer power of 2. Parallelism is M-fold on the first index of data.
INTEGER(I4B) :: n,i,istep,j,m,mmax,n2
REAL(DP) :: theta
COMPLEX(SPC), DIMENSION(size(data,1)) :: temp
COMPLEX(DPC) :: w,wp                        Double precision for the trigonometric recurrences.
COMPLEX(SPC) :: ws
n=size(data,2)
call assert(iand(n,n-1)==0, 'n must be a power of 2 in fourrow_sp')
n2=n/2
j=n2
  This is the bit-reversal section of the routine.
do i=1,n-2
    if (j > i) call swap(data(:,j+1),data(:,i+1))
    m=n2
    do
        if (m < 2 .or. j < m) exit
        j=j-m
        m=m/2
    end do
    j=j+m
end do
mmax=1
  Here begins the Danielson-Lanczos section of the routine.
do                                          Outer loop executed log2 N times.
    if (n <= mmax) exit
    istep=2*mmax
    theta=PI_D/(isign*mmax)                 Initialize for the trigonometric recurrence.
    wp=cmplx(-2.0_dp*sin(0.5_dp*theta)**2,sin(theta),kind=dpc)
    w=cmplx(1.0_dp,0.0_dp,kind=dpc)
    do m=1,mmax                             Here are the two nested inner loops.
        ws=w
        do i=m,n,istep
            j=i+mmax
            temp=ws*data(:,j)               This is the Danielson-Lanczos formula.
            data(:,j)=data(:,i)-temp
            data(:,i)=data(:,i)+temp
        end do
        w=w*wp+w                            Trigonometric recurrence.
    end do
    mmax=istep
```

```
end do
END SUBROUTINE fourrow_sp
```

`call assert(iand(n,n-1)==0 ...` All the Fourier routines in this chapter require the dimension N of the data to be a power of 2. This is easily tested for by AND'ing N and $N-1$: N should have the binary representation $10000\ldots$, in which case $N-1 = 01111\ldots$.

```
SUBROUTINE fourrow_dp(data,isign)
USE nrtype; USE nrutil, ONLY : assert,swap
IMPLICIT NONE
COMPLEX(DPC), DIMENSION(:,:), INTENT(INOUT) :: data
INTEGER(I4B), INTENT(IN) :: isign
INTEGER(I4B) :: n,i,istep,j,m,mmax,n2
REAL(DP) :: theta
COMPLEX(DPC), DIMENSION(size(data,1)) :: temp
COMPLEX(DPC) :: w,wp
COMPLEX(DPC) :: ws
n=size(data,2)
call assert(iand(n,n-1)==0, 'n must be a power of 2 in fourrow_dp')
n2=n/2
j=n2
do i=1,n-2
    if (j > i) call swap(data(:,j+1),data(:,i+1))
    m=n2
    do
        if (m < 2 .or. j < m) exit
        j=j-m
        m=m/2
    end do
    j=j+m
end do
mmax=1
do
    if (n <= mmax) exit
    istep=2*mmax
    theta=PI_D/(isign*mmax)
    wp=cmplx(-2.0_dp*sin(0.5_dp*theta)**2,sin(theta),kind=dpc)
    w=cmplx(1.0_dp,0.0_dp,kind=dpc)
    do m=1,mmax
        ws=w
        do i=m,n,istep
            j=i+mmax
            temp=ws*data(:,j)
            data(:,j)=data(:,i)-temp
            data(:,i)=data(:,i)+temp
        end do
        w=w*wp+w
    end do
    mmax=istep
end do
END SUBROUTINE fourrow_dp
```

```
SUBROUTINE fourrow_3d(data,isign)
USE nrtype; USE nrutil, ONLY : assert,swap
IMPLICIT NONE
COMPLEX(SPC), DIMENSION(:,:,:), INTENT(INOUT) :: data
INTEGER(I4B), INTENT(IN) :: isign
```

If `isign` is input as 1, replaces each third-index section (constant first and second indices) of `data(1:`L`,1:`M`,1:`N`)` by its discrete Fourier transform (transform on third index); or replaces each third-index section of data by N times its inverse discrete Fourier transform, if isign is input as -1. N must be an integer power of 2. Parallelism is $L \times M$-fold on the first and second indices of data.

```
INTEGER(I4B) :: n,i,istep,j,m,mmax,n2
REAL(DP) :: theta
COMPLEX(SPC), DIMENSION(size(data,1),size(data,2)) :: temp
COMPLEX(DPC) :: w,wp                    Double precision for the trigonometric recurrences.
COMPLEX(SPC) :: ws
n=size(data,3)
call assert(iand(n,n-1)==0, 'n must be a power of 2 in fourrow_3d')
n2=n/2
j=n2
  This is the bit-reversal section of the routine.
do i=1,n-2
    if (j > i) call swap(data(:,:,j+1),data(:,:,i+1))
    m=n2
    do
        if (m < 2 .or. j < m) exit
        j=j-m
        m=m/2
    end do
    j=j+m
end do
mmax=1
  Here begins the Danielson-Lanczos section of the routine.
do                                      Outer loop executed log2 N times.
    if (n <= mmax) exit
    istep=2*mmax
    theta=PI_D/(isign*mmax)             Initialize for the trigonometric recurrence.
    wp=cmplx(-2.0_dp*sin(0.5_dp*theta)**2,sin(theta),kind=dpc)
    w=cmplx(1.0_dp,0.0_dp,kind=dpc)
    do m=1,mmax                         Here are the two nested inner loops.
        ws=w
        do i=m,n,istep
            j=i+mmax
            temp=ws*data(:,:,j)         This is the Danielson-Lanczos formula.
            data(:,:,j)=data(:,:,i)-temp
            data(:,:,i)=data(:,:,i)+temp
        end do
        w=w*wp+w                        Trigonometric recurrence.
    end do
    mmax=istep
end do
END SUBROUTINE fourrow_3d
```

⋆ ⋆ ⋆

Exactly as in the preceding routines, we can take the FFT of each *column* of a two-dimensional matrix, and for each *first-index* section of a three-dimensional array.

```
SUBROUTINE fourcol(data,isign)
USE nrtype; USE nrutil, ONLY : assert,swap
IMPLICIT NONE
COMPLEX(SPC), DIMENSION(:,:), INTENT(INOUT) :: data
INTEGER(I4B), INTENT(IN) :: isign
```

Replaces each column (constant second index) of data$(1:N,1:M)$ by its discrete Fourier transform (transform on first index), if isign is input as 1; or replaces each row of data by N times its inverse discrete Fourier transform, if isign is input as -1. N must be an integer power of 2. Parallelism is M-fold on the second index of data.

```
INTEGER(I4B) :: n,i,istep,j,m,mmax,n2
REAL(DP) :: theta
COMPLEX(SPC), DIMENSION(size(data,2)) :: temp
COMPLEX(DPC) :: w,wp                    Double precision for the trigonometric recurrences.
COMPLEX(SPC) :: ws
n=size(data,1)
call assert(iand(n,n-1)==0, 'n must be a power of 2 in fourcol')
n2=n/2
j=n2
  This is the bit-reversal section of the routine.
do i=1,n-2
    if (j > i) call swap(data(j+1,:),data(i+1,:))
    m=n2
    do
        if (m < 2 .or. j < m) exit
        j=j-m
        m=m/2
    end do
    j=j+m
end do
mmax=1
  Here begins the Danielson-Lanczos section of the routine.
do                                      Outer loop executed log2 N times.
    if (n <= mmax) exit
    istep=2*mmax
    theta=PI_D/(isign*mmax)             Initialize for the trigonometric recurrence.
    wp=cmplx(-2.0_dp*sin(0.5_dp*theta)**2,sin(theta),kind=dpc)
    w=cmplx(1.0_dp,0.0_dp,kind=dpc)
    do m=1,mmax                         Here are the two nested inner loops.
        ws=w
        do i=m,n,istep
            j=i+mmax
            temp=ws*data(j,:)           This is the Danielson-Lanczos formula.
            data(j,:)=data(i,:)-temp
            data(i,:)=data(i,:)+temp
        end do
        w=w*wp+w                        Trigonometric recurrence.
    end do
    mmax=istep
end do
END SUBROUTINE fourcol
```

```
SUBROUTINE fourcol_3d(data,isign)
USE nrtype; USE nrutil, ONLY : assert,swap
IMPLICIT NONE
COMPLEX(SPC), DIMENSION(:,:,:), INTENT(INOUT) :: data
INTEGER(I4B), INTENT(IN) :: isign
```

If isign is input as 1, replaces each first-index section (constant second and third indices) of data(1:N,1:M,1:L) by its discrete Fourier transform (transform on first index); or replaces each first-index section of data by N times its inverse discrete Fourier transform, if isign is input as -1. N must be an integer power of 2. Parallelism is $M \times L$-fold on the second and third indices of data.

```
INTEGER(I4B) :: n,i,istep,j,m,mmax,n2
REAL(DP) :: theta
COMPLEX(SPC), DIMENSION(size(data,2),size(data,3)) :: temp
COMPLEX(DPC) :: w,wp                    Double precision for the trigonometric recurrences.
COMPLEX(SPC) :: ws
n=size(data,1)
call assert(iand(n,n-1)==0, 'n must be a power of 2 in fourcol_3d')
n2=n/2
```

```
j=n2
  This is the bit-reversal section of the routine.
do i=1,n-2
    if (j > i) call swap(data(j+1,:,:),data(i+1,:,:))
    m=n2
    do
        if (m < 2 .or. j < m) exit
        j=j-m
        m=m/2
    end do
    j=j+m
end do
mmax=1
  Here begins the Danielson-Lanczos section of the routine.
do                                      Outer loop executed log2 N times.
    if (n <= mmax) exit
    istep=2*mmax
    theta=PI_D/(isign*mmax)             Initialize for the trigonometric recurrence.
    wp=cmplx(-2.0_dp*sin(0.5_dp*theta)**2,sin(theta),kind=dpc)
    w=cmplx(1.0_dp,0.0_dp,kind=dpc)
    do m=1,mmax                         Here are the two nested inner loops.
        ws=w
        do i=m,n,istep
            j=i+mmax
            temp=ws*data(j,:,:)         This is the Danielson-Lanczos formula.
            data(j,:,:)=data(i,:,:)-temp
            data(i,:,:)=data(i,:,:)+temp
        end do
        w=w*wp+w                        Trigonometric recurrence.
    end do
    mmax=istep
end do
END SUBROUTINE fourcol_3d
```

⋆ ⋆ ⋆

Here now are implementations of the method of §22.4 for the FFT of one-dimensional single- and double-precision complex arrays:

```
SUBROUTINE four1_sp(data,isign)
USE nrtype; USE nrutil, ONLY : arth,assert
USE nr, ONLY : fourrow
IMPLICIT NONE
COMPLEX(SPC), DIMENSION(:), INTENT(INOUT) :: data
INTEGER(I4B), INTENT(IN) :: isign
   Replaces a complex array data by its discrete Fourier transform, if isign is input as 1;
   or replaces data by its inverse discrete Fourier transform times the size of data, if isign
   is input as -1. The size of data must be an integer power of 2. Parallelism is achieved
   by internally reshaping the input array to two dimensions. (Use this version if fourrow is
   faster than fourcol on your machine.)
COMPLEX(SPC), DIMENSION(:,:), ALLOCATABLE :: dat,temp
COMPLEX(DPC), DIMENSION(:), ALLOCATABLE :: w,wp
REAL(DP), DIMENSION(:), ALLOCATABLE :: theta
INTEGER(I4B) :: n,m1,m2,j
n=size(data)
call assert(iand(n,n-1)==0, 'n must be a power of 2 in four1_sp')
  Find dimensions as close to square as possible, allocate space, and reshape the input array.
m1=2**ceiling(0.5_sp*log(real(n,sp))/0.693147_sp)
m2=n/m1
allocate(dat(m1,m2),theta(m1),w(m1),wp(m1),temp(m2,m1))
dat=reshape(data,shape(dat))
call fourrow(dat,isign)                 Transform on second index.
```

```
theta=arth(0,isign,m1)*TWOPI_D/n        Set up recurrence.
wp=cmplx(-2.0_dp*sin(0.5_dp*theta)**2,sin(theta),kind=dpc)
w=cmplx(1.0_dp,0.0_dp,kind=dpc)
do j=2,m2                               Multiply by the extra phase factor.
    w=w*wp+w
    dat(:,j)=dat(:,j)*w
end do
temp=transpose(dat)                     Transpose, and transform on (original) first in-
call fourrow(temp,isign)                    dex.
data=reshape(temp,shape(data))          Reshape the result back to one dimension.
deallocate(dat,w,wp,theta,temp)
END SUBROUTINE four1_sp
```

```
SUBROUTINE four1_dp(data,isign)
USE nrtype; USE nrutil, ONLY : arth,assert
USE nr, ONLY : fourrow
IMPLICIT NONE
COMPLEX(DPC), DIMENSION(:), INTENT(INOUT) :: data
INTEGER(I4B), INTENT(IN) :: isign
COMPLEX(DPC), DIMENSION(:,:), ALLOCATABLE :: dat,temp
COMPLEX(DPC), DIMENSION(:), ALLOCATABLE :: w,wp
REAL(DP), DIMENSION(:), ALLOCATABLE :: theta
INTEGER(I4B) :: n,m1,m2,j
n=size(data)
call assert(iand(n,n-1)==0, 'n must be a power of 2 in four1_dp')
m1=2**ceiling(0.5_sp*log(real(n,sp))/0.693147_sp)
m2=n/m1
allocate(dat(m1,m2),theta(m1),w(m1),wp(m1),temp(m2,m1))
dat=reshape(data,shape(dat))
call fourrow(dat,isign)
theta=arth(0,isign,m1)*TWOPI_D/n
wp=cmplx(-2.0_dp*sin(0.5_dp*theta)**2,sin(theta),kind=dpc)
w=cmplx(1.0_dp,0.0_dp,kind=dpc)
do j=2,m2
    w=w*wp+w
    dat(:,j)=dat(:,j)*w
end do
temp=transpose(dat)
call fourrow(temp,isign)
data=reshape(temp,shape(data))
deallocate(dat,w,wp,theta,temp)
END SUBROUTINE four1_dp
```

The above routines use `fourrow` exclusively, on the assumption that it is faster than its sibling `fourcol`. When that is the case (as we typically find), it is likely that `four1_sp` is also faster than Volume 1's scalar `four1`. The reason, on scalar machines, is that `fourrow`'s parallelism is taking better advantage of cache memory locality.

If `fourrow` is *not* faster than `fourcol` on your machine, then you should instead try the following alternative FFT version that uses `fourcol` only.

```
SUBROUTINE four1_alt(data,isign)
USE nrtype; USE nrutil, ONLY : arth,assert
USE nr, ONLY : fourcol
IMPLICIT NONE
COMPLEX(SPC), DIMENSION(:), INTENT(INOUT) :: data
INTEGER(I4B), INTENT(IN) :: isign
    Replaces a complex array data by its discrete Fourier transform, if isign is input as 1; or
    replaces data by its inverse discrete Fourier transform times the size of data, if isign is
```

```
   input as -1. The size of data must be an integer power of 2. Parallelism is achieved by
   internally reshaping the input array to two dimensions. (Use this version only if fourcol
   is faster than fourrow on your machine.)
COMPLEX(SPC), DIMENSION(:,:), ALLOCATABLE :: dat,temp
COMPLEX(DPC), DIMENSION(:), ALLOCATABLE :: w,wp
REAL(DP), DIMENSION(:), ALLOCATABLE :: theta
INTEGER(I4B) :: n,m1,m2,j
n=size(data)
call assert(iand(n,n-1)==0, 'n must be a power of 2 in four1_alt')
  Find dimensions as close to square as possible, allocate space, and reshape the input array.
m1=2**ceiling(0.5_sp*log(real(n,sp))/0.693147_sp)
m2=n/m1
allocate(dat(m1,m2),theta(m1),w(m1),wp(m1),temp(m2,m1))
dat=reshape(data,shape(dat))
temp=transpose(dat)                    Transpose and transform on (original) second in-
call fourcol(temp,isign)                   dex.
theta=arth(0,isign,m1)*TWOPI_D/n       Set up recurrence.
wp=cmplx(-2.0_dp*sin(0.5_dp*theta)**2,sin(theta),kind=dpc)
w=cmplx(1.0_dp,0.0_dp,kind=dpc)
do j=2,m2                              Multiply by the extra phase factor.
    w=w*wp+w
    temp(j,:)=temp(j,:)*w
end do
dat=transpose(temp)                    Transpose, and transform on (original) first in-
call fourcol(dat,isign)                    dex.
temp=transpose(dat)                    Transpose and then reshape the result back to
data=reshape(temp,shape(data))             one dimension.
deallocate(dat,w,wp,theta,temp)
END SUBROUTINE four1_alt
```

⋆ ⋆ ⋆

With all the machinery of fourrow and fourcol, two-dimensional FFTs are extremely straightforward. Again there is an alternative version provided in case your hardware favors fourcol (which would be, we think, unusual).

```
SUBROUTINE four2(data,isign)
USE nrtype
USE nr, ONLY : fourrow
IMPLICIT NONE
COMPLEX(SPC), DIMENSION(:,:), INTENT(INOUT) :: data
INTEGER(I4B), INTENT(IN) :: isign
   Replaces a 2-d complex array data by its discrete 2-d Fourier transform, if isign is input
   as 1; or replaces data by its inverse 2-d discrete Fourier transform times the product of its
   two sizes, if isign is input as -1. Both of data's sizes must be integer powers of 2 (this
   is checked for in fourrow). Parallelism is by use of fourrow.
COMPLEX(SPC), DIMENSION(size(data,2),size(data,1)) :: temp
call fourrow(data,isign)          Transform in second dimension.
temp=transpose(data)              Tranpose.
call fourrow(temp,isign)          Transform in (original) first dimension.
data=transpose(temp)              Transpose into data.
END SUBROUTINE four2
```

```
SUBROUTINE four2_alt(data,isign)
USE nrtype
USE nr, ONLY : fourcol
IMPLICIT NONE
COMPLEX(SPC), DIMENSION(:,:), INTENT(INOUT) :: data
INTEGER(I4B), INTENT(IN) :: isign
```

Replaces a 2-d complex array `data` by its discrete 2-d Fourier transform, if `isign` is input as 1; or replaces `data` by its inverse 2-d discrete Fourier transform times the product of its two sizes, if `isign` is input as -1. Both of `data`'s sizes must be integer powers of 2 (this is checked for in `fourcol`). Parallelism is by use of `fourcol`. (Use this version *only* if `fourcol` is faster than `fourrow` on your machine.)

```
COMPLEX(SPC), DIMENSION(size(data,2),size(data,1)) :: temp
temp=transpose(data)           Tranpose.
call fourcol(temp,isign)       Transform in (original) second dimension.
data=transpose(temp)           Transpose.
call fourcol(data,isign)       Transform in (original) first dimension.
END SUBROUTINE four2_alt
```

⋆ ⋆ ⋆

Most of the remaining routines in this chapter simply call one or another of the above FFT routines, with a small amount of auxiliary computation, so they are fairly straightforward conversions from their Volume 1 counterparts.

```
SUBROUTINE twofft(data1,data2,fft1,fft2)
USE nrtype; USE nrutil, ONLY : assert,assert_eq
USE nr, ONLY : four1
IMPLICIT NONE
REAL(SP), DIMENSION(:), INTENT(IN) :: data1,data2
COMPLEX(SPC), DIMENSION(:), INTENT(OUT) :: fft1,fft2
```

Given two real input arrays `data1` and `data2` of length N, this routine calls `four1` and returns two complex output arrays, `fft1` and `fft2`, each of complex length N, that contain the discrete Fourier transforms of the respective `data` arrays. N must be an integer power of 2.

```
INTEGER(I4B) :: n,n2
COMPLEX(SPC), PARAMETER :: C1=(0.5_sp,0.0_sp), C2=(0.0_sp,-0.5_sp)
COMPLEX, DIMENSION(size(data1)/2+1) :: h1,h2
n=assert_eq(size(data1),size(data2),size(fft1),size(fft2),'twofft')
call assert(iand(n,n-1)==0, 'n must be a power of 2 in twofft')
fft1=cmplx(data1,data2,kind=spc)      Pack the two real arrays into one complex array.
call four1(fft1,1)                    Transform the complex array.
fft2(1)=cmplx(aimag(fft1(1)),0.0_sp,kind=spc)
fft1(1)=cmplx(real(fft1(1)),0.0_sp,kind=spc)
n2=n/2+1
h1(2:n2)=C1*(fft1(2:n2)+conjg(fft1(n:n2:-1)))     Use symmetries to separate the
h2(2:n2)=C2*(fft1(2:n2)-conjg(fft1(n:n2:-1)))        two transforms.
fft1(2:n2)=h1(2:n2)                   Ship them out in two complex arrays.
fft1(n:n2:-1)=conjg(h1(2:n2))
fft2(2:n2)=h2(2:n2)
fft2(n:n2:-1)=conjg(h2(2:n2))
END SUBROUTINE twofft
```

⋆ ⋆ ⋆

```
SUBROUTINE realft_sp(data,isign,zdata)
USE nrtype; USE nrutil, ONLY : assert,assert_eq,zroots_unity
USE nr, ONLY : four1
IMPLICIT NONE
REAL(SP), DIMENSION(:), INTENT(INOUT) :: data
INTEGER(I4B), INTENT(IN) :: isign
COMPLEX(SPC), DIMENSION(:), OPTIONAL, TARGET :: zdata
```

When isign $= 1$, calculates the Fourier transform of a set of N real-valued data points, input in the array data. If the optional argument zdata is not present, the data are replaced by the positive frequency half of its complex Fourier transform. The real-valued first and last components of the complex transform are returned as elements data(1) and data(2), respectively. If the complex array zdata of length $N/2$ is present, data is unchanged and the transform is returned in zdata. N must be a power of 2. If isign $= -1$, this routine calculates the inverse transform of a complex data array if it is the transform of real data. (Result in this case must be multiplied by $2/N$.) The data can be supplied either in data, with zdata absent, or in zdata.

```
INTEGER(I4B) :: n,ndum,nh,nq
COMPLEX(SPC), DIMENSION(size(data)/4) :: w
COMPLEX(SPC), DIMENSION(size(data)/4-1) :: h1,h2
COMPLEX(SPC), DIMENSION(:), POINTER :: cdata      Used for internal complex computa-
COMPLEX(SPC) :: z                                     tions.
REAL(SP) :: c1=0.5_sp,c2
n=size(data)
call assert(iand(n,n-1)==0, 'n must be a power of 2 in realft_sp')
nh=n/2
nq=n/4
if (present(zdata)) then
    ndum=assert_eq(n/2,size(zdata),'realft_sp')
    cdata=>zdata                                  Use zdata as cdata.
    if (isign == 1) cdata=cmplx(data(1:n-1:2),data(2:n:2),kind=spc)
else
    allocate(cdata(n/2))                          Have to allocate storage ourselves.
    cdata=cmplx(data(1:n-1:2),data(2:n:2),kind=spc)
end if
if (isign == 1) then
    c2=-0.5_sp
    call four1(cdata,+1)                          The forward transform is here.
else                                              Otherwise set up for an inverse trans-
    c2=0.5_sp                                         form.
end if
w=zroots_unity(sign(n,isign),n/4)
w=cmplx(-aimag(w),real(w),kind=spc)
h1=c1*(cdata(2:nq)+conjg(cdata(nh:nq+2:-1)))     The two separate transforms are sep-
h2=c2*(cdata(2:nq)-conjg(cdata(nh:nq+2:-1)))         arated out of cdata.
```

Next they are recombined to form the true transform of the original real data:

```
cdata(2:nq)=h1+w(2:nq)*h2
cdata(nh:nq+2:-1)=conjg(h1-w(2:nq)*h2)
z=cdata(1)                                        Squeeze the first and last data to-
if (isign == 1) then                                  gether to get them all within the
    cdata(1)=cmplx(real(z)+aimag(z),real(z)-aimag(z),kind=spc)      original array.
else
    cdata(1)=cmplx(c1*(real(z)+aimag(z)),c1*(real(z)-aimag(z)),kind=spc)
    call four1(cdata,-1)                          This is the inverse transform for the
end if                                                case isign=-1.
if (present(zdata)) then                          Ship out answer in data if required.
    if (isign /= 1) then
        data(1:n-1:2)=real(cdata)
        data(2:n:2)=aimag(cdata)
    end if
else
    data(1:n-1:2)=real(cdata)
    data(2:n:2)=aimag(cdata)
    deallocate(cdata)
end if
```

```
END SUBROUTINE realft_sp
```

```
SUBROUTINE realft_dp(data,isign,zdata)
USE nrtype; USE nrutil, ONLY : assert,assert_eq,zroots_unity
USE nr, ONLY : four1
IMPLICIT NONE
REAL(DP), DIMENSION(:), INTENT(INOUT) :: data
INTEGER(I4B), INTENT(IN) :: isign
COMPLEX(DPC), DIMENSION(:), OPTIONAL, TARGET :: zdata
INTEGER(I4B) :: n,ndum,nh,nq
COMPLEX(DPC), DIMENSION(size(data)/4) :: w
COMPLEX(DPC), DIMENSION(size(data)/4-1) :: h1,h2
COMPLEX(DPC), DIMENSION(:), POINTER :: cdata
COMPLEX(DPC) :: z
REAL(DP) :: c1=0.5_dp,c2
n=size(data)
call assert(iand(n,n-1)==0, 'n must be a power of 2 in realft_dp')
nh=n/2
nq=n/4
if (present(zdata)) then
    ndum=assert_eq(n/2,size(zdata),'realft_dp')
    cdata=>zdata
    if (isign == 1) cdata=cmplx(data(1:n-1:2),data(2:n:2),kind=spc)
else
    allocate(cdata(n/2))
    cdata=cmplx(data(1:n-1:2),data(2:n:2),kind=spc)
end if
if (isign == 1) then
    c2=-0.5_dp
    call four1(cdata,+1)
else
    c2=0.5_dp
end if
w=zroots_unity(sign(n,isign),n/4)
w=cmplx(-aimag(w),real(w),kind=dpc)
h1=c1*(cdata(2:nq)+conjg(cdata(nh:nq+2:-1)))
h2=c2*(cdata(2:nq)-conjg(cdata(nh:nq+2:-1)))
cdata(2:nq)=h1+w(2:nq)*h2
cdata(nh:nq+2:-1)=conjg(h1-w(2:nq)*h2)
z=cdata(1)
if (isign == 1) then
    cdata(1)=cmplx(real(z)+aimag(z),real(z)-aimag(z),kind=dpc)
else
    cdata(1)=cmplx(c1*(real(z)+aimag(z)),c1*(real(z)-aimag(z)),kind=dpc)
    call four1(cdata,-1)
end if
if (present(zdata)) then
    if (isign /= 1) then
        data(1:n-1:2)=real(cdata)
        data(2:n:2)=aimag(cdata)
    end if
else
    data(1:n-1:2)=real(cdata)
    data(2:n:2)=aimag(cdata)
    deallocate(cdata)
end if
END SUBROUTINE realft_dp
```

⋆ ⋆ ⋆

```
SUBROUTINE sinft(y)
USE nrtype; USE nrutil, ONLY : assert,cumsum,zroots_unity
USE nr, ONLY : realft
IMPLICIT NONE
REAL(SP), DIMENSION(:), INTENT(INOUT) :: y
    Calculates the sine transform of a set of N real-valued data points stored in array y. The
    number N must be a power of 2. On exit y is replaced by its transform. This program,
    without changes, also calculates the inverse sine transform, but in this case the output array
    should be multiplied by 2/N.
REAL(SP), DIMENSION(size(y)/2+1) :: wi
REAL(SP), DIMENSION(size(y)/2) :: y1,y2
INTEGER(I4B) :: n,nh
n=size(y)
call assert(iand(n,n-1)==0, 'n must be a power of 2 in sinft')
nh=n/2
wi=aimag(zroots_unity(n+n,nh+1))            Calculate the sine for the auxiliary array.
y(1)=0.0
y1=wi(2:nh+1)*(y(2:nh+1)+y(n:nh+1:-1))
  Construct the two pieces of the auxiliary array.
y2=0.5_sp*(y(2:nh+1)-y(n:nh+1:-1))           Put them together to make the auxiliary ar-
y(2:nh+1)=y1+y2                                  ray.
y(n:nh+1:-1)=y1-y2
call realft(y,+1)                           Transform the auxiliary array.
y(1)=0.5_sp*y(1)                            Initialize the sum used for odd terms.
y(2)=0.0
y1=cumsum(y(1:n-1:2))                       Odd terms are determined by this running sum.
y(1:n-1:2)=y(2:n:2)                         Even terms in the transform are determined di-
y(2:n:2)=y1                                     rectly.
END SUBROUTINE sinft
```

```
SUBROUTINE cosft1(y)
USE nrtype; USE nrutil, ONLY : assert,cumsum,zroots_unity
USE nr, ONLY : realft
IMPLICIT NONE
REAL(SP), DIMENSION(:), INTENT(INOUT) :: y
    Calculates the cosine transform of a set of N+1 real-valued data points y. The transformed
    data replace the original data in array y. N must be a power of 2. This program, without
    changes, also calculates the inverse cosine transform, but in this case the output array
    should be multiplied by 2/N.
COMPLEX(SPC), DIMENSION((size(y)-1)/2) :: w
REAL(SP), DIMENSION((size(y)-1)/2-1) :: y1,y2
REAL(SP) :: summ
INTEGER(I4B) :: n,nh
n=size(y)-1
call assert(iand(n,n-1)==0, 'n must be a power of 2 in cosft1')
nh=n/2
w=zroots_unity(n+n,nh)
summ=0.5_sp*(y(1)-y(n+1))
y(1)=0.5_sp*(y(1)+y(n+1))
y1=0.5_sp*(y(2:nh)+y(n:nh+2:-1))            Construct the two pieces of the auxiliary array.
y2=y(2:nh)-y(n:nh+2:-1)
summ=summ+sum(real(w(2:nh))*y2)             Carry along this sum for later use in unfolding
y2=y2*aimag(w(2:nh))                            the transform.
y(2:nh)=y1-y2                               Calculate the auxiliary function.
y(n:nh+2:-1)=y1+y2
call realft(y(1:n),1)                       Calculate the transform of the auxiliary function.
y(n+1)=y(2)
y(2)=summ                                   summ is the value of F1 in equation (12.3.21).
y(2:n:2)=cumsum(y(2:n:2))                   Equation (12.3.20).
END SUBROUTINE cosft1
```

```
SUBROUTINE cosft2(y,isign)
USE nrtype; USE nrutil, ONLY : assert,cumsum,zroots_unity
USE nr, ONLY : realft
IMPLICIT NONE
REAL(SP), DIMENSION(:), INTENT(INOUT) :: y
INTEGER(I4B), INTENT(IN) :: isign
```

Calculates the "staggered" cosine transform of a set of N real-valued data points `y`. The transformed data replace the original data in array `y`. N must be a power of 2. Set `isign` to $+1$ for a transform, and to -1 for an inverse transform. For an inverse transform, the output array should be multiplied by $2/N$.

```
COMPLEX(SPC), DIMENSION(size(y)) :: w
REAL(SP), DIMENSION(size(y)/2) :: y1,y2
REAL(SP) :: ytemp
INTEGER(I4B) :: n,nh
n=size(y)
call assert(iand(n,n-1)==0, 'n must be a power of 2 in cosft2')
nh=n/2
w=zroots_unity(4*n,n)
if (isign == 1) then                         Forward transform.
    y1=0.5_sp*(y(1:nh)+y(n:nh+1:-1))         Calculate the auxiliary function.
    y2=aimag(w(2:n:2))*(y(1:nh)-y(n:nh+1:-1))
    y(1:nh)=y1+y2
    y(n:nh+1:-1)=y1-y2
    call realft(y,1)                         Calculate transform of the auxiliary function.
    y1(1:nh-1)=y(3:n-1:2)*real(w(3:n-1:2)) &     Even terms.
        -y(4:n:2)*aimag(w(3:n-1:2))
    y2(1:nh-1)=y(4:n:2)*real(w(3:n-1:2)) &
        +y(3:n-1:2)*aimag(w(3:n-1:2))
    y(3:n-1:2)=y1(1:nh-1)
    y(4:n:2)=y2(1:nh-1)
    ytemp=0.5_sp*y(2)                        Initialize recurrence for odd terms with 1/2 R_{N/2}.
    y(n-2:2:-2)=cumsum(y(n:4:-2),ytemp)          Recurrence for odd terms.
    y(n)=ytemp
else if (isign == -1) then                   Inverse transform.
    ytemp=y(n)
    y(4:n:2)=y(2:n-2:2)-y(4:n:2)             Form difference of odd terms.
    y(2)=2.0_sp*ytemp
    y1(1:nh-1)=y(3:n-1:2)*real(w(3:n-1:2)) &     Calculate R_k and I_k.
        +y(4:n:2)*aimag(w(3:n-1:2))
    y2(1:nh-1)=y(4:n:2)*real(w(3:n-1:2)) &
        -y(3:n-1:2)*aimag(w(3:n-1:2))
    y(3:n-1:2)=y1(1:nh-1)
    y(4:n:2)=y2(1:nh-1)
    call realft(y,-1)
    y1=y(1:nh)+y(n:nh+1:-1)                  Invert auxiliary array.
    y2=(0.5_sp/aimag(w(2:n:2)))*(y(1:nh)-y(n:nh+1:-1))
    y(1:nh)=0.5_sp*(y1+y2)
    y(n:nh+1:-1)=0.5_sp*(y1-y2)
end if
END SUBROUTINE cosft2
```

⋆ ⋆ ⋆

```
SUBROUTINE four3(data,isign)
USE nrtype
USE nr, ONLY : fourrow_3d
IMPLICIT NONE
COMPLEX(SPC), DIMENSION(:,:,:), INTENT(INOUT) :: data
INTEGER(I4B), INTENT(IN) :: isign
```

Replaces a 3-d complex array `data` by its discrete 3-d Fourier transform, if `isign` is input as 1; or replaces `data` by its inverse 3-d discrete Fourier transform times the product of its

```
    three sizes, if isign is input as -1. All three of data's sizes must be integer powers of 2
    (this is checked for in fourrow_3d). Parallelism is by use of fourrow_3d.
COMPLEX(SPC), DIMENSION(:,:,:), ALLOCATABLE :: dat2,dat3
call fourrow_3d(data,isign)                         Transform in third dimension.
allocate(dat2(size(data,2),size(data,3),size(data,1)))
dat2=reshape(data,shape=shape(dat2),order=(/3,1,2/))    Transpose.
call fourrow_3d(dat2,isign)                         Transform in (original) first dimension.
allocate(dat3(size(data,3),size(data,1),size(data,2)))
dat3=reshape(dat2,shape=shape(dat3),order=(/3,1,2/))    Transpose.
deallocate(dat2)
call fourrow_3d(dat3,isign)                         Transform in (original) second dimension.
data=reshape(dat3,shape=shape(data),order=(/3,1,2/))    Transpose back to output or-
deallocate(dat3)                                        der.
END SUBROUTINE four3
```

The `reshape` intrinsic, used with an `order=` parameter, is the multidimensional generalization of the two-dimensional `transpose` operation. The line

```
dat2=reshape(data,shape=shape(dat2),order=(/3,1,2/))
```

is equivalent to the do-loop

```
do j=1,size(data,1)
   dat2(:,:,j)=data(j,:,:)
end do
```

Incidentally, we have found some Fortran 90 compilers that (for scalar machines) are significantly *slower* executing the `reshape` than executing the equivalent do-loop. This, of course, shouldn't happen, since the `reshape` basically *is* an implicit do-loop. If you find such inefficient behavior on your compiler, you should report it as a bug to your compiler vendor! (Only thus will Fortran 90 compilers be brought to mature states of efficiency.)

```
SUBROUTINE four3_alt(data,isign)
USE nrtype
USE nr, ONLY : fourcol_3d
IMPLICIT NONE
COMPLEX(SPC), DIMENSION(:,:,:), INTENT(INOUT) :: data
INTEGER(I4B), INTENT(IN) :: isign
    Replaces a 3-d complex array data by its discrete 2-d Fourier transform, if isign is input
    as 1; or replaces data by its inverse 3-d discrete Fourier transform times the product of
    its three sizes, if isign is input as -1. All three of data's sizes must be integer powers
    of 2 (this is checked for in fourcol_3d). Parallelism is by use of fourcol_3d. (Use this
    version only if fourcol_3d is faster than fourrow_3d on your machine.)
COMPLEX(SPC), DIMENSION(:,:,:), ALLOCATABLE :: dat2,dat3
call fourcol_3d(data,isign)                         Transform in first dimension.
allocate(dat2(size(data,2),size(data,3),size(data,1)))
dat2=reshape(data,shape=shape(dat2),order=(/3,1,2/))    Transpose.
call fourcol_3d(dat2,isign)                         Transform in (original) second dimension.
allocate(dat3(size(data,3),size(data,1),size(data,2)))
dat3=reshape(dat2,shape=shape(dat3),order=(/3,1,2/))    Transpose.
deallocate(dat2)
call fourcol_3d(dat3,isign)                         Transform in (original) third dimension.
data=reshape(dat3,shape=shape(data),order=(/3,1,2/))    Transpose back to output or-
deallocate(dat3)                                        der.
END SUBROUTINE four3_alt
```

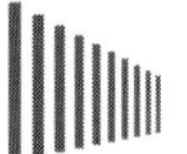

Note that `four3` uses `fourrow_3d`, the three-dimensional counterpart of `fourrow`, while `four3_alt` uses `fourcol_3d`, the three-dimensional counterpart of `fourcol`. You may want to time these programs to see which is faster on your machine.

⋆ ⋆ ⋆

In Volume 1, a single routine named `rlft3` was able to serve both as a three-dimensional real FFT, and as a two-dimensional real FFT. The trick is that the Fortran 77 version doesn't care whether the input array `data` is dimensioned as two- or three-dimensional. Fortran 90 is not so indifferent, and better programming practice is to have two separate versions of the algorithm:

```
SUBROUTINE rlft2(data,spec,speq,isign)
USE nrtype; USE nrutil, ONLY : assert,assert_eq
USE nr, ONLY : four2
REAL(SP), DIMENSION(:,:), INTENT(INOUT) :: data
COMPLEX(SPC), DIMENSION(:,:), INTENT(INOUT) :: spec
COMPLEX(SPC), DIMENSION(:), INTENT(INOUT) :: speq
INTEGER(I4B), INTENT(IN) :: isign
```

Given a two-dimensional real array `data(1:`M`,1:`N`)`, this routine returns (for `isign=1`) the complex fast Fourier transform as two complex arrays: On output, `spec(1:`$M/2$`,1:`N`)` contains the zero and positive frequency values of the first frequency component, while `speq(1:`N`)` contains the Nyquist critical frequency values of the first frequency component. The second frequency components are stored for zero, positive, and negative frequencies, in standard wrap-around order. For `isign=-1`, the inverse transform (times $M \times N/2$ as a constant multiplicative factor) is performed, with output `data` deriving from input `spec` and `speq`. For inverse transforms on data not generated first by a forward transform, make sure the complex input data array satisfies property (12.5.2). The size of all arrays must always be integer powers of 2.

```
INTEGER :: i1,j1,nn1,nn2
REAL(DP) :: theta
COMPLEX(SPC) :: c1=(0.5_sp,0.0_sp),c2,h1,h2,w
COMPLEX(SPC), DIMENSION(size(data,2)-1) :: h1a,h2a
COMPLEX(DPC) :: ww,wp
nn1=assert_eq(size(data,1),2*size(spec,1),'rlft2: nn1')
nn2=assert_eq(size(data,2),size(spec,2),size(speq),'rlft2: nn2')
call assert(iand((/nn1,nn2/),(/nn1,nn2/)-1)==0, &
    'dimensions must be powers of 2 in rlft2')
c2=cmplx(0.0_sp,-0.5_sp*isign,kind=spc)
theta=TWOPI_D/(isign*nn1)
wp=cmplx(-2.0_dp*sin(0.5_dp*theta)**2,sin(theta),kind=spc)
if (isign == 1) then                        Case of forward transform.
    spec(:,:)=cmplx(data(1:nn1:2,:),data(2:nn1:2,:),kind=spc)
    call four2(spec,isign)                  Here is where most all of the compute time
    speq=spec(1,:)                              is spent.
end if
h1=c1*(spec(1,1)+conjg(speq(1)))
h1a=c1*(spec(1,2:nn2)+conjg(speq(nn2:2:-1)))
h2=c2*(spec(1,1)-conjg(speq(1)))
h2a=c2*(spec(1,2:nn2)-conjg(speq(nn2:2:-1)))
spec(1,1)=h1+h2
spec(1,2:nn2)=h1a+h2a
speq(1)=conjg(h1-h2)
speq(nn2:2:-1)=conjg(h1a-h2a)
ww=cmplx(1.0_dp,0.0_dp,kind=dpc)            Initialize trigonometric recurrence.
do i1=2,nn1/4+1
    j1=nn1/2-i1+2                           Corresponding negative frequency.
    ww=ww*wp+ww                             Do the trig recurrence.
    w=ww
    h1=c1*(spec(i1,1)+conjg(spec(j1,1)))        Equation (12.3.5).
    h1a=c1*(spec(i1,2:nn2)+conjg(spec(j1,nn2:2:-1)))
```

```
    h2=c2*(spec(i1,1)-conjg(spec(j1,1)))
    h2a=c2*(spec(i1,2:nn2)-conjg(spec(j1,nn2:2:-1)))
    spec(i1,1)=h1+w*h2
    spec(i1,2:nn2)=h1a+w*h2a
    spec(j1,1)=conjg(h1-w*h2)
    spec(j1,nn2:2:-1)=conjg(h1a-w*h2a)
end do
if (isign == -1) then                       Case of reverse transform.
    call four2(spec,isign)
    data(1:nn1:2,:)=real(spec)
    data(2:nn1:2,:)=aimag(spec)
end if
END SUBROUTINE rlft2
```

f90 `call assert(iand((/nn1,nn2/),(/nn1,nn2/)-1)==0 ...` Here an overloaded version of `assert` that takes vector arguments is used to check that each dimension is a power of 2. Note that `iand` acts element-by-element on an array.

```
SUBROUTINE rlft3(data,spec,speq,isign)
USE nrtype; USE nrutil, ONLY : assert,assert_eq
USE nr, ONLY : four3
REAL(SP), DIMENSION(:,:,:), INTENT(INOUT) :: data
COMPLEX(SPC), DIMENSION(:,:,:), INTENT(INOUT) :: spec
COMPLEX(SPC), DIMENSION(:,:), INTENT(INOUT) :: speq
INTEGER(I4B), INTENT(IN) :: isign
```

Given a three-dimensional real array `data(1:`L`,1:`M`,1:`N`)`, this routine returns (for `isign=1`) the complex Fourier transform as two complex arrays: On output, the zero and positive frequency values of the first frequency component are in `spec(1:`$L/2$`,1:`M`,1:`N`)`, while `speq(1:`M`,1:`N`)` contains the Nyquist critical frequency values of the first frequency component. The second and third frequency components are stored for zero, positive, and negative frequencies, in standard wrap-around order. For `isign=-1`, the inverse transform (times $L \times M \times N/2$ as a constant multiplicative factor) is performed, with output `data` deriving from input `spec` and `speq`. For inverse transforms on data not generated first by a forward transform, make sure the complex input data array satisfies property (12.5.2). The size of all arrays must always be integer powers of 2.

```
INTEGER :: i1,i3,j1,j3,nn1,nn2,nn3
REAL(DP) :: theta
COMPLEX(SPC) :: c1=(0.5_sp,0.0_sp),c2,h1,h2,w
COMPLEX(SPC), DIMENSION(size(data,2)-1) :: h1a,h2a
COMPLEX(DPC) :: ww,wp
c2=cmplx(0.0_sp,-0.5_sp*isign,kind=spc)
nn1=assert_eq(size(data,1),2*size(spec,1),'rlft2: nn1')
nn2=assert_eq(size(data,2),size(spec,2),size(speq,1),'rlft2: nn2')
nn3=assert_eq(size(data,3),size(spec,3),size(speq,2),'rlft2: nn3')
call assert(iand((/nn1,nn2,nn3/),(/nn1,nn2,nn3/)-1)==0, &
    'dimensions must be powers of 2 in rlft3')
theta=TWOPI_D/(isign*nn1)
wp=cmplx(-2.0_dp*sin(0.5_dp*theta)**2,sin(theta),kind=dpc)
if (isign == 1) then                        Case of forward transform.
    spec(:,:,:)=cmplx(data(1:nn1:2,:,:),data(2:nn1:2,:,:),kind=spc)
    call four3(spec,isign)                  Here is where most all of the compute time
    speq=spec(1,:,:)                            is spent.
end if
do i3=1,nn3
    j3=1
    if (i3 /= 1) j3=nn3-i3+2
    h1=c1*(spec(1,1,i3)+conjg(speq(1,j3)))
    h1a=c1*(spec(1,2:nn2,i3)+conjg(speq(nn2:2:-1,j3)))
    h2=c2*(spec(1,1,i3)-conjg(speq(1,j3)))
    h2a=c2*(spec(1,2:nn2,i3)-conjg(speq(nn2:2:-1,j3)))
    spec(1,1,i3)=h1+h2
    spec(1,2:nn2,i3)=h1a+h2a
```

```
    speq(1,j3)=conjg(h1-h2)
    speq(nn2:2:-1,j3)=conjg(h1a-h2a)
    ww=cmplx(1.0_dp,0.0_dp,kind=dpc)          Initialize trigonometric recurrence.
    do i1=2,nn1/4+1
        j1=nn1/2-i1+2                         Corresponding negative frequency.
        ww=ww*wp+ww                           Do the trig recurrence.
        w=ww
        h1=c1*(spec(i1,1,i3)+conjg(spec(j1,1,j3)))       Equation (12.3.5).
        h1a=c1*(spec(i1,2:nn2,i3)+conjg(spec(j1,nn2:2:-1,j3)))
        h2=c2*(spec(i1,1,i3)-conjg(spec(j1,1,j3)))
        h2a=c2*(spec(i1,2:nn2,i3)-conjg(spec(j1,nn2:2:-1,j3)))
        spec(i1,1,i3)=h1+w*h2
        spec(i1,2:nn2,i3)=h1a+w*h2a
        spec(j1,1,j3)=conjg(h1-w*h2)
        spec(j1,nn2:2:-1,j3)=conjg(h1a-w*h2a)
    end do
end do
if (isign == -1) then                         Case of reverse transform.
    call four3(spec,isign)
    data(1:nn1:2,:,:)=real(spec)
    data(2:nn1:2,:,:)=aimag(spec)
end if
END SUBROUTINE rlft3
```

⋆ ⋆ ⋆

Referring back to the discussion of parallelism, §22.4, that led to four1's implementation with $\sqrt{N}$ parallelism, you might wonder whether Fortran 90 provides sufficiently powerful high-level constructs to enable an FFT routine with N-fold parallelism. The answer is, "*It does*, but you wouldn't want to use them!" Access to arbitrary interprocessor communication in Fortran 90 is through the mechanism of the "vector subscript" (one-dimensional array of indices in arbitrary order). When a vector subscript is on the right-hand side of an assignment statement, the operation performed is effectively a "gather"; when it is on the left-hand side, the operation is effectively a "scatter."

It is quite possible to write the classic FFT algorithm in terms of gather and scatter operations. In fact, we do so now. The problem is efficiency: The computations involved in constructing the vector subscripts for the scatter/gather operations, and the actual scatter/gather operations themselves, tend to swamp the underlying very lean FFT algorithm. The result is very slow, though theoretically perfectly parallelizable, code. Since small-scale parallel (SSP) machines can saturate their processors with $\sqrt{N}$ parallelism, while massively multiprocessor (MMP) machines inevitably come with architecture-optimized FFT library calls, there is really no niche for these routines, except as pedagogical demonstrations. We give here a one-dimensional routine, and also an arbitrary-dimensional routine modeled on Volume 1's fourn. Note the complete absence of do-loops of size N; the loops that remain are over $\log N$ stages, or over the number of dimensions.

```
SUBROUTINE four1_gather(data,isign)
USE nrtype; USE nrutil, ONLY : arth,assert
IMPLICIT NONE
COMPLEX(SPC), DIMENSION(:), INTENT(INOUT) :: data
INTEGER(I4B), INTENT(IN) :: isign
    Replaces a complex array data by its discrete Fourier transform, if isign is input as 1;
    or replaces data by size(data) times its inverse discrete Fourier transform, if isign is
    input as -1. The size of data must be an integer power of 2. This routine demonstrates
    coding the FFT algorithm in high-level Fortran 90 constructs. Generally the result is very
```

much slower than library routines coded for specific architectures, and also *significantly slower* than the parallelization-by-rows method used in the routine four1.

```
INTEGER(I4B) :: n,n2,m,mm
REAL(DP) :: theta
COMPLEX(SPC) :: wp
INTEGER(I4B), DIMENSION(size(data)) :: jarr
INTEGER(I4B), DIMENSION(:), ALLOCATABLE :: jrev
COMPLEX(SPC), DIMENSION(:), ALLOCATABLE :: wtab,dtemp
n=size(data)
call assert(iand(n,n-1)==0, 'n must be a power of 2 in four1_gather')
if (n <= 1) RETURN
allocate(jrev(n))                         Begin bit-reversal section of the routine.
jarr=arth(0,1,n)
jrev=0
n2=n/2
m=n2
do                                        Construct an array of pointers from an index
    where (iand(jarr,1) /= 0) jrev=jrev+m     to its bit-reverse.
    jarr=jarr/2
    m=m/2
    if (m == 0) exit
end do
data=data(jrev+1)                         Move all data to bit-reversed location by a
deallocate(jrev)                              single gather/scatter.
allocate(dtemp(n),wtab(n2))               Begin Danielson-Lanczos section of the rou-
jarr=arth(0,1,n)                              tine.
m=1
mm=n2
wtab(1)=(1.0_sp,0.0_sp)                   Seed the roots-of-unity table.
do                                        Outer loop executed log2 N times.
    where (iand(jarr,m) /= 0)
          The basic idea is to address the correct root-of-unity for each Danielson-Lanczos
          multiplication by tricky bit manipulations.
        dtemp=data*wtab(mm*iand(jarr,m-1)+1)
        data=eoshift(data,-m)-dtemp       This is half of Danielson-Lanczos.
    elsewhere
        data=data+eoshift(dtemp,m)        This is the other half. The referenced ele-
    end where                                 ments of dtemp will have been set in the
    m=m*2                                     where clause.
    if (m >= n) exit
    mm=mm/2
    theta=PI_D/(isign*m)                  Ready for trigonometry?
    wp=cmplx(-2.0_dp*sin(0.5_dp*theta)**2, sin(theta),kind=spc)
      Add entries to the table for the next iteration.
    wtab(mm+1:n2:2*mm)=wtab(1:n2-mm:2*mm)*wp+wtab(1:n2-mm:2*mm)
end do
deallocate(dtemp,wtab)
END SUBROUTINE four1_gather
```

```
SUBROUTINE fourn_gather(data,nn,isign)
USE nrtype; USE nrutil, ONLY : arth,assert
IMPLICIT NONE
COMPLEX(SPC), DIMENSION(:), INTENT(INOUT) :: data
INTEGER(I4B), DIMENSION(:) :: nn
INTEGER(I4B), INTENT(IN) :: isign
```

For data a one-dimensional complex array containing the values (in Fortran normal ordering) of an M-dimensional complex arrray, this routine replaces data by its M-dimensional discrete Fourier transform, if isign is input as 1. nn(1:M) is an integer array containing the lengths of each dimension (number of complex values), each of which must be a power of 2. If isign is input as -1, data is replaced by its inverse transform times the product of the lengths of all dimensions. This routine demonstrates coding the multidimensional FFT algorithm in high-level Fortran 90 constructs. Generally the result is *very much slower* than library routines coded for specific architectures, and *significantly slower* than routines `four2` and `four3` for the two- and three-dimensional cases.

```
INTEGER(I4B), DIMENSION(:), ALLOCATABLE :: jarr
INTEGER(I4B) :: ndim,idim,ntot,nprev,n,n2,msk0,msk1,msk2,m,mm,mn
REAL(DP) :: theta
COMPLEX(SPC) :: wp
COMPLEX(SPC), DIMENSION(:), ALLOCATABLE :: wtab,dtemp
call assert(iand(nn,nn-1)==0, &
    'each dimension must be a power of 2 in fourn_gather')
ndim=size(nn)
ntot=product(nn)
nprev=1
allocate(jarr(ntot))
do idim=1,ndim                          Loop over the dimensions.
    jarr=arth(0,1,ntot)                 We begin the bit-reversal section of the
    n=nn(idim)                              routine.
    n2=n/2
    msk0=nprev
    msk1=nprev*n2
    msk2=msk0+msk1
    do                                  Construct an array of pointers from an
        if (msk1 <= msk0) exit              index to its bit-reverse.
        where (iand(jarr,msk0) == 0 .neqv. iand(jarr,msk1) == 0) &
            jarr=ieor(jarr,msk2)
        msk0=msk0*2
        msk1=msk1/2
        msk2=msk0+msk1
    end do
    data=data(jarr+1)                   Move all data to bit-reversed location by
    allocate(dtemp(ntot),wtab(n2))          a single gather/scatter.
      We begin the Danielson-Lanczos section of the routine.
    jarr=iand(n-1,arth(0,1,ntot)/nprev)
    m=1
    mm=n2
    mn=m*nprev
    wtab(1)=(1.0_sp,0.0_sp)             Seed the roots-of-unity table.
    do                                  This loop executed log2 N times.
        if (mm == 0) exit
        where (iand(jarr,m) /= 0)
              The basic idea is to address the correct root-of-unity for each Danielson-Lanczos
              multiplication by tricky bit manipulations.
            dtemp=data*wtab(mm*iand(jarr,m-1)+1)
            data=eoshift(data,-mn)-dtemp    This is half of Danielson-Lanczos.
        elsewhere
            data=data+eoshift(dtemp,mn)     This is the other half. The referenced el-
        end where                               ements of dtemp will have been set
        m=m*2                                   in the where clause.
        if (m >= n) exit
        mn=m*nprev
        mm=mm/2
        theta=PI_D/(isign*m)            Ready for trigonometry?
        wp=cmplx(-2.0_dp*sin(0.5_dp*theta)**2,sin(theta),kind=dpc)
          Add entries to the table for the next iteration.
        wtab(mm+1:n2:2*mm)=wtab(1:n2-mm:2*mm)*wp &
            +wtab(1:n2-mm:2*mm)
    end do
    deallocate(dtemp,wtab)
    nprev=n*nprev
end do
deallocate(jarr)
END SUBROUTINE fourn_gather
```

`call assert(iand(nn,nn-1)==0 ...` Once again the vector version of `assert` is used to test all the dimensions stored in `nn` simultaneously.

Chapter B13. Fourier and Spectral Applications

```
FUNCTION convlv(data,respns,isign)
USE nrtype; USE nrutil, ONLY : assert,nrerror
USE nr, ONLY : realft
IMPLICIT NONE
REAL(SP), DIMENSION(:), INTENT(INOUT) :: data
REAL(SP), DIMENSION(:), INTENT(IN) :: respns
INTEGER(I4B), INTENT(IN) :: isign
REAL(SP), DIMENSION(size(data)) :: convlv
```

Convolves or deconvolves a real data set `data` (of length N, including any user-supplied zero padding) with a response function `respns`, stored in wrap-around order in a real array of length $M \le N$. (M should be an odd integer, N a power of 2.) Wrap-around order means that the first half of the array `respns` contains the impulse response function at positive times, while the second half of the array contains the impulse response function at negative times, counting down from the highest element `respns(`M`)`. On input `isign` is $+1$ for convolution, -1 for deconvolution. The answer is returned as the function `convlv`, an array of length N. `data` has `INTENT(INOUT)` for consistency with `realft`, but is actually unchanged.

```
INTEGER(I4B) :: no2,n,m
COMPLEX(SPC), DIMENSION(size(data)/2) :: tmpd,tmpr
n=size(data)
m=size(respns)
call assert(iand(n,n-1)==0, 'n must be a power of 2 in convlv')
call assert(mod(m,2)==1, 'm must be odd in convlv')
convlv(1:m)=respns(:)                    Put respns in array of length n.
convlv(n-(m-3)/2:n)=convlv((m+3)/2:m)
convlv((m+3)/2:n-(m-1)/2)=0.0            Pad with zeros.
no2=n/2
call realft(data,1,tmpd)                 FFT both arrays.
call realft(convlv,1,tmpr)
if (isign == 1) then                     Multiply FFTs to convolve.
    tmpr(1)=cmplx(real(tmpd(1))*real(tmpr(1))/no2, &
        aimag(tmpd(1))*aimag(tmpr(1))/no2, kind=spc)
    tmpr(2:)=tmpd(2:)*tmpr(2:)/no2
else if (isign == -1) then               Divide FFTs to deconvolve.
    if (any(abs(tmpr(2:)) == 0.0) .or. real(tmpr(1)) == 0.0 &
        .or. aimag(tmpr(1)) == 0.0) call nrerror &
        ('deconvolving at response zero in convlv')
    tmpr(1)=cmplx(real(tmpd(1))/real(tmpr(1))/no2, &
        aimag(tmpd(1))/aimag(tmpr(1))/no2, kind=spc)
    tmpr(2:)=tmpd(2:)/tmpr(2:)/no2
else
    call nrerror('no meaning for isign in convlv')
end if
call realft(convlv,-1,tmpr)              Inverse transform back to time domain.
END FUNCTION convlv
```

`tmpr(1)=cmplx(...kind=spc)` The intrinsic function `cmplx` returns a quantity of type default complex unless the `kind` argument is present. It is therefore a good idea always to include this argument. The intrinsic functions `real` and `aimag`, on the other hand, when called with a complex argument, return the same kind as their argument. So it is a good idea *not* to put in a `kind` argment for these. (In fact, `aimag` doesn't allow one.) Don't confuse these situations, regarding complex variables, with the completely unrelated use of `real` to convert a real or integer variable to a real value of specified kind. In this latter case, `kind` should be specified.

⋆ ⋆ ⋆

```
FUNCTION correl(data1,data2)
USE nrtype; USE nrutil, ONLY : assert,assert_eq
USE nr, ONLY : realft
IMPLICIT NONE
REAL(SP), DIMENSION(:), INTENT(INOUT) :: data1,data2
REAL(SP), DIMENSION(size(data1)) :: correl
```

Computes the correlation of two real data sets `data1` and `data2` of length N (including any user-supplied zero padding). N must be an integer power of 2. The answer is returned as the function `correl`, an array of length N. The answer is stored in wrap-around order, i.e., correlations at increasingly negative lags are in `correl(`N`)` on down to `correl(`$N/2+1$`)`, while correlations at increasingly positive lags are in `correl(1)` (zero lag) on up to `correl(`$N/2$`)`. Sign convention of this routine: if `data1` lags `data2`, i.e., is shifted to the right of it, then `correl` will show a peak at positive lags.

```
COMPLEX(SPC), DIMENSION(size(data1)/2) :: cdat1,cdat2
INTEGER(I4B) :: no2,n                          Normalization for inverse FFT.
n=assert_eq(size(data1),size(data2),'correl')
call assert(iand(n,n-1)==0, 'n must be a power of 2 in correl')
no2=n/2
call realft(data1,1,cdat1)                     Transform both data vectors.
call realft(data2,1,cdat2)
cdat1(1)=cmplx(real(cdat1(1))*real(cdat2(1))/no2, &     Multiply to find FFT of their
    aimag(cdat1(1))*aimag(cdat2(1))/no2, kind=spc)          correlation.
cdat1(2:)=cdat1(2:)*conjg(cdat2(2:))/no2
call realft(correl,-1,cdat1)                   Inverse transform gives correlation.
END FUNCTION correl
```

`cdat1(1)=cmplx(...kind=spc)` See just above for why we use the explicit kind type parameter `spc` for `cmplx`, but omit `sp` for `real`.

⋆ ⋆ ⋆

```
SUBROUTINE spctrm(p,k,ovrlap,unit,n_window)
USE nrtype; USE nrutil, ONLY : arth,nrerror
USE nr, ONLY : four1
IMPLICIT NONE
REAL(SP), DIMENSION(:), INTENT(OUT) :: p
INTEGER(I4B), INTENT(IN) :: k
LOGICAL(LGT), INTENT(IN) :: ovrlap             True for overlapping segments, false other-
INTEGER(I4B), OPTIONAL, INTENT(IN) :: n_window,unit     wise.
```

Reads data from input unit 9, or if the optional argument `unit` is present, from that input unit. The output is an array `p` of length M that contains the data's power (mean square amplitude) at frequency $(j-1)/2M$ cycles per grid point, for $j=1,2,\ldots,M$, based on `(2*k+1)*`M data points (if `ovrlap` is set `.true.`) or `4*k*`M data points (if `ovrlap` is set `.false.`). The number of segments of the data is `2*k` in both cases: The routine calls `four1` `k` times, each call with 2 partitions each of $2M$ real data points. If the optional argument `n_window` is present, the routine uses the Bartlett window, the square window, or the Welch window for n_window $= 1, 2, 3$ respectively. If n_window is not present, the Bartlett window is used.

```
INTEGER(I4B) :: j,joff,joffn,kk,m,m4,m43,m44,mm,iunit,nn_window
REAL(SP) :: den,facm,facp,sumw
REAL(SP), DIMENSION(2*size(p)) :: w
REAL(SP), DIMENSION(4*size(p)) :: w1
REAL(SP), DIMENSION(size(p)) :: w2
COMPLEX(SPC), DIMENSION(2*size(p)) :: cw1
m=size(p)
if (present(n_window)) then
    nn_window=n_window
else
    nn_window=1
end if
if (present(unit)) then
    iunit=unit
else
    iunit=9
end if
mm=m+m                                          Useful factors.
m4=mm+mm
m44=m4+4
m43=m4+3
den=0.0
facm=m                                          Factors used by the window function.
facp=1.0_sp/m
w1(1:mm)=window(arth(1,1,mm),facm,facp,nn_window)
sumw=dot_product(w1(1:mm),w1(1:mm))         Accumulate the squared sum of the weights.
p(:)=0.0                                        Initialize the spectrum to zero.
if (ovrlap) read (iunit,*) (w2(j),j=1,m)        Initialize the "save" half-buffer.
do kk=1,k                                       Loop over data segments in groups of two.
    do joff=-1,0,1                              Get two complete segments into workspace.
        if (ovrlap) then
            w1(joff+2:joff+mm:2)=w2(1:m)
            read (iunit,*) (w2(j),j=1,m)
            joffn=joff+mm
            w1(joffn+2:joffn+mm:2)=w2(1:m)
        else
            read (iunit,*) (w1(j),j=joff+2,m4,2)
        end if
    end do
    w=window(arth(1,1,mm),facm,facp,nn_window)          Apply the window to the data.
    w1(2:m4:2)=w1(2:m4:2)*w
    w1(1:m4:2)=w1(1:m4:2)*w
    cw1(1:mm)=cmplx(w1(1:m4:2),w1(2:m4:2),kind=spc)
    call four1(cw1(1:mm),1)                     Fourier transform the windowed data.
    w1(1:m4:2)=real(cw1(1:mm))
    w1(2:m4:2)=aimag(cw1(1:mm))
    p(1)=p(1)+w1(1)**2+w1(2)**2                 Sum results into previous segments.
    p(2:m)=p(2:m)+w1(4:2*m:2)**2+w1(3:2*m-1:2)**2+&
        w1(m44-4:m44-2*m:-2)**2+w1(m43-4:m43-2*m:-2)**2
    den=den+sumw
end do
p(:)=p(:)/(m4*den)                              Normalize the output.
CONTAINS

FUNCTION window(j,facm,facp,nn_window)
IMPLICIT NONE
INTEGER(I4B), DIMENSION(:), INTENT(IN) :: j
INTEGER(I4B), INTENT(IN) :: nn_window
REAL(SP), INTENT(IN) :: facm,facp
REAL(SP), DIMENSION(size(j)) :: window
select case(nn_window)
    case(1)
```

```
        window(j)=(1.0_sp-abs(((j-1)-facm)*facp))          Bartlett window.
    case(2)
        window(j)=1.0                                        Square window.
    case(3)
        window(j)=(1.0_sp-(((j-1)-facm)*facp)**2)           Welch window.
    case default
        call nrerror('unimplemented window function in spctrm')
end select
END FUNCTION window
END SUBROUTINE spctrm
```

f90 The Fortran 90 optional argument feature allows us to make unit 9 the default output unit in this routine, but leave the user the option of specifying a different output unit by supplying an actual argument for `unit`. We also use an optional argument to allow the user the option of overriding the default selection of the Bartlett window function.

`FUNCTION window(j,facm,facp,nn_window)` In Fortran 77 we coded this as a statement function. Here the internal function is equivalent, but allows full specification of the interface and so is preferred.

* * *

```
SUBROUTINE memcof(data,xms,d)
USE nrtype; USE nrutil, ONLY : nrerror
IMPLICIT NONE
REAL(SP), INTENT(OUT) :: xms
REAL(SP), DIMENSION(:), INTENT(IN) :: data
REAL(SP), DIMENSION(:), INTENT(OUT) :: d
   Given a real vector data of length N, this routine returns M linear prediction coefficients
   in a vector d of length M, and returns the mean square discrepancy as xms.
INTEGER(I4B) :: k,m,n
REAL(SP) :: denom,pneum
REAL(SP), DIMENSION(size(data)) :: wk1,wk2,wktmp
REAL(SP), DIMENSION(size(d)) :: wkm
m=size(d)
n=size(data)
xms=dot_product(data,data)/n
wk1(1:n-1)=data(1:n-1)
wk2(1:n-1)=data(2:n)
do k=1,m
    pneum=dot_product(wk1(1:n-k),wk2(1:n-k))
    denom=dot_product(wk1(1:n-k),wk1(1:n-k))+ &
        dot_product(wk2(1:n-k),wk2(1:n-k))
    d(k)=2.0_sp*pneum/denom
    xms=xms*(1.0_sp-d(k)**2)
    d(1:k-1)=wkm(1:k-1)-d(k)*wkm(k-1:1:-1)
      The algorithm is recursive, although it is implemented as an iteration. It builds up the
      answer for larger and larger values of m until the desired value is reached. At this point
      in the algorithm, one could return the vector d and scalar xms for a set of LP coefficients
      with k (rather than m) terms.
    if (k == m) RETURN
    wkm(1:k)=d(1:k)
    wktmp(2:n-k)=wk1(2:n-k)
    wk1(1:n-k-1)=wk1(1:n-k-1)-wkm(k)*wk2(1:n-k-1)
    wk2(1:n-k-1)=wk2(2:n-k)-wkm(k)*wktmp(2:n-k)
end do
call nrerror('never get here in memcof')
END SUBROUTINE memcof
```

* * *

```
SUBROUTINE fixrts(d)
USE nrtype
USE nr, ONLY : zroots
IMPLICIT NONE
REAL(SP), DIMENSION(:), INTENT(INOUT) :: d
```

Given the LP coefficients d, this routine finds all roots of the characteristic polynomial (13.6.14), reflects any roots that are outside the unit circle back inside, and then returns a modified set of coefficients in d.

```
INTEGER(I4B) :: i,m
LOGICAL(LGT) :: polish
COMPLEX(SPC), DIMENSION(size(d)+1) :: a
COMPLEX(SPC), DIMENSION(size(d)) :: roots
m=size(d)
a(m+1)=cmplx(1.0_sp,kind=spc)              Set up complex coefficients for polynomial
a(m:1:-1)=cmplx(-d(1:m),kind=spc)              root finder.
polish=.true.
call zroots(a(1:m+1),roots,polish)         Find all the roots.
where (abs(roots) > 1.0) roots=1.0_sp/conjg(roots)
  Reflect all roots outside the unit circle back inside.
a(1)=-roots(1)                             Now reconstruct the polynomial coefficients,
a(2:m+1)=cmplx(1.0_sp,kind=spc)
do i=2,m                                   by looping over the roots
    a(2:i)=a(1:i-1)-roots(i)*a(2:i)        and synthetically multiplying.
    a(1)=-roots(i)*a(1)
end do
d(m:1:-1)=-real(a(1:m))                    The polynomial coefficients are guaranteed
END SUBROUTINE fixrts                          to be real, so we need only return the
                                               real part as new LP coefficients.
```

f90 `a(m+1)=cmplx(1.0_sp,kind=spc)` See after `convlv` on p. 1254 to review why we use the explicit kind type parameter `spc` for `cmplx`.

⋆ ⋆ ⋆

```
FUNCTION predic(data,d,nfut)
USE nrtype
IMPLICIT NONE
REAL(SP), DIMENSION(:), INTENT(IN) :: data,d
INTEGER(I4B), INTENT(IN) :: nfut
REAL(SP), DIMENSION(nfut) :: predic
```

Given an array `data`, and given the data's LP coefficients d in an array of length M, this routine applies equation (13.6.11) to predict the next `nfut` data points, which it returns in an array as the function value `predic`. Note that the routine references only the last M values of `data`, as initial values for the prediction.

```
INTEGER(I4B) :: j,ndata,m
REAL(SP) :: discrp,sm
REAL(SP), DIMENSION(size(d)) :: reg
m=size(d)
ndata=size(data)
reg(1:m)=data(ndata:ndata+1-m:-1)
do j=1,nfut
    discrp=0.0
      This is where you would put in a known discrepancy if you were reconstructing a function
      by linear predictive coding rather than extrapolating a function by linear prediction. See
      text.
    sm=discrp+dot_product(d,reg)
    reg=eoshift(reg,-1,sm)            [If you want to implement circular arrays, you can
    predic(j)=sm                          avoid this shifting of coefficients!]
end do
END FUNCTION predic
```

⋆ ⋆ ⋆

```
FUNCTION evlmem(fdt,d,xms)
USE nrtype; USE nrutil, ONLY : poly
IMPLICIT NONE
REAL(SP), INTENT(IN) :: fdt,xms
REAL(SP), DIMENSION(:), INTENT(IN) :: d
REAL(SP) :: evlmem
```

Given d and xms as returned by memcof, this function returns the power spectrum estimate $P(f)$ as a function of fdt $= f\Delta$.

```
COMPLEX(SPC) :: z,zz
REAL(DP) :: theta                     Trigonometric recurrences in double precision.
theta=TWOPI_D*fdt
z=cmplx(cos(theta),sin(theta),kind=spc)
zz=1.0_sp-z*poly(z,d)
evlmem=xms/abs(zz)**2                 Equation (13.7.4).
END FUNCTION evlmem
```

f90 zz=...poly(z,d) The poly function in nrutil returns the value of the polynomial with coefficients d(:) at z. Here a version that takes real coefficients and a complex argument is actually invoked, but all the different versions have been overloaded onto the same name poly.

⋆ ⋆ ⋆

```
SUBROUTINE period(x,y,ofac,hifac,px,py,jmax,prob)
USE nrtype; USE nrutil, ONLY : assert_eq,imaxloc
USE nr, ONLY : avevar
IMPLICIT NONE
INTEGER(I4B), INTENT(OUT) :: jmax
REAL(SP), INTENT(IN) :: ofac,hifac
REAL(SP), INTENT(OUT) :: prob
REAL(SP), DIMENSION(:), INTENT(IN) :: x,y
REAL(SP), DIMENSION(:), POINTER :: px,py
```

Input is a set of N data points with abscissas x (which need not be equally spaced) and ordinates y, and a desired oversampling factor ofac (a typical value being 4 or larger). The routine returns pointers to internally allocated arrays px and py. px is filled with an increasing sequence of frequencies (not angular frequencies) up to hifac times the "average" Nyquist frequency, and py is filled with the values of the Lomb normalized periodogram at those frequencies. The length of these arrays is 0.5*ofac*hifac*N. The arrays x and y are not altered. The routine also returns jmax such that py(jmax) is the maximum element in py, and prob, an estimate of the significance of that maximum against the hypothesis of random noise. A small value of prob indicates that a significant periodic signal is present.

```
INTEGER(I4B) :: i,n,nout
REAL(SP) :: ave,cwtau,effm,expy,pnow,sumc,sumcy,&
    sums,sumsh,sumsy,swtau,var,wtau,xave,xdif,xmax,xmin
REAL(DP), DIMENSION(size(x)) :: tmp1,tmp2,wi,wpi,wpr,wr
LOGICAL(LGT), SAVE :: init=.true.
n=assert_eq(size(x),size(y),'period')
if (init) then
    init=.false.
    nullify(px,py)
else
    if (associated(px)) deallocate(px)
    if (associated(py)) deallocate(py)
end if
nout=0.5_sp*ofac*hifac*n
allocate(px(nout),py(nout))
call avevar(y(:),ave,var)             Get mean and variance of the input data.
xmax=maxval(x(:))                     Go through data to get the range of abscis-
xmin=minval(x(:))                         sas.
xdif=xmax-xmin
```

```
xave=0.5_sp*(xmax+xmin)
pnow=1.0_sp/(xdif*ofac)                   Starting frequency.
tmp1(:)=TWOPI_D*((x(:)-xave)*pnow)        Initialize values for the trigonometric recur-
wpr(:)=-2.0_dp*sin(0.5_dp*tmp1)**2            rences at each data point. The recur-
wpi(:)=sin(tmp1(:))                           rences are done in double precision.
wr(:)=cos(tmp1(:))
wi(:)=wpi(:)
do i=1,nout                               Main loop over the frequencies to be evalu-
    px(i)=pnow                                ated.
    sumsh=dot_product(wi,wr)              First, loop over the data to get τ and related
    sumc=dot_product(wr(:)-wi(:),wr(:)+wi(:))        quantities.
    wtau=0.5_sp*atan2(2.0_sp*sumsh,sumc)
    swtau=sin(wtau)
    cwtau=cos(wtau)
    tmp1(:)=wi(:)*cwtau-wr(:)*swtau       Then, loop over the data again to get the
    tmp2(:)=wr(:)*cwtau+wi(:)*swtau           periodogram value.
    sums=dot_product(tmp1,tmp1)
    sumc=dot_product(tmp2,tmp2)
    sumsy=dot_product(y(:)-ave,tmp1)
    sumcy=dot_product(y(:)-ave,tmp2)
    tmp1(:)=wr(:)                         Update the trigonometric recurrences.
    wr(:)=(wr(:)*wpr(:)-wi(:)*wpi(:))+wr(:)
    wi(:)=(wi(:)*wpr(:)+tmp1(:)*wpi(:))+wi(:)
    py(i)=0.5_sp*(sumcy**2/sumc+sumsy**2/sums)/var
    pnow=pnow+1.0_sp/(ofac*xdif)          The next frequency.
end do
jmax=imaxloc(py(1:nout))
expy=exp(-py(jmax))                       Evaluate statistical significance of the maxi-
effm=2.0_sp*nout/ofac                         mum.
prob=effm*expy
if (prob > 0.01_sp) prob=1.0_sp-(1.0_sp-expy)**effm
END SUBROUTINE period
```

f90 This routine shows another example of how to return arrays whose size is not known in advance (cf. zbrac in Chapter B9). The coding is explained in the subsection on pointers in §21.5. The size of the output arrays, nout in the code, is available as size(px).

jmax=imaxloc... See discussion of imaxloc on p. 1017.

```
SUBROUTINE fasper(x,y,ofac,hifac,px,py,jmax,prob)
USE nrtype; USE nrutil, ONLY : arth,assert_eq,imaxloc,nrerror
USE nr, ONLY : avevar,realft
IMPLICIT NONE
REAL(SP), DIMENSION(:), INTENT(IN) :: x,y
REAL(SP), INTENT(IN) :: ofac,hifac
INTEGER(I4B), INTENT(OUT) :: jmax
REAL(SP), INTENT(OUT) :: prob
REAL(SP), DIMENSION(:), POINTER :: px,py
INTEGER(I4B), PARAMETER :: MACC=4
```

Input is a set of N data points with abscissas x (which need not be equally spaced) and ordinates y, and a desired oversampling factor ofac (a typical value being 4 or larger). The routine returns pointers to internally allocated arrays px and py. px is filled with an increasing sequence of frequencies (not angular frequencies) up to hifac times the "average" Nyquist frequency, and py is filled with the values of the Lomb normalized periodogram at those frequencies. The length of these arrays is 0.5*ofac*hifac*N. The arrays x and y are not altered. The routine also returns jmax such that py(jmax) is the maximum element in py, and prob, an estimate of the significance of that maximum against the hypothesis of random noise. A small value of prob indicates that a significant

periodic signal is present.
Parameter: MACC is the number of interpolation points per 1/4 cycle of highest frequency.

```
INTEGER(I4B) :: j,k,n,ndim,nfreq,nfreqt,nout
REAL(SP) :: ave,ck,ckk,cterm,cwt,den,df,effm,expy,fac,fndim,hc2wt,&
    hs2wt,hypo,sterm,swt,var,xdif,xmax,xmin
REAL(SP), DIMENSION(:), ALLOCATABLE :: wk1,wk2
LOGICAL(LGT), SAVE :: init=.true.
n=assert_eq(size(x),size(y),'fasper')
if (init) then
    init=.false.
    nullify(px,py)
else
    if (associated(px)) deallocate(px)
    if (associated(py)) deallocate(py)
end if
nfreqt=ofac*hifac*n*MACC
nfreq=64
do                                  Size the FFT as next power of 2 above nfreqt.
    if (nfreq >= nfreqt) exit
    nfreq=nfreq*2
end do
ndim=2*nfreq
allocate(wk1(ndim),wk2(ndim))
call avevar(y(1:n),ave,var)         Compute the mean, variance, and range of the data.
xmax=maxval(x(:))
xmin=minval(x(:))
xdif=xmax-xmin
wk1(1:ndim)=0.0                     Zero the workspaces.
wk2(1:ndim)=0.0
fac=ndim/(xdif*ofac)
fndim=ndim
do j=1,n                            Extirpolate the data into the workspaces.
    ck=1.0_sp+mod((x(j)-xmin)*fac,fndim)
    ckk=1.0_sp+mod(2.0_sp*(ck-1.0_sp),fndim)
    call spreadval(y(j)-ave,wk1,ck,MACC)
    call spreadval(1.0_sp,wk2,ckk,MACC)
end do
call realft(wk1(1:ndim),1)          Take the fast Fourier transforms.
call realft(wk2(1:ndim),1)
df=1.0_sp/(xdif*ofac)
nout=0.5_sp*ofac*hifac*n
allocate(px(nout),py(nout))
k=3
do j=1,nout                         Compute the Lomb value for each frequency.
    hypo=sqrt(wk2(k)**2+wk2(k+1)**2)
    hc2wt=0.5_sp*wk2(k)/hypo
    hs2wt=0.5_sp*wk2(k+1)/hypo
    cwt=sqrt(0.5_sp+hc2wt)
    swt=sign(sqrt(0.5_sp-hc2wt),hs2wt)
    den=0.5_sp*n+hc2wt*wk2(k)+hs2wt*wk2(k+1)
    cterm=(cwt*wk1(k)+swt*wk1(k+1))**2/den
    sterm=(cwt*wk1(k+1)-swt*wk1(k))**2/(n-den)
    px(j)=j*df
    py(j)=(cterm+sterm)/(2.0_sp*var)
    k=k+2
end do
deallocate(wk1,wk2)
jmax=imaxloc(py(1:nout))
expy=exp(-py(jmax))                 Estimate significance of largest peak value.
effm=2.0_sp*nout/ofac
prob=effm*expy
if (prob > 0.01_sp) prob=1.0_sp-(1.0_sp-expy)**effm
CONTAINS
```

```
SUBROUTINE spreadval(y,yy,x,m)
IMPLICIT NONE
REAL(SP), INTENT(IN) :: y,x
REAL(SP), DIMENSION(:), INTENT(INOUT) :: yy
INTEGER(I4B), INTENT(IN) :: m
    Given an array yy of length N, extirpolate (spread) a value y into m actual array elements
    that best approximate the "fictional" (i.e., possibly noninteger) array element number x.
    The weights used are coefficients of the Lagrange interpolating polynomial.
INTEGER(I4B) :: ihi,ilo,ix,j,nden,n
REAL(SP) :: fac
INTEGER(I4B), DIMENSION(10) :: nfac = (/ &
    1,1,2,6,24,120,720,5040,40320,362880 /)
if (m > 10) call nrerror('factorial table too small in spreadval')
n=size(yy)
ix=x
if (x == real(ix,sp)) then
    yy(ix)=yy(ix)+y
else
    ilo=min(max(int(x-0.5_sp*m+1.0_sp),1),n-m+1)
    ihi=ilo+m-1
    nden=nfac(m)
    fac=product(x-arth(ilo,1,m))
    yy(ihi)=yy(ihi)+y*fac/(nden*(x-ihi))
    do j=ihi-1,ilo,-1
        nden=(nden/(j+1-ilo))*(j-ihi)
        yy(j)=yy(j)+y*fac/(nden*(x-j))
    end do
end if
END SUBROUTINE spreadval
END SUBROUTINE fasper
```

f90 This routine shows another example of how to return arrays whose size is not known in advance (cf. zbrac in Chapter B9). The coding is explained in the subsection on pointers in §21.5. The size of the output arrays, nout in the code, is available as size(px).

`jmax=imaxloc...` See discussion of imaxloc on p. 1017.

`if (x == real(ix,sp)) then` Without the explicit kind type parameter sp, real returns a value of type default real for an integer argument. This prevents automatic conversion of the routine from single to double precision. Here all you have to do is redefine sp in nrtype to get double precision.

* * *

```
SUBROUTINE dftcor(w,delta,a,b,endpts,corre,corim,corfac)
USE nrtype; USE nrutil, ONLY : assert
IMPLICIT NONE
REAL(SP), INTENT(IN) :: w,delta,a,b
REAL(SP), INTENT(OUT) :: corre,corim,corfac
REAL(SP), DIMENSION(:), INTENT(IN) :: endpts
    For an integral approximated by a discrete Fourier transform, this routine computes the
    correction factor that multiplies the DFT and the endpoint correction to be added. Input
    is the angular frequency w, stepsize delta, lower and upper limits of the integral a and
    b, while the array endpts of length 8 contains the first 4 and last 4 function values. The
```

```
    correction factor W(θ) is returned as corfac, while the real and imaginary parts of the
    endpoint correction are returned as corre and corim.
REAL(SP) :: a0i,a0r,a1i,a1r,a2i,a2r,a3i,a3r,arg,c,cl,cr,s,sl,sr,t,&
    t2,t4,t6
REAL(DP) :: cth,ctth,spth2,sth,sth4i,stth,th,th2,th4,&
    tmth2,tth4i
th=w*delta
call assert(a < b, th >= 0.0, th <= PI_D, 'dftcor args')
if (abs(th) < 5.0e-2_dp) then                    Use series.
    t=th
    t2=t*t
    t4=t2*t2
    t6=t4*t2
    corfac=1.0_sp-(11.0_sp/720.0_sp)*t4+(23.0_sp/15120.0_sp)*t6
    a0r=(-2.0_sp/3.0_sp)+t2/45.0_sp+(103.0_sp/15120.0_sp)*t4-&
        (169.0_sp/226800.0_sp)*t6
    a1r=(7.0_sp/24.0_sp)-(7.0_sp/180.0_sp)*t2+(5.0_sp/3456.0_sp)*t4&
        -(7.0_sp/259200.0_sp)*t6
    a2r=(-1.0_sp/6.0_sp)+t2/45.0_sp-(5.0_sp/6048.0_sp)*t4+t6/64800.0_sp
    a3r=(1.0_sp/24.0_sp)-t2/180.0_sp+(5.0_sp/24192.0_sp)*t4-t6/259200.0_sp
    a0i=t*(2.0_sp/45.0_sp+(2.0_sp/105.0_sp)*t2-&
        (8.0_sp/2835.0_sp)*t4+(86.0_sp/467775.0_sp)*t6)
    a1i=t*(7.0_sp/72.0_sp-t2/168.0_sp+(11.0_sp/72576.0_sp)*t4-&
        (13.0_sp/5987520.0_sp)*t6)
    a2i=t*(-7.0_sp/90.0_sp+t2/210.0_sp-(11.0_sp/90720.0_sp)*t4+&
        (13.0_sp/7484400.0_sp)*t6)
    a3i=t*(7.0_sp/360.0_sp-t2/840.0_sp+(11.0_sp/362880.0_sp)*t4-&
        (13.0_sp/29937600.0_sp)*t6)
else                                  Use trigonometric formulas in double precision.
    cth=cos(th)
    sth=sin(th)
    ctth=cth**2-sth**2
    stth=2.0_dp*sth*cth
    th2=th*th
    th4=th2*th2
    tmth2=3.0_dp-th2
    spth2=6.0_dp+th2
    sth4i=1.0_sp/(6.0_dp*th4)
    tth4i=2.0_dp*sth4i
    corfac=tth4i*spth2*(3.0_sp-4.0_dp*cth+ctth)
    a0r=sth4i*(-42.0_dp+5.0_dp*th2+spth2*(8.0_dp*cth-ctth))
    a0i=sth4i*(th*(-12.0_dp+6.0_dp*th2)+spth2*stth)
    a1r=sth4i*(14.0_dp*tmth2-7.0_dp*spth2*cth)
    a1i=sth4i*(30.0_dp*th-5.0_dp*spth2*sth)
    a2r=tth4i*(-4.0_dp*tmth2+2.0_dp*spth2*cth)
    a2i=tth4i*(-12.0_dp*th+2.0_dp*spth2*sth)
    a3r=sth4i*(2.0_dp*tmth2-spth2*cth)
    a3i=sth4i*(6.0_dp*th-spth2*sth)
end if
cl=a0r*endpts(1)+a1r*endpts(2)+a2r*endpts(3)+a3r*endpts(4)
sl=a0i*endpts(1)+a1i*endpts(2)+a2i*endpts(3)+a3i*endpts(4)
cr=a0r*endpts(8)+a1r*endpts(7)+a2r*endpts(6)+a3r*endpts(5)
sr=-a0i*endpts(8)-a1i*endpts(7)-a2i*endpts(6)-a3i*endpts(5)
arg=w*(b-a)
c=cos(arg)
s=sin(arg)
corre=cl+c*cr-s*sr
corim=sl+s*cr+c*sr
END SUBROUTINE dftcor
```

```
SUBROUTINE dftint(func,a,b,w,cosint,sinint)
USE nrtype; USE nrutil, ONLY : arth
USE nr, ONLY : dftcor,polint,realft
IMPLICIT NONE
REAL(SP), INTENT(IN) :: a,b,w
REAL(SP), INTENT(OUT) :: cosint,sinint
INTERFACE
    FUNCTION func(x)
    USE nrtype
    IMPLICIT NONE
    REAL(SP), DIMENSION(:), INTENT(IN) :: x
    REAL(SP), DIMENSION(size(x)) :: func
    END FUNCTION func
END INTERFACE
INTEGER(I4B), PARAMETER :: M=64,NDFT=1024,MPOL=6
```

Example subroutine illustrating how to use the routine dftcor. The user supplies an external function func that returns the quantity $h(t)$. The routine then returns $\int_a^b \cos(\omega t)h(t)\,dt$ as cosint and $\int_a^b \sin(\omega t)h(t)\,dt$ as sinint.

Parameters: The values of M, NDFT, and MPOL are merely illustrative and should be optimized for your particular application. M is the number of subintervals, NDFT is the length of the FFT (a power of 2), and MPOL is the degree of polynomial interpolation used to obtain the desired frequency from the FFT.

```
INTEGER(I4B) :: nn
INTEGER(I4B), SAVE :: init=0
INTEGER(I4B), DIMENSION(MPOL) :: nnmpol
REAL(SP) :: c,cdft,cerr,corfac,corim,corre,en,s,sdft,serr
REAL(SP), SAVE :: delta
REAL(SP), DIMENSION(MPOL) :: cpol,spol,xpol
REAL(SP), DIMENSION(NDFT), SAVE :: data
REAL(SP), DIMENSION(8), SAVE :: endpts
REAL(SP), SAVE :: aold=-1.0e30_sp,bold=-1.0e30_sp
if (init /= 1 .or. a /= aold .or. b /= bold) then      Do we need to initialize, or
                                                          is only ω changed?
    init=1
    aold=a
    bold=b
    delta=(b-a)/M
    data(1:M+1)=func(a+arth(0,1,M+1)*delta)
      Load the function values into the data array.
    data(M+2:NDFT)=0.0                       Zero pad the rest of the data array.
    endpts(1:4)=data(1:4)                    Load the endpoints.
    endpts(5:8)=data(M-2:M+1)
    call realft(data(1:NDFT),1)
      realft returns the unused value corresponding to ω_{N/2} in data(2). We actually want
      this element to contain the imaginary part corresponding to ω_0, which is zero.
    data(2)=0.0
end if
```

Now interpolate on the DFT result for the desired frequency. If the frequency is an ω_n, i.e., the quantity en is an integer, then cdft=data(2*en-1), sdft=data(2*en), and you could omit the interpolation.

```
en=w*delta*NDFT/TWOPI+1.0_sp
nn=min(max(int(en-0.5_sp*MPOL+1.0_sp),1),NDFT/2-MPOL+1)      Leftmost point for the in-
nnmpol=arth(nn,1,MPOL)                                          terpolation.
cpol(1:MPOL)=data(2*nnmpol(:)-1)
spol(1:MPOL)=data(2*nnmpol(:))
xpol(1:MPOL)=nnmpol(:)
call polint(xpol,cpol,en,cdft,cerr)
call polint(xpol,spol,en,sdft,serr)
call dftcor(w,delta,a,b,endpts,corre,corim,corfac)      Now get the endpoint cor-
cdft=cdft*corfac+corre                                     rection and the multiplica-
sdft=sdft*corfac+corim                                     tive factor W(θ).
c=delta*cos(w*a)                            Finally multiply by Δ and exp(iωa).
s=delta*sin(w*a)
cosint=c*cdft-s*sdft
```

```
sinint=s*cdft+c*sdft
END SUBROUTINE dftint
```

* * *

```
SUBROUTINE wt1(a,isign,wtstep)
USE nrtype; USE nrutil, ONLY : assert
IMPLICIT NONE
REAL(SP), DIMENSION(:), INTENT(INOUT) :: a
INTEGER(I4B), INTENT(IN) :: isign
INTERFACE
    SUBROUTINE wtstep(a,isign)
    USE nrtype
    IMPLICIT NONE
    REAL(SP), DIMENSION(:), INTENT(INOUT) :: a
    INTEGER(I4B), INTENT(IN) :: isign
    END SUBROUTINE wtstep
END INTERFACE
```

One-dimensional discrete wavelet transform. This routine implements the pyramid algorithm, replacing a by its wavelet transform (for isign=1), or performing the inverse operation (for isign=-1). The length of a is N, which must be an integer power of 2. The subroutine wtstep, whose actual name must be supplied in calling this routine, is the underlying wavelet filter. Examples of wtstep are daub4 and (preceded by pwtset) pwt.

```
INTEGER(I4B) :: n,nn
n=size(a)
call assert(iand(n,n-1)==0, 'n must be a power of 2 in wt1')
if (n < 4) RETURN
if (isign >= 0) then            Wavelet transform.
    nn=n                        Start at largest hierarchy,
    do
        if (nn < 4) exit
        call wtstep(a(1:nn),isign)
        nn=nn/2                 and work towards smallest.
    end do
else                            Inverse wavelet transform.
    nn=4                        Start at smallest hierarchy,
    do
        if (nn > n) exit
        call wtstep(a(1:nn),isign)
        nn=nn*2                 and work towards largest.
    end do
end if
END SUBROUTINE wt1
```

```
SUBROUTINE daub4(a,isign)
USE nrtype
IMPLICIT NONE
REAL(SP), DIMENSION(:), INTENT(INOUT) :: a
INTEGER(I4B), INTENT(IN) :: isign
```

Applies the Daubechies 4-coefficient wavelet filter to data vector a (for isign=1) or applies its transpose (for isign=-1). Used hierarchically by routines wt1 and wtn.

```
REAL(SP), DIMENSION(size(a)) :: wksp
REAL(SP), PARAMETER :: C0=0.4829629131445341_sp,&
    C1=0.8365163037378079_sp,C2=0.2241438680420134_sp,&
    C3=-0.1294095225512604_sp
INTEGER(I4B) :: n,nh,nhp,nhm
n=size(a)
if (n < 4) RETURN
nh=n/2
nhp=nh+1
nhm=nh-1
```

```
if (isign >= 0) then          Apply filter.
    wksp(1:nhm) = C0*a(1:n-3:2)+C1*a(2:n-2:2) &
        +C2*a(3:n-1:2)+C3*a(4:n:2)
    wksp(nh)=C0*a(n-1)+C1*a(n)+C2*a(1)+C3*a(2)
    wksp(nhp:n-1) = C3*a(1:n-3:2)-C2*a(2:n-2:2) &
        +C1*a(3:n-1:2)-C0*a(4:n:2)
    wksp(n)=C3*a(n-1)-C2*a(n)+C1*a(1)-C0*a(2)
else                          Apply transpose filter.
    wksp(1)=C2*a(nh)+C1*a(n)+C0*a(1)+C3*a(nhp)
    wksp(2)=C3*a(nh)-C0*a(n)+C1*a(1)-C2*a(nhp)
    wksp(3:n-1:2) = C2*a(1:nhm)+C1*a(nhp:n-1) &
        +C0*a(2:nh)+C3*a(nh+2:n)
    wksp(4:n:2) = C3*a(1:nhm)-C0*a(nhp:n-1) &
        +C1*a(2:nh)-C2*a(nh+2:n)
end if
a(1:n)=wksp(1:n)
END SUBROUTINE daub4
```

```
MODULE pwtcom
USE nrtype
INTEGER(I4B), SAVE :: ncof=0,ioff,joff          These module variables communicate the
REAL(SP), DIMENSION(:), ALLOCATABLE, SAVE :: cc,cr        filter to pwt.
END MODULE pwtcom
```

```
SUBROUTINE pwtset(n)
USE nrtype; USE nrutil, ONLY : nrerror
USE pwtcom
IMPLICIT NONE
INTEGER(I4B), INTENT(IN) :: n
    Initializing routine for pwt, here implementing the Daubechies wavelet filters with 4, 12,
    and 20 coefficients, as selected by the input value n. Further wavelet filters can be included
    in the obvious manner. This routine must be called (once) before the first use of pwt. (For
    the case n=4, the specific routine daub4 is considerably faster than pwt.)
REAL(SP) :: sig
REAL(SP), PARAMETER :: &
    c4(4)=(/&
    0.4829629131445341_sp, 0.8365163037378079_sp, &
    0.2241438680420134_sp,-0.1294095225512604_sp /), &
    c12(12)=(/&
    0.111540743350_sp, 0.494623890398_sp, 0.751133908021_sp, &
    0.315250351709_sp,-0.226264693965_sp,-0.129766867567_sp, &
    0.097501605587_sp, 0.027522865530_sp,-0.031582039318_sp, &
    0.000553842201_sp, 0.004777257511_sp,-0.001077301085_sp /), &
    c20(20)=(/&
    0.026670057901_sp, 0.188176800078_sp, 0.527201188932_sp, &
    0.688459039454_sp, 0.281172343661_sp,-0.249846424327_sp, &
    -0.195946274377_sp, 0.127369340336_sp, 0.093057364604_sp, &
    -0.071394147166_sp,-0.029457536822_sp, 0.033212674059_sp, &
    0.003606553567_sp,-0.010733175483_sp, 0.001395351747_sp, &
    0.001992405295_sp,-0.000685856695_sp,-0.000116466855_sp, &
    0.000093588670_sp,-0.000013264203_sp /)
if (allocated(cc)) deallocate(cc)
if (allocated(cr)) deallocate(cr)
allocate(cc(n),cr(n))
ncof=n
ioff=-n/2                    These values center the "support" of the wavelets at each
joff=-n/2                        level. Alternatively, the "peaks" of the wavelets can
sig=-1.0                         be approximately centered by the choices ioff=-2
select case(n)                   and joff=-n+2. Note that daub4 and pwtset with
    case(4)                      n=4 use different default centerings.
```

```
        cc=c4
    case(12)
        cc=c12
    case(20)
        cc=c20
    case default
        call nrerror('unimplemented value n in pwtset')
end select
cr(n:1:-1) = cc
cr(n:1:-2) = -cr(n:1:-2)
END SUBROUTINE pwtset
```

f90 Here we need to have as global variables arrays whose dimensions are known only at run time. At first sight the situation is the same as with the module fminln in newt on p. 1197. If you review the discussion there and in §21.5, you will recall that there are two good ways to implement this: with allocatable arrays ("Method 1") or with pointers ("Method 2"). There is a difference here that makes allocatable arrays simpler. We do not wish to deallocate the arrays on exiting pwtset. On the contrary, the values in cc and cr need to be preserved for use in pwt. Since allocatable arrays are born in the well-defined state of "not currently allocated," we can declare the arrays here as

```
REAL(SP), DIMENSION(:), ALLOCATABLE, SAVE :: cc,cr
```

and test whether they were used on a previous call with

```
if (allocated(cc)) deallocate(cc)
if (allocated(cr)) deallocate(cr)
```

We are then ready to allocate the new storage:

```
allocate(cc(n),cr(n))
```

With pointers, we would need the additional machinery of nullifying the pointers on the initial call, since pointers are born in an undefined state (see §21.5).

There is an additional important point in this example. The module variables need to be used by a "sibling" routine, pwt. We need to be sure that they do not become undefined when we exit pwtset. We could ensure this by putting a USE pwtcom in the main program that calls both pwtset and pwt, but it's easy to forget to do this. It is preferable to put explicit SAVEs on all the module variables.

```
SUBROUTINE pwt(a,isign)
USE nrtype; USE nrutil, ONLY : arth,nrerror
USE pwtcom
IMPLICIT NONE
REAL(SP), DIMENSION(:), INTENT(INOUT) :: a
INTEGER(I4B), INTENT(IN) :: isign
    Partial wavelet transform: applies an arbitrary wavelet filter to data vector a (for isign=1)
    or applies its transpose (for isign=-1). Used hierarchically by routines wt1 and wtn. The
    actual filter is determined by a preceding (and required) call to pwtset, which initializes
    the module pwtcom.
REAL(SP), DIMENSION(size(a)) :: wksp
INTEGER(I4B), DIMENSION(size(a)/2) :: jf,jr
INTEGER(I4B) :: k,n,nh,nmod
n=size(a)
if (n < 4) RETURN
if (ncof == 0) call nrerror('pwt: must call pwtset before pwt')
nmod=ncof*n                                  A positive constant equal to zero mod n.
```

```
nh=n/2
wksp(:)=0.0
jf=iand(n-1,arth(2+nmod+ioff,2,nh))       Use bitwise AND to wrap-around the point-
jr=iand(n-1,arth(2+nmod+joff,2,nh))           ers. n-1 is a mask of all bits, since n is
do k=1,ncof                                   a power of 2.
    if (isign >= 0) then                  Apply filter.
        wksp(1:nh)=wksp(1:nh)+cc(k)*a(jf+1)
        wksp(nh+1:n)=wksp(nh+1:n)+cr(k)*a(jr+1)
    else                                  Apply transpose filter.
        wksp(jf+1)=wksp(jf+1)+cc(k)*a(1:nh)
        wksp(jr+1)=wksp(jr+1)+cr(k)*a(nh+1:n)
    end if
    if (k == ncof) exit
    jf=iand(n-1,jf+1)
    jr=iand(n-1,jr+1)
end do
a(:)=wksp(:)                              Copy the results back from workspace.
END SUBROUTINE pwt
```

⋆ ⋆ ⋆

```
SUBROUTINE wtn(a,nn,isign,wtstep)
USE nrtype; USE nrutil, ONLY : arth,assert
IMPLICIT NONE
REAL(SP), DIMENSION(:), INTENT(INOUT) :: a
INTEGER(I4B), DIMENSION(:), INTENT(IN) :: nn
INTEGER(I4B), INTENT(IN) :: isign
INTERFACE
    SUBROUTINE wtstep(a,isign)
    USE nrtype
    IMPLICIT NONE
    REAL(SP), DIMENSION(:), INTENT(INOUT) :: a
    INTEGER(I4B), INTENT(IN) :: isign
    END SUBROUTINE wtstep
END INTERFACE
```

Replaces a by its N-dimensional discrete wavelet transform, if isign is input as 1. nn is an integer array of length N, containing the lengths of each dimension (number of real values), which must all be powers of 2. a is a real array of length equal to the product of these lengths, in which the data are stored as in a multidimensional real FORTRAN array. If isign is input as -1, a is replaced by its inverse wavelet transform. The subroutine wtstep, whose actual name must be supplied in calling this routine, is the underlying wavelet filter. Examples of wtstep are daub4 and (preceded by pwtset) pwt.

```
INTEGER(I4B) :: i1,i2,i3,idim,n,ndim,nnew,nprev,nt,ntot
REAL(SP), DIMENSION(:), ALLOCATABLE :: wksp
call assert(iand(nn,nn-1)==0, 'each dimension must be a power of 2 in wtn')
allocate(wksp(maxval(nn)))
ndim=size(nn)
ntot=product(nn(:))
nprev=1
do idim=1,ndim                                      Main loop over the dimensions.
    n=nn(idim)
    nnew=n*nprev
    if (n > 4) then
        do i2=0,ntot-1,nnew
            do i1=1,nprev
                i3=i1+i2
                wksp(1:n)=a(arth(i3,nprev,n))       Copy the relevant row or column
                i3=i3+n*nprev                           or etc. into workspace.
                if (isign >= 0) then                Do one-dimensional wavelet trans-
                    nt=n                                form.
                    do
```

```
                        if (nt < 4) exit
                        call wtstep(wksp(1:nt),isign)
                        nt=nt/2
                    end do
                else                                    Or inverse transform.
                    nt=4
                    do
                        if (nt > n) exit
                        call wtstep(wksp(1:nt),isign)
                        nt=nt*2
                    end do
                end if
                i3=i1+i2
                a(arth(i3,nprev,n))=wksp(1:n)           Copy back from workspace.
                i3=i3+n*nprev
            end do
        end do
    end if
    nprev=nnew
end do
deallocate(wksp)
END SUBROUTINE wtn
```

Chapter B14. Statistical Description of Data

```
SUBROUTINE moment(data,ave,adev,sdev,var,skew,curt)
USE nrtype; USE nrutil, ONLY : nrerror
IMPLICIT NONE
REAL(SP), INTENT(OUT) :: ave,adev,sdev,var,skew,curt
REAL(SP), DIMENSION(:), INTENT(IN) :: data
    Given an array of data, this routine returns its mean ave, average deviation adev, standard
    deviation sdev, variance var, skewness skew, and kurtosis curt.
INTEGER(I4B) :: n
REAL(SP) :: ep
REAL(SP), DIMENSION(size(data)) :: p,s
n=size(data)
if (n <= 1) call nrerror('moment: n must be at least 2')
ave=sum(data(:))/n                 First pass to get the mean.
s(:)=data(:)-ave                   Second pass to get the first (absolute), second, third, and
ep=sum(s(:))                           fourth moments of the deviation from the mean.
adev=sum(abs(s(:)))/n
p(:)=s(:)*s(:)
var=sum(p(:))
p(:)=p(:)*s(:)
skew=sum(p(:))
p(:)=p(:)*s(:)
curt=sum(p(:))
var=(var-ep**2/n)/(n-1)            Corrected two-pass formula.
sdev=sqrt(var)
if (var /= 0.0) then
    skew=skew/(n*sdev**3)
    curt=curt/(n*var**2)-3.0_sp
else
    call nrerror('moment: no skew or kurtosis when zero variance')
end if
END SUBROUTINE moment
```

⋆ ⋆ ⋆

```
SUBROUTINE ttest(data1,data2,t,prob)
USE nrtype
USE nr, ONLY : avevar,betai
IMPLICIT NONE
REAL(SP), DIMENSION(:), INTENT(IN) :: data1,data2
REAL(SP), INTENT(OUT) :: t,prob
    Given the arrays data1 and data2, which need not have the same length, this routine
    returns Student's t as t, and its significance as prob, small values of prob indicating that
```

the arrays have significantly different means. The data arrays are assumed to be drawn from populations with the same true variance.

```
INTEGER(I4B) :: n1,n2
REAL(SP) :: ave1,ave2,df,var,var1,var2
n1=size(data1)
n2=size(data2)
call avevar(data1,ave1,var1)
call avevar(data2,ave2,var2)
df=n1+n2-2                                          Degrees of freedom.
var=((n1-1)*var1+(n2-1)*var2)/df                    Pooled variance.
t=(ave1-ave2)/sqrt(var*(1.0_sp/n1+1.0_sp/n2))
prob=betai(0.5_sp*df,0.5_sp,df/(df+t**2))           See equation (6.4.9).
END SUBROUTINE ttest
```

⋆ ⋆ ⋆

```
SUBROUTINE avevar(data,ave,var)
USE nrtype
IMPLICIT NONE
REAL(SP), DIMENSION(:), INTENT(IN) :: data
REAL(SP), INTENT(OUT) :: ave,var
```

Given array data, returns its mean as ave and its variance as var.

```
INTEGER(I4B) :: n
REAL(SP), DIMENSION(size(data)) :: s
n=size(data)
ave=sum(data(:))/n
s(:)=data(:)-ave
var=dot_product(s,s)
var=(var-sum(s)**2/n)/(n-1)              Corrected two-pass formula (14.1.8).
END SUBROUTINE avevar
```

⋆ ⋆ ⋆

```
SUBROUTINE tutest(data1,data2,t,prob)
USE nrtype
USE nr, ONLY : avevar,betai
IMPLICIT NONE
REAL(SP), DIMENSION(:), INTENT(IN) :: data1,data2
REAL(SP), INTENT(OUT) :: t,prob
```

Given the arrays data1 and data2, which need not have the same length, this routine returns Student's t as t, and its significance as prob, small values of prob indicating that the arrays have significantly different means. The data arrays are allowed to be drawn from populations with unequal variances.

```
INTEGER(I4B) :: n1,n2
REAL(SP) :: ave1,ave2,df,var1,var2
n1=size(data1)
n2=size(data2)
call avevar(data1,ave1,var1)
call avevar(data2,ave2,var2)
t=(ave1-ave2)/sqrt(var1/n1+var2/n2)
df=(var1/n1+var2/n2)**2/((var1/n1)**2/(n1-1)+(var2/n2)**2/(n2-1))
prob=betai(0.5_sp*df,0.5_sp,df/(df+t**2))
END SUBROUTINE tutest
```

⋆ ⋆ ⋆

```
SUBROUTINE tptest(data1,data2,t,prob)
USE nrtype; USE nrutil, ONLY : assert_eq
USE nr, ONLY : avevar,betai
IMPLICIT NONE
REAL(SP), DIMENSION(:), INTENT(IN) :: data1,data2
REAL(SP), INTENT(OUT) :: t,prob
```

Given the paired arrays data1 and data2 of the same length, this routine returns Student's t for paired data as t, and its significance as prob, small values of prob indicating a significant difference of means.

```
INTEGER(I4B) :: n
REAL(SP) :: ave1,ave2,cov,df,sd,var1,var2
n=assert_eq(size(data1),size(data2),'tptest')
call avevar(data1,ave1,var1)
call avevar(data2,ave2,var2)
cov=dot_product(data1(:)-ave1,data2(:)-ave2)
df=n-1
cov=cov/df
sd=sqrt((var1+var2-2.0_sp*cov)/n)
t=(ave1-ave2)/sd
prob=betai(0.5_sp*df,0.5_sp,df/(df+t**2))
END SUBROUTINE tptest
```

⋆ ⋆ ⋆

```
SUBROUTINE ftest(data1,data2,f,prob)
USE nrtype
USE nr, ONLY : avevar,betai
IMPLICIT NONE
REAL(SP), INTENT(OUT) :: f,prob
REAL(SP), DIMENSION(:), INTENT(IN) :: data1,data2
```

Given the arrays data1 and data2, which need not have the same length, this routine returns the value of f, and its significance as prob. Small values of prob indicate that the two arrays have significantly different variances.

```
INTEGER(I4B) :: n1,n2
REAL(SP) :: ave1,ave2,df1,df2,var1,var2
n1=size(data1)
n2=size(data2)
call avevar(data1,ave1,var1)
call avevar(data2,ave2,var2)
if (var1 > var2) then        Make F the ratio of the larger variance to the smaller one.
    f=var1/var2
    df1=n1-1
    df2=n2-1
else
    f=var2/var1
    df1=n2-1
    df2=n1-1
end if
prob=2.0_sp*betai(0.5_sp*df2,0.5_sp*df1,df2/(df2+df1*f))
if (prob > 1.0) prob=2.0_sp-prob
END SUBROUTINE ftest
```

⋆ ⋆ ⋆

```
SUBROUTINE chsone(bins,ebins,knstrn,df,chsq,prob)
USE nrtype; USE nrutil, ONLY : assert_eq,nrerror
USE nr, ONLY : gammq
IMPLICIT NONE
INTEGER(I4B), INTENT(IN) :: knstrn
REAL(SP), INTENT(OUT) :: df,chsq,prob
REAL(SP), DIMENSION(:), INTENT(IN) :: bins,ebins
```

Given the same-size arrays `bins` containing the observed numbers of events, and `ebins` containing the expected numbers of events, and given the number of constraints `knstrn` (normally one), this routine returns (trivially) the number of degrees of freedom `df`, and (nontrivially) the chi-square `chsq` and the significance `prob`. A small value of `prob` indicates a significant difference between the distributions `bins` and `ebins`. Note that `bins` and `ebins` are both real arrays, although `bins` will normally contain integer values.

```
INTEGER(I4B) :: ndum
ndum=assert_eq(size(bins),size(ebins),'chsone')
if (any(ebins(:) <= 0.0)) call nrerror('bad expected number in chsone')
df=size(bins)-knstrn
chsq=sum((bins(:)-ebins(:))**2/ebins(:))
prob=gammq(0.5_sp*df,0.5_sp*chsq)          Chi-square probability function. See §6.2.
END SUBROUTINE chsone
```

```
SUBROUTINE chstwo(bins1,bins2,knstrn,df,chsq,prob)
USE nrtype; USE nrutil, ONLY : assert_eq
USE nr, ONLY : gammq
IMPLICIT NONE
INTEGER(I4B), INTENT(IN) :: knstrn
REAL(SP), INTENT(OUT) :: df,chsq,prob
REAL(SP), DIMENSION(:), INTENT(IN) :: bins1,bins2
```

Given the same-size arrays `bins1` and `bins2`, containing two sets of binned data, and given the number of constraints `knstrn` (normally 1 or 0), this routine returns the number of degrees of freedom `df`, the chi-square `chsq`, and the significance `prob`. A small value of `prob` indicates a significant difference between the distributions `bins1` and `bins2`. Note that `bins1` and `bins2` are both real arrays, although they will normally contain integer values.

```
INTEGER(I4B) :: ndum
LOGICAL(LGT), DIMENSION(size(bins1)) :: nzeromask
ndum=assert_eq(size(bins1),size(bins2),'chstwo')
nzeromask = bins1(:) /= 0.0 .or. bins2(:) /= 0.0
chsq=sum((bins1(:)-bins2(:))**2/(bins1(:)+bins2(:)),mask=nzeromask)
df=count(nzeromask)-knstrn                 No data means one less degree of freedom.
prob=gammq(0.5_sp*df,0.5_sp*chsq)          Chi-square probability function. See §6.2.
END SUBROUTINE chstwo
```

f90 `nzeromask=...chisq=sum(...mask=nzeromask)` We use the optional argument `mask` in `sum` to select out the elements to be summed over. In this case, at least one of the elements of `bins1` or `bins2` is not zero for each term in the sum.

⋆ ⋆ ⋆

```
SUBROUTINE ksone(data,func,d,prob)
USE nrtype; USE nrutil, ONLY : arth
USE nr, ONLY : probks,sort
IMPLICIT NONE
REAL(SP), INTENT(OUT) :: d,prob
REAL(SP), DIMENSION(:), INTENT(INOUT) :: data
INTERFACE
    FUNCTION func(x)
    USE nrtype
    REAL(SP), DIMENSION(:), INTENT(IN) :: x
    REAL(SP), DIMENSION(size(x)) :: func
    END FUNCTION func
END INTERFACE
```

Given an array data, and given a user-supplied function of a single variable func which is a cumulative distribution function ranging from 0 (for smallest values of its argument) to 1 (for largest values of its argument), this routine returns the K–S statistic d, and the significance level prob. Small values of prob show that the cumulative distribution function of data is significantly different from func. The array data is modified by being sorted into ascending order.

```
INTEGER(I4B) :: n
REAL(SP) :: en
REAL(SP), DIMENSION(size(data)) :: fvals
REAL(SP), DIMENSION(size(data)+1) :: temp
call sort(data)                              If the data are already sorted into as-
n=size(data)                                     cending order, then this call can be
en=n                                             omitted.
fvals(:)=func(data(:))
temp=arth(0,1,n+1)/en
d=maxval(max(abs(temp(1:n)-fvals(:)), &      Compute the maximum distance between
    abs(temp(2:n+1)-fvals(:))))                  the data's c.d.f. and the user-supplied
en=sqrt(en)                                      function.
prob=probks((en+0.12_sp+0.11_sp/en)*d)       Compute significance.
END SUBROUTINE ksone
```

f90 `d=maxval(max...` Note the difference between max and maxval: max takes two or more arguments and returns the maximum. If the arguments are two arrays, it returns an array each of whose elements is the maximum of the corresponding elements in the two arrays. maxval takes a single array argument and returns its maximum value.

```
SUBROUTINE kstwo(data1,data2,d,prob)
USE nrtype; USE nrutil, ONLY : cumsum
USE nr, ONLY : probks,sort2
IMPLICIT NONE
REAL(SP), INTENT(OUT) :: d,prob
REAL(SP), DIMENSION(:), INTENT(IN) :: data1,data2
```

Given arrays data1 and data2, which can be of different length, this routine returns the K–S statistic d, and the significance level prob for the null hypothesis that the data sets are drawn from the same distribution. Small values of prob show that the cumulative distribution function of data1 is significantly different from that of data2. The arrays data1 and data2 are not modified.

```
INTEGER(I4B) :: n1,n2
REAL(SP) :: en1,en2,en
REAL(SP), DIMENSION(size(data1)+size(data2)) :: dat,org
n1=size(data1)
n2=size(data2)
en1=n1
en2=n2
dat(1:n1)=data1                              Copy the two data sets into a single ar-
dat(n1+1:)=data2                                 ray.
```

```
org(1:n1)=0.0                                  Define an array that contains 0 when the
org(n1+1:)=1.0                                     corresponding element comes from
call sort2(dat,org)                                data1, 1 from data2.
  Sort the array of 1's and 0's into the order of the merged data sets.
d=maxval(abs(cumsum(org)/en2-cumsum(1.0_sp-org)/en1))
  Now use cumsum to get the c.d.f. corresponding to each set of data.
en=sqrt(en1*en2/(en1+en2))
prob=probks((en+0.12_sp+0.11_sp/en)*d)         Compute significance.
END SUBROUTINE kstwo
```

The problem here is how to compute the cumulative distribution function (c.d.f.) corresponding to each set of data, and then find the corresponding KS statistic, without a serial loop over the data. The trick is to define an array that contains 0 when the corresponding element comes from the first data set and 1 when it's from the second data set. Sort the array of 1's and 0's into the same order as the merged data sets. Now tabulate the partial sums of the array. Every time you encounter a 1, the partial sum increases by 1. So if you normalize the partial sums by dividing by the number of elements in the second data set, you have the c.d.f. of the second data set.

If you subtract the array of 1's and 0's from an array of all 1's, you get an array where 1 corresponds to an element in the first data set, 0 the second data set. So tabulating its partial sums and normalizing gives the c.d.f. of the first data set. As we've seen before, tabulating partial sums can be done with a parallel algorithm (`cumsum` in `nrutil`). The KS statistic is just the maximum absolute difference of the c.d.f.'s, computed in parallel with Fortran 90's `maxval` function.

```
FUNCTION probks(alam)
USE nrtype
IMPLICIT NONE
REAL(SP), INTENT(IN) :: alam
REAL(SP) :: probks
REAL(SP), PARAMETER :: EPS1=0.001_sp,EPS2=1.0e-8_sp
INTEGER(I4B), PARAMETER :: NITER=100
  Kolmogorov-Smirnov probability function.
INTEGER(I4B) :: j
REAL(SP) :: a2,fac,term,termbf
a2=-2.0_sp*alam**2
fac=2.0
probks=0.0
termbf=0.0                      Previous term in sum.
do j=1,NITER
    term=fac*exp(a2*j**2)
    probks=probks+term
    if (abs(term) <= EPS1*termbf .or. abs(term) <= EPS2*probks) RETURN
    fac=-fac                    Alternating signs in sum.
    termbf=abs(term)
end do
probks=1.0                      Get here only by failing to converge, which implies the func-
END FUNCTION probks                 tion is very close to 1.
```

⋆ ⋆ ⋆

```
SUBROUTINE cntab1(nn,chisq,df,prob,cramrv,ccc)
USE nrtype; USE nrutil, ONLY : outerprod
USE nr, ONLY : gammq
IMPLICIT NONE
INTEGER(I4B), DIMENSION(:,:), INTENT(IN) :: nn
REAL(SP), INTENT(OUT) :: chisq,df,prob,cramrv,ccc
REAL(SP), PARAMETER :: TINY=1.0e-30_sp
```

Given a two-dimensional contingency table in the form of a rectangular integer array nn, this routine returns the chi-square chisq, the number of degrees of freedom df, the significance level prob (small values indicating a significant association), and two measures of association, Cramer's V (cramrv), and the contingency coefficient C (ccc).

```
INTEGER(I4B) :: nni,nnj
REAL(SP) :: sumn
REAL(SP), DIMENSION(size(nn,1)) :: sumi
REAL(SP), DIMENSION(size(nn,2)) :: sumj
REAL(SP), DIMENSION(size(nn,1),size(nn,2)) :: expctd
sumi(:)=sum(nn(:,:),dim=2)                  Get the row totals.
sumj(:)=sum(nn(:,:),dim=1)                  Get the column totals.
sumn=sum(sumi(:))                           Get the grand total.
nni=size(sumi)-count(sumi(:) == 0.0)
  Eliminate any zero rows by reducing the number of rows.
nnj=size(sumj)-count(sumj(:) == 0.0)        Eliminate any zero columns.
df=nni*nnj-nni-nnj+1                        Corrected number of degrees of freedom.
expctd(:,:)=outerprod(sumi(:),sumj(:))/sumn
chisq=sum((nn(:,:)-expctd(:,:))**2/(expctd(:,:)+TINY))
  Do the chi-square sum. Here TINY guarantees that any eliminated row or column will not
  contribute to the sum.
prob=gammq(0.5_sp*df,0.5_sp*chisq)          Chi-square probability function.
cramrv=sqrt(chisq/(sumn*min(nni-1,nnj-1)))
ccc=sqrt(chisq/(chisq+sumn))
END SUBROUTINE cntab1
```

f90 `sumi(:)=sum(...dim=2)...sumj(:)=sum(...dim=1)` We use the optional argument dim of sum to sum first over the columns (dim=2) to get the row totals, and then to sum over the rows (dim=1) to get the column totals.

`expctd(:,:)=...` This is a direct implementation of equation (14.4.2) using outerprod from nrutil.

`chisq=...` And here is a direct implementation of equation (14.4.3).

```
SUBROUTINE cntab2(nn,h,hx,hy,hygx,hxgy,uygx,uxgy,uxy)
USE nrtype
IMPLICIT NONE
INTEGER(I4B), DIMENSION(:,:), INTENT(IN) :: nn
REAL(SP), INTENT(OUT) :: h,hx,hy,hygx,hxgy,uygx,uxgy,uxy
REAL(SP), PARAMETER :: TINY=1.0e-30_sp
```

Given a two-dimensional contingency table in the form of a rectangular integer array nn, where the first index labels the x-variable and the second index labels the y variable, this routine returns the entropy h of the whole table, the entropy hx of the x-distribution, the entropy hy of the y-distribution, the entropy hygx of y given x, the entropy hxgy of x given y, the dependency uygx of y on x (eq. 14.4.15), the dependency uxgy of x on y (eq. 14.4.16), and the symmetrical dependency uxy (eq. 14.4.17).

```
REAL(SP) :: sumn
REAL(SP), DIMENSION(size(nn,1)) :: sumi
REAL(SP), DIMENSION(size(nn,2)) :: sumj
sumi(:)=sum(nn(:,:),dim=2)                  Get the row totals.
sumj(:)=sum(nn(:,:),dim=1)                  Get the column totals.
sumn=sum(sumi(:))
hx=-sum(sumi(:)*log(sumi(:)/sumn), mask=(sumi(:) /= 0.0) )/sumn
  Entropy of the x distribution,
hy=-sum(sumj(:)*log(sumj(:)/sumn), mask=(sumj(:) /= 0.0) )/sumn
```

```
  and of the y distribution.
h=-sum(nn(:,:)*log(nn(:,:)/sumn), mask=(nn(:,:) /= 0) )/sumn
  Total entropy: loop over both x and y.
hygx=h-hx                                 Uses equation (14.4.18),
hxgy=h-hy                                 as does this.
uygx=(hy-hygx)/(hy+TINY)                  Equation (14.4.15).
uxgy=(hx-hxgy)/(hx+TINY)                  Equation (14.4.16).
uxy=2.0_sp*(hx+hy-h)/(hx+hy+TINY)         Equation (14.4.17).
END SUBROUTINE cntab2
```

This code exploits both the dim feature of sum (see discussion after cntab1) and the mask feature to restrict the elements to be summed over.

⋆ ⋆ ⋆

```
SUBROUTINE pearsn(x,y,r,prob,z)
USE nrtype; USE nrutil, ONLY : assert_eq
USE nr, ONLY : betai
IMPLICIT NONE
REAL(SP), INTENT(OUT) :: r,prob,z
REAL(SP), DIMENSION(:), INTENT(IN) :: x,y
REAL(SP), PARAMETER :: TINY=1.0e-20_sp
```

Given two arrays x and y of the same size, this routine computes their correlation coefficient r (returned as r), the significance level at which the null hypothesis of zero correlation is disproved (prob whose small value indicates a significant correlation), and Fisher's z (returned as z), whose value can be used in further statistical tests as described above the routine in Volume 1.

Parameter: TINY will regularize the unusual case of complete correlation.

```
REAL(SP), DIMENSION(size(x)) :: xt,yt
REAL(SP) :: ax,ay,df,sxx,sxy,syy,t
INTEGER(I4B) :: n
n=assert_eq(size(x),size(y),'pearsn')
ax=sum(x)/n                                              Find the means.
ay=sum(y)/n
xt(:)=x(:)-ax                                            Compute the correlation co-
yt(:)=y(:)-ay                                               efficient.
sxx=dot_product(xt,xt)
syy=dot_product(yt,yt)
sxy=dot_product(xt,yt)
r=sxy/(sqrt(sxx*syy)+TINY)
z=0.5_sp*log(((1.0_sp+r)+TINY)/((1.0_sp-r)+TINY))        Fisher's z transformation.
df=n-2
t=r*sqrt(df/(((1.0_sp-r)+TINY)*((1.0_sp+r)+TINY)))       Equation (14.5.5).
prob=betai(0.5_sp*df,0.5_sp,df/(df+t**2))                Student's t probability.
prob=erfcc(abs(z*sqrt(n-1.0_sp))/SQRT2)
  For large n, this easier computation of prob, using the short routine erfcc, would give
  approximately the same value.
END SUBROUTINE pearsn
```

⋆ ⋆ ⋆

```
SUBROUTINE spear(data1,data2,d,zd,probd,rs,probrs)
USE nrtype; USE nrutil, ONLY : assert_eq
USE nr, ONLY : betai,erfcc,sort2
IMPLICIT NONE
REAL(SP), DIMENSION(:), INTENT(IN) :: data1,data2
REAL(SP), INTENT(OUT) :: d,zd,probd,rs,probrs
```

Given two data arrays of the same size, data1 and data2, this routine returns their sum-squared difference of ranks as D, the number of standard deviations by which D deviates from its null-hypothesis expected value as zd, the two-sided significance level of this deviation as probd, Spearman's rank correlation r_s as rs, and the two-sided significance level of its deviation from zero as probrs. data1 and data2 are not modified. A small value of either probd or probrs indicates a significant correlation (rs positive) or anticorrelation (rs negative).

```
INTEGER(I4B) :: n
REAL(SP) :: aved,df,en,en3n,fac,sf,sg,t,vard
REAL(SP), DIMENSION(size(data1)) :: wksp1,wksp2
n=assert_eq(size(data1),size(data2),'spear')
wksp1(:)=data1(:)
wksp2(:)=data2(:)
call sort2(wksp1,wksp2)
call crank(wksp1,sf)
call sort2(wksp2,wksp1)
call crank(wksp2,sg)
```

Sort each of the data arrays, and convert the entries to ranks. The values sf and sg return the sums $\sum(f_k^3 - f_k)$ and $\sum(g_m^3 - g_m)$, respectively.

```
wksp1(:)=wksp1(:)-wksp2(:)
d=dot_product(wksp1,wksp1)
```

Sum the squared difference of ranks.

```
en=n
en3n=en**3-en
aved=en3n/6.0_sp-(sf+sg)/12.0_sp
```

Expectation value of D,

```
fac=(1.0_sp-sf/en3n)*(1.0_sp-sg/en3n)
vard=((en-1.0_sp)*en**2*(en+1.0_sp)**2/36.0_sp)*fac
zd=(d-aved)/sqrt(vard)
probd=erfcc(abs(zd)/SQRT2)
```

and variance of D give number of standard deviations, and significance.

```
rs=(1.0_sp-(6.0_sp/en3n)*(d+(sf+sg)/12.0_sp))/sqrt(fac)
fac=(1.0_sp+rs)*(1.0_sp-rs)
```

Rank correlation coefficient,

```
if (fac > 0.0) then
    t=rs*sqrt((en-2.0_sp)/fac)
```

and its t value,

```
    df=en-2.0_sp
    probrs=betai(0.5_sp*df,0.5_sp,df/(df+t**2))
```

give its significance.

```
else
    probrs=0.0
end if
CONTAINS

SUBROUTINE crank(w,s)
USE nrtype; USE nrutil, ONLY : arth,array_copy
IMPLICIT NONE
REAL(SP), INTENT(OUT) :: s
REAL(SP), DIMENSION(:), INTENT(INOUT) :: w
```

Given a sorted array w, replaces the elements by their rank, including midranking of ties, and returns as s the sum of $f^3 - f$, where f is the number of elements in each tie.

```
INTEGER(I4B) :: i,n,ndum,nties
INTEGER(I4B), DIMENSION(size(w)) :: tstart,tend,tie,idx
n=size(w)
idx(:)=arth(1,1,n)
```

Index vector.

```
tie(:)=merge(1,0,w == eoshift(w,-1))
```

Look for ties: Compare each element to the one before. If it's equal, it's part of a tie, and we put 1 into tie. Otherwise we put 0.

```
tie(1)=0
```

Boundary; the first element must be zero.

```
w(:)=idx(:)
```

Assign ranks ignoring possible ties.

```
if (all(tie == 0)) then
```

No ties—we're done.

```
    s=0.0
    RETURN
end if
call array_copy(pack(idx(:),tie(:)<eoshift(tie(:),1)),tstart,nties,ndum)
```

```
  Look for 0 → 1 transitions in tie, which mean that the 0 element is the start of a tie run.
  Store index of each transition in tstart. nties is the number of ties found.
tend(1:nties)=pack(idx(:),tie(:)>eoshift(tie(:),1))
  Look for 1 → 0 transitions in tie, which mean that the 1 element is the end of a tie run.
do i=1,nties                                          Midrank assignments.
    w(tstart(i):tend(i))=(tstart(i)+tend(i))/2.0_sp
end do
tend(1:nties)=tend(1:nties)-tstart(1:nties)+1         Now calculate s.
s=sum(tend(1:nties)**3-tend(1:nties))
END SUBROUTINE crank
END SUBROUTINE spear
```

To understand how the parallel version of crank works, let's consider an example of 9 elements in the array w, which is input in sorted order to crank. The elements in our example are given in the second line of the following table:

index	1	2	3	4	5	6	7	8	9	
data in w	0	0	1	1	1	2	3	4	4	
shift right	0	0	0	1	1	1	2	3	4	
compare	1	1	0	1	1	0	0	0	1	
tie array	0	1	0	1	1	0	0	0	1	
shift left	1	0	1	1	0	0	0	1	0	
$0 \to 1$	1		3					8		start index
$1 \to 0$		2			5				9	stop index

We look for ties by comparing this array with itself, right shifted by one element ("shift right" in table). We record a 1 for each element that is the same, a 0 for each element that is different ("compare"). A 1 indicates the element is part of a tie with the *preceding* element, so we always set the first element to 0, even if it was a 1 as in our example. This gives the "tie array." Now wherever the tie array makes a transition $0 \to 1$ indicates the start of a tie run, while a $1 \to 0$ transition indicates the end of a tie run. We find these transitions by comparing the tie array to itself left shifted by one ("shift left"). If the tie array element is smaller than the shifted array element, we have a $0 \to 1$ transition and we record the corresponding index as the start of a tie. Similarly if the tie array element is larger we record the index as the end of a tie. Note that the shifts must be end-off shifts with zeros inserted in the gaps for the boundary conditions to work.

f90

```
call array_copy(pack(idx(:),tie(:)<eoshift(tie(:),1)),
    tstart,nties,ndum)
```

The start indices (1, 3, and 8 in our example above) are here packed into the first few elements of tstart. array_copy is a useful routine in nrutil for copying elements from one array to another, when the number of elements to be copied is not known in advance. This line of code is equivalent to

```
tstart(:)=0
tstart(:)=pack(idx(:), tie(:) < eoshift(tie(:),1),tstart(:))
nties=count(tstart(:) > 0)
```

The point is that we don't know how many elements pack is going to select. We have to make sure the dimensions of both sides of the pack statement are the same, so we set the optional third argument of pack to tstart. We then make a separate pass through tstart to count how many elements we copied. Alternatively, we could have used an additional logical array mask and coded this as

```
mask(:)=tie(:) < eoshift(tie(:),1)
nties=count(mask)
tstart(1:nties)=pack(idx(:),mask)
```

But we still need two passes through the mask array. The beauty of the array_copy routine is that nties is determined from the *size* of the first argument, without the necessity for a second pass through the array.

⋆ ⋆ ⋆

```
SUBROUTINE kendl1(data1,data2,tau,z,prob)
USE nrtype; USE nrutil, ONLY : assert_eq
USE nr, ONLY : erfcc
IMPLICIT NONE
REAL(SP), INTENT(OUT) :: tau,z,prob
REAL(SP), DIMENSION(:), INTENT(IN) :: data1,data2
    Given same-size data arrays data1 and data2, this program returns Kendall's τ as tau, its
    number of standard deviations from zero as z, and its two-sided significance level as prob.
    Small values of prob indicate a significant correlation (tau positive) or anticorrelation
    (tau negative).
INTEGER(I4B) :: is,j,n,n1,n2
REAL(SP) :: var
REAL(SP), DIMENSION(size(data1)) :: a1,a2
n=assert_eq(size(data1),size(data2),'kendl1')
n1=0                                This will be the argument of one square root in (14.6.8),
n2=0                                and this the other.
is=0                                This will be the numerator in (14.6.8).
do j=1,n-1                                         For each first member of pair,
    a1(j+1:n)=data1(j)-data1(j+1:n)                loop over second member.
    a2(j+1:n)=data2(j)-data2(j+1:n)
    n1=n1+count(a1(j+1:n) /= 0.0)
    n2=n2+count(a2(j+1:n) /= 0.0)
      Now accumulate the numerator in (14.6.8):
    is=is+count((a1(j+1:n) > 0.0 .and. a2(j+1:n) > 0.0) &
        .or. (a1(j+1:n) < 0.0 .and. a2(j+1:n) < 0.0)) - &
        count((a1(j+1:n) > 0.0 .and. a2(j+1:n) < 0.0) &
        .or. (a1(j+1:n) < 0.0 .and. a2(j+1:n) > 0.0))
end do
tau=real(is,sp)/sqrt(real(n1,sp)*real(n2,sp))      Equation (14.6.8).
var=(4.0_sp*n+10.0_sp)/(9.0_sp*n*(n-1.0_sp))       Equation (14.6.9).
z=tau/sqrt(var)
prob=erfcc(abs(z)/SQRT2)                           Significance.
END SUBROUTINE kendl1
```

```
SUBROUTINE kendl2(tab,tau,z,prob)
USE nrtype; USE nrutil, ONLY : cumsum
USE nr, ONLY : erfcc
IMPLICIT NONE
REAL(SP), DIMENSION(:,:), INTENT(IN) :: tab
REAL(SP), INTENT(OUT) :: tau,z,prob
    Given a two-dimensional table tab such that tab(k,l) contains the number of events falling
    in bin k of one variable and bin l of another, this program returns Kendall's τ as tau, its
    number of standard deviations from zero as z, and its two-sided significance level as prob.
    Small values of prob indicate a significant correlation (tau positive) or anticorrelation (tau
```

negative) between the two variables. Although tab is a real array, it will normally contain integral values.

```
REAL(SP), DIMENSION(size(tab,1),size(tab,2)) :: cum,cumt
INTEGER(I4B) :: i,j,ii,jj
REAL(SP) :: sc,sd,en1,en2,points,var
ii=size(tab,1)
jj=size(tab,2)
do i=1,ii                                          Get cumulative sums leftward along
    cumt(i,jj:1:-1)=cumsum(tab(i,jj:1:-1))             rows.
end do
en2=sum(tab(1:ii,1:jj-1)*cumt(1:ii,2:jj))          Tally the extra-y pairs.
do j=1,jj                                          Get counts of points to lower-right
    cum(ii:1:-1,j)=cumsum(cumt(ii:1:-1,j))             of each cell in cum.
end do
points=cum(1,1)                                    Total number of entries in table.
sc=sum(tab(1:ii-1,1:jj-1)*cum(2:ii,2:jj))          Tally the concordant pairs.
do j=1,jj                                          Now get counts of points to upper-
    cum(1:ii,j)=cumsum(cumt(1:ii,j))                   right of each cell in cum,
end do
sd=sum(tab(2:ii,1:jj-1)*cum(1:ii-1,2:jj))          giving tally of discordant points.
do j=1,jj                                          Finally, get cumulative sums upward
    cumt(ii:1:-1,j)=cumsum(tab(ii:1:-1,j))             along columns,
end do
en1=sum(tab(1:ii-1,1:jj)*cumt(2:ii,1:jj))          giving the count of extra-x pairs,
tau=(sc-sd)/sqrt((en1+sc+sd)*(en2+sc+sd))          and compute desired results.
var=(4.0_sp*points+10.0_sp)/(9.0_sp*points*(points-1.0_sp))
z=tau/sqrt(var)
prob=erfcc(abs(z)/SQRT2)
END SUBROUTINE kendl2
```

The underlying algorithm in kendl2 might seem to require looping over all *pairs* of cells in the two-dimensional table tab. Actually, however, clever use of the cumsum utility function reduces this to a simple loop over all the cells; moreover this "loop" parallelizes into a simple parallel product and call to the sum intrinsic. The basic idea is shown in the following table:

		d	d
	t	y	y
	x	c	c
	x	c	c
	x	c	c

Relative to the cell marked t (which we use to denote the numerical value it contains), the cells marked d contribute to the "discordant" tally in Volume 1's equation (14.6.8), while the cells marked c contribute to the "concordant" tally. Likewise, the cells marked x and y contribute, respectively, to the "extra-x" and "extra-y" tallies. What about the cells left blank? Since we want to count pairs of cells only *once*, without duplication, these cells will be counted, relative to the location shown as t, when t itself moves into the blank-cell area.

Symbolically we have

$$\begin{aligned}
\text{concordant} &= \sum_n t_n \left(\sum_{\text{lower right}} c_m \right) \\
\text{discordant} &= \sum_n t_n \left(\sum_{\text{upper right}} d_m \right) \\
\text{extra-}x &= \sum_n t_n \left(\sum_{\text{below}} x_m \right) \\
\text{extra-}y &= \sum_n t_n \left(\sum_{\text{to the right}} y_m \right)
\end{aligned} \tag{B14.1}$$

Here n varies over all the positions in the table, while the limits of the inner sums are relative to the position of n. (The letters t_n, c_m, d_m, x_m, y_m all represent the value in a cell; we use different letters only to make the relation with the above table clear.) Now the final trick is to recognize that the inner sums, over cells to the lower- or upper-right, below, and to the right can be done in parallel by cumulative sums (cumsum) sweeping to the right and up. The routine does these in a nonintuitive order merely to be able to reuse maximally the scratch spaces cum and cumt.

⋆ ⋆ ⋆

```
SUBROUTINE ks2d1s(x1,y1,quadvl,d1,prob)
USE nrtype; USE nrutil, ONLY : assert_eq
USE nr, ONLY : pearsn,probks,quadct
IMPLICIT NONE
REAL(SP), DIMENSION(:), INTENT(IN) :: x1,y1
REAL(SP), INTENT(OUT) :: d1,prob
INTERFACE
    SUBROUTINE quadvl(x,y,fa,fb,fc,fd)
    USE nrtype
    IMPLICIT NONE
    REAL(SP), INTENT(IN) :: x,y
    REAL(SP), INTENT(OUT) :: fa,fb,fc,fd
    END SUBROUTINE quadvl
END INTERFACE
```

Two-dimensional Kolmogorov-Smirnov test of one sample against a model. Given the x- and y-coordinates of a set of data points in arrays x1 and y1 of the same length, and given a user-supplied function quadvl that exemplifies the model, this routine returns the two-dimensional K-S statistic as d1, and its significance level as prob. Small values of prob show that the sample is significantly different from the model. Note that the test is slightly distribution-dependent, so prob is only an estimate.

```
INTEGER(I4B) :: j,n1
REAL(SP) :: dum,dumm,fa,fb,fc,fd,ga,gb,gc,gd,r1,rr,sqen
n1=assert_eq(size(x1),size(y1),'ks2d1s')
d1=0.0
```

```
do j=1,n1                                       Loop over the data points.
    call quadct(x1(j),y1(j),x1,y1,fa,fb,fc,fd)
    call quadvl(x1(j),y1(j),ga,gb,gc,gd)
    d1=max(d1,abs(fa-ga),abs(fb-gb),abs(fc-gc),abs(fd-gd))
      For both the sample and the model, the distribution is integrated in each of four quad-
      rants, and the maximum difference is saved.
end do
call pearsn(x1,y1,r1,dum,dumm)          Get the linear correlation coefficient r1.
sqen=sqrt(real(n1,sp))
rr=sqrt(1.0_sp-r1**2)
  Estimate the probability using the K-S probability function probks.
prob=probks(d1*sqen/(1.0_sp+rr*(0.25_sp-0.75_sp/sqen)))
END SUBROUTINE ks2d1s
```

```
SUBROUTINE quadct(x,y,xx,yy,fa,fb,fc,fd)
USE nrtype; USE nrutil, ONLY : assert_eq
IMPLICIT NONE
REAL(SP), INTENT(IN) :: x,y
REAL(SP), DIMENSION(:), INTENT(IN) :: xx,yy
REAL(SP), INTENT(OUT) :: fa,fb,fc,fd
    Given an origin (x, y), and an array of points with coordinates xx and yy, count how many of
    them are in each quadrant around the origin, and return the normalized fractions. Quadrants
    are labeled alphabetically, counterclockwise from the upper right. Used by ks2d1s and
    ks2d2s.
INTEGER(I4B) :: na,nb,nc,nd,nn
REAL(SP) :: ff
nn=assert_eq(size(xx),size(yy),'quadct')
na=count(yy(:) > y .and. xx(:) > x)
nb=count(yy(:) > y .and. xx(:) <= x)
nc=count(yy(:) <= y .and. xx(:) <= x)
nd=nn-na-nb-nc
ff=1.0_sp/nn
fa=ff*na
fb=ff*nb
fc=ff*nc
fd=ff*nd
END SUBROUTINE quadct
```

```
SUBROUTINE quadvl(x,y,fa,fb,fc,fd)
USE nrtype
IMPLICIT NONE
REAL(SP), INTENT(IN) :: x,y
REAL(SP), INTENT(OUT) :: fa,fb,fc,fd
    This is a sample of a user-supplied routine to be used with ks2d1s. In this case, the model
    distribution is uniform inside the square -1 < x < 1, -1 < y < 1. In general this routine
    should return, for any point (x, y), the fraction of the total distribution in each of the
    four quadrants around that point. The fractions, fa, fb, fc, and fd, must add up to 1.
    Quadrants are alphabetical, counterclockwise from the upper right.
REAL(SP) :: qa,qb,qc,qd
qa=min(2.0_sp,max(0.0_sp,1.0_sp-x))
qb=min(2.0_sp,max(0.0_sp,1.0_sp-y))
qc=min(2.0_sp,max(0.0_sp,x+1.0_sp))
qd=min(2.0_sp,max(0.0_sp,y+1.0_sp))
fa=0.25_sp*qa*qb
fb=0.25_sp*qb*qc
fc=0.25_sp*qc*qd
fd=0.25_sp*qd*qa
END SUBROUTINE quadvl
```

```
SUBROUTINE ks2d2s(x1,y1,x2,y2,d,prob)
USE nrtype; USE nrutil, ONLY : assert_eq
USE nr, ONLY : pearsn,probks,quadct
IMPLICIT NONE
REAL(SP), DIMENSION(:), INTENT(IN) :: x1,y1,x2,y2
REAL(SP), INTENT(OUT) :: d,prob
```

Compute two-dimensional Kolmogorov-Smirnov test on two samples. Input are the x- and y-coordinates of the first sample in arrays x1 and y1 of the same length, and of the second sample in arrays x2 and y2 of the same length (possibly different from the length of the first sample). The routine returns the two-dimensional, two-sample K-S statistic as d, and its significance level as prob. Small values of prob show that the two samples are significantly different. Note that the test is slightly distribution-dependent, so prob is only an estimate.

```
INTEGER(I4B) :: j,n1,n2
REAL(SP) :: d1,d2,dum,dumm,fa,fb,fc,fd,ga,gb,gc,gd,r1,r2,rr,sqen
n1=assert_eq(size(x1),size(y1),'ks2d2s: n1')
n2=assert_eq(size(x2),size(y2),'ks2d2s: n2')
d1=0.0
do j=1,n1                                   First, use points in the first sample as origins.
    call quadct(x1(j),y1(j),x1,y1,fa,fb,fc,fd)
    call quadct(x1(j),y1(j),x2,y2,ga,gb,gc,gd)
    d1=max(d1,abs(fa-ga),abs(fb-gb),abs(fc-gc),abs(fd-gd))
end do
d2=0.0
do j=1,n2                                   Then, use points in the second sample as ori-
    call quadct(x2(j),y2(j),x1,y1,fa,fb,fc,fd)       gins.
    call quadct(x2(j),y2(j),x2,y2,ga,gb,gc,gd)
    d2=max(d2,abs(fa-ga),abs(fb-gb),abs(fc-gc),abs(fd-gd))
end do
d=0.5_sp*(d1+d2)                            Average the K-S statistics.
sqen=sqrt(real(n1,sp)*real(n2,sp)/real(n1+n2,sp))
call pearsn(x1,y1,r1,dum,dumm)              Get the linear correlation coefficient for each sam-
call pearsn(x2,y2,r2,dum,dumm)                  ple.
rr=sqrt(1.0_sp-0.5_sp*(r1**2+r2**2))
  Estimate the probability using the K-S probability function probks.
prob=probks(d*sqen/(1.0_sp+rr*(0.25_sp-0.75_sp/sqen)))
END SUBROUTINE ks2d2s
```

⋆ ⋆ ⋆

```
FUNCTION savgol(nl,nrr,ld,m)
USE nrtype; USE nrutil, ONLY : arth,assert,poly
USE nr, ONLY : lubksb,ludcmp
IMPLICIT NONE
INTEGER(I4B), INTENT(IN) :: nl,nrr,ld,m
```

Returns in array c, in wrap-around order (N.B.!) consistent with the argument respns in routine convlv, a set of Savitzky-Golay filter coefficients. nl is the number of leftward (past) data points used, while nrr is the number of rightward (future) data points, making the total number of data points used nl $+$ nrr $+1$. ld is the order of the derivative desired (e.g., ld $=0$ for smoothed function). m is the order of the smoothing polynomial, also equal to the highest conserved moment; usual value is m $=2$ or m $=4$.

```
REAL(SP), DIMENSION(nl+nrr+1) :: savgol
INTEGER(I4B) :: imj,ipj,mm,np
INTEGER(I4B), DIMENSION(m+1) :: indx
REAL(SP) :: d,sm
REAL(SP), DIMENSION(m+1) :: b
REAL(SP), DIMENSION(m+1,m+1) :: a
INTEGER(I4B) :: irng(nl+nrr+1)
call assert(nl >= 0, nrr >= 0, ld <= m, nl+nrr >= m, 'savgol args')
do ipj=0,2*m                                Set up the normal equations of the desired least
    sm=sum(arth(1.0_sp,1.0_sp,nrr)**ipj)+&          squares fit.
        sum(arth(-1.0_sp,-1.0_sp,nl)**ipj)
```

```
    if (ipj == 0) sm=sm+1.0_sp
    mm=min(ipj,2*m-ipj)
    do imj=-mm,mm,2
        a(1+(ipj+imj)/2,1+(ipj-imj)/2)=sm
    end do
end do
call ludcmp(a(:,:),indx(:),d)              Solve them: LU decomposition.
b(:)=0.0
b(ld+1)=1.0                                Right-hand-side vector is unit vector, depending
call lubksb(a(:,:),indx(:),b(:))               on which derivative we want.
  Backsubstitute, giving one row of the inverse matrix.
savgol(:)=0.0                              Zero the output array (it may be bigger than
irng(:)=arth(-nl,1,nrr+nl+1)                   number of coefficients).
np=nl+nrr+1
savgol(mod(np-irng(:),np)+1)=poly(real(irng(:),sp),b(:))
  Each Savitzky-Golay coefficient is the value of the polynomial in (14.8.6) at the corresponding
  integer. The polynomial coefficients are a row of the inverse matrix. The mod function takes
  care of the wrap-around order.
END FUNCTION savgol
```

`do imj=-mm,mm,2` Here is an example of a loop that cannot be parallelized in the framework of Fortran 90: We need to access "skew" sections of the matrix `a`.

`savgol...=poly(real(irng(:),sp),b(:)))` The `poly` function in `nrutil` returns the value of a polynomial, here the one in equation (14.8.6). We need the explicit kind type parameter `sp` in the `real` function, otherwise it would return type default real for the integer argument and would not automatically convert to double precision if desired.

Chapter B15. Modeling of Data

```
SUBROUTINE fit(x,y,a,b,siga,sigb,chi2,q,sig)
USE nrtype; USE nrutil, ONLY : assert_eq
USE nr, ONLY : gammq
IMPLICIT NONE
REAL(SP), DIMENSION(:), INTENT(IN) :: x,y
REAL(SP), INTENT(OUT) :: a,b,siga,sigb,chi2,q
REAL(SP), DIMENSION(:), OPTIONAL, INTENT(IN) :: sig
```

Given a set of data points in same-size arrays x and y, fit them to a straight line $y = a + bx$ by minimizing χ^2. sig is an optional array of the same length containing the individual standard deviations. If it is present, then a,b are returned with their respective probable uncertainties siga and sigb, the chi-square chi2, and the goodness-of-fit probability q (that the fit would have χ^2 this large or larger). If sig is not present, then q is returned as 1.0 and the normalization of chi2 is to unit standard deviation on all points.

```
INTEGER(I4B) :: ndata
REAL(SP) :: sigdat,ss,sx,sxoss,sy,st2
REAL(SP), DIMENSION(size(x)), TARGET :: t
REAL(SP), DIMENSION(:), POINTER :: wt
if (present(sig)) then
    ndata=assert_eq(size(x),size(y),size(sig),'fit')
    wt=>t                                   Use temporary variable t to store weights.
    wt(:)=1.0_sp/(sig(:)**2)
    ss=sum(wt(:))                           Accumulate sums with weights.
    sx=dot_product(wt,x)
    sy=dot_product(wt,y)
else
    ndata=assert_eq(size(x),size(y),'fit')
    ss=real(size(x),sp)                     Accumulate sums without weights.
    sx=sum(x)
    sy=sum(y)
end if
sxoss=sx/ss
t(:)=x(:)-sxoss
if (present(sig)) then
    t(:)=t(:)/sig(:)
    b=dot_product(t/sig,y)
else
    b=dot_product(t,y)
end if
st2=dot_product(t,t)
b=b/st2                                     Solve for a, b, σa, and σb.
a=(sy-sx*b)/ss
siga=sqrt((1.0_sp+sx*sx/(ss*st2))/ss)
sigb=sqrt(1.0_sp/st2)
t(:)=y(:)-a-b*x(:)
q=1.0
if (present(sig)) then
    t(:)=t(:)/sig(:)
    chi2=dot_product(t,t)                   Calculate χ^2.
    if (ndata > 2) q=gammq(0.5_sp*(size(x)-2),0.5_sp*chi2)      Equation (15.2.1:
else
    chi2=dot_product(t,t)
```

```
    sigdat=sqrt(chi2/(size(x)-2))            For unweighted data evaluate typical
    siga=siga*sigdat                            sig using chi2, and adjust the
    sigb=sigb*sigdat                            standard deviations.
end if
END SUBROUTINE fit
```

f90 REAL(SP), DIMENSION(:), POINTER :: wt...wt=>t When standard deviations are supplied in sig, we need to compute the weights for the least squares fit in a temporary array wt. Later in the routine, we need another temporary array, which we call t to correspond to the variable in equation (15.2.15). It would be confusing to use the same name for both arrays. In Fortran 77 the arrays could share storage with an EQUIVALENCE declaration, but that is a deprecated feature in Fortran 90. We accomplish the same thing by making wt a pointer alias to t.

⋆ ⋆ ⋆

```
SUBROUTINE fitexy(x,y,sigx,sigy,a,b,siga,sigb,chi2,q)
USE nrtype; USE nrutil, ONLY : assert_eq,swap
USE nr, ONLY : avevar,brent,fit,gammq,mnbrak,zbrent
USE chixyfit
IMPLICIT NONE
REAL(SP), DIMENSION(:), INTENT(IN) :: x,y,sigx,sigy
REAL(SP), INTENT(OUT) :: a,b,siga,sigb,chi2,q
REAL(SP), PARAMETER :: POTN=1.571000_sp,BIG=1.0e30_sp,ACC=1.0e-3_sp
    Straight-line fit to input data x and y with errors in both x and y, the respective standard
    deviations being the input quantities sigx and sigy. x, y, sigx, and sigy are all arrays of
    the same length. Output quantities are a and b such that y = a + bx minimizes χ^2, whose
    value is returned as chi2. The χ^2 probability is returned as q, a small value indicating
    a poor fit (sometimes indicating underestimated errors). Standard errors on a and b are
    returned as siga and sigb. These are not meaningful if either (i) the fit is poor, or (ii) b
    is so large that the data are consistent with a vertical (infinite b) line. If siga and sigb
    are returned as BIG, then the data are consistent with all values of b.
INTEGER(I4B) :: j,n
REAL(SP), DIMENSION(size(x)), TARGET :: xx,yy,sx,sy,ww
REAL(SP), DIMENSION(6) :: ang,ch
REAL(SP) :: amx,amn,varx,vary,scale,bmn,bmx,d1,d2,r2,&
    dum1,dum2,dum3,dum4,dum5
n=assert_eq(size(x),size(y),size(sigx),size(sigy),'fitexy')
xxp=>xx                                     Set up communication with function chixy
yyp=>yy                                         through global variables in the module
sxp=>sx                                         chixyfit.
syp=>sy
wwp=>ww
call avevar(x,dum1,varx)                    Find the x and y variances, and scale the
call avevar(y,dum1,vary)                        data.
scale=sqrt(varx/vary)
xx(:)=x(:)
yy(:)=y(:)*scale
sx(:)=sigx(:)
sy(:)=sigy(:)*scale
ww(:)=sqrt(sx(:)**2+sy(:)**2)               Use both x and y weights in first trial fit.
call fit(xx,yy,dum1,b,dum2,dum3,dum4,dum5,ww)      Trial fit for b.
offs=0.0
ang(1)=0.0                                  Construct several angles for reference points.
ang(2)=atan(b)                              Make b an angle.
ang(4)=0.0
ang(5)=ang(2)
ang(6)=POTN
do j=4,6
    ch(j)=chixy(ang(j))
```

```
end do
call mnbrak(ang(1),ang(2),ang(3),ch(1),ch(2),ch(3),chixy)
  Bracket the χ^2 minimum and then locate it with brent.
chi2=brent(ang(1),ang(2),ang(3),chixy,ACC,b)
chi2=chixy(b)
a=aa
q=gammq(0.5_sp*(n-2),0.5_sp*chi2)              Compute χ^2 probability.
r2=1.0_sp/sum(ww(:))                           Save inverse sum of weights at the minimum.
bmx=BIG                                        Now, find standard errors for b as points where
bmn=BIG                                            Δχ^2 = 1.
offs=chi2+1.0_sp
do j=1,6                                       Go through saved values to bracket the de-
    if (ch(j) > offs) then                         sired roots. Note periodicity in slope an-
        d1=mod(abs(ang(j)-b),PI)                   gles.
        d2=PI-d1
        if (ang(j) < b) call swap(d1,d2)
        if (d1 < bmx) bmx=d1
        if (d2 < bmn) bmn=d2
    end if
end do
if (bmx < BIG) then                            Call zbrent to find the roots.
    bmx=zbrent(chixy,b,b+bmx,ACC)-b
    amx=aa-a
    bmn=zbrent(chixy,b,b-bmn,ACC)-b
    amn=aa-a
    sigb=sqrt(0.5_sp*(bmx**2+bmn**2))/(scale*cos(b)**2)
    siga=sqrt(0.5_sp*(amx**2+amn**2)+r2)/scale     Error in a has additional piece
else                                                   r2.
    sigb=BIG
    siga=BIG
end if
a=a/scale                                      Unscale the answers.
b=tan(b)/scale
END SUBROUTINE fitexy
```

f90 `USE chixyfit` We need to pass arrays and other variables to `chixy`, but not as arguments. See §21.5 and the discussion of `fminln` on p. 1197 for two good ways to do this. The pointer construction here is analogous to the one used in `fminln`.

```
MODULE chixyfit
USE nrtype; USE nrutil, ONLY : nrerror
REAL(SP), DIMENSION(:), POINTER :: xxp,yyp,sxp,syp,wwp
REAL(SP) :: aa,offs
CONTAINS

FUNCTION chixy(bang)
IMPLICIT NONE
REAL(SP), INTENT(IN) :: bang
REAL(SP) :: chixy
REAL(SP), PARAMETER :: BIG=1.0e30_sp
  Captive function of fitexy, returns the value of (χ^2 − offs) for the slope b=tan(bang).
  Scaled data and offs are communicated via the module chixyfit.
REAL(SP) :: avex,avey,sumw,b
if (.not. associated(wwp)) call nrerror("chixy: bad pointers")
b=tan(bang)
wwp(:)=(b*sxp(:))**2+syp(:)**2
where (wwp(:) < 1.0/BIG)
    wwp(:)=BIG
elsewhere
    wwp(:)=1.0_sp/wwp(:)
end where
```

```
sumw=sum(wwp)
avex=dot_product(wwp,xxp)/sumw
avey=dot_product(wwp,yyp)/sumw
aa=avey-b*avex
chixy=sum(wwp(:)*(yyp(:)-aa-b*xxp(:))**2)-offs
END FUNCTION chixy
END MODULE chixyfit
```

⋆ ⋆ ⋆

```
SUBROUTINE lfit(x,y,sig,a,maska,covar,chisq,funcs)
USE nrtype; USE nrutil, ONLY : assert_eq,diagmult,nrerror
USE nr, ONLY :covsrt,gaussj
IMPLICIT NONE
REAL(SP), DIMENSION(:), INTENT(IN) :: x,y,sig
REAL(SP), DIMENSION(:), INTENT(INOUT) :: a
LOGICAL(LGT), DIMENSION(:), INTENT(IN) :: maska
REAL(SP), DIMENSION(:,:), INTENT(INOUT) :: covar
REAL(SP), INTENT(OUT) :: chisq
INTERFACE
    SUBROUTINE funcs(x,arr)
    USE nrtype
    IMPLICIT NONE
    REAL(SP),INTENT(IN) :: x
    REAL(SP), DIMENSION(:), INTENT(OUT) :: arr
    END SUBROUTINE funcs
END INTERFACE
```

Given a set of N data points x, y with individual standard deviations sig, all arrays of length N, use χ^2 minimization to fit for some or all of the M coefficients a of a function that depends linearly on a, $y = \sum_{i=1}^{M} \mathrm{a}_i \times \mathtt{afunc}_i(x)$. The input logical array maska of length M indicates by true entries those components of a that should be fitted for, and by false entries those components that should be held fixed at their input values. The program returns values for a, $\chi^2 =$ chisq, and the $M \times M$ covariance matrix covar. (Parameters held fixed will return zero covariances.) The user supplies a subroutine funcs(x,afunc) that returns the M basis functions evaluated at $x =$ x in the array afunc.

```
INTEGER(I4B) :: i,j,k,l,ma,mfit,n
REAL(SP) :: sig2i,wt,ym
REAL(SP), DIMENSION(size(maska)) :: afunc
REAL(SP), DIMENSION(size(maska),1) :: beta
n=assert_eq(size(x),size(y),size(sig),'lfit: n')
ma=assert_eq(size(maska),size(a),size(covar,1),size(covar,2),'lfit: ma')
mfit=count(maska)                         Number of parameters to fit for.
if (mfit == 0) call nrerror('lfit: no parameters to be fitted')
covar(1:mfit,1:mfit)=0.0                  Initialize the (symmetric) matrix.
beta(1:mfit,1)=0.0
do i=1,n                                  Loop over data to accumulate coefficients of
    call funcs(x(i),afunc)                    the normal equations.
    ym=y(i)
    if (mfit < ma) ym=ym-sum(a(1:ma)*afunc(1:ma), mask=.not. maska)
      Subtract off dependences on known pieces of the fitting function.
    sig2i=1.0_sp/sig(i)**2
    j=0
    do l=1,ma
        if (maska(l)) then
            j=j+1
            wt=afunc(l)*sig2i
            k=count(maska(1:l))
            covar(j,1:k)=covar(j,1:k)+wt*pack(afunc(1:l),maska(1:l))
            beta(j,1)=beta(j,1)+ym*wt
        end if
    end do
```

```
end do
call diagmult(covar(1:mfit,1:mfit),0.5_sp)
covar(1:mfit,1:mfit)= &                        Fill in above the diagonal from symmetry.
    covar(1:mfit,1:mfit)+transpose(covar(1:mfit,1:mfit))
call gaussj(covar(1:mfit,1:mfit),beta(1:mfit,1:1))       Matrix solution.
a(1:ma)=unpack(beta(1:ma,1),maska,a(1:ma))
  Partition solution to appropriate coefficients a.
chisq=0.0                                      Evaluate χ^2 of the fit.
do i=1,n
    call funcs(x(i),afunc)
    chisq=chisq+((y(i)-dot_product(a(1:ma),afunc(1:ma)))/sig(i))**2
end do
call covsrt(covar,maska)                       Sort covariance matrix to true order of fitting
END SUBROUTINE lfit                                coefficients.
```

f90 `if (mfit < ma) ym=ym-sum(a(1:ma)*afunc(1:ma), mask=.not. maska)` This is the first of several uses of maska in this routine to control which elements of an array are to be used. Here we include in the sum only elements for which maska is false, i.e., elements corresponding to parameters that are not being fitted for.

`covar(j,1:k)=covar(j,1:k)+wt*pack(afunc(1:l),maska(1:l))` Here maska controls which elements of afunc get packed into the covariance matrix.

`call diagmult(covar(1:mfit,1:mfit),0.5_sp)` See discussion of diagadd after hqr on p. 1234.

`a(1:ma)=unpack(beta(1:ma,1),maska,a(1:ma))` And here maska controls which elements of beta get unpacked into the appropriate slots in a. Where maska is false, corresponding elements are selected from the third argument of unpack, here a itself. The net effect is that those elements remain unchanged.

⋆ ⋆ ⋆

```
SUBROUTINE covsrt(covar,maska)
USE nrtype; USE nrutil, ONLY : assert_eq,swap
IMPLICIT NONE
REAL(SP), DIMENSION(:,:), INTENT(INOUT) :: covar
LOGICAL(LGT), DIMENSION(:), INTENT(IN) :: maska
    Expand in storage the covariance matrix covar, so as to take into account parameters that
    are being held fixed. (For the latter, return zero covariances.)
INTEGER(I4B) :: ma,mfit,j,k
ma=assert_eq(size(covar,1),size(covar,2),size(maska),'covsrt')
mfit=count(maska)
covar(mfit+1:ma,1:ma)=0.0
covar(1:ma,mfit+1:ma)=0.0
k=mfit
do j=ma,1,-1
    if (maska(j)) then
        call swap(covar(1:ma,k),covar(1:ma,j))
        call swap(covar(k,1:ma),covar(j,1:ma))
        k=k-1
    end if
end do
END SUBROUTINE covsrt
```

⋆ ⋆ ⋆

```
SUBROUTINE svdfit(x,y,sig,a,v,w,chisq,funcs)
USE nrtype; USE nrutil, ONLY : assert_eq,vabs
USE nr, ONLY : svbksb,svdcmp
IMPLICIT NONE
REAL(SP), DIMENSION(:), INTENT(IN) :: x,y,sig
REAL(SP), DIMENSION(:), INTENT(OUT) :: a,w
REAL(SP), DIMENSION(:,:), INTENT(OUT) :: v
REAL(SP), INTENT(OUT) :: chisq
INTERFACE
    FUNCTION funcs(x,n)
    USE nrtype
    IMPLICIT NONE
    REAL(SP), INTENT(IN) :: x
    INTEGER(I4B), INTENT(IN) :: n
    REAL(SP), DIMENSION(n) :: funcs
    END FUNCTION funcs
END INTERFACE
REAL(SP), PARAMETER :: TOL=1.0e-5_sp
```

Given a set of N data points x, y with individual standard deviations sig, all arrays of length N, use χ^2 minimization to determine the M coefficients a of a function that depends linearly on a, $y = \sum_{i=1}^{M} \mathtt{a}_i \times \mathtt{afunc}_i(x)$. Here we solve the fitting equations using singular value decomposition of the $N \times M$ matrix, as in §2.6. On output, the $M \times M$ array v and the vector w of length M define part of the singular value decomposition, and can be used to obtain the covariance matrix. The program returns values for the M fit parameters a, and χ^2, chisq. The user supplies a subroutine funcs(x,afunc) that returns the M basis functions evaluated at $x = X$ in the array afunc.

```
INTEGER(I4B) :: i,ma,n
REAL(SP), DIMENSION(size(x)) :: b,sigi
REAL(SP), DIMENSION(size(x),size(a)) :: u,usav
n=assert_eq(size(x),size(y),size(sig),'svdfit: n')
ma=assert_eq(size(a),size(v,1),size(v,2),size(w),'svdfit: ma')
sigi=1.0_sp/sig                          Accumulate coefficients of the fitting matrix in
b=y*sigi                                     u.
do i=1,n
    usav(i,:)=funcs(x(i),ma)
end do
u=usav*spread(sigi,dim=2,ncopies=ma)
usav=u
call svdcmp(u,w,v)                       Singular value decomposition.
where (w < TOL*maxval(w)) w=0.0          Edit the singular values, given TOL from the pa-
call svbksb(u,w,v,b,a)                       rameter statement.
chisq=vabs(matmul(usav,a)-b)**2          Evaluate chi-square.
END SUBROUTINE svdfit
```

f90 u=usav*spread(sigi,dim=2,ncopies=ma) Remember how spread works: the vector sigi is copied *along* the dimension 2, making a matrix whose columns are each a copy of sigi. The multiplication here is element by element, so each row of usav is multiplied by the corresponding element of sigi.

chisq=vabs(matmul(usav,a)-b)**2 Fortran 90's matmul intrinsic allows us to evaluate χ^2 from the mathematical definition in terms of matrices. vabs in nrutil returns the length of a vector (L_2 norm).

```
SUBROUTINE svdvar(v,w,cvm)
USE nrtype; USE nrutil, ONLY : assert_eq
IMPLICIT NONE
REAL(SP), DIMENSION(:,:), INTENT(IN) :: v
REAL(SP), DIMENSION(:), INTENT(IN) :: w
REAL(SP), DIMENSION(:,:), INTENT(OUT) :: cvm
```

```
   To evaluate the covariance matrix cvm of the fit for M parameters obtained by svdfit,
   call this routine with matrices v,w as returned from svdfit. The dimensions are M for
   w and M × M for v and cvm.
INTEGER(I4B) :: ma
REAL(SP), DIMENSION(size(w)) :: wti
ma=assert_eq((/size(v,1),size(v,2),size(w),size(cvm,1),size(cvm,2)/),&
    'svdvar')
where (w /= 0.0)
    wti=1.0_sp/(w*w)
elsewhere
    wti=0.0
end where
cvm=v*spread(wti,dim=1,ncopies=ma)
cvm=matmul(cvm,transpose(v))          Covariance matrix is given by (15.4.20).
END SUBROUTINE svdvar
```

f90 `where (w /= 0.0)...elsewhere...end where` This is the standard Fortran 90 construction for doing different things to a matrix depending on some condition. Here we want to avoid inverting elements of `w` that are zero.

`cvm=v*spread(wti,dim=1,ncopies=ma)` Each column of `v` gets multiplied by the corresponding element of `wti`. Contrast the construction `spread(...dim=2...)` in `svdfit`.

⋆ ⋆ ⋆

```
FUNCTION fpoly(x,n)
USE nrtype; USE nrutil, ONLY : geop
IMPLICIT NONE
REAL(SP), INTENT(IN) :: x
INTEGER(I4B), INTENT(IN) :: n
REAL(SP), DIMENSION(n) :: fpoly
   Fitting routine for a polynomial of degree n − 1, returning n coefficients in fpoly.
fpoly=geop(1.0_sp,x,n)
END FUNCTION fpoly
```

⋆ ⋆ ⋆

```
FUNCTION fleg(x,nl)
USE nrtype
IMPLICIT NONE
REAL(SP), INTENT(IN) :: x
INTEGER(I4B), INTENT(IN) :: nl
REAL(SP), DIMENSION(nl) :: fleg
   Fitting routine for an expansion with nl Legendre polynomials evaluated at x and returned
   in the array fleg of length nl. The evaluation uses the recurrence relation as in §5.5.
INTEGER(I4B) :: j
REAL(SP) :: d,f1,f2,twox
fleg(1)=1.0
fleg(2)=x
if (nl > 2) then
    twox=2.0_sp*x
    f2=x
    d=1.0
    do j=3,nl
        f1=d
        f2=f2+twox
        d=d+1.0_sp
```

```
        fleg(j)=(f2*fleg(j-1)-f1*fleg(j-2))/d
    end do
end if
END FUNCTION fleg
```

⋆ ⋆ ⋆

```
SUBROUTINE mrqmin(x,y,sig,a,maska,covar,alpha,chisq,funcs,alamda)
USE nrtype; USE nrutil, ONLY : assert_eq,diagmult
USE nr, ONLY : covsrt,gaussj
IMPLICIT NONE
REAL(SP), DIMENSION(:), INTENT(IN) :: x,y,sig
REAL(SP), DIMENSION(:), INTENT(INOUT) :: a
REAL(SP), DIMENSION(:,:), INTENT(OUT) :: covar,alpha
REAL(SP), INTENT(OUT) :: chisq
REAL(SP), INTENT(INOUT) :: alamda
LOGICAL(LGT), DIMENSION(:), INTENT(IN) :: maska
INTERFACE
    SUBROUTINE funcs(x,a,yfit,dyda)
    USE nrtype
    REAL(SP), DIMENSION(:), INTENT(IN) :: x,a
    REAL(SP), DIMENSION(:), INTENT(OUT) :: yfit
    REAL(SP), DIMENSION(:,:), INTENT(OUT) :: dyda
    END SUBROUTINE funcs
END INTERFACE
```

Levenberg-Marquardt method, attempting to reduce the value χ^2 of a fit between a set of N data points x, y with individual standard deviations sig, and a nonlinear function dependent on M coefficients a. The input logical array maska of length M indicates by true entries those components of a that should be fitted for, and by false entries those components that should be held fixed at their input values. The program returns current best-fit values for the parameters a, and $\chi^2 =$ chisq. The $M \times M$ arrays covar and alpha are used as working space during most iterations. Supply a subroutine funcs(x,a,yfit,dyda) that evaluates the fitting function yfit, and its derivatives dyda with respect to the fitting parameters a at x. On the first call provide an initial guess for the parameters a, and set alamda<0 for initialization (which then sets alamda=.001). If a step succeeds chisq becomes smaller and alamda decreases by a factor of 10. If a step fails alamda grows by a factor of 10. You must call this routine repeatedly until convergence is achieved. Then, make one final call with alamda=0, so that covar returns the covariance matrix, and alpha the curvature matrix. (Parameters held fixed will return zero covariances.)

```
INTEGER(I4B) :: ma,ndata
INTEGER(I4B), SAVE :: mfit
call mrqmin_private
CONTAINS

SUBROUTINE mrqmin_private
REAL(SP), SAVE :: ochisq
REAL(SP), DIMENSION(:), ALLOCATABLE, SAVE :: atry,beta
REAL(SP), DIMENSION(:,:), ALLOCATABLE, SAVE :: da
ndata=assert_eq(size(x),size(y),size(sig),'mrqmin: ndata')
ma=assert_eq((/size(a),size(maska),size(covar,1),size(covar,2),&
    size(alpha,1),size(alpha,2)/),'mrqmin: ma')
mfit=count(maska)
if (alamda < 0.0) then                          Initialization.
    allocate(atry(ma),beta(ma),da(ma,1))
    alamda=0.001_sp
    call mrqcof(a,alpha,beta)
    ochisq=chisq
    atry=a
end if
covar(1:mfit,1:mfit)=alpha(1:mfit,1:mfit)
call diagmult(covar(1:mfit,1:mfit),1.0_sp+alamda)
```

```
    Alter linearized fitting matrix, by augmenting diagonal elements.
da(1:mfit,1)=beta(1:mfit)
call gaussj(covar(1:mfit,1:mfit),da(1:mfit,1:1))    Matrix solution.
if (alamda == 0.0) then                           Once converged, evaluate covariance ma-
    call covsrt(covar,maska)                          trix.
    call covsrt(alpha,maska)                      Spread out alpha to its full size too.
    deallocate(atry,beta,da)
    RETURN
end if
atry=a+unpack(da(1:mfit,1),maska,0.0_sp)          Did the trial succeed?
call mrqcof(atry,covar,da(1:mfit,1))
if (chisq < ochisq) then                          Success, accept the new solution.
    alamda=0.1_sp*alamda
    ochisq=chisq
    alpha(1:mfit,1:mfit)=covar(1:mfit,1:mfit)
    beta(1:mfit)=da(1:mfit,1)
    a=atry
else                                              Failure, increase alamda and return.
    alamda=10.0_sp*alamda
    chisq=ochisq
end if
END SUBROUTINE mrqmin_private

SUBROUTINE mrqcof(a,alpha,beta)
REAL(SP), DIMENSION(:), INTENT(IN) :: a
REAL(SP), DIMENSION(:), INTENT(OUT) :: beta
REAL(SP), DIMENSION(:,:), INTENT(OUT) :: alpha
    Used by mrqmin to evaluate the linearized fitting matrix alpha, and vector beta as in
    (15.5.8), and calculate χ².
INTEGER(I4B) :: j,k,l,m
REAL(SP), DIMENSION(size(x),size(a)) :: dyda
REAL(SP), DIMENSION(size(x)) :: dy,sig2i,wt,ymod
call funcs(x,a,ymod,dyda)                         Loop over all the data.
sig2i=1.0_sp/(sig**2)
dy=y-ymod
j=0
do l=1,ma
    if (maska(l)) then
        j=j+1
        wt=dyda(:,l)*sig2i
        k=0
        do m=1,l
            if (maska(m)) then
                k=k+1
                alpha(j,k)=dot_product(wt,dyda(:,m))
                alpha(k,j)=alpha(j,k)             Fill in the symmetric side.
            end if
        end do
        beta(j)=dot_product(dy,wt)
    end if
end do
chisq=dot_product(dy**2,sig2i)                    Find χ².
END SUBROUTINE mrqcof
END SUBROUTINE mrqmin
```

f90 The organization of this routine is similar to that of amoeba, discussed on p. 1209. We want to keep the argument list of mrqcof to a minimum, but we want to make clear what global variables it accesses, and protect mrqmin_private's name space.

`REAL(SP), DIMENSION(:), ALLOCATABLE, SAVE :: atry,beta` These arrays, as well as da, are allocated with the correct dimensions on the first call to mrqmin.

They need to retain their values between calls, so they are declared with the SAVE attribute. They get deallocated only on the final call when alamda=0.

call diagmult(...) See discussion of diagadd after hqr on p. 1234.

atry=a+unpack(da(1:mfit,1),maska,0.0_sp) maska controls which elements of a get incremented by da and which by 0.

⋆ ⋆ ⋆

```
SUBROUTINE fgauss(x,a,y,dyda)
USE nrtype; USE nrutil, ONLY : assert_eq
IMPLICIT NONE
REAL(SP), DIMENSION(:), INTENT(IN) :: x,a
REAL(SP), DIMENSION(:), INTENT(OUT) :: y
REAL(SP), DIMENSION(:,:), INTENT(OUT) :: dyda
```

$y(x; a)$ is the sum of $N/3$ Gaussians (15.5.16). Here N is the length of the vectors x, y and a, while dyda is an $N \times N$ matrix. The amplitude, center, and width of the Gaussians are stored in consecutive locations of a: a(i) $= B_k$, a(i+1) $= E_k$, a(i+2) $= G_k$, $k = 1, \ldots, N/3$.

```
INTEGER(I4B) :: i,na,nx
REAL(SP), DIMENSION(size(x)) :: arg,ex,fac
nx=assert_eq(size(x),size(y),size(dyda,1),'fgauss: nx')
na=assert_eq(size(a),size(dyda,2),'fgauss: na')
y(:)=0.0
do i=1,na-1,3
    arg(:)=(x(:)-a(i+1))/a(i+2)
    ex(:)=exp(-arg(:)**2)
    fac(:)=a(i)*ex(:)*2.0_sp*arg(:)
    y(:)=y(:)+a(i)*ex(:)
    dyda(:,i)=ex(:)
    dyda(:,i+1)=fac(:)/a(i+2)
    dyda(:,i+2)=fac(:)*arg(:)/a(i+2)
end do
END SUBROUTINE fgauss
```

⋆ ⋆ ⋆

```
SUBROUTINE medfit(x,y,a,b,abdev)
USE nrtype; USE nrutil, ONLY : assert_eq
USE nr, ONLY : select
IMPLICIT NONE
REAL(SP), DIMENSION(:), INTENT(IN) :: x,y
REAL(SP), INTENT(OUT) :: a,b,abdev
```

Fits $y = a + bx$ by the criterion of least absolute deviations. The same-size arrays x and y are the input experimental points. The fitted parameters a and b are output, along with abdev, which is the mean absolute deviation (in y) of the experimental points from the fitted line.

```
INTEGER(I4B) :: ndata
REAL(SP) :: aa
call medfit_private
CONTAINS

SUBROUTINE medfit_private
IMPLICIT NONE
REAL(SP) :: b1,b2,bb,chisq,del,f,f1,f2,sigb,sx,sxx,sxy,sy
REAL(SP), DIMENSION(size(x)) :: tmp
ndata=assert_eq(size(x),size(y),'medfit')
sx=sum(x)                           As a first guess for a and b, we will find the least
sy=sum(y)                              squares fitting line.
sxy=dot_product(x,y)
```

```
sxx=dot_product(x,x)
del=ndata*sxx-sx**2
aa=(sxx*sy-sx*sxy)/del                Least squares solutions.
bb=(ndata*sxy-sx*sy)/del
tmp(:)=y(:)-(aa+bb*x(:))
chisq=dot_product(tmp,tmp)
sigb=sqrt(chisq/del)                  The standard deviation will give some idea of how
b1=bb                                     big an iteration step to take.
f1=rofunc(b1)
b2=bb+sign(3.0_sp*sigb,f1)            Guess bracket as 3-σ away, in the downhill direction
f2=rofunc(b2)                             known from f1.
if (b2 == b1) then
    a=aa
    b=bb
    RETURN
endif
do                                    Bracketing.
    if (f1*f2 <= 0.0) exit
    bb=b2+1.6_sp*(b2-b1)
    b1=b2
    f1=f2
    b2=bb
    f2=rofunc(b2)
end do
sigb=0.01_sp*sigb                     Refine until error a negligible number of standard de-
do                                        viations.
    if (abs(b2-b1) <= sigb) exit
    bb=b1+0.5_sp*(b2-b1)              Bisection.
    if (bb == b1 .or. bb == b2) exit
    f=rofunc(bb)
    if (f*f1 >= 0.0) then
        f1=f
        b1=bb
    else
        f2=f
        b2=bb
    end if
end do
a=aa
b=bb
abdev=abdev/ndata
END SUBROUTINE medfit_private

FUNCTION rofunc(b)
IMPLICIT NONE
REAL(SP), INTENT(IN) :: b
REAL(SP) :: rofunc
REAL(SP), PARAMETER :: EPS=epsilon(b)
   Evaluates the right-hand side of equation (15.7.16) for a given value of b.
INTEGER(I4B) :: j
REAL(SP), DIMENSION(size(x)) :: arr,d
arr(:)=y(:)-b*x(:)
if (mod(ndata,2) == 0) then
    j=ndata/2
    aa=0.5_sp*(select(j,arr)+select(j+1,arr))
else
    aa=select((ndata+1)/2,arr)
end if
d(:)=y(:)-(b*x(:)+aa)
abdev=sum(abs(d))
where (y(:) /= 0.0) d(:)=d(:)/abs(y(:))
rofunc=sum(x(:)*sign(1.0_sp,d(:)), mask=(abs(d(:)) > EPS) )
END FUNCTION rofunc
END SUBROUTINE medfit
```

f90 The organization of this routine is similar to that of amoeba discussed on p. 1209. We want to keep the argument list of rofunc to a minimum, but we want to make clear what global variables it accesses and protect medfit_private's name space. In the Fortran 77 version, we kept the only argument as b by passing the global variables in a common block. This required us to make copies of the arrays x and y. An alternative Fortran 90 implementation would be to use a module with pointers to the arguments of medfit like x and y that need to be passed to rofunc. We think the medfit_private construction is simpler.

Chapter B16. Integration of Ordinary Differential Equations

```
SUBROUTINE rk4(y,dydx,x,h,yout,derivs)
USE nrtype; USE nrutil, ONLY : assert_eq
IMPLICIT NONE
REAL(SP), DIMENSION(:), INTENT(IN) :: y,dydx
REAL(SP), INTENT(IN) :: x,h
REAL(SP), DIMENSION(:), INTENT(OUT) :: yout
INTERFACE
    SUBROUTINE derivs(x,y,dydx)
    USE nrtype
    IMPLICIT NONE
    REAL(SP), INTENT(IN) :: x
    REAL(SP), DIMENSION(:), INTENT(IN) :: y
    REAL(SP), DIMENSION(:), INTENT(OUT) :: dydx
    END SUBROUTINE derivs
END INTERFACE
```

Given values for the N variables y and their derivatives dydx known at x, use the fourth-order Runge-Kutta method to advance the solution over an interval h and return the incremented variables as yout, which need not be a distinct array from y. y, dydx and yout are all of length N. The user supplies the subroutine derivs(x,y,dydx), which returns derivatives dydx at x.

```
INTEGER(I4B) :: ndum
REAL(SP) :: h6,hh,xh
REAL(SP), DIMENSION(size(y)) :: dym,dyt,yt
ndum=assert_eq(size(y),size(dydx),size(yout),'rk4')
hh=h*0.5_sp
h6=h/6.0_sp
xh=x+hh
yt=y+hh*dydx                            First step.
call derivs(xh,yt,dyt)                  Second step.
yt=y+hh*dyt
call derivs(xh,yt,dym)                  Third step.
yt=y+h*dym
dym=dyt+dym
call derivs(x+h,yt,dyt)                 Fourth step.
yout=y+h6*(dydx+dyt+2.0_sp*dym)         Accumulate increments with proper weights.
END SUBROUTINE rk4
```

⋆ ⋆ ⋆

```
MODULE rkdumb_path                      Storage of results.
USE nrtype
REAL(SP), DIMENSION(:), ALLOCATABLE:: xx
REAL(SP), DIMENSION(:,:), ALLOCATABLE :: y
END MODULE rkdumb_path
```

```
SUBROUTINE rkdumb(vstart,x1,x2,nstep,derivs)
USE nrtype; USE nrutil, ONLY : nrerror
USE nr, ONLY : rk4
USE rkdumb_path
IMPLICIT NONE
REAL(SP), DIMENSION(:), INTENT(IN) :: vstart
REAL(SP), INTENT(IN) :: x1,x2
INTEGER(I4B), INTENT(IN) :: nstep
INTERFACE
    SUBROUTINE derivs(x,y,dydx)
    USE nrtype
    IMPLICIT NONE
    REAL(SP), INTENT(IN) :: x
    REAL(SP), DIMENSION(:), INTENT(IN) :: y
    REAL(SP), DIMENSION(:), INTENT(OUT) :: dydx
    END SUBROUTINE derivs
END INTERFACE
```

Starting from N initial values `vstart` known at `x1`, use fourth-order Runge-Kutta to advance `nstep` equal increments to `x2`. The user-supplied subroutine `derivs(x,y,dydx)` evaluates derivatives. Results are stored in the module variables `xx` and `y`.

```
INTEGER(I4B) :: k
REAL(SP) :: h,x
REAL(SP), DIMENSION(size(vstart)) :: dv,v
v(:)=vstart(:)                               Load starting values.
if (allocated(xx)) deallocate(xx)            Clear out old stored variables if necessary.
if (allocated(y)) deallocate(y)
allocate(xx(nstep+1))                        Allocate storage for saved values.
allocate(y(size(vstart),nstep+1))
y(:,1)=v(:)
xx(1)=x1
x=x1
h=(x2-x1)/nstep
do k=1,nstep                                 Take nstep steps.
    call derivs(x,v,dv)
    call rk4(v,dv,x,h,v,derivs)
    if (x+h == x) call nrerror('stepsize not significant in rkdumb')
    x=x+h
    xx(k+1)=x                                Store intermediate steps.
    y(:,k+1)=v(:)
end do
END SUBROUTINE rkdumb
```

f90 `MODULE rkdumb_path` This routine needs straightforward communication of arrays with the calling program. The dimension of the arrays is not known in advance, and if the routine is called a second time we need to throw away the old array information. The Fortran 90 construction for this is to declare allocatable arrays in a module, and then test them at the beginning of the routine with `if (allocated...)`.

⋆ ⋆ ⋆

```
SUBROUTINE rkqs(y,dydx,x,htry,eps,yscal,hdid,hnext,derivs)
USE nrtype; USE nrutil, ONLY : assert_eq,nrerror
USE nr, ONLY : rkck
IMPLICIT NONE
REAL(SP), DIMENSION(:), INTENT(INOUT) :: y
REAL(SP), DIMENSION(:), INTENT(IN) :: dydx,yscal
REAL(SP), INTENT(INOUT) :: x
REAL(SP), INTENT(IN) :: htry,eps
REAL(SP), INTENT(OUT) :: hdid,hnext
```

```
INTERFACE
    SUBROUTINE derivs(x,y,dydx)
    USE nrtype
    IMPLICIT NONE
    REAL(SP), INTENT(IN) :: x
    REAL(SP), DIMENSION(:), INTENT(IN) :: y
    REAL(SP), DIMENSION(:), INTENT(OUT) :: dydx
    END SUBROUTINE derivs
END INTERFACE
```

Fifth order Runge-Kutta step with monitoring of local truncation error to ensure accuracy and adjust stepsize. Input are the dependent variable vector y and its derivative dydx at the starting value of the independent variable x. Also input are the stepsize to be attempted htry, the required accuracy eps, and the vector yscal against which the error is scaled. y, dydx, and yscal are all of the same length. On output, y and x are replaced by their new values, hdid is the stepsize that was actually accomplished, and hnext is the estimated next stepsize. derivs is the user-supplied subroutine that computes the right-hand-side derivatives.

```
INTEGER(I4B) :: ndum
REAL(SP) :: errmax,h,htemp,xnew
REAL(SP), DIMENSION(size(y)) :: yerr,ytemp
REAL(SP), PARAMETER :: SAFETY=0.9_sp,PGROW=-0.2_sp,PSHRNK=-0.25_sp,&
    ERRCON=1.89e-4
      The value ERRCON equals (5/SAFETY)**(1/PGROW), see use below.
ndum=assert_eq(size(y),size(dydx),size(yscal),'rkqs')
h=htry                                          Set stepsize to the initial trial value.
do
    call rkck(y,dydx,x,h,ytemp,yerr,derivs)         Take a step.
    errmax=maxval(abs(yerr(:)/yscal(:)))/eps        Evaluate accuracy.
    if (errmax <= 1.0) exit                     Step succeeded.
    htemp=SAFETY*h*(errmax**PSHRNK)             Truncation error too large, reduce stepsize.
    h=sign(max(abs(htemp),0.1_sp*abs(h)),h)         No more than a factor of 10.
    xnew=x+h
    if (xnew == x) call nrerror('stepsize underflow in rkqs')
end do                                          Go back for another try.
if (errmax > ERRCON) then                       Compute size of next step.
    hnext=SAFETY*h*(errmax**PGROW)
else                                            No more than a factor of 5 increase.
    hnext=5.0_sp*h
end if
hdid=h
x=x+h
y(:)=ytemp(:)
END SUBROUTINE rkqs
```

* * *

```
SUBROUTINE rkck(y,dydx,x,h,yout,yerr,derivs)
USE nrtype; USE nrutil, ONLY : assert_eq
IMPLICIT NONE
REAL(SP), DIMENSION(:), INTENT(IN) :: y,dydx
REAL(SP), INTENT(IN) :: x,h
REAL(SP), DIMENSION(:), INTENT(OUT) :: yout,yerr
INTERFACE
    SUBROUTINE derivs(x,y,dydx)
    USE nrtype
    IMPLICIT NONE
    REAL(SP), INTENT(IN) :: x
    REAL(SP), DIMENSION(:), INTENT(IN) :: y
    REAL(SP), DIMENSION(:), INTENT(OUT) :: dydx
    END SUBROUTINE derivs
END INTERFACE
```

Given values for N variables y and their derivatives dydx known at x, use the fifth order Cash-Karp Runge-Kutta method to advance the solution over an interval h and return the incremented variables as yout. Also return an estimate of the local truncation error in yout using the embedded fourth order method. The user supplies the subroutine derivs(x,y,dydx), which returns derivatives dydx at x.

```
INTEGER(I4B) :: ndum
REAL(SP), DIMENSION(size(y)) :: ak2,ak3,ak4,ak5,ak6,ytemp
REAL(SP), PARAMETER :: A2=0.2_sp,A3=0.3_sp,A4=0.6_sp,A5=1.0_sp,&
    A6=0.875_sp,B21=0.2_sp,B31=3.0_sp/40.0_sp,B32=9.0_sp/40.0_sp,&
    B41=0.3_sp,B42=-0.9_sp,B43=1.2_sp,B51=-11.0_sp/54.0_sp,&
    B52=2.5_sp,B53=-70.0_sp/27.0_sp,B54=35.0_sp/27.0_sp,&
    B61=1631.0_sp/55296.0_sp,B62=175.0_sp/512.0_sp,&
    B63=575.0_sp/13824.0_sp,B64=44275.0_sp/110592.0_sp,&
    B65=253.0_sp/4096.0_sp,C1=37.0_sp/378.0_sp,&
    C3=250.0_sp/621.0_sp,C4=125.0_sp/594.0_sp,&
    C6=512.0_sp/1771.0_sp,DC1=C1-2825.0_sp/27648.0_sp,&
    DC3=C3-18575.0_sp/48384.0_sp,DC4=C4-13525.0_sp/55296.0_sp,&
    DC5=-277.0_sp/14336.0_sp,DC6=C6-0.25_sp
ndum=assert_eq(size(y),size(dydx),size(yout),size(yerr),'rkck')
ytemp=y+B21*h*dydx                            First step.
call derivs(x+A2*h,ytemp,ak2)                 Second step.
ytemp=y+h*(B31*dydx+B32*ak2)
call derivs(x+A3*h,ytemp,ak3)                 Third step.
ytemp=y+h*(B41*dydx+B42*ak2+B43*ak3)
call derivs(x+A4*h,ytemp,ak4)                 Fourth step.
ytemp=y+h*(B51*dydx+B52*ak2+B53*ak3+B54*ak4)
call derivs(x+A5*h,ytemp,ak5)                 Fifth step.
ytemp=y+h*(B61*dydx+B62*ak2+B63*ak3+B64*ak4+B65*ak5)
call derivs(x+A6*h,ytemp,ak6)                 Sixth step.
yout=y+h*(C1*dydx+C3*ak3+C4*ak4+C6*ak6)       Accumulate increments with proper weights.
yerr=h*(DC1*dydx+DC3*ak3+DC4*ak4+DC5*ak5+DC6*ak6)
  Estimate error as difference between fourth and fifth order methods.
END SUBROUTINE rkck
```

* * *

```
MODULE ode_path
USE nrtype
INTEGER(I4B) :: nok,nbad,kount
LOGICAL(LGT), SAVE :: save_steps=.false.
REAL(SP) :: dxsav
REAL(SP), DIMENSION(:), POINTER :: xp
REAL(SP), DIMENSION(:,:), POINTER :: yp
END MODULE ode_path
```

On output nok and nbad are the number of good and bad (but retried and fixed) steps taken. If save_steps is set to true in the calling program, then intermediate values are stored in xp and yp at intervals greater than dxsav. kount is the total number of saved steps.

```
SUBROUTINE odeint(ystart,x1,x2,eps,h1,hmin,derivs,rkqs)
USE nrtype; USE nrutil, ONLY : nrerror,reallocate
USE ode_path
IMPLICIT NONE
REAL(SP), DIMENSION(:), INTENT(INOUT) :: ystart
REAL(SP), INTENT(IN) :: x1,x2,eps,h1,hmin
INTERFACE
    SUBROUTINE derivs(x,y,dydx)
    USE nrtype
    IMPLICIT NONE
    REAL(SP), INTENT(IN) :: x
    REAL(SP), DIMENSION(:), INTENT(IN) :: y
    REAL(SP), DIMENSION(:), INTENT(OUT) :: dydx
    END SUBROUTINE derivs

    SUBROUTINE rkqs(y,dydx,x,htry,eps,yscal,hdid,hnext,derivs)
    USE nrtype
```

```
        IMPLICIT NONE
        REAL(SP), DIMENSION(:), INTENT(INOUT) :: y
        REAL(SP), DIMENSION(:), INTENT(IN) :: dydx,yscal
        REAL(SP), INTENT(INOUT) :: x
        REAL(SP), INTENT(IN) :: htry,eps
        REAL(SP), INTENT(OUT) :: hdid,hnext
        INTERFACE
            SUBROUTINE derivs(x,y,dydx)
            USE nrtype
            IMPLICIT NONE
            REAL(SP), INTENT(IN) :: x
            REAL(SP), DIMENSION(:), INTENT(IN) :: y
            REAL(SP), DIMENSION(:), INTENT(OUT) :: dydx
            END SUBROUTINE derivs
        END INTERFACE
        END SUBROUTINE rkqs
    END INTERFACE
    REAL(SP), PARAMETER :: TINY=1.0e-30_sp
    INTEGER(I4B), PARAMETER :: MAXSTP=10000
```

Runge-Kutta driver with adaptive stepsize control. Integrate the array of starting values `ystart` from `x1` to `x2` with accuracy `eps`, storing intermediate results in the module variables in `ode_path`. `h1` should be set as a guessed first stepsize, `hmin` as the minimum allowed stepsize (can be zero). On output `ystart` is replaced by values at the end of the integration interval. `derivs` is the user-supplied subroutine for calculating the right-hand-side derivative, while `rkqs` is the name of the stepper routine to be used.

```
    INTEGER(I4B) :: nstp
    REAL(SP) :: h,hdid,hnext,x,xsav
    REAL(SP), DIMENSION(size(ystart)) :: dydx,y,yscal
    x=x1
    h=sign(h1,x2-x1)
    nok=0
    nbad=0
    kount=0
    y(:)=ystart(:)
    if (save_steps) then
        xsav=x-2.0_sp*dxsav                       Assures storage of first step.
        nullify(xp,yp)                            Pointers nullified here, but memory not
        allocate(xp(256))                             deallocated. If odeint is called mul-
        allocate(yp(size(ystart),size(xp)))           tiple times, calling program should
    end if                                            deallocate xp and yp between calls.
    do nstp=1,MAXSTP                              Take at most MAXSTP steps.
        call derivs(x,y,dydx)
        yscal(:)=abs(y(:))+abs(h*dydx(:))+TINY
```

Scaling used to monitor accuracy. This general purpose choice can be modified if need be.

```
        if (save_steps .and. (abs(x-xsav) > abs(dxsav))) &      Store intermediate results.
            call save_a_step
        if ((x+h-x2)*(x+h-x1) > 0.0) h=x2-x       If stepsize can overshoot, decrease.
        call rkqs(y,dydx,x,h,eps,yscal,hdid,hnext,derivs)
        if (hdid == h) then
            nok=nok+1
        else
            nbad=nbad+1
        end if
        if ((x-x2)*(x2-x1) >= 0.0) then           Are we done?
            ystart(:)=y(:)
            if (save_steps) call save_a_step      Save final step.
            RETURN                                Normal exit.
        end if
        if (abs(hnext) < hmin)&
            call nrerror('stepsize smaller than minimum in odeint')
        h=hnext
    end do
    call nrerror('too many steps in odeint')
```

```
CONTAINS

SUBROUTINE save_a_step
kount=kount+1
if (kount > size(xp)) then
    xp=>reallocate(xp,2*size(xp))
    yp=>reallocate(yp,size(yp,1),size(xp))
end if
xp(kount)=x
yp(:,kount)=y(:)
xsav=x
END SUBROUTINE save_a_step
END SUBROUTINE odeint
```

f90 MODULE ode_path The situation here is similar to rkdumb_path, except we don't know at run time how much storage to allocate. We may need to use reallocate from nrutil to increase the storage. The solution is pointers to arrays, with a nullify to be sure the pointer status is well-defined at the beginning of the routine.

SUBROUTINE save_a_step An internal subprogram with no arguments is like a macro in C: you could imagine just copying its code wherever it is called in the parent routine.

⋆ ⋆ ⋆

```
SUBROUTINE mmid(y,dydx,xs,htot,nstep,yout,derivs)
USE nrtype; USE nrutil, ONLY : assert_eq,swap
IMPLICIT NONE
INTEGER(I4B), INTENT(IN) :: nstep
REAL(SP), INTENT(IN) :: xs,htot
REAL(SP), DIMENSION(:), INTENT(IN) :: y,dydx
REAL(SP), DIMENSION(:), INTENT(OUT) :: yout
INTERFACE
    SUBROUTINE derivs(x,y,dydx)
    USE nrtype
    IMPLICIT NONE
    REAL(SP), INTENT(IN) :: x
    REAL(SP), DIMENSION(:), INTENT(IN) :: y
    REAL(SP), DIMENSION(:), INTENT(OUT) :: dydx
    END SUBROUTINE derivs
END INTERFACE
    Modified midpoint step. Dependent variable vector y and its derivative vector dydx are
    input at xs. Also input is htot, the total step to be taken, and nstep, the number of
    substeps to be used. The output is returned as yout, which need not be a distinct array
    from y; if it is distinct, however, then y and dydx are returned undamaged. y, dydx, and
    yout must all have the same length.
INTEGER(I4B) :: n,ndum
REAL(SP) :: h,h2,x
REAL(SP), DIMENSION(size(y)) :: ym,yn
ndum=assert_eq(size(y),size(dydx),size(yout),'mmid')
h=htot/nstep                          Stepsize this trip.
ym=y
yn=y+h*dydx                           First step.
x=xs+h
call derivs(x,yn,yout)                Will use yout for temporary storage of derivatives.
h2=2.0_sp*h
do n=2,nstep                          General step.
    call swap(ym,yn)
    yn=yn+h2*yout
```

```
    x=x+h
    call derivs(x,yn,yout)
end do
yout=0.5_sp*(ym+yn+h*yout)          Last step.
END SUBROUTINE mmid
```

* * *

```
SUBROUTINE bsstep(y,dydx,x,htry,eps,yscal,hdid,hnext,derivs)
USE nrtype; USE nrutil, ONLY : arth,assert_eq,cumsum,iminloc,nrerror,&
    outerdiff,outerprod,upper_triangle
USE nr, ONLY : mmid,pzextr
IMPLICIT NONE
REAL(SP), DIMENSION(:), INTENT(INOUT) :: y
REAL(SP), DIMENSION(:), INTENT(IN) :: dydx,yscal
REAL(SP), INTENT(INOUT) :: x
REAL(SP), INTENT(IN) :: htry,eps
REAL(SP), INTENT(OUT) :: hdid,hnext
INTERFACE
    SUBROUTINE derivs(x,y,dydx)
    USE nrtype
    IMPLICIT NONE
    REAL(SP), INTENT(IN) :: x
    REAL(SP), DIMENSION(:), INTENT(IN) :: y
    REAL(SP), DIMENSION(:), INTENT(OUT) :: dydx
    END SUBROUTINE derivs
END INTERFACE
INTEGER(I4B), PARAMETER :: IMAX=9, KMAXX=IMAX-1
REAL(SP), PARAMETER :: SAFE1=0.25_sp,SAFE2=0.7_sp,REDMAX=1.0e-5_sp,&
    REDMIN=0.7_sp,TINY=1.0e-30_sp,SCALMX=0.1_sp
```

Bulirsch-Stoer step with monitoring of local truncation error to ensure accuracy and adjust stepsize. Input are the dependent variable vector y and its derivative dydx at the starting value of the independent variable x. Also input are the stepsize to be attempted htry, the required accuracy eps, and the vector yscal against which the error is scaled. On output, y and x are replaced by their new values, hdid is the stepsize that was actually accomplished, and hnext is the estimated next stepsize. derivs is the user-supplied subroutine that computes the right-hand-side derivatives. y, dydx, and yscal must all have the same length. Be sure to set htry on successive steps to the value of hnext returned from the previous step, as is the case if the routine is called by odeint.

Parameters: KMAXX is the maximum row number used in the extrapolation; IMAX is the next row number; SAFE1 and SAFE2 are safety factors; REDMAX is the maximum factor used when a stepsize is reduced, REDMIN the minimum; TINY prevents division by zero; 1/SCALMX is the maximum factor by which a stepsize can be increased.

```
INTEGER(I4B) :: k,km,ndum
INTEGER(I4B), DIMENSION(IMAX) :: nseq = (/ 2,4,6,8,10,12,14,16,18 /)
INTEGER(I4B), SAVE :: kopt,kmax
REAL(SP), DIMENSION(KMAXX,KMAXX), SAVE :: alf
REAL(SP), DIMENSION(KMAXX) :: err
REAL(SP), DIMENSION(IMAX), SAVE :: a
REAL(SP), SAVE :: epsold = -1.0_sp,xnew
REAL(SP) :: eps1,errmax,fact,h,red,scale,wrkmin,xest
REAL(SP), DIMENSION(size(y)) :: yerr,ysav,yseq
LOGICAL(LGT) :: reduct
LOGICAL(LGT), SAVE :: first=.true.
ndum=assert_eq(size(y),size(dydx),size(yscal),'bsstep')
if (eps /= epsold) then                      A new tolerance, so reinitialize.
    hnext=-1.0e29_sp                         "Impossible" values.
    xnew=-1.0e29_sp
    eps1=SAFE1*eps
    a(:)=cumsum(nseq,1)
      Compute α(k,q):
    where (upper_triangle(KMAXX,KMAXX)) alf=eps1** &
        (outerdiff(a(2:),a(2:))/outerprod(arth( &
```

```
            3.0_sp,2.0_sp,KMAXX),(a(2:)-a(1)+1.0_sp)))
        epsold=eps
        do kopt=2,KMAXX-1                           Determine optimal row number for con-
            if (a(kopt+1) > a(kopt)*alf(kopt-1,kopt)) exit     vergence.
        end do
        kmax=kopt
    end if
    h=htry
    ysav(:)=y(:)                                    Save the starting values.
    if (h /= hnext .or. x /= xnew) then             A new stepsize or a new integration: Re-
        first=.true.                                    establish the order window.
        kopt=kmax
    end if
    reduct=.false.
    main_loop: do
        do k=1,kmax                                 Evaluate the sequence of modified mid-
            xnew=x+h                                    point integrations.
            if (xnew == x) call nrerror('step size underflow in bsstep')
            call mmid(ysav,dydx,x,h,nseq(k),yseq,derivs)
            xest=(h/nseq(k))**2                     Squared, since error series is even.
            call pzextr(k,xest,yseq,y,yerr)         Perform extrapolation.
            if (k /= 1) then                        Compute normalized error estimate ε(k).
                errmax=maxval(abs(yerr(:)/yscal(:)))
                errmax=max(TINY,errmax)/eps         Scale error relative to tolerance.
                km=k-1
                err(km)=(errmax/SAFE1)**(1.0_sp/(2*km+1))
            end if
            if (k /= 1 .and. (k >= kopt-1 .or. first)) then     In order window.
                if (errmax < 1.0) exit main_loop                Converged.
                if (k == kmax .or. k == kopt+1) then            Check for possible step-
                    red=SAFE2/err(km)                               size reduction.
                    exit
                else if (k == kopt) then
                    if (alf(kopt-1,kopt) < err(km)) then
                        red=1.0_sp/err(km)
                        exit
                    end if
                else if (kopt == kmax) then
                    if (alf(km,kmax-1) < err(km)) then
                        red=alf(km,kmax-1)*SAFE2/err(km)
                        exit
                    end if
                else if (alf(km,kopt) < err(km)) then
                    red=alf(km,kopt-1)/err(km)
                    exit
                end if
            end if
        end do
        red=max(min(red,REDMIN),REDMAX)             Reduce stepsize by at least REDMIN and
        h=h*red                                         at most REDMAX.
        reduct=.true.
    end do main_loop                                Try again.
    x=xnew                                          Successful step taken.
    hdid=h
    first=.false.
    kopt=1+iminloc(a(2:km+1)*max(err(1:km),SCALMX))
      Compute optimal row for convergence and corresponding stepsize.
    scale=max(err(kopt-1),SCALMX)
    wrkmin=scale*a(kopt)
    hnext=h/scale
    if (kopt >= k .and. kopt /= kmax .and. .not. reduct) then     Check for possible or-
        fact=max(scale/alf(kopt-1,kopt),SCALMX)                       der increase, but
        if (a(kopt+1)*fact <= wrkmin) then                            not if stepsize was
            hnext=h/fact                                              just reduced.
```

```
        kopt=kopt+1
    end if
end if
END SUBROUTINE bsstep
```

f90 `a(:)=cumsum(nseq,1)` The function cumsum in nrutil with the optional argument seed=1 gives a direct implementation of equation (16.4.6).

`where (upper_triangle(KMAXX,KMAXX))...` The upper_triangle function in nrutil returns an upper triangular logical mask. As used here, the mask is true everywhere in the upper triangle of a KMAXX $\times$ KMAXX matrix, excluding the diagonal. An optional integer argument extra allows additional diagonals to be set to true. With extra=1 the upper triangle including the diagonal would be true.

`main_loop: do` Using a named do-loop provides clear structured code that required goto's in the Fortran 77 version.

`kopt=1+iminloc(...)` See the discussion of imaxloc on p. 1017.

* * *

```
SUBROUTINE pzextr(iest,xest,yest,yz,dy)
USE nrtype; USE nrutil, ONLY : assert_eq,nrerror
IMPLICIT NONE
INTEGER(I4B), INTENT(IN) :: iest
REAL(SP), INTENT(IN) :: xest
REAL(SP), DIMENSION(:), INTENT(IN) :: yest
REAL(SP), DIMENSION(:), INTENT(OUT) :: yz,dy
    Use polynomial extrapolation to evaluate N functions at x = 0 by fitting a polynomial to
    a sequence of estimates with progressively smaller values x = xest, and corresponding
    function vectors yest. This call is number iest in the sequence of calls. Extrapolated
    function values are output as yz, and their estimated error is output as dy. yest, yz, and
    dy are arrays of length N.
INTEGER(I4B), PARAMETER :: IEST_MAX=16
INTEGER(I4B) :: j,nv
INTEGER(I4B), SAVE :: nvold=-1
REAL(SP) :: delta,f1,f2
REAL(SP), DIMENSION(size(yz)) :: d,tmp,q
REAL(SP), DIMENSION(IEST_MAX), SAVE :: x
REAL(SP), DIMENSION(:,:), ALLOCATABLE, SAVE :: qcol
nv=assert_eq(size(yz),size(yest),size(dy),'pzextr')
if (iest > IEST_MAX) call &
    nrerror('pzextr: probable misuse, too much extrapolation')
if (nv /= nvold) then                Set up internal storage.
    if (allocated(qcol)) deallocate(qcol)
    allocate(qcol(nv,IEST_MAX))
    nvold=nv
end if
x(iest)=xest                         Save current independent variable.
dy(:)=yest(:)
yz(:)=yest(:)
if (iest == 1) then                  Store first estimate in first column.
    qcol(:,1)=yest(:)
else
    d(:)=yest(:)
    do j=1,iest-1
        delta=1.0_sp/(x(iest-j)-xest)
        f1=xest*delta
        f2=x(iest-j)*delta
        q(:)=qcol(:,j)               Propagate tableau 1 diagonal more.
```

```
        qcol(:,j)=dy(:)
        tmp(:)=d(:)-q(:)
        dy(:)=f1*tmp(:)
        d(:)=f2*tmp(:)
        yz(:)=yz(:)+dy(:)
    end do
    qcol(:,iest)=dy(:)
end if
END SUBROUTINE pzextr
```

f90 `REAL(SP), DIMENSION(:,:), ALLOCATABLE, SAVE :: qcol` The second dimension of `qcol` is known at compile time to be `IEST_MAX`, but the first dimension is known only at run time, from `size(yz)`. The language requires us to have all dimensions allocatable if any one of them is.

`if (nv /= nvold) then...` This routine generally gets called many times with `iest` cycling repeatedly through the values $1, 2, \ldots,$ up to some value less than `IEST_MAX`. The number of variables, `nv`, is fixed during the solution of the problem. The routine might be called again in solving a different problem with a new value of `nv`. This if block ensures that `qcol` is dimensioned correctly both for the first and subsequent problems, if any.

```
SUBROUTINE rzextr(iest,xest,yest,yz,dy)
USE nrtype; USE nrutil, ONLY : assert_eq,nrerror
IMPLICIT NONE
INTEGER(I4B), INTENT(IN) :: iest
REAL(SP), INTENT(IN) :: xest
REAL(SP), DIMENSION(:), INTENT(IN) :: yest
REAL(SP), DIMENSION(:), INTENT(OUT) :: yz,dy
    Exact substitute for pzextr, but uses diagonal rational function extrapolation instead of
    polynomial extrapolation.
INTEGER(I4B), PARAMETER :: IEST_MAX=16
INTEGER(I4B) :: k,nv
INTEGER(I4B), SAVE :: nvold=-1
REAL(SP), DIMENSION(size(yz)) :: yy,v,c,b,b1,ddy
REAL(SP), DIMENSION(:,:), ALLOCATABLE, SAVE :: d
REAL(SP), DIMENSION(IEST_MAX), SAVE :: fx,x
nv=assert_eq(size(yz),size(dy),size(yest),'rzextr')
if (iest > IEST_MAX) call &
    nrerror('rzextr: probable misuse, too much extrapolation')
if (nv /= nvold) then
    if (allocated(d)) deallocate(d)
    allocate(d(nv,IEST_MAX))
    nvold=nv
end if
x(iest)=xest                        Save current independent variable.
if (iest == 1) then
    yz=yest
    d(:,1)=yest
    dy=yest
else
    fx(2:iest)=x(iest-1:1:-1)/xest
    yy=yest                         Evaluate next diagonal in tableau.
    v=d(1:nv,1)
    c=yy
    d(1:nv,1)=yy
    do k=2,iest
        b1=fx(k)*v
        b=b1-c
        where (b /= 0.0)
```

```
            b=(c-v)/b
            ddy=c*b
            c=b1*b
        elsewhere                Care needed to avoid division by 0.
            ddy=v
        end where
        if (k /= iest) v=d(1:nv,k)
        d(1:nv,k)=ddy
        yy=yy+ddy
    end do
    dy=ddy
    yz=yy
end if
END SUBROUTINE rzextr
```

*   *   *

```
SUBROUTINE stoerm(y,d2y,xs,htot,nstep,yout,derivs)
USE nrtype; USE nrutil, ONLY : assert_eq
IMPLICIT NONE
REAL(SP), DIMENSION(:), INTENT(IN) :: y,d2y
REAL(SP), INTENT(IN) :: xs,htot
INTEGER(I4B), INTENT(IN) :: nstep
REAL(SP), DIMENSION(:), INTENT(OUT) :: yout
INTERFACE
    SUBROUTINE derivs(x,y,dydx)
    USE nrtype
    IMPLICIT NONE
    REAL(SP), INTENT(IN) :: x
    REAL(SP), DIMENSION(:), INTENT(IN) :: y
    REAL(SP), DIMENSION(:), INTENT(OUT) :: dydx
    END SUBROUTINE derivs
END INTERFACE
```

Stoermer's rule for integrating $y'' = f(x, y)$ for a system of n equations. On input y contains y in its first n elements and y' in its second n elements, all evaluated at xs. d2y contains the right-hand-side function f (also evaluated at xs) in its first n elements. Its second n elements are not referenced. Also input is htot, the total step to be taken, and nstep, the number of substeps to be used. The output is returned as yout, with the same storage arrangement as y. derivs is the user-supplied subroutine that calculates f.

```
INTEGER(I4B) :: neqn,neqn1,nn,nv
REAL(SP) :: h,h2,halfh,x
REAL(SP), DIMENSION(size(y)) :: ytemp
nv=assert_eq(size(y),size(d2y),size(yout),'stoerm')
neqn=nv/2                                         Number of equations.
neqn1=neqn+1
h=htot/nstep                                      Stepsize this trip.
halfh=0.5_sp*h                                    First step.
ytemp(neqn1:nv)=h*(y(neqn1:nv)+halfh*d2y(1:neqn))
ytemp(1:neqn)=y(1:neqn)+ytemp(neqn1:nv)
x=xs+h
call derivs(x,ytemp,yout)                         Use yout for temporary storage of deriva-
h2=h*h                                                tives.
do nn=2,nstep                                     General step.
    ytemp(neqn1:nv)=ytemp(neqn1:nv)+h2*yout(1:neqn)
    ytemp(1:neqn)=ytemp(1:neqn)+ytemp(neqn1:nv)
    x=x+h
    call derivs(x,ytemp,yout)
end do
yout(neqn1:nv)=ytemp(neqn1:nv)/h+halfh*yout(1:neqn)      Last step.
yout(1:neqn)=ytemp(1:neqn)
END SUBROUTINE stoerm
```

*   *   *

```
SUBROUTINE stiff(y,dydx,x,htry,eps,yscal,hdid,hnext,derivs)
USE nrtype; USE nrutil, ONLY : assert_eq,diagadd,nrerror
USE nr, ONLY : lubksb,ludcmp
IMPLICIT NONE
REAL(SP), DIMENSION(:), INTENT(INOUT) :: y
REAL(SP), DIMENSION(:), INTENT(IN) :: dydx,yscal
REAL(SP), INTENT(INOUT) :: x
REAL(SP), INTENT(IN) :: htry,eps
REAL(SP), INTENT(OUT) :: hdid,hnext
INTERFACE
    SUBROUTINE derivs(x,y,dydx)
    USE nrtype
    IMPLICIT NONE
    REAL(SP), INTENT(IN) :: x
    REAL(SP), DIMENSION(:), INTENT(IN) :: y
    REAL(SP), DIMENSION(:), INTENT(OUT) :: dydx
    END SUBROUTINE derivs

    SUBROUTINE jacobn(x,y,dfdx,dfdy)
    USE nrtype
    IMPLICIT NONE
    REAL(SP), INTENT(IN) :: x
    REAL(SP), DIMENSION(:), INTENT(IN) :: y
    REAL(SP), DIMENSION(:), INTENT(OUT) :: dfdx
    REAL(SP), DIMENSION(:,:), INTENT(OUT) :: dfdy
    END SUBROUTINE jacobn
END INTERFACE
INTEGER(I4B), PARAMETER :: MAXTRY=40
REAL(SP), PARAMETER :: SAFETY=0.9_sp,GROW=1.5_sp,PGROW=-0.25_sp,&
    SHRNK=0.5_sp,PSHRNK=-1.0_sp/3.0_sp,ERRCON=0.1296_sp,&
    GAM=1.0_sp/2.0_sp,&
    A21=2.0_sp,A31=48.0_sp/25.0_sp,A32=6.0_sp/25.0_sp,C21=-8.0_sp,&
    C31=372.0_sp/25.0_sp,C32=12.0_sp/5.0_sp,&
    C41=-112.0_sp/125.0_sp,C42=-54.0_sp/125.0_sp,&
    C43=-2.0_sp/5.0_sp,B1=19.0_sp/9.0_sp,B2=1.0_sp/2.0_sp,&
    B3=25.0_sp/108.0_sp,B4=125.0_sp/108.0_sp,E1=17.0_sp/54.0_sp,&
    E2=7.0_sp/36.0_sp,E3=0.0_sp,E4=125.0_sp/108.0_sp,&
    C1X=1.0_sp/2.0_sp,C2X=-3.0_sp/2.0_sp,C3X=121.0_sp/50.0_sp,&
    C4X=29.0_sp/250.0_sp,A2X=1.0_sp,A3X=3.0_sp/5.0_sp
```

Fourth order Rosenbrock step for integrating stiff ODEs, with monitoring of local truncation error to adjust stepsize. Input are the dependent variable vector `y` and its derivative `dydx` at the starting value of the independent variable `x`. Also input are the stepsize to be attempted `htry`, the required accuracy `eps`, and the vector `yscal` against which the error is scaled. On output, `y` and `x` are replaced by their new values, `hdid` is the stepsize that was actually accomplished, and `hnext` is the estimated next stepsize. `derivs` is a user-supplied subroutine that computes the derivatives of the right-hand side with respect to `x`, while `jacobn` (a fixed name) is a user-supplied subroutine that computes the Jacobi matrix of derivatives of the right-hand side with respect to the components of `y`. `y`, `dydx`, and `yscal` must have the same length.

Parameters: `GROW` and `SHRNK` are the largest and smallest factors by which stepsize can change in one step; `ERRCON=(GROW/SAFETY)**(1/PGROW)` and handles the case when `errmax` $\simeq 0$.

```
INTEGER(I4B) :: jtry,ndum
INTEGER(I4B), DIMENSION(size(y)) :: indx
REAL(SP), DIMENSION(size(y)) :: dfdx,dytmp,err,g1,g2,g3,g4,ysav
REAL(SP), DIMENSION(size(y),size(y)) :: a,dfdy
REAL(SP) :: d,errmax,h,xsav
ndum=assert_eq(size(y),size(dydx),size(yscal),'stiff')
xsav=x                                          Save initial values.
ysav(:)=y(:)
call jacobn(xsav,ysav,dfdx,dfdy)
```

The user must supply this subroutine to return the $n \times n$ matrix `dfdy` and the vector `dfdx`.

```
h=htry                                          Set stepsize to the initial trial value.
do jtry=1,MAXTRY
```

```
    a(:,:)=-dfdy(:,:)                        Set up the matrix 1 − γhf'.
    call diagadd(a,1.0_sp/(GAM*h))
    call ludcmp(a,indx,d)                    LU decomposition of the matrix.
    g1=dydx+h*C1X*dfdx                       Set up right-hand side for g1.
    call lubksb(a,indx,g1)                   Solve for g1.
    y=ysav+A21*g1                            Compute intermediate values of y and x.
    x=xsav+A2X*h
    call derivs(x,y,dytmp)                   Compute dydx at the intermediate values.
    g2=dytmp+h*C2X*dfdx+C21*g1/h             Set up right-hand side for g2.
    call lubksb(a,indx,g2)                   Solve for g2.
    y=ysav+A31*g1+A32*g2                     Compute intermediate values of y and x.
    x=xsav+A3X*h
    call derivs(x,y,dytmp)                   Compute dydx at the intermediate values.
    g3=dytmp+h*C3X*dfdx+(C31*g1+C32*g2)/h              Set up right-hand side for g3.
    call lubksb(a,indx,g3)                             Solve for g3.
    g4=dytmp+h*C4X*dfdx+(C41*g1+C42*g2+C43*g3)/h       Set up right-hand side for g4.
    call lubksb(a,indx,g4)                             Solve for g4.
    y=ysav+B1*g1+B2*g2+B3*g3+B4*g4           Get fourth order estimate of y and error es-
    err=E1*g1+E2*g2+E3*g3+E4*g4                  timate.
    x=xsav+h
    if (x == xsav) call &
        nrerror('stepsize not significant in stiff')
    errmax=maxval(abs(err/yscal))/eps        Evaluate accuracy.
    if (errmax <= 1.0) then                  Step succeeded. Compute size of next step
        hdid=h                                   and return.
        hnext=merge(SAFETY*h*errmax**PGROW, GROW*h, &
            errmax > ERRCON)
        RETURN
    else                                     Truncation error too large, reduce stepsize.
        hnext=SAFETY*h*errmax**PSHRNK
        h=sign(max(abs(hnext),SHRNK*abs(h)),h)
    end if
end do                                       Go back and retry step.
call nrerror('exceeded MAXTRY in stiff')
END SUBROUTINE stiff
```

`call diagadd(...)` See discussion of diagadd after hqr on p. 1234.

```
SUBROUTINE jacobn(x,y,dfdx,dfdy)
USE nrtype
IMPLICIT NONE
REAL(SP), INTENT(IN) :: x
REAL(SP), DIMENSION(:), INTENT(IN) :: y
REAL(SP), DIMENSION(:), INTENT(OUT) :: dfdx
REAL(SP), DIMENSION(:,:), INTENT(OUT) :: dfdy
   Routine for Jacobi matrix corresponding to example in equations (16.6.27).
dfdx(:)=0.0
dfdy(1,1)=-0.013_sp-1000.0_sp*y(3)
dfdy(1,2)=0.0
dfdy(1,3)=-1000.0_sp*y(1)
dfdy(2,1)=0.0
dfdy(2,2)=-2500.0_sp*y(3)
dfdy(2,3)=-2500.0_sp*y(2)
dfdy(3,1)=-0.013_sp-1000.0_sp*y(3)
dfdy(3,2)=-2500.0_sp*y(3)
dfdy(3,3)=-1000.0_sp*y(1)-2500.0_sp*y(2)
END SUBROUTINE jacobn
```

```
SUBROUTINE derivs(x,y,dydx)
USE nrtype
IMPLICIT NONE
REAL(SP), INTENT(IN) :: x
REAL(SP), DIMENSION(:), INTENT(IN) :: y
REAL(SP), DIMENSION(:), INTENT(OUT) :: dydx
   Routine for right-hand side of example in equations (16.6.27).
dydx(1)=-0.013_sp*y(1)-1000.0_sp*y(1)*y(3)
dydx(2)=-2500.0_sp*y(2)*y(3)
dydx(3)=-0.013_sp*y(1)-1000.0_sp*y(1)*y(3)-2500.0_sp*y(2)*y(3)
END SUBROUTINE derivs
```

* * *

```
SUBROUTINE simpr(y,dydx,dfdx,dfdy,xs,htot,nstep,yout,derivs)
USE nrtype; USE nrutil, ONLY : assert_eq,diagadd
USE nr, ONLY : lubksb,ludcmp
IMPLICIT NONE
REAL(SP), INTENT(IN) :: xs,htot
REAL(SP), DIMENSION(:), INTENT(IN) :: y,dydx,dfdx
REAL(SP), DIMENSION(:,:), INTENT(IN) :: dfdy
INTEGER(I4B), INTENT(IN) :: nstep
REAL(SP), DIMENSION(:), INTENT(OUT) :: yout
INTERFACE
    SUBROUTINE derivs(x,y,dydx)
    USE nrtype
    IMPLICIT NONE
    REAL(SP), INTENT(IN) :: x
    REAL(SP), DIMENSION(:), INTENT(IN) :: y
    REAL(SP), DIMENSION(:), INTENT(OUT) :: dydx
    END SUBROUTINE derivs
END INTERFACE
   Performs one step of semi-implicit midpoint rule. Input are the dependent variable y, its
   derivative dydx, the derivative of the right-hand side with respect to x, dfdx, which are all
   vectors of length N, and the N × N Jacobian dfdy at xs. Also input are htot, the total
   step to be taken, and nstep, the number of substeps to be used. The output is returned as
   yout, a vector of length N. derivs is the user-supplied subroutine that calculates dydx.
INTEGER(I4B) :: ndum,nn
INTEGER(I4B), DIMENSION(size(y)) :: indx
REAL(SP) :: d,h,x
REAL(SP), DIMENSION(size(y)) :: del,ytemp
REAL(SP), DIMENSION(size(y),size(y)) :: a
ndum=assert_eq((/size(y),size(dydx),size(dfdx),size(dfdy,1),&
    size(dfdy,2),size(yout)/),'simpr')
h=htot/nstep                          Stepsize this trip.
a(:,:)=-h*dfdy(:,:)                   Set up the matrix 1 - hf'.
call diagadd(a,1.0_sp)
call ludcmp(a,indx,d)                 LU decomposition of the matrix.
yout=h*(dydx+h*dfdx)                  Set up right-hand side for first step. Use yout for
call lubksb(a,indx,yout)                  temporary storage.
del=yout                              First step.
ytemp=y+del
x=xs+h
call derivs(x,ytemp,yout)             Use yout for temporary storage of derivatives.
do nn=2,nstep                         General step.
    yout=h*yout-del                   Set up right-hand side for general step.
    call lubksb(a,indx,yout)
    del=del+2.0_sp*yout
    ytemp=ytemp+del
    x=x+h
    call derivs(x,ytemp,yout)
```

```
end do
yout=h*yout-del                         Set up right-hand side for last step.
call lubksb(a,indx,yout)
yout=ytemp+yout                         Take last step.
END SUBROUTINE simpr
```

`call diagadd(...)` See discussion of diagadd after hqr on p. 1234.

⋆ ⋆ ⋆

```
SUBROUTINE stifbs(y,dydx,x,htry,eps,yscal,hdid,hnext,derivs)
USE nrtype; USE nrutil, ONLY : arth,assert_eq,cumsum,iminloc,nrerror,&
    outerdiff,outerprod,upper_triangle
USE nr, ONLY : simpr,pzextr
IMPLICIT NONE
REAL(SP), DIMENSION(:), INTENT(INOUT) :: y
REAL(SP), DIMENSION(:), INTENT(IN) :: dydx,yscal
REAL(SP), INTENT(IN) :: htry,eps
REAL(SP), INTENT(INOUT) :: x
REAL(SP), INTENT(OUT) :: hdid,hnext
INTERFACE
    SUBROUTINE derivs(x,y,dydx)
    USE nrtype
    IMPLICIT NONE
    REAL(SP), INTENT(IN) :: x
    REAL(SP), DIMENSION(:), INTENT(IN) :: y
    REAL(SP), DIMENSION(:), INTENT(OUT) :: dydx
    END SUBROUTINE derivs

    SUBROUTINE jacobn(x,y,dfdx,dfdy)
    USE nrtype
    IMPLICIT NONE
    REAL(SP), INTENT(IN) :: x
    REAL(SP), DIMENSION(:), INTENT(IN) :: y
    REAL(SP), DIMENSION(:), INTENT(OUT) :: dfdx
    REAL(SP), DIMENSION(:,:), INTENT(OUT) :: dfdy
    END SUBROUTINE jacobn
END INTERFACE
INTEGER(I4B), PARAMETER :: IMAX=8, KMAXX=IMAX-1
REAL(SP), PARAMETER :: SAFE1=0.25_sp,SAFE2=0.7_sp,REDMAX=1.0e-5_sp,&
    REDMIN=0.7_sp,TINY=1.0e-30_sp,SCALMX=0.1_sp
```

Semi-implicit extrapolation step for integrating stiff ODEs, with monitoring of local truncation error to adjust stepsize. Input are the dependent variable vector `y` and its derivative `dydx` at the starting value of the independent variable `x`. Also input are the stepsize to be attempted `htry`, the required accuracy `eps`, and the vector `yscal` against which the error is scaled. On output, `y` and `x` are replaced by their new values, `hdid` is the stepsize that was actually accomplished, and `hnext` is the estimated next stepsize. `derivs` is a user-supplied subroutine that computes the derivatives of the right-hand side with respect to `x`, while `jacobn` (a fixed name) is a user-supplied subroutine that computes the Jacobi matrix of derivatives of the right-hand side with respect to the components of `y`. `y`, `dydx`, and `yscal` must all have the same length. Be sure to set `htry` on successive steps to the value of `hnext` returned from the previous step, as is the case if the routine is called by `odeint`.

```
INTEGER(I4B) :: k,km,ndum
INTEGER(I4B), DIMENSION(IMAX) :: nseq = (/ 2,6,10,14,22,34,50,70 /)
```

Sequence is different from `bsstep`.

```
INTEGER(I4B), SAVE :: kopt,kmax,nvold=-1
REAL(SP), DIMENSION(KMAXX,KMAXX), SAVE :: alf
REAL(SP), DIMENSION(KMAXX) :: err
REAL(SP), DIMENSION(IMAX), SAVE :: a
REAL(SP), SAVE :: epsold = -1.0
REAL(SP) :: eps1,errmax,fact,h,red,scale,wrkmin,xest
```

```
REAL(SP), SAVE :: xnew
REAL(SP), DIMENSION(size(y)) :: dfdx,yerr,ysav,yseq
REAL(SP), DIMENSION(size(y),size(y)) :: dfdy
LOGICAL(LGT) :: reduct
LOGICAL(LGT), SAVE :: first=.true.
ndum=assert_eq(size(y),size(dydx),size(yscal),'stifbs')
if (eps /= epsold .or. nvold /= size(y)) then        Reinitialize also if number of vari-
    hnext=-1.0e29_sp                                      ables has changed.
    xnew=-1.0e29_sp
    eps1=SAFE1*eps
    a(:)=cumsum(nseq,1)
    where (upper_triangle(KMAXX,KMAXX)) alf=eps1** &
        (outerdiff(a(2:),a(2:))/outerprod(arth( &
        3.0_sp,2.0_sp,KMAXX),(a(2:)-a(1)+1.0_sp)))
    epsold=eps
    nvold=size(y)                            Save number of variables.
    a(:)=cumsum(nseq,1+nvold)                Add cost of Jacobian evaluations to work co-
    do kopt=2,KMAXX-1                            efficients.
        if (a(kopt+1) > a(kopt)*alf(kopt-1,kopt)) exit
    end do
    kmax=kopt
end if
h=htry
ysav(:)=y(:)
call jacobn(x,y,dfdx,dfdy)                   Evaluate Jacobian.
if (h /= hnext .or. x /= xnew) then
    first=.true.
    kopt=kmax
end if
reduct=.false.
main_loop: do
    do k=1,kmax
        xnew=x+h
        if (xnew == x) call nrerror('step size underflow in stifbs')
        call simpr(ysav,dydx,dfdx,dfdy,x,h,nseq(k),yseq,derivs)
          Here is the call to the semi-implicit midpoint rule.
        xest=(h/nseq(k))**2                  The rest of the routine is identical to bsstep.
        call pzextr(k,xest,yseq,y,yerr)
        if (k /= 1) then
            errmax=maxval(abs(yerr(:)/yscal(:)))
            errmax=max(TINY,errmax)/eps
            km=k-1
            err(km)=(errmax/SAFE1)**(1.0_sp/(2*km+1))
        end if
        if (k /= 1 .and. (k >= kopt-1 .or. first)) then
            if (errmax < 1.0) exit main_loop
            if (k == kmax .or. k == kopt+1) then
                red=SAFE2/err(km)
                exit
            else if (k == kopt) then
                if (alf(kopt-1,kopt) < err(km)) then
                    red=1.0_sp/err(km)
                    exit
                end if
            else if (kopt == kmax) then
                if (alf(km,kmax-1) < err(km)) then
                    red=alf(km,kmax-1)*SAFE2/err(km)
                    exit
                end if
            else if (alf(km,kopt) < err(km)) then
                red=alf(km,kopt-1)/err(km)
                exit
            end if
        end if
```

```
    end do
    red=max(min(red,REDMIN),REDMAX)
    h=h*red
    reduct=.true.
end do main_loop
x=xnew
hdid=h
first=.false.
kopt=1+iminloc(a(2:km+1)*max(err(1:km),SCALMX))
scale=max(err(kopt-1),SCALMX)
wrkmin=scale*a(kopt)
hnext=h/scale
if (kopt >= k .and. kopt /= kmax .and. .not. reduct) then
    fact=max(scale/alf(kopt-1,kopt),SCALMX)
    if (a(kopt+1)*fact <= wrkmin) then
        hnext=h/fact
        kopt=kopt+1
    end if
end if
END SUBROUTINE stifbs
```

This routine is very similar to `bsstep`, and the same remarks about Fortran 90 constructions on p. 1305 apply here.

Chapter B17. Two Point Boundary Value Problems

```
!   FUNCTION shoot(v) is named "funcv" for use with "newt"
    FUNCTION funcv(v)
    USE nrtype
    USE nr, ONLY : odeint,rkqs
    USE sphoot_caller, ONLY : nvar,x1,x2; USE ode_path, ONLY : xp,yp
    IMPLICIT NONE
    REAL(SP), DIMENSION(:), INTENT(IN) :: v
    REAL(SP), DIMENSION(size(v)) :: funcv
    REAL(SP), PARAMETER :: EPS=1.0e-6_sp
```

Routine for use with `newt` to solve a two point boundary value problem for N coupled ODEs by shooting from `x1` to `x2`. Initial values for the ODEs at `x1` are generated from the n_2 input coefficients `v`, using the user-supplied routine `load`. The routine integrates the ODEs to `x2` using the Runge-Kutta method with tolerance EPS, initial stepsize `h1`, and minimum stepsize `hmin`. At `x2` it calls the user-supplied subroutine `score` to evaluate the n_2 functions `funcv` that ought to be zero to satisfy the boundary conditions at `x2`. The functions `funcv` are returned on output. `newt` uses a globally convergent Newton's method to adjust the values of `v` until the functions `funcv` are zero. The user-supplied subroutine `derivs(x,y,dydx)` supplies derivative information to the ODE integrator (see Chapter 16). The module `sphoot_caller` receives its values from the main program so that `funcv` can have the syntax required by `newt`. Set `nvar` $= N$ in the main program.

```
    REAL(SP) :: h1,hmin
    REAL(SP), DIMENSION(nvar) :: y
    INTERFACE
        SUBROUTINE derivs(x,y,dydx)
        USE nrtype
        IMPLICIT NONE
        REAL(SP), INTENT(IN) :: x
        REAL(SP), DIMENSION(:), INTENT(IN) :: y
        REAL(SP), DIMENSION(:), INTENT(OUT) :: dydx
        END SUBROUTINE derivs

        SUBROUTINE load(x1,v,y)
        USE nrtype
        IMPLICIT NONE
        REAL(SP), INTENT(IN) :: x1
        REAL(SP), DIMENSION(:), INTENT(IN) :: v
        REAL(SP), DIMENSION(:), INTENT(OUT) :: y
        END SUBROUTINE load

        SUBROUTINE score(x2,y,f)
        USE nrtype
        IMPLICIT NONE
        REAL(SP), INTENT(IN) :: x2
        REAL(SP), DIMENSION(:), INTENT(IN) :: y
        REAL(SP), DIMENSION(:), INTENT(OUT) :: f
        END SUBROUTINE score
    END INTERFACE
    h1=(x2-x1)/100.0_sp
```

```
hmin=0.0
call load(x1,v,y)
if (associated(xp)) deallocate(xp,yp)          Prevent memory leak if save_steps set
call odeint(y,x1,x2,EPS,h1,hmin,derivs,rkqs)     to .true.
call score(x2,y,funcv)
END FUNCTION funcv
```

⋆ ⋆ ⋆

```
! FUNCTION shootf(v) is named "funcv" for use with "newt"
FUNCTION funcv(v)
USE nrtype
USE nr, ONLY : odeint,rkqs
USE sphfpt_caller, ONLY : x1,x2,xf,nn2; USE ode_path, ONLY : xp,yp
IMPLICIT NONE
REAL(SP), DIMENSION(:), INTENT(IN) :: v
REAL(SP), DIMENSION(size(v)) :: funcv
REAL(SP), PARAMETER :: EPS=1.0e-6_sp
```

Routine for use with `newt` to solve a two point boundary value problem for N coupled ODEs by shooting from `x1` and `x2` to a fitting point `xf`. Initial values for the ODEs at `x1` (`x2`) are generated from the n_2 (n_1) coefficients V_1 (V_2), using the user-supplied routine `load1` (`load2`). The coefficients V_1 and V_2 should be stored in a single array `v` of length N in the main program, and referenced by pointers as `v1=>v(1:`n_2`)`, `v2=>v(`n_2+1`:`N`)`. Here $N = n_1 + n_2$. The routine integrates the ODEs to `xf` using the Runge-Kutta method with tolerance EPS, initial stepsize `h1`, and minimum stepsize `hmin`. At `xf` it calls the user-supplied subroutine `score` to evaluate the N functions `f1` and `f2` that ought to match at `xf`. The differences `funcv` are returned on output. `newt` uses a globally convergent Newton's method to adjust the values of `v` until the functions `funcv` are zero. The user-supplied subroutine `derivs(x,y,dydx)` supplies derivative information to the ODE integrator (see Chapter 16). The module `sphfpt_caller` receives its values from the main program so that `funcv` can have the syntax required by `newt`. Set `nn2` $= n_2$ in the main program.

```
REAL(SP) :: h1,hmin
REAL(SP), DIMENSION(size(v)) :: f1,f2,y
INTERFACE
    SUBROUTINE derivs(x,y,dydx)
    USE nrtype
    IMPLICIT NONE
    REAL(SP), INTENT(IN) :: x
    REAL(SP), DIMENSION(:), INTENT(IN) :: y
    REAL(SP), DIMENSION(:), INTENT(OUT) :: dydx
    END SUBROUTINE derivs

    SUBROUTINE load1(x1,v1,y)
    USE nrtype
    IMPLICIT NONE
    REAL(SP), INTENT(IN) :: x1
    REAL(SP), DIMENSION(:), INTENT(IN) :: v1
    REAL(SP), DIMENSION(:), INTENT(OUT) :: y
    END SUBROUTINE load1

    SUBROUTINE load2(x2,v2,y)
    USE nrtype
    IMPLICIT NONE
    REAL(SP), INTENT(IN) :: x2
    REAL(SP), DIMENSION(:), INTENT(IN) :: v2
    REAL(SP), DIMENSION(:), INTENT(OUT) :: y
    END SUBROUTINE load2

    SUBROUTINE score(x2,y,f)
    USE nrtype
    IMPLICIT NONE
    REAL(SP), INTENT(IN) :: x2
    REAL(SP), DIMENSION(:), INTENT(IN) :: y
```

```
    REAL(SP), DIMENSION(:), INTENT(OUT) :: f
    END SUBROUTINE score
END INTERFACE
h1=(x2-x1)/100.0_sp
hmin=0.0
call load1(x1,v,y)                              Path from x1 to xf with best trial values V1.
if (associated(xp)) deallocate(xp,yp)              Prevent memory leak if save_steps set
call odeint(y,x1,xf,EPS,h1,hmin,derivs,rkqs)          to .true.
call score(xf,y,f1)
call load2(x2,v(nn2+1:),y)                      Path from x2 to xf with best trial values V2.
call odeint(y,x2,xf,EPS,h1,hmin,derivs,rkqs)
call score(xf,y,f2)
funcv(:)=f1(:)-f2(:)
END FUNCTION funcv
```

⋆ ⋆ ⋆

```
SUBROUTINE solvde(itmax,conv,slowc,scalv,indexv,nb,y)
USE nrtype; USE nrutil, ONLY : assert_eq,imaxloc,nrerror
USE nr, ONLY : difeq
IMPLICIT NONE
INTEGER(I4B), INTENT(IN) :: itmax,nb
REAL(SP), INTENT(IN) :: conv,slowc
REAL(SP), DIMENSION(:), INTENT(IN) :: scalv
INTEGER(I4B), DIMENSION(:), INTENT(IN) :: indexv
REAL(SP), DIMENSION(:,:), INTENT(INOUT) :: y
```

Driver routine for solution of two point boundary value problems with N equations by relaxation. `itmax` is the maximum number of iterations. `conv` is the convergence criterion (see text). `slowc` controls the fraction of corrections actually used after each iteration. `scalv`, a vector of length N, contains typical sizes for each dependent variable, used to weight errors. `indexv`, also of length N, lists the column ordering of variables used to construct the matrix `s` of derivatives. (The `nb` boundary conditions at the first mesh point must contain some dependence on the first `nb` variables listed in `indexv`.) There are a total of M mesh points. `y` is the $N \times M$ array that contains the initial guess for all the dependent variables at each mesh point. On each iteration, it is updated by the calculated correction.

```
INTEGER(I4B) :: ic1,ic2,ic3,ic4,it,j,j1,j2,j3,j4,j5,j6,j7,j8,&
    j9,jc1,jcf,jv,k,k1,k2,km,kp,m,ne,nvars
INTEGER(I4B), DIMENSION(size(scalv)) :: kmax
REAL(SP) :: err,fac
REAL(SP), DIMENSION(size(scalv)) :: ermax
REAL(SP), DIMENSION(size(scalv),2*size(scalv)+1) :: s
REAL(SP), DIMENSION(size(scalv),size(scalv)-nb+1,size(y,2)+1) :: c
ne=assert_eq(size(scalv),size(indexv),size(y,1),'solvde: ne')
m=size(y,2)
k1=1                                             Set up row and column markers.
k2=m
nvars=ne*m
j1=1
j2=nb
j3=nb+1
j4=ne
j5=j4+j1
j6=j4+j2
j7=j4+j3
j8=j4+j4
j9=j8+j1
ic1=1
ic2=ne-nb
ic3=ic2+1
ic4=ne
jc1=1
jcf=ic3
do it=1,itmax                                    Primary iteration loop.
    k=k1                                         Boundary conditions at first point
```

```
    call difeq(k,k1,k2,j9,ic3,ic4,indexv,s,y)
    call pinvs(ic3,ic4,j5,j9,jc1,k1,c,s)
    do k=k1+1,k2                                Finite difference equations at all point
        kp=k-1                                      pairs.
        call difeq(k,k1,k2,j9,ic1,ic4,indexv,s,y)
        call red(ic1,ic4,j1,j2,j3,j4,j9,ic3,jc1,jcf,kp,c,s)
        call pinvs(ic1,ic4,j3,j9,jc1,k,c,s)
    end do
    k=k2+1                                      Final boundary conditions.
    call difeq(k,k1,k2,j9,ic1,ic2,indexv,s,y)
    call red(ic1,ic2,j5,j6,j7,j8,j9,ic3,jc1,jcf,k2,c,s)
    call pinvs(ic1,ic2,j7,j9,jcf,k2+1,c,s)
    call bksub(ne,nb,jcf,k1,k2,c)               Backsubstitution.
    do j=1,ne                                   Convergence check, accumulate average
        jv=indexv(j)                                error.
        km=imaxloc(abs(c(jv,1,k1:k2)))+k1-1
          Find point with largest error, for each dependent variable.
        ermax(j)=c(jv,1,km)
        kmax(j)=km
    end do
    ermax(:)=ermax(:)/scalv(:)                  Weighting for each dependent variable.
    err=sum(sum(abs(c(indexv(:),1,k1:k2)),dim=2)/scalv(:))/nvars
    fac=slowc/max(slowc,err)
      Reduce correction applied when error is large.
    y(:,k1:k2)=y(:,k1:k2)-fac*c(indexv(:),1,k1:k2)      Apply corrections.
    write(*,'(1x,i4,2f12.6)') it,err,fac
      Summary of corrections for this step. Point with largest error for each variable can be
      monitored by writing out kmax and ermax.
    if (err < conv) RETURN
end do
call nrerror('itmax exceeded in solvde')        Convergence failed.
CONTAINS

SUBROUTINE bksub(ne,nb,jf,k1,k2,c)
IMPLICIT NONE
INTEGER(I4B), INTENT(IN) :: ne,nb,jf,k1,k2
REAL(SP), DIMENSION(:,:,:), INTENT(INOUT) :: c
   Backsubstitution, used internally by solvde.
INTEGER(I4B) :: im,k,nbf
nbf=ne-nb
im=1
do k=k2,k1,-1
      Use recurrence relations to eliminate remaining dependences.
    if (k == k1) im=nbf+1                       Special handling of first point.
    c(im:ne,jf,k)=c(im:ne,jf,k)-matmul(c(im:ne,1:nbf,k),c(1:nbf,jf,k+1))
end do
c(1:nb,1,k1:k2)=c(1+nbf:nb+nbf,jf,k1:k2)        Reorder corrections to be in column 1.
c(1+nb:nbf+nb,1,k1:k2)=c(1:nbf,jf,k1+1:k2+1)
END SUBROUTINE bksub

SUBROUTINE pinvs(ie1,ie2,je1,jsf,jc1,k,c,s)
IMPLICIT NONE
INTEGER(I4B), INTENT(IN) :: ie1,ie2,je1,jsf,jc1,k
REAL(SP), DIMENSION(:,:,:), INTENT(OUT) :: c
REAL(SP), DIMENSION(:,:), INTENT(INOUT) :: s
   Diagonalize the square subsection of the s matrix, and store the recursion coefficients in
   c; used internally by solvde.
INTEGER(I4B) :: i,icoff,id,ipiv,jcoff,je2,jp,jpiv,js1
INTEGER(I4B), DIMENSION(ie2) :: indxr
REAL(SP) :: big,piv,pivinv
REAL(SP), DIMENSION(ie2) :: pscl
je2=je1+ie2-ie1
js1=je2+1
pscl(ie1:ie2)=maxval(abs(s(ie1:ie2,je1:je2)),dim=2)
  Implicit pivoting, as in §2.1.
```

```
if (any(pscl(ie1:ie2) == 0.0)) &
    call nrerror('singular matrix, row all 0 in pinvs')
pscl(ie1:ie2)=1.0_sp/pscl(ie1:ie2)
indxr(ie1:ie2)=0
do id=ie1,ie2
    piv=0.0
    do i=ie1,ie2                                Find pivot element.
        if (indxr(i) == 0) then
            jp=imaxloc(abs(s(i,je1:je2)))+je1-1
            big=abs(s(i,jp))
            if (big*pscl(i) > piv) then
                ipiv=i
                jpiv=jp
                piv=big*pscl(i)
            end if
        end if
    end do
    if (s(ipiv,jpiv) == 0.0) call nrerror('singular matrix in pinvs')
    indxr(ipiv)=jpiv                            In place reduction. Save column order-
    pivinv=1.0_sp/s(ipiv,jpiv)                     ing.
    s(ipiv,je1:jsf)=s(ipiv,je1:jsf)*pivinv         Normalize pivot row.
    s(ipiv,jpiv)=1.0
    do i=ie1,ie2                                Reduce nonpivot elements in column.
        if (indxr(i) /= jpiv .and. s(i,jpiv) /= 0.0) then
            s(i,je1:jsf)=s(i,je1:jsf)-s(i,jpiv)*s(ipiv,je1:jsf)
            s(i,jpiv)=0.0
        end if
    end do
end do
jcoff=jc1-js1                                   Sort and store unreduced coefficients.
icoff=ie1-je1
c(indxr(ie1:ie2)+icoff,js1+jcoff:jsf+jcoff,k)=s(ie1:ie2,js1:jsf)
END SUBROUTINE pinvs

SUBROUTINE red(iz1,iz2,jz1,jz2,jm1,jm2,jmf,ic1,jc1,jcf,kc,c,s)
IMPLICIT NONE
INTEGER(I4B), INTENT(IN) :: iz1,iz2,jz1,jz2,jm1,jm2,jmf,ic1,jc1,jcf,kc
REAL(SP), DIMENSION(:,:), INTENT(INOUT) :: s
REAL(SP), DIMENSION(:,:,:), INTENT(IN) :: c
   Reduce columns jz1-jz2 of the s matrix, using previous results as stored in the c matrix.
   Only columns jm1-jm2,jmf are affected by the prior results. red is used internally by
   solvde.
INTEGER(I4B) :: ic,l,loff
loff=jc1-jm1
ic=ic1
do j=jz1,jz2                                    Loop over columns to be zeroed.
    do l=jm1,jm2                                Loop over columns altered.
        s(iz1:iz2,l)=s(iz1:iz2,l)-s(iz1:iz2,j)*c(ic,l+loff,kc)
          Loop over rows.
    end do
    s(iz1:iz2,jmf)=s(iz1:iz2,jmf)-s(iz1:iz2,j)*c(ic,jcf,kc)    Plus final element.
    ic=ic+1
end do
END SUBROUTINE red
END SUBROUTINE solvde
```

`km=imaxloc...` See discussion of imaxloc on p. 1017.

⋆ ⋆ ⋆

```
MODULE sfroid_data
```
Communicates with difeq.
```
USE nrtype
INTEGER(I4B), PARAMETER :: M=41
INTEGER(I4B) :: mm,n
REAL(SP) :: anorm,c2,h
REAL(SP), DIMENSION(M) :: x
END MODULE sfroid_data
```

```
PROGRAM sfroid
USE nrtype; USE nrutil, ONLY : arth
USE nr, ONLY : plgndr,solvde
USE sfroid_data
IMPLICIT NONE
INTEGER(I4B), PARAMETER :: NE=3,NB=1
```
Sample program using solvde. Computes eigenvalues of spheroidal harmonics $S_{mn}(x;c)$ for $m \geq 0$ and $n \geq m$. In the program, m is mm, c^2 is c2, and γ of equation (17.4.20) is anorm.
```
INTEGER(I4B) :: itmax
INTEGER(I4B), DIMENSION(NE) :: indexv
REAL(SP) :: conv,slowc
REAL(SP), DIMENSION(M) :: deriv,fac1,fac2
REAL(SP), DIMENSION(NE) :: scalv
REAL(SP), DIMENSION(NE,M) :: y
itmax=100
conv=5.0e-6_sp
slowc=1.0
h=1.0_sp/(M-1)
c2=0.0
write(*,*) 'ENTER M,N'
read(*,*) mm,n
indexv(1:3)=merge( (/ 1, 2, 3 /), (/ 2, 1, 3 /), (mod(n+mm,2) == 1) )
```
No interchanges necessary if n+mm is odd; otherwise interchange y_1 and y_2.
```
anorm=1.0
```
Compute γ.
```
if (mm /= 0) then
    anorm=(-0.5_sp)**mm*product(&
        arth(n+1,1,mm)*arth(real(n,sp),-1.0_sp,mm)/arth(1,1,mm))
end if
x(1:M-1)=arth(0,1,M-1)*h
fac1(1:M-1)=1.0_sp-x(1:M-1)**2
```
Compute initial guess.
```
fac2(1:M-1)=fac1(1:M-1)**(-mm/2.0_sp)
y(1,1:M-1)=plgndr(n,mm,x(1:M-1))*fac2(1:M-1)
```
P_n^m from §6.8.
```
deriv(1:M-1)=-((n-mm+1)*plgndr(n+1,mm,x(1:M-1))-(n+1)*&
    x(1:M-1)*plgndr(n,mm,x(1:M-1)))/fac1(1:M-1)
```
Derivative of P_n^m from a recurrence relation.
```
y(2,1:M-1)=mm*x(1:M-1)*y(1,1:M-1)/fac1(1:M-1)+deriv(1:M-1)*fac2(1:M-1)
y(3,1:M-1)=n*(n+1)-mm*(mm+1)
x(M)=1.0
```
Initial guess at $x = 1$ done separately.
```
y(1,M)=anorm
y(3,M)=n*(n+1)-mm*(mm+1)
y(2,M)=(y(3,M)-c2)*y(1,M)/(2.0_sp*(mm+1.0_sp))
scalv(1:3)=(/ abs(anorm), max(abs(anorm),y(2,M)), max(1.0_sp,y(3,M)) /)
do
    write (*,*) 'ENTER C**2 OR 999 TO END'
    read (*,*) c2
    if (c2 == 999.0) exit
    call solvde(itmax,conv,slowc,scalv,indexv,NB,y)
    write (*,*) ' M = ',mm,' N = ',n,&
        ' C**2 = ',c2,' LAMBDA = ',y(3,1)+mm*(mm+1)
end do
```
Go back for another value of c^2.
```
END PROGRAM sfroid
```

`MODULE sfroid_data` This module functions just like a common block to communicate variables with `difeq`. The advantage of a module is that it allows complete specification of the variables.

`anorm=(-0.5_sp)**mm*product(...` This statement computes equation (17.4.20) by direct multiplication.

⋆ ⋆ ⋆

```
SUBROUTINE difeq(k,k1,k2,jsf,is1,isf,indexv,s,y)
USE nrtype
USE sfroid_data
IMPLICIT NONE
INTEGER(I4B), INTENT(IN) :: is1,isf,jsf,k,k1,k2
INTEGER(I4B), DIMENSION(:), INTENT(IN) :: indexv
REAL(SP), DIMENSION(:,:), INTENT(OUT) :: s
REAL(SP), DIMENSION(:,:), INTENT(IN) :: y
    Returns matrix s(i,j) for solvde.
REAL(SP) :: temp,temp2
INTEGER(I4B), DIMENSION(3) :: indexv3
indexv3(1:3)=3+indexv(1:3)
if (k == k1) then                Boundary condition at first point.
    if (mod(n+mm,2) == 1) then
        s(3,indexv3(1:3))= (/ 1.0_sp, 0.0_sp, 0.0_sp /)        Equation (17.4.32).
        s(3,jsf)=y(1,1)                                         Equation (17.4.31).
    else
        s(3,indexv3(1:3))= (/ 0.0_sp, 1.0_sp, 0.0_sp /)        Equation (17.4.32).
        s(3,jsf)=y(2,1)                                         Equation (17.4.31).
    end if
else if (k > k2) then            Boundary conditions at last point.
    s(1,indexv3(1:3))= (/ -(y(3,M)-c2)/(2.0_sp*(mm+1.0_sp)),&
        1.0_sp, -y(1,M)/(2.0_sp*(mm+1.0_sp)) /)                Equation (17.4.35).
    s(1,jsf)=y(2,M)-(y(3,M)-c2)*y(1,M)/(2.0_sp*(mm+1.0_sp))    Equation (17.4.33).
    s(2,indexv3(1:3))=(/ 1.0_sp, 0.0_sp, 0.0_sp /)             Equation (17.4.36).
    s(2,jsf)=y(1,M)-anorm                                       Equation (17.4.34).
else                             Interior point.
    s(1,indexv(1:3))=(/ -1.0_sp, -0.5_sp*h, 0.0_sp /)          Equation (17.4.28).
    s(1,indexv3(1:3))=(/ 1.0_sp, -0.5_sp*h, 0.0_sp /)
    temp=h/(1.0_sp-(x(k)+x(k-1))**2*0.25_sp)
    temp2=0.5_sp*(y(3,k)+y(3,k-1))-c2*0.25_sp*(x(k)+x(k-1))**2
    s(2,indexv(1:3))=(/ temp*temp2*0.5_sp,&                     Equation (17.4.29).
        -1.0_sp-0.5_sp*temp*(mm+1.0_sp)*(x(k)+x(k-1)),&
        0.25_sp*temp*(y(1,k)+y(1,k-1)) /)
    s(2,indexv3(1:3))=s(2,indexv(1:3))
    s(2,indexv3(2))=s(2,indexv3(2))+2.0_sp
    s(3,indexv(1:3))=(/ 0.0_sp, 0.0_sp, -1.0_sp /)             Equation (17.4.30).
    s(3,indexv3(1:3))=(/ 0.0_sp, 0.0_sp, 1.0_sp /)
    s(1,jsf)=y(1,k)-y(1,k-1)-0.5_sp*h*(y(2,k)+y(2,k-1))        Equation (17.4.23).
    s(2,jsf)=y(2,k)-y(2,k-1)-temp*((x(k)+x(k-1))*&              Equation (17.4.24).
        0.5_sp*(mm+1.0_sp)*(y(2,k)+y(2,k-1))-temp2*&
        0.5_sp*(y(1,k)+y(1,k-1)))
    s(3,jsf)=y(3,k)-y(3,k-1)                                    Equation (17.4.27).
end if
END SUBROUTINE difeq
```

⋆ ⋆ ⋆

```
MODULE sphoot_data                      Communicates with load, score, and derivs.
USE nrtype
INTEGER(I4B) :: m,n
REAL(SP) :: c2,dx,gamma
END MODULE sphoot_data
```

```
MODULE sphoot_caller                    Communicates with shoot.
USE nrtype
INTEGER(I4B) :: nvar
REAL(SP) :: x1,x2
END MODULE sphoot_caller
```

```
PROGRAM sphoot
```

Sample program using shoot. Computes eigenvalues of spheroidal harmonics $S_{mn}(x;c)$ for $m \geq 0$ and $n \geq m$. Be sure that routine funcv for newt is provided by shoot (§17.1).

```
USE nrtype; USE nrutil, ONLY : arth
USE nr, ONLY : newt
USE sphoot_data
USE sphoot_caller
IMPLICIT NONE
INTEGER(I4B), PARAMETER :: NV=3,N2=1
REAL(SP), DIMENSION(N2) :: v
LOGICAL(LGT) :: check
nvar=NV                                 Number of equations.
dx=1.0e-4_sp                            Avoid evaluating derivatives exactly at x = -1.
do
    write(*,*) 'input m,n,c-squared (999 to end)'
    read(*,*) m,n,c2
    if (c2 == 999.0) exit
    if ((n < m) .or. (m < 0)) cycle
    gamma=(-0.5_sp)**m*product(&        Compute γ of equation (17.4.20).
        arth(n+1,1,m)*(arth(real(n,sp),-1.0_sp,m)/arth(1,1,m)))
    v(1)=n*(n+1)-m*(m+1)+c2/2.0_sp      Initial guess for eigenvalue.
    x1=-1.0_sp+dx                       Set range of integration.
    x2=0.0
    call newt(v,check)                  Find v that zeros function f in score.
    if (check) then
        write(*,*)'shoot failed; bad initial guess'
        exit
    else
        write(*,'(1x,t6,a)') 'mu(m,n)'
        write(*,'(1x,f12.6)') v(1)
    end if
end do
END PROGRAM sphoot
```

```
SUBROUTINE load(x1,v,y)
USE nrtype
USE sphoot_data
IMPLICIT NONE
REAL(SP), INTENT(IN) :: x1
REAL(SP), DIMENSION(:), INTENT(IN) :: v
REAL(SP), DIMENSION(:), INTENT(OUT) :: y
```

Supplies starting values for integration at $x = -1 + dx$.

```
REAL(SP) :: y1
y(3)=v(1)
y1=merge(gamma,-gamma, mod(n-m,2) == 0 )
y(2)=-(y(3)-c2)*y1/(2*(m+1))
y(1)=y1+y(2)*dx
END SUBROUTINE load
```

```
SUBROUTINE score(x2,y,f)
USE nrtype
USE sphoot_data
IMPLICIT NONE
REAL(SP), INTENT(IN) :: x2
REAL(SP), DIMENSION(:), INTENT(IN) :: y
REAL(SP), DIMENSION(:), INTENT(OUT) :: f
    Tests whether boundary condition at x = 0 is satisfied.
f(1)=merge(y(2),y(1), mod(n-m,2) == 0 )
END SUBROUTINE score
```

f90 `MODULE sphoot_data...MODULE sphoot_caller` These modules function just like common blocks to communicate variables from `sphoot` to the various subsidiary routines. The advantage of a module is that it allows complete specification of the variables.

```
SUBROUTINE derivs(x,y,dydx)
USE nrtype
USE sphoot_data
IMPLICIT NONE
REAL(SP), INTENT(IN) :: x
REAL(SP), DIMENSION(:), INTENT(IN) :: y
REAL(SP), DIMENSION(:), INTENT(OUT) :: dydx
    Evaluates derivatives for odeint.
dydx(1)=y(2)
dydx(2)=(2.0_sp*x*(m+1.0_sp)*y(2)-(y(3)-c2*x*x)*y(1))/(1.0_sp-x*x)
dydx(3)=0.0
END SUBROUTINE derivs
```

* * *

```
MODULE sphfpt_data                        Communicates with load1, load2, score,
USE nrtype                                    and derivs.
INTEGER(I4B) :: m,n
REAL(SP) :: c2,dx,gamma
END MODULE sphfpt_data
```

```
MODULE sphfpt_caller                      Communicates with shootf.
USE nrtype
INTEGER(I4B) :: nn2
REAL(SP) :: x1,x2,xf
END MODULE sphfpt_caller
```

```
PROGRAM sphfpt
   Sample program using shootf. Computes eigenvalues of spheroidal harmonics Smn(x;c)
   for m >= 0 and n >= m. Be sure that routine funcv for newt is provided by shootf (§17.2).
   The routine derivs is the same as for sphoot.
USE nrtype; USE nrutil, ONLY : arth
USE nr, ONLY : newt
USE sphfpt_data
USE sphfpt_caller
IMPLICIT NONE
INTEGER(I4B), PARAMETER :: N1=2,N2=1,NTOT=N1+N2
REAL(SP), PARAMETER :: DXX=1.0e-4_sp
REAL(SP), DIMENSION(:), POINTER :: v1,v2
REAL(SP), DIMENSION(NTOT), TARGET :: v
LOGICAL(LGT) :: check
v1=>v(1:N2)
v2=>v(N2+1:NTOT)
nn2=N2
dx=DXX                                   Avoid evaluating derivatives exactly at x =
do                                          ±1.
   write(*,*) 'input m,n,c-squared (999 to end)'
   read(*,*) m,n,c2
   if (c2 == 999.0) exit
   if ((n < m) .or. (m < 0)) cycle
   gamma=(-0.5_sp)**m*product(&          Compute γ of equation (17.4.20).
      arth(n+1,1,m)*(arth(real(n,sp),-1.0_sp,m)/arth(1,1,m)))
   v1(1)=n*(n+1)-m*(m+1)+c2/2.0_sp       Initial guess for eigenvalue and function value.
   v2(2)=v1(1)
   v2(1)=gamma*(1.0_sp-(v2(2)-c2)*dx/(2*(m+1)))
   x1=-1.0_sp+dx                         Set range of integration.
   x2=1.0_sp-dx
   xf=0.0                                Fitting point.
   call newt(v,check)                    Find v that zeros function f in score.
   if (check) then
      write(*,*) 'shootf failed; bad initial guess'
      exit
   else
      write(*,'(1x,t6,a)') 'mu(m,n)'
      write(*,'(1x,f12.6)') v1(1)
   end if
end do
END PROGRAM sphfpt
```

```
SUBROUTINE load1(x1,v1,y)
USE nrtype
USE sphfpt_data
IMPLICIT NONE
REAL(SP), INTENT(IN) :: x1
REAL(SP), DIMENSION(:), INTENT(IN) :: v1
REAL(SP), DIMENSION(:), INTENT(OUT) :: y
   Supplies starting values for integration at x = -1 + dx.
REAL(SP) :: y1
y(3)=v1(1)
y1=merge(gamma,-gamma,mod(n-m,2) == 0)
y(2)=-(y(3)-c2)*y1/(2*(m+1))
y(1)=y1+y(2)*dx
END SUBROUTINE load1
```

```
SUBROUTINE load2(x2,v2,y)
USE nrtype
USE sphfpt_data
IMPLICIT NONE
REAL(SP), INTENT(IN) :: x2
REAL(SP), DIMENSION(:), INTENT(IN) :: v2
REAL(SP), DIMENSION(:), INTENT(OUT) :: y
   Supplies starting values for integration at x = 1 - dx.
y(3)=v2(2)
y(1)=v2(1)
y(2)=(y(3)-c2)*y(1)/(2*(m+1))
END SUBROUTINE load2
```

```
SUBROUTINE score(xf,y,f)
USE nrtype
USE sphfpt_data
IMPLICIT NONE
REAL(SP), INTENT(IN) :: xf
REAL(SP), DIMENSION(:), INTENT(IN) :: y
REAL(SP), DIMENSION(:), INTENT(OUT) :: f
   Tests whether solutions match at fitting point x = 0.
f(1:3)=y(1:3)
END SUBROUTINE score
```

f90 `MODULE sphfpt_data...MODULE sphfpt_caller` These modules function just like common blocks to communicate variables from `sphfpt` to the various subsidiary routines. The advantage of a module is that it allows complete specification of the variables.

Chapter B18. Integral Equations and Inverse Theory

```
SUBROUTINE fred2(a,b,t,f,w,g,ak)
USE nrtype; USE nrutil, ONLY : assert_eq,unit_matrix
USE nr, ONLY : gauleg,lubksb,ludcmp
IMPLICIT NONE
REAL(SP), INTENT(IN) :: a,b
REAL(SP), DIMENSION(:), INTENT(OUT) :: t,f,w
INTERFACE
    FUNCTION g(t)
    USE nrtype
    IMPLICIT NONE
    REAL(SP), DIMENSION(:), INTENT(IN) :: t
    REAL(SP), DIMENSION(size(t)) :: g
    END FUNCTION g

    FUNCTION ak(t,s)
    USE nrtype
    IMPLICIT NONE
    REAL(SP), DIMENSION(:), INTENT(IN) :: t,s
    REAL(SP), DIMENSION(size(t),size(s)) :: ak
    END FUNCTION ak
END INTERFACE
```

Solves a linear Fredholm equation of the second kind by N-point Gaussian quadrature. On input, a and b are the limits of integration. g and ak are user-supplied external functions. g returns $g(t)$ as a vector of length N for a vector of N arguments, while ak returns $\lambda K(t,s)$ as an $N \times N$ matrix. The routine returns arrays t and f of length N containing the abscissas t_i of the Gaussian quadrature and the solution f at these abscissas. Also returned is the array w of length N of Gaussian weights for use with the Nystrom interpolation routine fredin.

```
INTEGER(I4B) :: n
INTEGER(I4B), DIMENSION(size(f)) :: indx
REAL(SP) :: d
REAL(SP), DIMENSION(size(f),size(f)) :: omk
n=assert_eq(size(f),size(t),size(w),'fred2')
call gauleg(a,b,t,w)                         Replace gauleg with another routine if not
call unit_matrix(omk)                          using Gauss-Legendre quadrature.
omk=omk-ak(t,t)*spread(w,dim=1,ncopies=n)      Form 1 - λK̃.
f=g(t)
call ludcmp(omk,indx,d)                        Solve linear equations.
call lubksb(omk,indx,f)
END SUBROUTINE fred2
```

call unit_matrix(omk) The unit_matrix routine in nrutil does exactly what its name suggests.

`omk=omk-ak(t,t)*spread(w,dim=1,ncopies=n)` By now this idiom should be second nature: the first column of `ak` gets multiplied by the first element of `w`, and so on.

⋆ ⋆ ⋆

```
FUNCTION fredin(x,a,b,t,f,w,g,ak)
USE nrtype; USE nrutil, ONLY : assert_eq
IMPLICIT NONE
REAL(SP), INTENT(IN) :: a,b
REAL(SP), DIMENSION(:), INTENT(IN) :: x,t,f,w
REAL(SP), DIMENSION(size(x)) :: fredin
INTERFACE
    FUNCTION g(t)
    USE nrtype
    IMPLICIT NONE
    REAL(SP), DIMENSION(:), INTENT(IN) :: t
    REAL(SP), DIMENSION(size(t)) :: g
    END FUNCTION g

    FUNCTION ak(t,s)
    USE nrtype
    IMPLICIT NONE
    REAL(SP), DIMENSION(:), INTENT(IN) :: t,s
    REAL(SP), DIMENSION(size(t),size(s)) :: ak
    END FUNCTION ak
END INTERFACE
```

Input are arrays `t` and `w` of length N containing the abscissas and weights of the N-point Gaussian quadrature, and the solution array `f` of length N from `fred2`. The function `fredin` returns the array of values of f at an array of points `x` using the Nystrom interpolation formula. On input, `a` and `b` are the limits of integration. `g` and `ak` are user-supplied external functions. `g` returns $g(t)$ as a vector of length N for a vector of N arguments, while `ak` returns $\lambda K(t,s)$ as an $N \times N$ matrix.

```
INTEGER(I4B) :: n
n=assert_eq(size(f),size(t),size(w),'fredin')
fredin=g(x)+matmul(ak(x,t),w*f)
END FUNCTION fredin
```

`fredin=g(x)+matmul...` Fortran 90 allows very concise coding here, which also happens to be much closer to the mathematical formulation than the loops required in Fortran 77.

⋆ ⋆ ⋆

```
SUBROUTINE voltra(t0,h,t,f,g,ak)
USE nrtype; USE nrutil, ONLY : array_copy,assert_eq,unit_matrix
USE nr, ONLY : lubksb,ludcmp
IMPLICIT NONE
REAL(SP), INTENT(IN) :: t0,h
REAL(SP), DIMENSION(:), INTENT(OUT) :: t
REAL(SP), DIMENSION(:,:), INTENT(OUT) :: f
INTERFACE
    FUNCTION g(t)
    USE nrtype
    IMPLICIT NONE
    REAL(SP), INTENT(IN) :: t
    REAL(SP), DIMENSION(:), POINTER :: g
    END FUNCTION g

    FUNCTION ak(t,s)
```

```
    USE nrtype
    IMPLICIT NONE
    REAL(SP), INTENT(IN) :: t,s
    REAL(SP), DIMENSION(:,:), POINTER :: ak
    END FUNCTION ak
END INTERFACE
```

Solves a set of M linear Volterra equations of the second kind using the extended trapezoidal rule. On input, `t0` is the starting point of the integration. The routine takes $N-1$ steps of size `h` and returns the abscissas in `t`, a vector of length N. The solution at these points is returned in the $M \times N$ matrix `f`. `g` is a user-supplied external function that returns a pointer to the M-dimensional vector of functions $g_k(t)$, while `ak` is another user-supplied external function that returns a pointer to the $M \times M$ matrix $K(t,s)$.

```
INTEGER(I4B) :: i,j,n,ncop,nerr,m
INTEGER(I4B), DIMENSION(size(f,1)) :: indx
REAL(SP) :: d
REAL(SP), DIMENSION(size(f,1)) :: b
REAL(SP), DIMENSION(size(f,1),size(f,1)) :: a
n=assert_eq(size(f,2),size(t),'voltra: n')
t(1)=t0                                                  Initialize.
call array_copy(g(t(1)),f(:,1),ncop,nerr)
m=assert_eq(size(f,1),ncop,ncop+nerr,'voltra: m')
do i=2,n                                                 Take a step h.
    t(i)=t(i-1)+h
    b=g(t(i))+0.5_sp*h*matmul(ak(t(i),t(1)),f(:,1))      Accumulate right-hand side
    do j=2,i-1                                              of linear equations in b.
        b=b+h*matmul(ak(t(i),t(j)),f(:,j))
    end do
    call unit_matrix(a)                                  Left-hand side goes in ma-
    a=a-0.5_sp*h*ak(t(i),t(i))                              trix a.
    call ludcmp(a,indx,d)                                Solve linear equations.
    call lubksb(a,indx,b)
    f(:,i)=b(:)
end do
END SUBROUTINE voltra
```

f90 `FUNCTION g(t)...REAL(SP), DIMENSION(:), POINTER :: g` The routine `voltra` requires an argument that is a function returning a vector, but we don't know the dimension of the vector at compile time. The solution is to make the function return a *pointer* to the vector. This is not the same thing as a pointer to a function, which is not allowed in Fortran 90. When you use the pointer in the routine, Fortran 90 figures out from the context that you want the vector of values, so the code remains highly readable. Similarly, the argument `ak` is a function returning a pointer to a matrix.

The coding of the user-supplied functions `g` and `ak` deserves some comment: functions returning pointers to arrays are potential memory leaks if the arrays are allocated dynamically in the functions. Here the user knows in advance the dimension of the problem, and so there is no need to use dynamical allocation in the functions. For example, in a two-dimensional problem, you can code `g` as follows:

```
FUNCTION g(t)
USE nrtype
IMPLICIT NONE
REAL(SP), INTENT(IN) :: t
REAL(SP), DIMENSION(:), POINTER :: g
REAL(SP), DIMENSION(2), TARGET, SAVE :: gg
g=>gg
g(1)=...
g(2)=...
END FUNCTION g
```

and similarly for ak.

Suppose, however, we coded g with dynamical allocation:

```
FUNCTION g(t)
USE nrtype
IMPLICIT NONE
REAL(SP), INTENT(IN) :: t
REAL(SP), DIMENSION(:), POINTER :: g
allocate(g(2))
g(1)=...
g(2)=...
END FUNCTION g
```

Now g never gets deallocated; each time we call the function fresh memory gets consumed. If you have a problem that really does require dynamical allocation in a pointer function, you have to be sure to deallocate the pointer in the calling routine. In voltra, for example, we would declare pointers gtemp and aktemp. Then instead of writing simply

```
b=g(t(i))+...
```

we would write

```
gtemp=>g(t(i))
b=gtemp+...
deallocate(gtemp)
```

and similarly for each pointer function invocation.

call array_copy(g(t(1)),f(:,1),ncop,nerr) The routine would work if we replaced this statement with simply f(:,1)=g(t(1)). The purpose of using array_copy from nrutil is that we can check that f and g have consistent dimensions with a call to assert_eq.

* * *

```
FUNCTION wwghts(n,h,kermom)
USE nrtype; USE nrutil, ONLY : geop
IMPLICIT NONE
INTEGER(I4B), INTENT(IN) :: n
REAL(SP), INTENT(IN) :: h
REAL(SP), DIMENSION(n) :: wwghts
INTERFACE
    FUNCTION kermom(y,m)
    USE nrtype
    IMPLICIT NONE
    REAL(DP), INTENT(IN) :: y
    INTEGER(I4B), INTENT(IN) :: m
    REAL(DP), DIMENSION(m) :: kermom
    END FUNCTION kermom
END INTERFACE
    Returns in wwghts(1:n) weights for the n-point equal-interval quadrature from 0 to (n-
    1)h of a function f(x) times an arbitrary (possibly singular) weight function w(x) whose
    indefinite-integral moments F_n(y) are provided by the user-supplied function kermom.
INTEGER(I4B) :: j
REAL(DP) :: hh,hi,c,a,b
REAL(DP), DIMENSION(4) :: wold,wnew,w
hh=h                                Double precision on internal calculations even though
hi=1.0_dp/hh                            the interface is in single precision.
wwghts(1:n)=0.0                     Zero all the weights so we can sum into them.
wold(1:4)=kermom(0.0_dp,4)          Evaluate indefinite integrals at lower end.
```

```
if (n >= 4) then                        Use highest available order.
    b=0.0                               For another problem, you might change this lower
                                            limit.
    do j=1,n-3
        c=j-1                           This is called k in equation (18.3.5).
        a=b                             Set upper and lower limits for this step.
        b=a+hh
        if (j == n-3) b=(n-1)*hh        Last interval: go all the way to end.
        wnew(1:4)=kermom(b,4)
        w(1:4)=(wnew(1:4)-wold(1:4))*geop(1.0_dp,hi,4)        Equation (18.3.4).
        wwghts(j:j+3)=wwghts(j:j+3)+(/&                        Equation (18.3.5).
            ((c+1.0_dp)*(c+2.0_dp)*(c+3.0_dp)*w(1)&
            -(11.0_dp+c*(12.0_dp+c*3.0_dp))*w(2)&
                +3.0_dp*(c+2.0_dp)*w(3)-w(4))/6.0_dp,&
            (-c*(c+2.0_dp)*(c+3.0_dp)*w(1)&
            +(6.0_dp+c*(10.0_dp+c*3.0_dp))*w(2)&
                -(3.0_dp*c+5.0_dp)*w(3)+w(4))*0.50_dp,&
            (c*(c+1.0_dp)*(c+3.0_dp)*w(1)&
            -(3.0_dp+c*(8.0_dp+c*3.0_dp))*w(2)&
                +(3.0_dp*c+4.0_dp)*w(3)-w(4))*0.50_dp,&
            (-c*(c+1.0_dp)*(c+2.0_dp)*w(1)&
            +(2.0_dp+c*(6.0_dp+c*3.0_dp))*w(2)&
            -3.0_dp*(c+1.0_dp)*w(3)+w(4))/6.0_dp /)
        wold(1:4)=wnew(1:4)             Reset lower limits for moments.
    end do
else if (n == 3) then                   Lower-order cases; not recommended.
    wnew(1:3)=kermom(hh+hh,3)
    w(1:3)= (/ wnew(1)-wold(1), hi*(wnew(2)-wold(2)),&
        hi**2*(wnew(3)-wold(3)) /)
    wwghts(1:3)= (/ w(1)-1.50_dp*w(2)+0.50_dp*w(3),&
        2.0_dp*w(2)-w(3), 0.50_dp*(w(3)-w(2)) /)
else if (n == 2) then
    wnew(1:2)=kermom(hh,2)
    wwghts(2)=hi*(wnew(2)-wold(2))
    wwghts(1)=wnew(1)-wold(1)-wwghts(2)
end if
END FUNCTION wwghts
```

⋆ ⋆ ⋆

```
MODULE kermom_info
USE nrtype
REAL(DP) :: kermom_x
END MODULE kermom_info
```

```
FUNCTION kermom(y,m)
USE nrtype
USE kermom_info
IMPLICIT NONE
REAL(DP), INTENT(IN) :: y
INTEGER(I4B), INTENT(IN) :: m
REAL(DP), DIMENSION(m) :: kermom
```

Returns in kermom(1:m) the first m indefinite-integral moments of one row of the singular part of the kernel. (For this example, m is hard-wired to be 4.) The input variable y labels the column, while kermom_x (in the module kermom_info) is the row.

```
REAL(DP) :: x,d,df,clog,x2,x3,x4
x=kermom_x                      We can take x as the lower limit of integration. Thus, we
if (y >= x) then                    return the moment integrals either purely to the left or
    d=y-x                           purely to the right of the diagonal.
    df=2.0_dp*sqrt(d)*d
    kermom(1:4) = (/ df/3.0_dp, df*(x/3.0_dp+d/5.0_dp),&
```

```
        df*((x/3.0_dp + 0.4_dp*d)*x + d**2/7.0_dp),&
        df*(((x/3.0_dp + 0.6_dp*d)*x + 3.0_dp*d**2/7.0_dp)*x&
            + d**3/9.0_dp) /)
else
    x2=x**2
    x3=x2*x
    x4=x2*x2
    d=x-y
    clog=log(d)
    kermom(1:4) = (/ d*(clog-1.0_dp),&
        -0.25_dp*(3.0_dp*x+y-2.0_dp*clog*(x+y))*d,&
        (-11.0_dp*x3+y*(6.0_dp*x2+y*(3.0_dp*x+2.0_dp*y))&
            +6.0_dp*clog*(x3-y**3))/18.0_dp,&
        (-25.0_dp*x4+y*(12.0_dp*x3+y*(6.0_dp*x2+y*&
            (4.0_dp*x+3.0_dp*y)))+12.0_dp*clog*(x4-y**4))/48.0_dp /)
end if
END FUNCTION kermom
```

`MODULE kermom_info` This module functions just like a common block to share the variable `kermom_x` with the routine `quadmx`.

★ ★ ★

```
SUBROUTINE quadmx(a)
USE nrtype; USE nrutil, ONLY : arth,assert_eq,diagadd,outerprod
USE nr, ONLY : wwghts,kermom
USE kermom_info
IMPLICIT NONE
REAL(SP), DIMENSION(:,:), INTENT(OUT) :: a
```

Constructs in the $N \times N$ array a the quadrature matrix for an example Fredholm equation of the second kind. The nonsingular part of the kernel is computed within this routine, while the quadrature weights that integrate the singular part of the kernel are obtained via calls to wwghts. An external routine kermom, which supplies indefinite-integral moments of the singular part of the kernel, is passed to wwghts.

```
INTEGER(I4B) :: j,n
REAL(SP) :: h,x
REAL(SP), DIMENSION(size(a,1)) :: wt
n=assert_eq(size(a,1),size(a,2),'quadmx')
h=PI/(n-1)
do j=1,n
    x=(j-1)*h
    kermom_x=x                      Put x in the module kermom_info for use by kermom.
    wt(:)=wwghts(n,h,kermom)        Part of nonsingular kernel.
    a(j,:)=wt(:)                    Put together all the pieces of the kernel.
end do
wt(:)=cos(arth(0,1,n)*h)
a(:,:)=a(:,:)*outerprod(wt(:),wt(:))
call diagadd(a,1.0_sp)              Since equation of the second kind, there is diagonal
END SUBROUTINE quadmx                   piece independent of h.
```

`call diagadd...` See discussion of diagadd after hqr on p. 1234.

★ ★ ★

```
PROGRAM fredex
USE nrtype; USE nrutil, ONLY : arth
USE nr, ONLY : quadmx,ludcmp,lubksb
IMPLICIT NONE
INTEGER(I4B), PARAMETER :: N=40
INTEGER(I4B) :: j
INTEGER(I4B), DIMENSION(N) :: indx
REAL(SP) :: d
REAL(SP), DIMENSION(N) :: g,x
REAL(SP), DIMENSION(N,N) :: a
    This sample program shows how to solve a Fredholm equation of the second kind using
    the product Nystrom method and a quadrature rule especially constructed for a particular,
    singular, kernel.
    Parameter: N is the size of the grid.
call quadmx(a)                    Make the quadrature matrix; all the action is here.
call ludcmp(a,indx,d)             Decompose the matrix.
x(:)=arth(0,1,n)*PI/(n-1)
g(:)=sin(x(:))                    Construct the right-hand side, here sin x.
call lubksb(a,indx,g)             Backsubstitute.
do j=1,n                          Write out the solution.
    write (*,*) j,x(j),g(j)
end do
write (*,*) 'normal completion'
END PROGRAM fredex
```

Chapter B19. Partial Differential Equations

```
SUBROUTINE sor(a,b,c,d,e,f,u,rjac)
USE nrtype; USE nrutil, ONLY : assert_eq,nrerror
IMPLICIT NONE
REAL(DP), DIMENSION(:,:), INTENT(IN) :: a,b,c,d,e,f
REAL(DP), DIMENSION(:,:), INTENT(INOUT) :: u
REAL(DP), INTENT(IN) :: rjac
INTEGER(I4B), PARAMETER :: MAXITS=1000
REAL(DP), PARAMETER :: EPS=1.0e-5_dp
```

Successive overrelaxation solution of equation (19.5.25) with Chebyshev acceleration. a, b, c, d, e, and f are input as the coefficients of the equation, each dimensioned to the grid size $J \times J$. u is input as the initial guess to the solution, usually zero, and returns with the final value. `rjac` is input as the spectral radius of the Jacobi iteration, or an estimate of it. Double precision is a good idea for J bigger than about 25.

```
REAL(DP), DIMENSION(size(a,1),size(a,1)) :: resid
INTEGER(I4B) :: jmax,jm1,jm2,jm3,n
REAL(DP) :: anorm,anormf,omega
jmax=assert_eq((/size(a,1),size(a,2),size(b,1),size(b,2), &
    size(c,1),size(c,2),size(d,1),size(d,2),size(e,1), &
    size(e,2),size(f,1),size(f,2),size(u,1),size(u,2)/),'sor')
jm1=jmax-1
jm2=jmax-2
jm3=jmax-3
anormf=sum(abs(f(2:jm1,2:jm1)))
```

Compute initial norm of residual and terminate iteration when norm has been reduced by a factor EPS. This computation assumes initial u is zero.

```
omega=1.0
do n=1,MAXITS
```

First do the even-even and odd-odd squares of the grid, i.e., the red squares of the checkerboard:

```
    resid(2:jm1:2,2:jm1:2)=a(2:jm1:2,2:jm1:2)*u(3:jmax:2,2:jm1:2)+&
        b(2:jm1:2,2:jm1:2)*u(1:jm2:2,2:jm1:2)+&
        c(2:jm1:2,2:jm1:2)*u(2:jm1:2,3:jmax:2)+&
        d(2:jm1:2,2:jm1:2)*u(2:jm1:2,1:jm2:2)+&
        e(2:jm1:2,2:jm1:2)*u(2:jm1:2,2:jm1:2)-f(2:jm1:2,2:jm1:2)
    u(2:jm1:2,2:jm1:2)=u(2:jm1:2,2:jm1:2)-omega*&
        resid(2:jm1:2,2:jm1:2)/e(2:jm1:2,2:jm1:2)
    resid(3:jm2:2,3:jm2:2)=a(3:jm2:2,3:jm2:2)*u(4:jm1:2,3:jm2:2)+&
        b(3:jm2:2,3:jm2:2)*u(2:jm3:2,3:jm2:2)+&
        c(3:jm2:2,3:jm2:2)*u(3:jm2:2,4:jm1:2)+&
        d(3:jm2:2,3:jm2:2)*u(3:jm2:2,2:jm3:2)+&
        e(3:jm2:2,3:jm2:2)*u(3:jm2:2,3:jm2:2)-f(3:jm2:2,3:jm2:2)
    u(3:jm2:2,3:jm2:2)=u(3:jm2:2,3:jm2:2)-omega*&
        resid(3:jm2:2,3:jm2:2)/e(3:jm2:2,3:jm2:2)
    omega=merge(1.0_dp/(1.0_dp-0.5_dp*rjac**2), &
        1.0_dp/(1.0_dp-0.25_dp*rjac**2*omega), n == 1)
```

Now do even-odd and odd-even squares of the grid, i.e., the black squares of the checkerboard:

```
    resid(3:jm2:2,2:jm1:2)=a(3:jm2:2,2:jm1:2)*u(4:jm1:2,2:jm1:2)+&
        b(3:jm2:2,2:jm1:2)*u(2:jm3:2,2:jm1:2)+&
```

```
        c(3:jm2:2,2:jm1:2)*u(3:jm2:2,3:jmax:2)+&
        d(3:jm2:2,2:jm1:2)*u(3:jm2:2,1:jm2:2)+&
        e(3:jm2:2,2:jm1:2)*u(3:jm2:2,2:jm1:2)-f(3:jm2:2,2:jm1:2)
    u(3:jm2:2,2:jm1:2)=u(3:jm2:2,2:jm1:2)-omega*&
        resid(3:jm2:2,2:jm1:2)/e(3:jm2:2,2:jm1:2)
    resid(2:jm1:2,3:jm2:2)=a(2:jm1:2,3:jm2:2)*u(3:jmax:2,3:jm2:2)+&
        b(2:jm1:2,3:jm2:2)*u(1:jm2:2,3:jm2:2)+&
        c(2:jm1:2,3:jm2:2)*u(2:jm1:2,4:jm1:2)+&
        d(2:jm1:2,3:jm2:2)*u(2:jm1:2,2:jm3:2)+&
        e(2:jm1:2,3:jm2:2)*u(2:jm1:2,3:jm2:2)-f(2:jm1:2,3:jm2:2)
    u(2:jm1:2,3:jm2:2)=u(2:jm1:2,3:jm2:2)-omega*&
        resid(2:jm1:2,3:jm2:2)/e(2:jm1:2,3:jm2:2)
    omega=1.0_dp/(1.0_dp-0.25_dp*rjac**2*omega)
    anorm=sum(abs(resid(2:jm1,2:jm1)))
    if (anorm < EPS*anormf) exit
end do
if (n > MAXITS) call nrerror('MAXITS exceeded in sor')
END SUBROUTINE sor
```

Red-black iterative schemes like the one used in sor are easily parallelizable. Updating the red grid points requires information only from the black grid points, so they can all be updated independently. Similarly the black grid points can all be updated independently. Since nearest neighbors are involved in the updating, communication costs can be kept to a minimum.

f90 There are several possibilities for coding the red-black iteration in a data parallel way using only Fortran 90 and no parallel language extensions. One way is to define an $N \times N$ logical mask red that is true on the red grid points and false on the black. Then each iteration consists of an update governed by a `where(red)...end where` block and a `where(.not. red)...end where` block. We have chosen a more direct coding that avoids the need for storage of the array red. The red update corresponds to the even-even and odd-odd grid points, the black to the even-odd and odd-even points. We can code each of these four cases directly with array sections, as in the routine above.

The array section notation used in sor is rather dense and hard to read. We could use pointer aliases to try to simplify things, but since each array section is different, we end up merely giving names to each term that was there all along. Pointer aliases do help if we code sor using a logical mask. Since there may be machines on which this version is faster, and since it is of some pedagogic interest, we give the alternative code:

```
SUBROUTINE sor_mask(a,b,c,d,e,f,u,rjac)
USE nrtype; USE nrutil, ONLY : assert_eq,nrerror
IMPLICIT NONE
REAL(DP), DIMENSION(:,:), TARGET, INTENT(IN) :: a,b,c,d,e,f
REAL(DP), DIMENSION(:,:), TARGET, INTENT(INOUT) :: u
REAL(DP), INTENT(IN) :: rjac
INTEGER(I4B), PARAMETER :: MAXITS=1000
REAL(DP), PARAMETER :: EPS=1.0e-5_dp
REAL(DP), DIMENSION(:,:), ALLOCATABLE :: resid
REAL(DP), DIMENSION(:,:), POINTER :: u_int,u_down,u_up,u_left,&
    u_right,a_int,b_int,c_int,d_int,e_int,f_int
INTEGER(I4B) :: jmax,jm1,jm2,jm3,n
REAL(DP) anorm,anormf,omega
LOGICAL, DIMENSION(:,:), ALLOCATABLE :: red
jmax=assert_eq((/size(a,1),size(a,2),size(b,1),size(b,2), &
```

```
    size(c,1),size(c,2),size(d,1),size(d,2),size(e,1), &
    size(e,2),size(f,1),size(f,2),size(u,1),size(u,2)/),'sor')
jm1=jmax-1
jm2=jmax-2
jm3=jmax-3
allocate(resid(jm2,jm2),red(jm2,jm2))          Interior is (jmax − 2) × (jmax − 2).
red=.false.
red(1:jm2:2,1:jm2:2)=.true.
red(2:jm3:2,2:jm3:2)=.true.
u_int=>u(2:jm1,2:jm1)
u_down=>u(3:jmax,2:jm1)
u_up=>u(1:jm2,2:jm1)
u_left=>u(2:jm1,1:jm2)
u_right=>u(2:jm1,3:jmax)
a_int=>a(2:jm1,2:jm1)
b_int=>b(2:jm1,2:jm1)
c_int=>c(2:jm1,2:jm1)
d_int=>d(2:jm1,2:jm1)
e_int=>e(2:jm1,2:jm1)
f_int=>f(2:jm1,2:jm1)
anormf=sum(abs(f_int))
omega=1.0
do n=1,MAXITS
    where(red)
        resid=a_int*u_down+b_int*u_up+c_int*u_right+&
            d_int*u_left+e_int*u_int-f_int
        u_int=u_int-omega*resid/e_int
    end where
    omega=merge(1.0_dp/(1.0_dp-0.5_dp*rjac**2), &
        1.0_dp/(1.0_dp-0.25_dp*rjac**2*omega), n == 1)
    where(.not.red)
        resid=a_int*u_down+b_int*u_up+c_int*u_right+&
            d_int*u_left+e_int*u_int-f_int
        u_int=u_int-omega*resid/e_int
    end where
    omega=1.0_dp/(1.0_dp-0.25_dp*rjac**2*omega)
    anorm=sum(abs(resid))
    if(anorm < EPS*anormf)exit
end do
deallocate(resid,red)
if (n > MAXITS) call nrerror('MAXITS exceeded in sor')
END SUBROUTINE sor_mask
```

⋆ ⋆ ⋆

```
SUBROUTINE mglin(u,ncycle)
USE nrtype; USE nrutil, ONLY : assert_eq,nrerror
USE nr, ONLY : interp,rstrct,slvsml
IMPLICIT NONE
REAL(DP), DIMENSION(:,:), INTENT(INOUT) :: u
INTEGER(I4B), INTENT(IN) :: ncycle
```

Full Multigrid Algorithm for solution of linear elliptic equation, here the model problem (19.0.6). On input u contains the right-hand side ρ in an $N \times N$ array, while on output it returns the solution. The dimension N is related to the number of grid levels used in the solution, ng below, by $N =$ 2**ng+1. ncycle is the number of V-cycles to be used at each level.

```
INTEGER(I4B) :: j,jcycle,n,ng,ngrid,nn
TYPE ptr2d                             Define a type so we can have an array of pointers
    REAL(DP), POINTER :: a(:,:)            to arrays of grid variables.
END TYPE ptr2d
TYPE(ptr2d), ALLOCATABLE :: rho(:)
```

```
REAL(DP), DIMENSION(:,:), POINTER :: uj,uj_1
n=assert_eq(size(u,1),size(u,2),'mglin')
ng=nint(log(n-1.0)/log(2.0))
if (n /= 2**ng+1) call nrerror('n-1 must be a power of 2 in mglin')
allocate(rho(ng))
nn=n
ngrid=ng
allocate(rho(ngrid)%a(nn,nn))             Allocate storage for r.h.s. on grid ng,
rho(ngrid)%a=u                            and fill it with the input r.h.s.
do                                        Similarly allocate storage and fill r.h.s. on all coarse
    if (nn <= 3) exit                         grids by restricting from finer grids.
    nn=nn/2+1
    ngrid=ngrid-1
    allocate(rho(ngrid)%a(nn,nn))
    rho(ngrid)%a=rstrct(rho(ngrid+1)%a)
end do
nn=3
allocate(uj(nn,nn))
call slvsml(uj,rho(1)%a)                  Initial solution on coarsest grid.
do j=2,ng                                 Nested iteration loop.
    nn=2*nn-1
    uj_1=>uj
    allocate(uj(nn,nn))
    uj=interp(uj_1)                       Interpolate from grid j-1 to next finer grid j.
    deallocate(uj_1)
    do jcycle=1,ncycle                    V-cycle loop.
        call mg(j,uj,rho(j)%a)
    end do
end do
u=uj                                      Return solution in u.
deallocate(uj)
do j=1,ng
    deallocate(rho(j)%a)
end do
deallocate(rho)
CONTAINS

RECURSIVE SUBROUTINE mg(j,u,rhs)
USE nrtype
USE nr, ONLY : interp,relax,resid,rstrct,slvsml
IMPLICIT NONE
INTEGER(I4B), INTENT(IN) :: j
REAL(DP), DIMENSION(:,:), INTENT(INOUT) :: u
REAL(DP), DIMENSION(:,:), INTENT(IN) :: rhs
INTEGER(I4B), PARAMETER :: NPRE=1,NPOST=1
```

Recursive multigrid iteration. On input, j is the current level, u is the current value of the solution, and rhs is the right-hand side. On output u contains the improved solution at the current level.

Parameters: NPRE and NPOST are the number of relaxation sweeps before and after the coarse-grid correction is computed.

```
INTEGER(I4B) :: jpost,jpre
REAL(DP), DIMENSION((size(u,1)+1)/2,(size(u,1)+1)/2) :: res,v
if (j == 1) then                          Bottom of V: Solve on coarsest grid.
    call slvsml(u,rhs)
else                                      On downward stoke of the V.
    do jpre=1,NPRE                        Pre-smoothing.
        call relax(u,rhs)
    end do
    res=rstrct(resid(u,rhs))              Restriction of the residual is the next r.h.s.
    v=0.0                                 Zero for initial guess in next relaxation.
    call mg(j-1,v,res)                    Recursive call for the coarse grid correction.
    u=u+interp(v)                         On upward stroke of V.
    do jpost=1,NPOST                      Post-smoothing.
        call relax(u,rhs)
```

```
    end do
end if
END SUBROUTINE mg
END SUBROUTINE mglin
```

f90 The Fortran 90 version of mglin (and of mgfas below) is quite different from the Fortran 77 version, although the algorithm is identical. First, we use a recursive implementation. This makes the code much more transparent. It also makes the memory management much better: we simply define the new arrays res and v as automatic arrays of the appropriate dimension on each recursive call to a coarser level. And a third benefit is that it is trivial to change the code to increase the number of multigrid iterations done at level $j-1$ by each iteration at level j, i.e., to set the quantity γ in §19.6 to a value greater than one. (Recall that $\gamma = 1$ as chosen in mglin gives V-cycles, $\gamma = 2$ gives W-cycles.) Simply enclose the recursive call in a do-loop:

```
do i=1,merge(gamma,1,j /= 2)
   call mg(j-1,v,res)
end do
```

The merge expression ensures that there is no more than one call to the coarsest level, where the problem is solved exactly.

A second improvement in the Fortran 90 version is to make the procedures resid, interp, and rstrct functions instead of subroutines. This allows us to code the algorithm exactly as written mathematically.

TYPE ptr2d... The right-hand-side quantity ρ is supplied initially on the finest grid in the argument u. It has to be defined on the coarser grids by restriction, and then supplied as the right-hand side to mg in the nested iteration loop. This loop starts at the coarsest level and progresses up to the finest level. We thus need a data structure to store ρ on all the grid levels. A convenient way to implement this in Fortran 90 is to define a type ptr2d, a pointer to a two-dimensional array a that represents a grid. (In three dimensions, a would of course be three-dimensional.) We then declare the variable ρ as an allocatable array of type ptr2d:

```
TYPE(ptr2d), ALLOCATABLE :: rho(:)
```

Next we allocate storage for ρ on each level. The number of levels or grids, ng, is known only at run time:

```
allocate(rho(ng))
```

Then we allocate storage as needed on particular sized grids. For example,

```
allocate(rho(ngrid)%a(nn,nn))
```

allocates an nn $\times$ nn grid for rho on grid number ngrid.

The various subsidiary routines of mglin such as rstrct and interp are written to accept two-dimensional arrays as arguments. With the data structure we've employed, using these routines is simple. For example,

```
rho(ngrid)%a=rstrct(rho(ngrid+1)%a)
```

will restrict rho from the grid ngrid+1 to the grid ngrid. The statement is even more readable if we mentally ignore the %a that is tagged onto each variable. (If we actually did omit %a in the code, the compiler would think we meant the array of type ptr2d instead of the grid array.)

Note that while Fortran 90 does not allow you to declare an array of pointers directly, you can achieve the same effect by declaring your own type, as we have done with ptr2d in this example.

```
FUNCTION rstrct(uf)
USE nrtype; USE nrutil, ONLY : assert_eq
IMPLICIT NONE
REAL(DP), DIMENSION(:,:), INTENT(IN) :: uf
REAL(DP), DIMENSION((size(uf,1)+1)/2,(size(uf,1)+1)/2) :: rstrct
```

Half-weighting restriction. If N_c is the coarse-grid dimension, the fine-grid solution is input in the $(2N_c-1)\times(2N_c-1)$ array uf, the coarse-grid solution is returned in the $N_c \times N_c$ array rstrct.

```
INTEGER(I4B) :: nc,nf
nf=assert_eq(size(uf,1),size(uf,2),'rstrct')
nc=(nf+1)/2
rstrct(2:nc-1,2:nc-1)=0.5_dp*uf(3:nf-2:2,3:nf-2:2)+0.125_dp*(&     Interior points.
    uf(4:nf-1:2,3:nf-2:2)+uf(2:nf-3:2,3:nf-2:2)+&
    uf(3:nf-2:2,4:nf-1:2)+uf(3:nf-2:2,2:nf-3:2))
rstrct(1:nc,1)=uf(1:nf:2,1)                                        Boundary points.
rstrct(1:nc,nc)=uf(1:nf:2,nf)
rstrct(1,1:nc)=uf(1,1:nf:2)
rstrct(nc,1:nc)=uf(nf,1:nf:2)
END FUNCTION rstrct
```

```
FUNCTION interp(uc)
USE nrtype; USE nrutil, ONLY : assert_eq
IMPLICIT NONE
REAL(DP), DIMENSION(:,:), INTENT(IN) :: uc
REAL(DP), DIMENSION(2*size(uc,1)-1,2*size(uc,1)-1) :: interp
```

Coarse-to-fine prolongation by bilinear interpolation. If N_f is the fine-grid dimension and N_c the coarse-grid dimension, then $N_f = 2N_c - 1$. The coarse-grid solution is input as uc, the fine-grid solution is returned in interp.

```
INTEGER(I4B) :: nc,nf
nc=assert_eq(size(uc,1),size(uc,2),'interp')
nf=2*nc-1
interp(1:nf:2,1:nf:2)=uc(1:nc,1:nc)
  Do elements that are copies.
interp(2:nf-1:2,1:nf:2)=0.5_dp*(interp(3:nf:2,1:nf:2)+ &
    interp(1:nf-2:2,1:nf:2))
      Do odd-numbered columns, interpolating vertically.
interp(1:nf,2:nf-1:2)=0.5_dp*(interp(1:nf,3:nf:2)+interp(1:nf,1:nf-2:2))
  Do even-numbered columns, interpolating horizontally.
END FUNCTION interp
```

```
SUBROUTINE slvsml(u,rhs)
USE nrtype
IMPLICIT NONE
REAL(DP), DIMENSION(3,3), INTENT(OUT) :: u
REAL(DP), DIMENSION(3,3), INTENT(IN) :: rhs
```

Solution of the model problem on the coarsest grid, where $h = \frac{1}{2}$. The right-hand side is input in rhs(1:3,1:3) and the solution is returned in u(1:3,1:3).

```
REAL(DP) :: h
u=0.0
h=0.5_dp
u(2,2)=-h*h*rhs(2,2)/4.0_dp
END SUBROUTINE slvsml
```

```
SUBROUTINE relax(u,rhs)
USE nrtype; USE nrutil, ONLY : assert_eq
IMPLICIT NONE
REAL(DP), DIMENSION(:,:), INTENT(INOUT) :: u
REAL(DP), DIMENSION(:,:), INTENT(IN) :: rhs
```

Red-black Gauss-Seidel relaxation for model problem. The current value of the solution `u` is updated, using the right-hand-side function `rhs`. `u` and `rhs` are square arrays of the same odd dimension.

```
INTEGER(I4B) :: n
REAL(DP) :: h,h2
n=assert_eq(size(u,1),size(u,2),size(rhs,1),size(rhs,2),'relax')
h=1.0_dp/(n-1)
h2=h*h
```

First do the even-even and odd-odd squares of the grid, i.e., the red squares of the checkerboard:

```
u(2:n-1:2,2:n-1:2)=0.25_dp*(u(3:n:2,2:n-1:2)+u(1:n-2:2,2:n-1:2)+&
    u(2:n-1:2,3:n:2)+u(2:n-1:2,1:n-2:2)-h2*rhs(2:n-1:2,2:n-1:2))
u(3:n-2:2,3:n-2:2)=0.25_dp*(u(4:n-1:2,3:n-2:2)+u(2:n-3:2,3:n-2:2)+&
    u(3:n-2:2,4:n-1:2)+u(3:n-2:2,2:n-3:2)-h2*rhs(3:n-2:2,3:n-2:2))
```

Now do even-odd and odd-even squares of the grid, i.e., the black squares of the checkerboard:

```
u(3:n-2:2,2:n-1:2)=0.25_dp*(u(4:n-1:2,2:n-1:2)+u(2:n-3:2,2:n-1:2)+&
    u(3:n-2:2,3:n:2)+u(3:n-2:2,1:n-2:2)-h2*rhs(3:n-2:2,2:n-1:2))
u(2:n-1:2,3:n-2:2)=0.25_dp*(u(3:n:2,3:n-2:2)+u(1:n-2:2,3:n-2:2)+&
    u(2:n-1:2,4:n-1:2)+u(2:n-1:2,2:n-3:2)-h2*rhs(2:n-1:2,3:n-2:2))
END SUBROUTINE relax
```

See the discussion of red-black relaxation after `sor` on p. 1333.

```
FUNCTION resid(u,rhs)
USE nrtype; USE nrutil, ONLY : assert_eq
IMPLICIT NONE
REAL(DP), DIMENSION(:,:), INTENT(IN) :: u,rhs
REAL(DP), DIMENSION(size(u,1),size(u,1)) :: resid
```

Returns *minus* the residual for the model problem. Input quantities are `u` and `rhs`, while the residual is returned in `resid`. All three quantities are square arrays with the same odd dimension.

```
INTEGER(I4B) :: n
REAL(DP) :: h,h2i
n=assert_eq((/size(u,1),size(u,2),size(rhs,1),size(rhs,2)/),'resid')
n=size(u,1)
h=1.0_dp/(n-1)
h2i=1.0_dp/(h*h)
resid(2:n-1,2:n-1)=-h2i*(u(3:n,2:n-1)+u(1:n-2,2:n-1)+u(2:n-1,3:n)+&
    u(2:n-1,1:n-2)-4.0_dp*u(2:n-1,2:n-1))+rhs(2:n-1,2:n-1)        Interior points.
resid(1:n,1)=0.0                                                  Boundary points.
resid(1:n,n)=0.0
resid(1,1:n)=0.0
resid(n,1:n)=0.0
END FUNCTION resid
```

⋆ ⋆ ⋆

```
SUBROUTINE mgfas(u,maxcyc)
USE nrtype; USE nrutil, ONLY : assert_eq,nrerror
USE nr, ONLY : interp,lop,rstrct,slvsm2
IMPLICIT NONE
REAL(DP), DIMENSION(:,:), INTENT(INOUT) :: u
INTEGER(I4B), INTENT(IN) :: maxcyc
```

Full Multigrid Algorithm for FAS solution of nonlinear elliptic equation, here equation (19.6.44). On input u contains the right-hand side ρ in an $N \times N$ array, while on output it returns the solution. The dimension N is related to the number of grid levels used in the solution, ng below, by $N =$ 2**ng+1. maxcyc is the maximum number of V-cycles to be used at each level.

```
INTEGER(I4B) :: j,jcycle,n,ng,ngrid,nn
REAL(DP) :: res,trerr
TYPE ptr2d                                  Define a type so we can have an array of
    REAL(DP), POINTER :: a(:,:)                 pointers to arrays of grid variables.
END TYPE ptr2d
TYPE(ptr2d), ALLOCATABLE :: rho(:)
REAL(DP), DIMENSION(:,:), POINTER :: uj,uj_1
n=assert_eq(size(u,1),size(u,2),'mgfas')
ng=nint(log(n-1.0)/log(2.0))
if (n /= 2**ng+1) call nrerror('n-1 must be a power of 2 in mgfas')
allocate(rho(ng))
nn=n
ngrid=ng
allocate(rho(ngrid)%a(nn,nn))               Allocate storage for r.h.s. on grid ng,
rho(ngrid)%a=u                              and fill it with ρ from the fine grid.
do                                          Similarly allocate storage and fill r.h.s. by re-
    if (nn <= 3) exit                           striction on all coarse grids.
    nn=nn/2+1
    ngrid=ngrid-1
    allocate(rho(ngrid)%a(nn,nn))
    rho(ngrid)%a=rstrct(rho(ngrid+1)%a)
end do
nn=3
allocate(uj(nn,nn))
call slvsm2(uj,rho(1)%a)                    Initial solution on coarsest grid.
do j=2,ng                                   Nested iteration loop.
    nn=2*nn-1
    uj_1=>uj
    allocate(uj(nn,nn))
    uj=interp(uj_1)                         Interpolate from grid j-1 to next finer grid
    deallocate(uj_1)                            j.
    do jcycle=1,maxcyc                      V-cycle loop.
        call mg(j,uj,trerr=trerr)
        res=sqrt(sum((lop(uj)-rho(j)%a)**2))/nn      Form residual ||d_h||.
        if (res < trerr) exit               No more V-cycles needed if residual small
    end do                                      enough.
end do
u=uj                                        Return solution in u.
deallocate(uj)
do j=1,ng
    deallocate(rho(j)%a)
end do
deallocate(rho)
CONTAINS

RECURSIVE SUBROUTINE mg(j,u,rhs,trerr)
USE nrtype
USE nr, ONLY : interp,lop,relax2,rstrct,slvsm2
IMPLICIT NONE
INTEGER(I4B), INTENT(IN) :: j
REAL(DP), DIMENSION(:,:), INTENT(INOUT) :: u
REAL(DP), DIMENSION(:,:), INTENT(IN), OPTIONAL :: rhs
REAL(DP), INTENT(OUT), OPTIONAL :: trerr
```

```
INTEGER(I4B), PARAMETER :: NPRE=1,NPOST=1
REAL(DP), PARAMETER :: ALPHA=0.33_dp
    Recursive multigrid iteration. On input, j is the current level and u is the current value
    of the solution. For the first call on a given level, the right-hand side is zero, and the
    optional argument rhs is not present. Subsequent recursive calls supply a nonzero rhs as
    in equation (19.6.33). On output u contains the improved solution at the current level.
    When the first call on a given level is made, the relative truncation error τ is returned in
    the optional argument trerr.
    Parameters: NPRE and NPOST are the number of relaxation sweeps before and after the
    coarse-grid correction is computed; ALPHA relates the estimated truncation error to the
    norm of the residual.
INTEGER(I4B) :: jpost,jpre
REAL(DP), DIMENSION((size(u,1)+1)/2,(size(u,1)+1)/2) :: v,ut,tau
if (j == 1) then                              Bottom of V: Solve on coarsest grid.
    call slvsm2(u,rhs+rho(j)%a)
else                                          On downward stoke of the V.
    do jpre=1,NPRE                            Pre-smoothing.
        if (present(rhs)) then
            call relax2(u,rhs+rho(j)%a)
        else
            call relax2(u,rho(j)%a)
        end if
    end do
    ut=rstrct(u)                              R ũ_h.
    v=ut                                      Make a copy in v.
    if (present(rhs)) then
        tau=lop(ut)-rstrct(lop(u)-rhs)        Form τ̃_h + f_H = L_H(R ũ_h) − R L_h(ũ_h) +
    else                                        f_H.
        tau=lop(ut)-rstrct(lop(u))
        trerr=ALPHA*sqrt(sum(tau**2))/size(tau,1)      Estimate truncation error τ.
    end if
    call mg(j-1,v,tau)                        Recursive call for the coarse-grid correction.
    u=u+interp(v-ut)                          ũ_h^new = ũ_h + P(ũ_H − R ũ_h)
    do jpost=1,NPOST                          Post-smoothing.
        if (present(rhs)) then
            call relax2(u,rhs+rho(j)%a)
        else
            call relax2(u,rho(j)%a)
        end if
    end do
end if
END SUBROUTINE mg
END SUBROUTINE mgfas
```

See the discussion after mglin on p. 1336 for the changes made in the Fortran 90 versions of the multigrid routines from the Fortran 77 versions.

TYPE ptr2d... See discussion after mglin on p. 1336.

RECURSIVE SUBROUTINE mg(j,u,rhs,trerr) Recall that mgfas solves the problem $\mathcal{L}u = 0$, but that nonzero right-hand sides appear during the solution. We implement this by having rhs be an optional argument to mg. On the first call at a given level j, the right-hand side is zero and so you just omit it from the calling sequence. On the other hand, the truncation error trerr is computed only on the first call at a given level, so it is also an optional argument that does get supplied on the first call:

```
call mg(j,uj,trerr=trerr)
```

The second and subsequent calls at a given level supply rhs=tau but omit trerr:

```
call mg(j-1,v,tau)
```

Note that we can omit the keyword rhs from this call because the variable tau appears in the correct order of arguments. However, in the other call above, the keyword trerr must be supplied because rhs has been omitted.

The example equation that is solved in mgfas, equation (19.6.44), is almost linear, and the code is set up so that ρ is supplied as part of the right-hand side instead of pulling it over to the left-hand side. The variable rho is visible to mg by host association. Note also that the function lop does not include rho, but that the statement

```
tau=lop(ut)-rstrct(lop(u))
```

is nevertheless correct, since rho would cancel out if it were included in lop. This feature is also true in the Fortran 77 code.

```
SUBROUTINE relax2(u,rhs)
USE nrtype; USE nrutil, ONLY : assert_eq
IMPLICIT NONE
REAL(DP), DIMENSION(:,:), INTENT(INOUT) :: u
REAL(DP), DIMENSION(:,:), INTENT(IN) :: rhs
    Red-black Gauss-Seidel relaxation for equation (19.6.44). The current value of the solution
    u is updated, using the right-hand-side function rhs. u and rhs are square arrays of the
    same odd dimension.
INTEGER(I4B) :: n
REAL(DP) :: foh2,h,h2i
REAL(DP) :: res(size(u,1),size(u,1))
n=assert_eq(size(u,1),size(u,2),size(rhs,1),size(rhs,2),'relax2')
h=1.0_dp/(n-1)
h2i=1.0_dp/(h*h)
foh2=-4.0_dp*h2i
  First do the even-even and odd-odd squares of the grid, i.e., the red squares of the checker-
  board:
res(2:n-1:2,2:n-1:2)=h2i*(u(3:n:2,2:n-1:2)+u(1:n-2:2,2:n-1:2)+&
    u(2:n-1:2,3:n:2)+u(2:n-1:2,1:n-2:2)-4.0_dp*u(2:n-1:2,2:n-1:2))&
    +u(2:n-1:2,2:n-1:2)**2-rhs(2:n-1:2,2:n-1:2)
u(2:n-1:2,2:n-1:2)=u(2:n-1:2,2:n-1:2)-res(2:n-1:2,2:n-1:2)/&
    (foh2+2.0_dp*u(2:n-1:2,2:n-1:2))
res(3:n-2:2,3:n-2:2)=h2i*(u(4:n-1:2,3:n-2:2)+u(2:n-3:2,3:n-2:2)+&
    u(3:n-2:2,4:n-1:2)+u(3:n-2:2,2:n-3:2)-4.0_dp*u(3:n-2:2,3:n-2:2))&
    +u(3:n-2:2,3:n-2:2)**2-rhs(3:n-2:2,3:n-2:2)
u(3:n-2:2,3:n-2:2)=u(3:n-2:2,3:n-2:2)-res(3:n-2:2,3:n-2:2)/&
    (foh2+2.0_dp*u(3:n-2:2,3:n-2:2))
  Now do even-odd and odd-even squares of the grid, i.e., the black squares of the checker-
  board:
res(3:n-2:2,2:n-1:2)=h2i*(u(4:n-1:2,2:n-1:2)+u(2:n-3:2,2:n-1:2)+&
    u(3:n-2:2,3:n:2)+u(3:n-2:2,1:n-2:2)-4.0_dp*u(3:n-2:2,2:n-1:2))&
    +u(3:n-2:2,2:n-1:2)**2-rhs(3:n-2:2,2:n-1:2)
u(3:n-2:2,2:n-1:2)=u(3:n-2:2,2:n-1:2)-res(3:n-2:2,2:n-1:2)/&
    (foh2+2.0_dp*u(3:n-2:2,2:n-1:2))
res(2:n-1:2,3:n-2:2)=h2i*(u(3:n:2,3:n-2:2)+u(1:n-2:2,3:n-2:2)+&
    u(2:n-1:2,4:n-1:2)+u(2:n-1:2,2:n-3:2)-4.0_dp*u(2:n-1:2,3:n-2:2))&
    +u(2:n-1:2,3:n-2:2)**2-rhs(2:n-1:2,3:n-2:2)
u(2:n-1:2,3:n-2:2)=u(2:n-1:2,3:n-2:2)-res(2:n-1:2,3:n-2:2)/&
    (foh2+2.0_dp*u(2:n-1:2,3:n-2:2))
END SUBROUTINE relax2
```

See the discussion of red-black relaxation after sor on p. 1333.

```
SUBROUTINE slvsm2(u,rhs)
USE nrtype
IMPLICIT NONE
REAL(DP), DIMENSION(3,3), INTENT(OUT) :: u
REAL(DP), DIMENSION(3,3), INTENT(IN) :: rhs
```

Solution of equation (19.6.44) on the coarsest grid, where $h = \frac{1}{2}$. The right-hand side is input in `rhs(1:3,1:3)` and the solution is returned in `u(1:3,1:3)`.

```
REAL(DP) :: disc,fact,h
u=0.0
h=0.5_dp
fact=2.0_dp/h**2
disc=sqrt(fact**2+rhs(2,2))
u(2,2)=-rhs(2,2)/(fact+disc)
END SUBROUTINE slvsm2
```

```
FUNCTION lop(u)
USE nrtype; USE nrutil, ONLY : assert_eq
IMPLICIT NONE
REAL(DP), DIMENSION(:,:), INTENT(IN) :: u
REAL(DP), DIMENSION(size(u,1),size(u,1)) :: lop
```

Given `u`, returns $\mathcal{L}_h(\widetilde{u}_h)$ for equation (19.6.44). `u` and `lop` are square arrays of the same odd dimension.

```
INTEGER(I4B) :: n
REAL(DP) :: h,h2i
n=assert_eq(size(u,1),size(u,2),'lop')
h=1.0_dp/(n-1)
h2i=1.0_dp/(h*h)
lop(2:n-1,2:n-1)=h2i*(u(3:n,2:n-1)+u(1:n-2,2:n-1)+u(2:n-1,3:n)+&
    u(2:n-1,1:n-2)-4.0_dp*u(2:n-1,2:n-1))+u(2:n-1,2:n-1)**2      Interior points.
lop(1:n,1)=0.0                                                    Boundary points.
lop(1:n,n)=0.0
lop(1,1:n)=0.0
lop(n,1:n)=0.0
END FUNCTION lop
```

Chapter B20. Less-Numerical Algorithms

Volume 1's Fortran 77 routine machar performed various clever contortions (due to Cody, Malcolm, and others) to discover the underlying properties of a machine's floating-point representation. Fortran 90, by contrast, provides a built-in set of "numeric inquiry functions" that accomplish the same goal. The routine machar included here makes use of these and is included largely for compatibility with the previous version.

```
SUBROUTINE machar(ibeta,it,irnd,ngrd,machep,negep,iexp,minexp,&
    maxexp,eps,epsneg,xmin,xmax)
USE nrtype
IMPLICIT NONE
INTEGER(I4B), INTENT(OUT) :: ibeta,iexp,irnd,it,machep,maxexp,minexp,negep,ngrd
REAL(SP), INTENT(OUT) :: eps,epsneg,xmax,xmin
REAL(SP), PARAMETER :: RX=1.0
    Determines and returns machine-specific parameters affecting floating-point arithmetic. Re-
    turned values include ibeta, the floating-point radix; it, the number of base-ibeta digits
    in the floating-point mantissa; eps, the smallest positive number that, added to 1.0, is
    not equal to 1.0; epsneg, the smallest positive number that, subtracted from 1.0, is not
    equal to 1.0; xmin, the smallest representable positive number; and xmax, the largest rep-
    resentable positive number. See text for description of other returned parameters. Change
    all REAL(SP) declarations to REAL(DP) to find double-precision parameters.
REAL(SP) :: a,beta,betah,one,temp,tempa,two,zero
ibeta=radix(RX)                             Most of the parameters are easily determined
it=digits(RX)                                   from intrinsic functions.
machep=exponent(nearest(RX,RX)-RX)-1
negep=exponent(nearest(RX,-RX)-RX)-1
minexp=minexponent(RX)-1
maxexp=maxexponent(RX)
iexp=nint(log(real(maxexp-minexp+2,sp))/log(2.0_sp))
eps=real(ibeta,sp)**machep
epsneg=real(ibeta,sp)**negep
xmax=huge(RX)
xmin=tiny(RX)
one=RX                                      Determine irnd.
two=one+one
zero=one-one
beta=real(ibeta,sp)
a=beta**(-negep)
irnd=0
betah=beta/two
temp=a+betah
if (temp-a /= zero) irnd=1
tempa=a+beta
temp=tempa+betah
if ((irnd == 0) .and. (temp-tempa /= zero)) irnd=2
ngrd=0                                      Determine ngrd.
```

```
temp=one+eps
if ((irnd == 0) .and. (temp*one-one /= zero)) ngrd=1
temp=xmin/two
if (temp /= zero) irnd=irnd+3          Adjust irnd to reflect partial underflow.
END SUBROUTINE machar
```

* * *

```
FUNCTION igray(n,is)
USE nrtype
IMPLICIT NONE
INTEGER(I4B), INTENT(IN) :: n,is
INTEGER(I4B) :: igray
   For zero or positive values of is, return the Gray code of n; if is is negative, return the
   inverse Gray code of n.
INTEGER(I4B) :: idiv,ish
if (is >= 0) then                This is the easy direction!
    igray=ieor(n,n/2)
else                             This is the more complicated direction: In hierarchical stages,
    ish=-1                           starting with a one-bit right shift, cause each bit to be
    igray=n                          XORed with all more significant bits.
    do
        idiv=ishft(igray,ish)
        igray=ieor(igray,idiv)
        if (idiv <= 1 .or. ish == -16) RETURN
        ish=ish+ish              Double the amount of shift on the next cycle.
    end do
end if
END FUNCTION igray
```

* * *

```
FUNCTION icrc(crc,buf,jinit,jrev)
USE nrtype
IMPLICIT NONE
CHARACTER(1), DIMENSION(:), INTENT(IN) :: buf
INTEGER(I2B), INTENT(IN) :: crc,jinit
INTEGER(I4B), INTENT(IN) :: jrev
INTEGER(I2B) :: icrc
   Computes a 16-bit Cyclic Redundancy Check for an array buf of bytes, using any of several
   conventions as determined by the settings of jinit and jrev (see accompanying table).
   The result is returned both as an integer icrc and as a 2-byte array crc. If jinit is neg-
   ative, then crc is used on input to initialize the remainder register, in effect concatenating
   buf to the previous call.
INTEGER(I4B), SAVE :: init=0
INTEGER(I2B) :: j,cword,ich
INTEGER(I2B), DIMENSION(0:255), SAVE :: icrctb,rchr
INTEGER(I2B), DIMENSION(0:15) :: it = &       Table of 4-bit bit-reverses.
    (/ 0,8,4,12,2,10,6,14,1,9,5,13,3,11,7,15 /)
if (init == 0) then                           Do we need to initialize tables?
    init=1
    do j=0,255                                The two tables are: CRCs of all characters,
        icrctb(j)=icrc1(ishft(j,8),char(0))       and bit-reverses of all characters.
        rchr(j)=ishft(it(iand(j,15_I2B)),4)+it(ishft(j,-4))
    end do
end if
cword=crc
if (jinit >= 0) then                          Initialize the remainder register.
    cword=ior(jinit,ishft(jinit,8))
```

```
else if (jrev < 0) then                  If not initializing, do we reverse the register?
    cword=ior(rchr(hibyte()),ishft(rchr(lobyte()),8))
end if
do j=1,size(buf)                         Main loop over the characters in the array.
    ich=ichar(buf(j))
    if (jrev < 0) ich=rchr(ich)
    cword=ieor(icrctb(ieor(ich,hibyte())),ishft(lobyte(),8))
end do
icrc=merge(cword, &                      Do we need to reverse the output?
    ior(rchr(hibyte()),ishft(rchr(lobyte()),8)), jrev >= 0)
CONTAINS

FUNCTION hibyte()
INTEGER(I2B) :: hibyte
  Extracts the high byte of the 2-byte integer cword.
hibyte = ishft(cword,-8)
END FUNCTION hibyte

FUNCTION lobyte()
INTEGER(I2B) :: lobyte
  Extracts the low byte of the 2-byte integer cword.
lobyte = iand(cword,255_I2B)
END FUNCTION lobyte

FUNCTION icrc1(crc,onech)
INTEGER(I2B), INTENT(IN) :: crc
CHARACTER(1), INTENT(IN) :: onech
INTEGER(I2B) :: icrc1
  Given a remainder up to now, return the new CRC after one character is added. This routine is
  functionally equivalent to icrc(,,-1,1), but slower. It is used by icrc to initialize its table.
INTEGER(I2B) :: i,ich, bit16, ccitt
DATA bit16,ccitt /Z'8000', Z'1021'/
ich=ichar(onech)                         Here is where the character is folded into the
icrc1=ieor(crc,ishft(ich,8))                 register.
do i=1,8                                 Here is where 8 one-bit shifts, and some XORs
    icrc1=merge(ieor(ccitt,ishft(icrc1,1)), &      with the generator polynomial,
        ishft(icrc1,1), iand(icrc1,bit16) /= 0)    are done.
end do
END FUNCTION icrc1
END FUNCTION icrc
```

The embedded functions hibyte and lobyte always act on the same variable, cword. Thus they don't need any explicit argument.

⋆ ⋆ ⋆

```
FUNCTION decchk(string,ch)
USE nrtype; USE nrutil, ONLY : ifirstloc
IMPLICIT NONE
CHARACTER(1), DIMENSION(:), INTENT(IN) :: string
CHARACTER(1), INTENT(OUT) :: ch
LOGICAL(LGT) :: decchk
    Decimal check digit computation or verification. Returns as ch a check digit for appending
    to string. In this mode, ignore the returned logical value. If string already ends with
    a check digit, returns the function value .true. if the check digit is valid, otherwise
    .false. In this mode, ignore the returned value of ch. Note that string and ch contain
    ASCII characters corresponding to the digits 0-9, not byte values in that range. Other ASCII
    characters are allowed in string, and are ignored in calculating the check digit.
INTEGER(I4B) :: i,j,k,m
INTEGER(I4B) :: ip(0:9,0:7) = reshape((/ &       Group multiplication and permuta-
    0,1,2,3,4,5,6,7,8,9,1,5,7,6,2,8,3,0,9,4,&        tion tables.
    5,8,0,3,7,9,6,1,4,2,8,9,1,6,0,4,3,5,2,7,9,4,5,3,1,2,6,8,7,0,&
    4,2,8,6,5,7,3,9,0,1,2,7,9,3,8,0,6,4,1,5,7,0,4,6,9,1,3,2,5,8 /),&
```

```
    (/ 10,8 /) )
INTEGER(I4B) :: ij(0:9,0:9) = reshape((/ &
    0,1,2,3,4,5,6,7,8,9,1,2,3,4,0,9,5,6,7,8,2,3,4,0,1,8,9,5,6,&
    7,3,4,0,1,2,7,8,9,5,6,4,0,1,2,3,6,7,8,9,5,5,6,7,8,9,0,1,2,3,&
    4,6,7,8,9,5,4,0,1,2,3,7,8,9,5,6,3,4,0,1,2,8,9,5,6,7,2,3,4,0,&
    1,9,5,6,7,8,1,2,3,4,0 /),(/ 10,10 /))
k=0
m=0
do j=1,size(string)                         Look at successive characters.
    i=ichar(string(j))
    if (i >= 48 .and. i <= 57) then          Ignore everything except digits.
        k=ij(k,ip(mod(i+2,10),mod(m,8)))
        m=m+1
    end if
end do
decchk=logical(k == 0,kind=lgt)
i=mod(m,8)                                  Find which appended digit will check prop-
i=ifirstloc(ij(k,ip(0:9,i)) == 0)-1              erly.
ch=char(i+48)                               Convert to ASCII.
END FUNCTION decchk
```

f90 Note the use of the utility function `ifirstloc` to find the first (in this case, the only) correct check digit.

⋆ ⋆ ⋆

f90 The Huffman and arithmetic coding routines exemplify the use of modules to encapsulate user-defined data types. In these algorithms, "the code" is a fairly complicated construct containing scalar and array data. We define types `huffcode` and `arithcode`, then can pass "the code" from the routine that constructs it to the routine that uses it as a single variable.

```
MODULE huf_info
USE nrtype
IMPLICIT NONE
TYPE huffcode
    INTEGER(I4B) :: nch,nodemax
    INTEGER(I4B), DIMENSION(:), POINTER :: icode,left,iright,ncode
END TYPE huffcode
CONTAINS
SUBROUTINE huff_allocate(hcode,mc)
USE nrtype
IMPLICIT NONE
TYPE(huffcode) :: hcode
INTEGER(I4B) :: mc
INTEGER(I4B) :: mq
mq=2*mc-1
allocate(hcode%icode(mq),hcode%ncode(mq),hcode%left(mq),hcode%iright(mq))
hcode%icode(:)=0
hcode%ncode(:)=0
END SUBROUTINE huff_allocate

SUBROUTINE huff_deallocate(hcode)
USE nrtype
IMPLICIT NONE
TYPE(huffcode) :: hcode
deallocate(hcode%iright,hcode%left,hcode%ncode,hcode%icode)
nullify(hcode%icode)
nullify(hcode%ncode)
nullify(hcode%left)
nullify(hcode%iright)
```

```
END SUBROUTINE huff_deallocate
END MODULE huf_info


SUBROUTINE hufmak(nfreq,ilong,nlong,hcode)
USE nrtype; USE nrutil, ONLY : array_copy,arth,imaxloc,nrerror
USE huf_info
IMPLICIT NONE
INTEGER(I4B), INTENT(OUT) :: ilong,nlong
INTEGER(I4B), DIMENSION(:), INTENT(IN) :: nfreq
TYPE(huffcode) :: hcode
   Given the frequency of occurrence table nfreq of size(nfreq) characters, return the
   Huffman code hcode. Returned values ilong and nlong are the character number that
   produced the longest code symbol, and the length of that symbol.
INTEGER(I4B) :: ibit,j,k,n,node,nused,nerr
INTEGER(I4B), DIMENSION(2*size(nfreq)-1) :: indx,iup,nprob
hcode%nch=size(nfreq)               Initialization.
call huff_allocate(hcode,size(nfreq))
nused=0
nprob(1:hcode%nch)=nfreq(1:hcode%nch)
call array_copy(pack(arth(1,1,hcode%nch), nfreq(1:hcode%nch) /= 0 ),&
    indx,nused,nerr)
do j=nused,1,-1                     Sort nprob into a heap structure in indx.
    call hufapp(j)
end do
k=hcode%nch
do                                  Combine heap nodes, remaking the heap at each stage.
    if (nused <= 1) exit
    node=indx(1)
    indx(1)=indx(nused)
    nused=nused-1
    call hufapp(1)
    k=k+1
    nprob(k)=nprob(indx(1))+nprob(node)
    hcode%left(k)=node              Store left and right children of a node.
    hcode%iright(k)=indx(1)
    iup(indx(1))=-k                 Indicate whether a node is a left or right child of its par-
    iup(node)=k                         ent.
    indx(1)=k
    call hufapp(1)
end do
hcode%nodemax=k
iup(hcode%nodemax)=0
do j=1,hcode%nch                    Make the Huffman code from the tree.
    if (nprob(j) /= 0) then
        n=0
        ibit=0
        node=iup(j)
        do
            if (node == 0) exit
            if (node < 0) then
                n=ibset(n,ibit)
                node=-node
            end if
            node=iup(node)
            ibit=ibit+1
        end do
        hcode%icode(j)=n
        hcode%ncode(j)=ibit
    end if
end do
ilong=imaxloc(hcode%ncode(1:hcode%nch))
nlong=hcode%ncode(ilong)
```

```
if (nlong > bit_size(1_i4b)) call &        Check nlong not larger than word length.
    nrerror('hufmak: Number of possible bits for code exceeded')
CONTAINS

SUBROUTINE hufapp(l)
IMPLICIT NONE
INTEGER(I4B), INTENT(IN) :: l
  Used by hufmak to maintain a heap structure in the array indx(1:l).
INTEGER(I4B) :: i,j,k,n
n=nused
i=1
k=indx(i)
do
    if (i > n/2) exit
    j=i+i
    if (j < n .and. nprob(indx(j)) > nprob(indx(j+1))) &
        j=j+1
    if (nprob(k) <= nprob(indx(j))) exit
    indx(i)=indx(j)
    i=j
end do
indx(i)=k
END SUBROUTINE hufapp
END SUBROUTINE hufmak
```

```
SUBROUTINE hufenc(ich,codep,nb,hcode)
USE nrtype; USE nrutil, ONLY : nrerror,reallocate
USE huf_info
IMPLICIT NONE
INTEGER(I4B), INTENT(IN) :: ich
INTEGER(I4B), INTENT(INOUT) :: nb
CHARACTER(1), DIMENSION(:), POINTER :: codep
TYPE(huffcode) :: hcode
    Huffman encode the single character ich (in the range 0..nch-1) using the code in hcode,
    write the result to the character array pointed to by codep starting at bit nb (whose smallest
    valid value is zero), and increment nb appropriately. This routine is called repeatedly to
    encode consecutive characters in a message, but must be preceded by a single initializing
    call to hufmak.
INTEGER(I4B) :: k,l,n,nc,ntmp
k=ich+1                                    Convert character range 0..nch-1 to ar-
if (k > hcode%nch .or. k < 1) call &           ray index range 1..nch.
    nrerror('hufenc: ich out of range')
do n=hcode%ncode(k),1,-1                   Loop over the bits in the stored Huffman
    nc=nb/8+1                                  code for ich.
    if (nc > size(codep)) codep=>reallocate(codep,2*size(codep))
    l=mod(nb,8)
    if (l == 0) codep(nc)=char(0)
    if (btest(hcode%icode(k),n-1)) then    Set appropriate bits in codep.
        ntmp=ibset(ichar(codep(nc)),l)
        codep(nc)=char(ntmp)
    end if
    nb=nb+1
end do
END SUBROUTINE hufenc
```

```
SUBROUTINE hufdec(ich,code,nb,hcode)
USE nrtype
USE huf_info
IMPLICIT NONE
INTEGER(I4B), INTENT(OUT) :: ich
INTEGER(I4B), INTENT(INOUT) :: nb
CHARACTER(1), DIMENSION(:), INTENT(IN) :: code
TYPE(huffcode) :: hcode
```

Starting at bit number nb in the character array code, use the Huffman code in hcode to decode a single character (returned as ich in the range 0..nch-1) and increment nb appropriately. Repeated calls, starting with nb = 0, will return successive characters in a compressed message. The returned value ich=nch indicates end-of-message. This routine must be preceded by a single initializing call to hufmak.

```
INTEGER(I4B) :: l,nc,node
node=hcode%nodemax                          Set node to the top of the decoding tree.
do                                          Loop until a valid character is obtained.
    nc=nb/8+1
    if (nc > size(code)) then               Ran out of input; return with ich=nch
        ich=hcode%nch                           indicating end of message.
        RETURN
    end if
    l=mod(nb,8)                             Now decoding this bit.
    nb=nb+1
    if (btest(ichar(code(nc)),l)) then      Branch left or right in tree, depending on
        node=hcode%iright(node)                 its value.
    else
        node=hcode%left(node)
    end if
    if (node <= hcode%nch) then             If we reach a terminal node, we have a
        ich=node-1                              complete character and can return.
        RETURN
    end if
end do
END SUBROUTINE hufdec
```

*   *   *

```
MODULE arcode_info
USE nrtype
IMPLICIT NONE
INTEGER(I4B), PARAMETER :: NWK=20
```

NWK is the number of working digits (see text).

```
TYPE arithcode
    INTEGER(I4B), DIMENSION(:), POINTER :: ilob,iupb,ncumfq
    INTEGER(I4B) :: jdif,nc,minint,nch,ncum,nrad
END TYPE arithcode
CONTAINS
SUBROUTINE arcode_allocate(acode,mc)
USE nrtype
IMPLICIT NONE
TYPE(arithcode) :: acode
INTEGER(I4B) :: mc
allocate(acode%ilob(NWK),acode%iupb(NWK),acode%ncumfq(mc+2))
END SUBROUTINE arcode_allocate

SUBROUTINE arcode_deallocate(acode)
USE nrtype
IMPLICIT NONE
TYPE(arithcode) :: acode
deallocate(acode%ncumfq,acode%iupb,acode%ilob)
nullify(acode%ilob)
nullify(acode%iupb)
```

```
nullify(acode%ncumfq)
END SUBROUTINE arcode_deallocate
END MODULE arcode_info
```

```
SUBROUTINE arcmak(nfreq,nradd,acode)
USE nrtype; USE nrutil, ONLY : cumsum,nrerror
USE arcode_info
IMPLICIT NONE
INTEGER(I4B), INTENT(IN) :: nradd
INTEGER(I4B), DIMENSION(:), INTENT(IN) :: nfreq
TYPE(arithcode) :: acode
INTEGER(I4B), PARAMETER :: MAXINT=huge(nradd)
```

Given a table nfreq of the frequency of occurrence of size(nfreq) symbols, and given a desired output radix nradd, initialize the cumulative frequency table and other variables for arithmetic compression. Store the code in acode.

MAXINT is a large positive integer that does not overflow.

```
if (nradd > 256) call nrerror('output radix may not exceed 256 in arcmak')
acode%minint=MAXINT/nradd
acode%nch=size(nfreq)
acode%nrad=nradd
call arcode_allocate(acode,acode%nch)
acode%ncumfq(1)=0
acode%ncumfq(2:acode%nch+1)=cumsum(max(nfreq(1:acode%nch),1))
acode%ncumfq(acode%nch+2)=acode%ncumfq(acode%nch+1)+1
acode%ncum=acode%ncumfq(acode%nch+2)
END SUBROUTINE arcmak
```

```
SUBROUTINE arcode(ich,codep,lcd,isign,acode)
USE nrtype; USE nrutil, ONLY : nrerror,reallocate
USE arcode_info
IMPLICIT NONE
INTEGER(I4B), INTENT(INOUT) :: ich,lcd
INTEGER(I4B), INTENT(IN) :: isign
CHARACTER(1), DIMENSION(:), POINTER :: codep
TYPE(arithcode) :: acode
```

Compress ($isign = 1$) or decompress ($isign = -1$) the single character ich into or out of the character array pointed to by codep, starting with byte codep(lcd) and (if necessary) incrementing lcd so that, on return, lcd points to the first unused byte in codep. Note that this routine saves the result of previous calls until a new byte of code is produced, and only then increments lcd. An initializing call with isign=0 is required for each different array codep. The routine arcmak must have previously been called to initialize the code acode. A call with ich=arcode%nch (as set in arcmak) has the reserved meaning "end of message."

```
INTEGER(I4B) :: ihi,j,ja,jh,jl,m
if (isign == 0) then                        Initialize enough digits of the upper and lower
    acode%jdif=acode%nrad-1                     bounds.
    acode%ilob(:)=0
    acode%iupb(:)=acode%nrad-1
    do j=NWK,1,-1
        acode%nc=j
        if (acode%jdif > acode%minint) RETURN       Initialization complete.
        acode%jdif=(acode%jdif+1)*acode%nrad-1
    end do
    call nrerror('NWK too small in arcode')
else
    if (isign > 0) then                     If encoding, check for valid input character.
        if (ich > acode%nch .or. ich < 0) call nrerror('bad ich in arcode')
    else                                    If decoding, locate the character ich by bi-
        ja=ichar(codep(lcd))-acode%ilob(acode%nc)       section.
        do j=acode%nc+1,NWK
```

```
            ja=ja*acode%nrad+(ichar(codep(j+lcd-acode%nc))-acode%ilob(j))
        end do
        ich=0
        ihi=acode%nch+1
        do
            if (ihi-ich <= 1) exit
            m=(ich+ihi)/2
            if (ja >= jtry(acode%jdif,acode%ncumfq(m+1),acode%ncum)) then
                ich=m
            else
                ihi=m
            end if
        end do
        if (ich == acode%nch) RETURN        Detected end of message.
    end if
      Following code is common for encoding and decoding. Convert character ich to a new
      subrange [ilob,iupb).
    jh=jtry(acode%jdif,acode%ncumfq(ich+2),acode%ncum)
    jl=jtry(acode%jdif,acode%ncumfq(ich+1),acode%ncum)
    acode%jdif=jh-jl
    call arcsum(acode%ilob,acode%iupb,jh,NWK,acode%nrad,acode%nc)
      How many leading digits to output (if encoding) or skip over?
    call arcsum(acode%ilob,acode%ilob,jl,NWK,acode%nrad,acode%nc)
    do j=acode%nc,NWK
        if (ich /= acode%nch .and. acode%iupb(j) /= acode%ilob(j)) exit
        if (acode%nc > size(codep)) codep=>reallocate(codep,2*size(codep))
        if (isign > 0) codep(lcd)=char(acode%ilob(j))
        lcd=lcd+1
    end do
    if (j > NWK) RETURN                     Ran out of message. Did someone forget to
    acode%nc=j                                  encode a terminating ncd?
    j=0                                     How many digits to shift?
    do
        if (acode%jdif >= acode%minint) exit
        j=j+1
        acode%jdif=acode%jdif*acode%nrad
    end do
    if (acode%nc-j < 1) call nrerror('NWK too small in arcode')
    if (j /= 0) then                        Shift them.
        acode%iupb((acode%nc-j):(NWK-j))=acode%iupb(acode%nc:NWK)
        acode%ilob((acode%nc-j):(NWK-j))=acode%ilob(acode%nc:NWK)
    end if
    acode%nc=acode%nc-j
    acode%iupb((NWK-j+1):NWK)=0
    acode%ilob((NWK-j+1):NWK)=0
end if                                      Normal return.
CONTAINS

FUNCTION jtry(m,n,k)
USE nrtype
IMPLICIT NONE
INTEGER(I4B), INTENT(IN) :: m,n,k
INTEGER(I4B) :: jtry
  Calculate (m*n)/k without overflow. Program efficiency can be improved by substituting an
  assembly language routine that does integer multiply to a double register.
jtry=int((real(m,dp)*real(n,dp))/real(k,dp))
END FUNCTION jtry

SUBROUTINE arcsum(iin,iout,ja,nwk,nrad,nc)
USE nrtype
IMPLICIT NONE
INTEGER(I4B), DIMENSION(:), INTENT(IN) :: iin
INTEGER(I4B), DIMENSION(:), INTENT(OUT) :: iout
INTEGER(I4B), INTENT(IN) :: nwk,nrad,nc
INTEGER(I4B), INTENT(INOUT) :: ja
```

```
  Add the integer ja to the radix nrad multiple-precision integer iin(nc..nwk). Return the
  result in iout(nc..nwk).
INTEGER(I4B) :: j,jtmp,karry
karry=0
do j=nwk,nc+1,-1
    jtmp=ja
    ja=ja/nrad
    iout(j)=iin(j)+(jtmp-ja*nrad)+karry
    if (iout(j) >= nrad) then
        iout(j)=iout(j)-nrad
        karry=1
    else
        karry=0
    end if
end do
iout(nc)=iin(nc)+ja+karry
END SUBROUTINE arcsum
END SUBROUTINE arcode
```

⋆ ⋆ ⋆

```
MODULE mpops
USE nrtype
INTEGER(I4B), PARAMETER :: NPAR_ICARRY=64
CONTAINS

SUBROUTINE icarry(karry,isum,nbits)
IMPLICIT NONE
INTEGER(I4B), INTENT(OUT) :: karry
    Perform deferred carry operation on an array isum of multiple-precision digits. Nonzero bits
    of higher order than nbits (typically 8) are carried to the next-lower (leftward) component
    of isum. The final (most leftward) carry value is returned as karry.
INTEGER(I2B), DIMENSION(:), INTENT(INOUT) :: isum
INTEGER(I4B), INTENT(IN) :: nbits
INTEGER(I4B) :: n,j
INTEGER(I2B), DIMENSION(size(isum)) :: ihi
INTEGER(I2B) :: mb,ihh
n=size(isum)
mb=ishft(1,nbits)-1                       Make mask for low-order bits.
karry=0
if (n < NPAR_ICARRY ) then
    do j=n,2,-1                           Keep going until all carries have cascaded.
        ihh=ishft(isum(j),-nbits)
        if (ihh /= 0) then
            isum(j)=iand(isum(j),mb)
            isum(j-1)=isum(j-1)+ihh
        end if
    end do
    ihh=ishft(isum(1),-nbits)
    isum(1)=iand(isum(1),mb)
    karry=karry+ihh
else
    do
        ihi=ishft(isum,-nbits)            Get high bits.
        if (all(ihi == 0)) exit           Check if done.
        where (ihi /= 0) isum=iand(isum,mb)       Remove bits to be carried and add
        where (ihi(2:n) /= 0) isum(1:n-1)=isum(1:n-1)+ihi(2:n)     them to left.
        karry=karry+ihi(1)                Final carry.
    end do
end if
END SUBROUTINE icarry
```

```
SUBROUTINE mpadd(w,u,v,n)
IMPLICIT NONE
CHARACTER(1), DIMENSION(:), INTENT(OUT) :: w
CHARACTER(1), DIMENSION(:), INTENT(IN) :: u,v
INTEGER(I4B), INTENT(IN) :: n
```

Adds the unsigned radix 256 integers u(1:n) and v(1:n) yielding the unsigned integer w(1:n+1).

```
INTEGER(I2B), DIMENSION(n) :: isum
INTEGER(I4B) :: karry
isum=ichar(u(1:n))+ichar(v(1:n))
call icarry(karry,isum,8_I4B)
w(2:n+1)=char(isum)
w(1)=char(karry)
END SUBROUTINE mpadd
```

```
SUBROUTINE mpsub(is,w,u,v,n)
IMPLICIT NONE
INTEGER(I4B), INTENT(OUT) :: is
CHARACTER(1), DIMENSION(:), INTENT(OUT) :: w
CHARACTER(1), DIMENSION(:), INTENT(IN) :: u,v
INTEGER(I4B), INTENT(IN) :: n
```

Subtracts the unsigned radix 256 integer v(1:n) from u(1:n) yielding the unsigned integer w(1:n). If the result is negative (wraps around), is is returned as -1; otherwise it is returned as 0.

```
INTEGER(I4B) :: karry
INTEGER(I2B), DIMENSION(n) :: isum
isum=255+ichar(u(1:n))-ichar(v(1:n))
isum(n)=isum(n)+1
call icarry(karry,isum,8_I4B)
w(1:n)=char(isum)
is=karry-1
END SUBROUTINE mpsub
```

```
SUBROUTINE mpsad(w,u,n,iv)
IMPLICIT NONE
CHARACTER(1), DIMENSION(:), INTENT(OUT) :: w
CHARACTER(1), DIMENSION(:), INTENT(IN) :: u
INTEGER(I4B), INTENT(IN) :: n,iv
```

Short addition: The integer iv (in the range $0 \le$ iv ≤ 255) is added to the unsigned radix 256 integer u(1:n), yielding w(1:n+1).

```
INTEGER(I4B) :: karry
INTEGER(I2B), DIMENSION(n) :: isum
isum=ichar(u(1:n))
isum(n)=isum(n)+iv
call icarry(karry,isum,8_I4B)
w(2:n+1)=char(isum)
w(1)=char(karry)
END SUBROUTINE mpsad
```

```
SUBROUTINE mpsmu(w,u,n,iv)
IMPLICIT NONE
CHARACTER(1), DIMENSION(:), INTENT(OUT) :: w
CHARACTER(1), DIMENSION(:), INTENT(IN) :: u
INTEGER(I4B), INTENT(IN) :: n,iv
```

Short multiplication: The unsigned radix 256 integer u(1:n) is multiplied by the integer iv (in the range $0 \le$ iv ≤ 255), yielding w(1:n+1).

```
INTEGER(I4B) :: karry
INTEGER(I2B), DIMENSION(n) :: isum
isum=ichar(u(1:n))*iv
call icarry(karry,isum,8_I4B)
w(2:n+1)=char(isum)
w(1)=char(karry)
END SUBROUTINE mpsmu
```

```
SUBROUTINE mpneg(u,n)
IMPLICIT NONE
```

```
CHARACTER(1), DIMENSION(:), INTENT(INOUT) :: u
INTEGER(I4B), INTENT(IN) :: n
    Ones-complement negate the unsigned radix 256 integer u(1:n).
INTEGER(I4B) :: karry
INTEGER(I2B), DIMENSION(n) :: isum
isum=255-ichar(u(1:n))
isum(n)=isum(n)+1
call icarry(karry,isum,8_I4B)
u(1:n)=char(isum)
END SUBROUTINE mpneg

SUBROUTINE mplsh(u,n)
IMPLICIT NONE
CHARACTER(1), DIMENSION(:), INTENT(INOUT) :: u
INTEGER(I4B), INTENT(IN) :: n
    Left shift u(2..n+1) onto u(1:n).
u(1:n)=u(2:n+1)
END SUBROUTINE mplsh

SUBROUTINE mpmov(u,v,n)
IMPLICIT NONE
CHARACTER(1), DIMENSION(:), INTENT(OUT) :: u
CHARACTER(1), DIMENSION(:), INTENT(IN) :: v
INTEGER(I4B), INTENT(IN) :: n
    Move v(1:n) onto u(1:n).
u(1:n)=v(1:n)
END SUBROUTINE mpmov

SUBROUTINE mpsdv(w,u,n,iv,ir)
IMPLICIT NONE
CHARACTER(1), DIMENSION(:), INTENT(OUT) :: w
CHARACTER(1), DIMENSION(:), INTENT(IN) :: u
INTEGER(I4B), INTENT(IN) :: n,iv
INTEGER(I4B), INTENT(OUT) :: ir
    Short division: The unsigned radix 256 integer u(1:n) is divided by the integer iv (in the
    range 0 ≤ iv ≤ 255), yielding a quotient w(1:n) and a remainder ir (with 0 ≤ ir ≤ 255).
    Note: Your Numerical Recipes authors don't know how to parallelize this routine in Fortran
    90!
INTEGER(I4B) :: i,j
ir=0
do j=1,n
    i=256*ir+ichar(u(j))
    w(j)=char(i/iv)
    ir=mod(i,iv)
end do
END SUBROUTINE mpsdv
END MODULE mpops
```

```
SUBROUTINE mpmul(w,u,v,n,m)
USE nrtype; USE nrutil, ONLY : nrerror
USE nr, ONLY : realft
IMPLICIT NONE
INTEGER(I4B), INTENT(IN) :: n,m
CHARACTER(1), DIMENSION(:), INTENT(IN) :: u,v
CHARACTER(1), DIMENSION(:), INTENT(OUT) :: w
The logical dimensions are: CHARACTER(1) :: w(n+m),u(n),v(m)
REAL(DP), PARAMETER :: RX=256.0
    Uses fast Fourier transform to multiply the unsigned radix 256 integers u(1:n) and v(1:m),
    yielding a product w(1:n+m).
INTEGER(I4B) :: j,mn,nn
REAL(DP) :: cy,t
REAL(DP), DIMENSION(:), ALLOCATABLE :: a,b,tb
mn=max(m,n)
nn=1                                Find the smallest useable power of two for the transform.
```

```
do
    if (nn >= mn) exit
    nn=nn+nn
end do
nn=nn+nn
allocate(a(nn),b(nn),tb((nn-1)/2))
a(1:n)=ichar(u(1:n))                  Move U to a double-precision floating array.
a(n+1:nn)=0.0
b(1:m)=ichar(v(1:m))                  Move V to a double-precision floating array.
b(m+1:nn)=0.0
call realft(a(1:nn),1)                Perform the convolution: First, the two Fourier trans-
call realft(b(1:nn),1)                    forms.
b(1)=b(1)*a(1)                        Then multiply the complex results (real and imaginary
b(2)=b(2)*a(2)                            parts).
tb=b(3:nn:2)
b(3:nn:2)=tb*a(3:nn:2)-b(4:nn:2)*a(4:nn:2)
b(4:nn:2)=tb*a(4:nn:2)+b(4:nn:2)*a(3:nn:2)
call realft(b(1:nn),-1)               Then do the inverse Fourier transform.
b(:)=b(:)/(nn/2)
cy=0.0                                Make a final pass to do all the carries.
do j=nn,1,-1
    t=b(j)+cy+0.5_dp                  The 0.5 allows for roundoff error.
    b(j)=mod(t,RX)
    cy=int(t/RX)
end do
if (cy >= RX) call nrerror('mpmul: sanity check failed in fftmul')
w(1)=char(int(cy))                    Copy answer to output.
w(2:(n+m))=char(int(b(1:(n+m-1))))
deallocate(a,b,tb)
END SUBROUTINE mpmul
```

```
SUBROUTINE mpinv(u,v,n,m)
USE nrtype; USE nrutil, ONLY : poly
USE nr, ONLY : mpmul
USE mpops, ONLY : mpmov,mpneg
IMPLICIT NONE
CHARACTER(1), DIMENSION(:), INTENT(OUT) :: u
CHARACTER(1), DIMENSION(:), INTENT(IN) :: v
INTEGER(I4B), INTENT(IN) :: n,m
INTEGER(I4B), PARAMETER :: MF=4
REAL(SP), PARAMETER :: BI=1.0_sp/256.0_sp
    Character string v(1:m) is interpreted as a radix 256 number with the radix point after
    (nonzero) v(1); u(1:n) is set to the most significant digits of its reciprocal, with the radix
    point after u(1).
INTEGER(I4B) :: i,j,mm
REAL(SP) :: fu
CHARACTER(1), DIMENSION(:), ALLOCATABLE :: rr,s
allocate(rr(max(n,m)+n+1),s(n))
mm=min(MF,m)
fu=1.0_sp/poly(BI,real(ichar(v(:)),sp))       Use ordinary floating arithmetic to get an
do j=1,n                                          initial approximation.
    i=int(fu)
    u(j)=char(i)
    fu=256.0_sp*(fu-i)
end do
do                                            Iterate Newton's rule to convergence.
    call mpmul(rr,u,v,n,m)                    Construct 2 - UV in S.
    call mpmov(s,rr(2:),n)
    call mpneg(s,n)
    s(1)=char(ichar(s(1))-254)                Multiply SU into U.
    call mpmul(rr,s,u,n,n)
    call mpmov(u,rr(2:),n)
```

```
    if (all(ichar(s(2:n-1)) == 0)) exit
end do
deallocate(rr,s)
END SUBROUTINE mpinv
```

If fractional part of S is not zero, it has not converged to 1.

```
SUBROUTINE mpdiv(q,r,u,v,n,m)
USE nrtype; USE nrutil, ONLY : nrerror
USE nr, ONLY : mpinv,mpmul
USE mpops, ONLY : mpsad,mpmov,mpsub
IMPLICIT NONE
CHARACTER(1), DIMENSION(:), INTENT(OUT) :: q,r
CHARACTER(1), DIMENSION(:), INTENT(IN) :: u,v
! The logical dimensions are: CHARACTER(1) :: q(n-m+1),r(m),u(n),v(m)
INTEGER(I4B), INTENT(IN) :: n,m
```

Divides unsigned radix 256 integers `u(1:n)` by `v(1:m)` (with `m` $\le$ `n` required), yielding a quotient `q(1:n-m+1)` and a remainder `r(1:m)`.

```
INTEGER(I4B), PARAMETER :: MACC=6
INTEGER(I4B) :: is
CHARACTER(1), DIMENSION(:), ALLOCATABLE, TARGET :: rr,s
CHARACTER(1), DIMENSION(:), POINTER :: rr2,s3
allocate(rr(2*(n+MACC)),s(2*(n+MACC)))
rr2=>rr(2:)
s3=>s(3:)
call mpinv(s,v,n+MACC,m)                Set S = 1/V.
call mpmul(rr,s,u,n+MACC,n)             Set Q = SU.
call mpsad(s,rr,n+n+MACC/2,1)
call mpmov(q,s3,n-m+1)
call mpmul(rr,q,v,n-m+1,m)              Multiply and subtract to get the remainder.
call mpsub(is,rr2,u,rr2,n)
if (is /= 0) call nrerror('MACC too small in mpdiv')
call mpmov(r,rr(n-m+2:),m)
deallocate(rr,s)
END SUBROUTINE mpdiv
```

```
SUBROUTINE mpsqrt(w,u,v,n,m)
USE nrtype; USE nrutil, ONLY : poly
USE nr, ONLY : mpmul
USE mpops, ONLY : mplsh,mpmov,mpneg,mpsdv
IMPLICIT NONE
CHARACTER(1), DIMENSION(:), INTENT(OUT) :: w,u
CHARACTER(1), DIMENSION(:), INTENT(IN) :: v
INTEGER(I4B), INTENT(IN) :: n,m
INTEGER(I4B), PARAMETER :: MF=3
REAL(SP), PARAMETER :: BI=1.0_sp/256.0_sp
```

Character string `v(1:m)` is interpreted as a radix 256 number with the radix point after `v(1)`; `w(1:n)` is set to its square root (radix point after `w(1)`), and `u(1:n)` is set to the reciprocal thereof (radix point before `u(1)`). `w` and `u` need not be distinct, in which case they are set to the square root.

```
INTEGER(I4B) :: i,ir,j,mm
REAL(SP) :: fu
CHARACTER(1), DIMENSION(:), ALLOCATABLE :: r,s
allocate(r(2*n),s(2*n))
mm=min(m,MF)
fu=1.0_sp/sqrt(poly(BI,real(ichar(v(:)),sp)))   Use ordinary floating arithmetic
do j=1,n                                            to get an initial approxima-
    i=int(fu)                                       tion.
    u(j)=char(i)
    fu=256.0_sp*(fu-i)
end do
do                                       Iterate Newton's rule to convergence.
    call mpmul(r,u,u,n,n)                Construct S = (3 - VU^2)/2.
```

```
        call mplsh(r,n)
        call mpmul(s,r,v,n,min(m,n))
        call mplsh(s,n)
        call mpneg(s,n)
        s(1)=char(ichar(s(1))-253)
        call mpsdv(s,s,n,2,ir)
        if (any(ichar(s(2:n-1)) /= 0)) then
            If fractional part of S is not zero, it has not converged to 1.
            call mpmul(r,s,u,n,n)              Replace U by SU.
            call mpmov(u,r(2:),n)
            cycle
        end if
        call mpmul(r,u,v,n,min(m,n))           Get square root from reciprocal and return.
        call mpmov(w,r(2:),n)
        deallocate(r,s)
        RETURN
    end do
    END SUBROUTINE mpsqrt
```

```
    SUBROUTINE mp2dfr(a,s,n,m)
    USE nrtype
    USE mpops, ONLY : mplsh,mpsmu
    IMPLICIT NONE
    INTEGER(I4B), INTENT(IN) :: n
    INTEGER(I4B), INTENT(OUT) :: m
    CHARACTER(1), DIMENSION(:), INTENT(INOUT) :: a
    CHARACTER(1), DIMENSION(:), INTENT(OUT) :: s
    INTEGER(I4B), PARAMETER :: IAZ=48
        Converts a radix 256 fraction a(1:n) (radix point before a(1)) to a decimal fraction
        represented as an ascii string s(1:m), where m is a returned value. The input array a(1:n)
        is destroyed. NOTE: For simplicity, this routine implements a slow (∝ N^2) algorithm. Fast
        (∝ N ln N), more complicated, radix conversion algorithms do exist.
    INTEGER(I4B) :: j
    m=int(2.408_sp*n)
    do j=1,m
        call mpsmu(a,a,n,10)
        s(j)=char(ichar(a(1))+IAZ)
        call mplsh(a,n)
    end do
    END SUBROUTINE mp2dfr
```

```
    SUBROUTINE mppi(n)
    USE nrtype
    USE nr, ONLY : mp2dfr,mpinv,mpmul,mpsqrt
    USE mpops, ONLY : mpadd,mplsh,mpmov,mpsdv
    IMPLICIT NONE
    INTEGER(I4B), INTENT(IN) :: n
    INTEGER(I4B), PARAMETER :: IAOFF=48
        Demonstrate multiple precision routines by calculating and printing the first n bytes of π.
    INTEGER(I4B) :: ir,j,m
    CHARACTER(1), DIMENSION(n) :: sx,sxi
    CHARACTER(1), DIMENSION(2*n) :: t,y
    CHARACTER(1), DIMENSION(3*n) :: s
    CHARACTER(1), DIMENSION(n+1) :: x,bigpi
    t(1)=char(2)                               Set T = 2.
    t(2:n)=char(0)
    call mpsqrt(x,x,t,n,n)                     Set X_0 = √2.
    call mpadd(bigpi,t,x,n)                    Set π_0 = 2 + √2.
    call mplsh(bigpi,n)
    call mpsqrt(sx,sxi,x,n,n)                  Set Y_0 = 2^(1/4).
```

```
call mpmov(y,sx,n)
do
    call mpadd(x,sx,sxi,n)                  Set X_{i+1} = (X_i^{1/2} + X_i^{-1/2})/2.
    call mpsdv(x,x(2:),n,2,ir)
    call mpsqrt(sx,sxi,x,n,n)               Form the temporary T = Y_i X_{i+1}^{1/2} + X_{i+1}^{-1/2}.
    call mpmul(t,y,sx,n,n)
    call mpadd(t(2:),t(2:),sxi,n)
    x(1)=char(ichar(x(1))+1)                Increment X_{i+1} and Y_i by 1.
    y(1)=char(ichar(y(1))+1)
    call mpinv(s,y,n,n)                     Set Y_{i+1} = T/(Y_i + 1).
    call mpmul(y,t(3:),s,n,n)
    call mplsh(y,n)
    call mpmul(t,x,s,n,n)                   Form temporary T = (X_{i+1} + 1)/(Y_i + 1).
    m=mod(255+ichar(t(2)),256)              If T = 1 then we have converged.
    if (abs(ichar(t(n+1))-m) > 1 .or. any(ichar(t(3:n)) /= m)) then
        call mpmul(s,bigpi,t(2:),n,n)       Set π_{i+1} = T π_i.
        call mpmov(bigpi,s(2:),n)
        cycle
    end if
    write (*,*) 'pi='
    s(1)=char(ichar(bigpi(1))+IAOFF)
    s(2)='.'
    call mp2dfr(bigpi(2:),s(3:),n-1,m)
      Convert to decimal for printing. NOTE: The conversion routine, for this demonstration
      only, is a slow (∝ N^2) algorithm. Fast (∝ N ln N), more complicated, radix conversion
      algorithms do exist.
    write (*,'(1x,64a1)') (s(j),j=1,m+1)
    RETURN
end do
END SUBROUTINE mppi
```

References

The references collected here are those of general usefulness, cited in this volume. For references to the material in Volume 1, see the References section of that volume.

A first group of references relates to the Fortran 90 language itself:

Metcalf, M., and Reid, J. 1996, *Fortran 90/95 Explained* (Oxford: Oxford University Press).

Kerrigan, J.F. 1993, *Migrating to Fortran 90* (Sebastopol, CA: O'Reilly).

Brainerd, W.S., Goldberg, C.H., and Adams, J.C. 1996, *Programmer's Guide to Fortran 90*, 3rd ed. (New York: Springer-Verlag).

A second group of references relates to, or includes material on, parallel programming and algorithms:

Akl, S.G. 1989, *The Design and Analysis of Parallel Algorithms* (Englewood Cliffs, NJ: Prentice Hall).

Bertsekas, D.P., and Tsitsiklis, J.N. 1989, *Parallel and Distributed Computation: Numerical Methods* (Englewood Cliffs, NJ: Prentice Hall).

Carey, G.F. 1989, *Parallel Supercomputing: Methods, Algorithms, and Applications* (New York: Wiley).

Fountain, T.J. 1994, *Parallel Computing: Principles and Practice* (New York: Cambridge University Press).

Fox, G.C., et al. 1988, *Solving Problems on Concurrent Processors*, Volume I (Englewood Cliffs, NJ: Prentice Hall).

Golub, G., and Ortega, J.M. 1993, *Scientific Computing: An Introduction with Parallel Computing* (San Diego, CA: Academic Press).

Golub, G.H., and Van Loan, C.F. 1989, *Matrix Computations*, 2nd ed. (Baltimore: Johns Hopkins University Press).

Hockney, R.W., and Jesshope, C.R. 1988, *Parallel Computers 2* (Bristol and Philadelphia: Adam Hilger).

Kumar, V., et al. 1994, *Introduction to Parallel Computing: Design and Analysis of Parallel Algorithms* (Redwood City, CA: Benjamin/Cummings).

Lewis, T.G., and El-Rewini, H. 1992, *Introduction to Parallel Computing* (Englewood Cliffs, NJ: Prentice Hall).

Modi, J.J. 1988, *Parallel Algorithms and Matrix Computation* (New York: Oxford University Press).

Smith, J.R. 1993, *The Design and Analysis of Parallel Algorithms* (New York: Oxford University Press).

Van de Velde, E. 1994, *Concurrent Scientific Computing* (New York: Springer-Verlag).

Van Loan, C.F. 1992, *Computational Frameworks for the Fast Fourier Transform* (Philadelphia: S.I.A.M.).

C1. Listing of Utility Modules (nrtype and nrutil)

C1.1 Numerical Recipes Types (nrtype)

The file supplied as nrtype.f90 contains a single module named nrtype, which in turn contains definitions for a number of named constants (that is, PARAMETERs), and a couple of elementary derived data types used by the sparse matrix routines in this book. Of the named constants, by far the most important are those that define the KIND types of virtually all the variables used in this book: I4B, I2B, and I1B for integer variables, SP and DP for real variables (and SPC and DPC for the corresponding complex cases), and LGT for the default logical type.

```
MODULE nrtype
    Symbolic names for kind types of 4-, 2-, and 1-byte integers:
INTEGER, PARAMETER :: I4B = SELECTED_INT_KIND(9)
INTEGER, PARAMETER :: I2B = SELECTED_INT_KIND(4)
INTEGER, PARAMETER :: I1B = SELECTED_INT_KIND(2)
    Symbolic names for kind types of single- and double-precision reals:
INTEGER, PARAMETER :: SP = KIND(1.0)
INTEGER, PARAMETER :: DP = KIND(1.0D0)
    Symbolic names for kind types of single- and double-precision complex:
INTEGER, PARAMETER :: SPC = KIND((1.0,1.0))
INTEGER, PARAMETER :: DPC = KIND((1.0D0,1.0D0))
    Symbolic name for kind type of default logical:
INTEGER, PARAMETER :: LGT = KIND(.true.)
    Frequently used mathematical constants (with precision to spare):
REAL(SP), PARAMETER :: PI=3.141592653589793238462643383279502884197_sp
REAL(SP), PARAMETER :: PIO2=1.57079632679489661923132169163975144209858_sp
REAL(SP), PARAMETER :: TWOPI=6.283185307179586476925286766559005768394_sp
REAL(SP), PARAMETER :: SQRT2=1.41421356237309504880168872420969807856967_sp
REAL(SP), PARAMETER :: EULER=0.5772156649015328606065120900824024310422_sp
REAL(DP), PARAMETER :: PI_D=3.141592653589793238462643383279502884197_dp
REAL(DP), PARAMETER :: PIO2_D=1.57079632679489661923132169163975144209858_dp
REAL(DP), PARAMETER :: TWOPI_D=6.283185307179586476925286766559005768394_dp
    Derived data types for sparse matrices, single and double precision (see use in Chapter B2):
TYPE sprs2_sp
    INTEGER(I4B) :: n,len
    REAL(SP), DIMENSION(:), POINTER :: val
    INTEGER(I4B), DIMENSION(:), POINTER :: irow
    INTEGER(I4B), DIMENSION(:), POINTER :: jcol
END TYPE sprs2_sp
TYPE sprs2_dp
    INTEGER(I4B) :: n,len
    REAL(DP), DIMENSION(:), POINTER :: val
```

```
    INTEGER(I4B), DIMENSION(:), POINTER :: irow
    INTEGER(I4B), DIMENSION(:), POINTER :: jcol
END TYPE sprs2_dp
END MODULE nrtype
```

About Converting to Higher Precision

You might hope that changing all the Numerical Recipes routines from single precision to double precision would be as simple as redefining the values of SP and DP in nrtype. Well ... not quite.

Converting algorithms to a higher precision is not a purely mechanical task because of the distinction between "roundoff error" and "truncation error." (Please see Volume 1, §1.2, if you are not familiar with these concepts.) While increasing the precision implied by the kind values SP and DP will indeed reduce a routine's roundoff error, it will not reduce any truncation error that may be intrinsic to the algorithm. Sometimes, a routine contains "accuracy parameters" that can be adjusted to reduce the truncation error to the new, desired level. In other cases, however, the truncation error cannot be so easily reduced; then, a whole new algorithm is needed. Clearly such new algorithms are beyond the scope of a simple mechanical "conversion."

If, despite these cautionary words, you want to proceed with converting some routines to a higher precision, here are some hints:

If your machine has a kind type that is distinct from, and has equal or greater precision than, the kind type that we use for DP, then, in nrtype, you can simply redefine DP to this new highest precision and redefine SP to what was previously DP. For example, DEC machines usually have a "quadruple precision" real type available, which can be used in this way. You should not need to make any further edits of nrtype or nrutil.

If, on the other hand, the kind type that we already use for DP is the highest precision available, then you must leave DP defined as it is, and redefine SP in nrtype to be this same kind type. Now, however, you will also have to edit nrutil, because some overloaded routines that were previously distinguishable (by the different kind types) will now be seen by the compiler as indistinguishable — and it will object strenuously. Simply delete all the "_dp" function names from the list of overloaded procedures (i.e., from the MODULE PROCEDURE statements). Note that it is not necessary to delete the routines from the MODULE itself. Similarly, in the interface file nr.f90 you must delete the "_dp" interfaces, *except* for the sprs... routines. (Since they have TYPE(sprs2_dp) or TYPE(sprs2_sp), they are treated as distinct even though they have functionally equivalent kind types.)

Finally, the following table gives some suggestions for changing the accuracy parameters, or constants, in some of the routines. Please note that this table is not necessarily complete, and that higher-precision performance is not guaranteed for all the routines, *even if* you make all the changes indicated. The above edits, and these suggestions, do, however, work in the majority of cases.

In routine...	*change...*	*to...*
beschb	NUSE1=5,NUSE2=5	NUSE1=7,NUSE2=8
bessi	IACC=40	IACC=200
bessik	EPS=1.0e-10_dp	EPS=epsilon(x)
bessj	IACC=40	IACC=160
bessjy	EPS=1.0e-10_dp	EPS=epsilon(x)
broydn	TOLF=1.0e-4_sp	TOLF=1.0e-8_sp
	TOLMIN=1.0e-6_sp	TOLMIN=1.0e-12_sp
fdjac	EPS=1.0e-4_sp	EPS=1.0e-8_sp
frprmn	EPS=1.0e-10_sp	EPS=1.0e-18_sp
gauher	EPS=3.0e-13_dp	EPS=1.0e-14_dp
gaujac	EPS=3.0e-14_dp	EPS=1.0e-14_dp
gaulag	EPS=3.0e-13_dp	EPS=1.0e-14_dp
gauleg	EPS=3.0e-14_dp	EPS=1.0e-14_dp
hypgeo	EPS=1.0e-6_sp	EPS=1.0e-14_sp
linmin	TOL=1.0e-4_sp	TOL=1.0e-8_sp
newt	TOLF=1.0e-4_sp	TOLF=1.0e-8_sp
	TOLMIN=1.0e-6_sp	TOLMIN=1.0e-12_sp
probks	EPS1=0.001_sp	EPS1=1.0e-6_sp
	EPS2=1.0e-8_sp	EPS2=1.0e-16_sp
qromb	EPS=1.0e-6_sp	EPS=1.0e-10_sp
qromo	EPS=1.0e-6_sp	EPS=1.0e-10_sp
qroot	TINY=1.0e-6_sp	TINY=1.0e-14_sp
qsimp	EPS=1.0e-6_sp	EPS=1.0e-10_sp
qtrap	EPS=1.0e-6_sp	EPS=1.0e-10_sp
rc	ERRTOL=0.04_sp	ERRTOL=0.0012_sp
rd	ERRTOL=0.05_sp	ERRTOL=0.0015_sp
rf	ERRTOL=0.08_sp	ERRTOL=0.0025_sp
rj	ERRTOL=0.05_sp	ERRTOL=0.0015_sp
sfroid	conv=5.0e-6_sp	conv=1.0e-14_sp
shoot	EPS=1.0e-6_sp	EPS=1.0e-14_sp
shootf	EPS=1.0e-6_sp	EPS=1.0e-14_sp
simplx	EPS=1.0e-6_sp	EPS=1.0e-14_sp
sncndn	CA=0.0003_sp	CA=1.0e-8_sp
sor	EPS=1.0e-5_dp	EPS=1.0e-13_dp
sphfpt	DXX=1.0e-4_sp	DXX=1.0e-8_sp
sphoot	dx=1.0e-4_sp	dx=1.0e-8_sp
svdfit	TOL=1.0e-5_sp	TOL=1.0e-13_sp
zroots	EPS=1.0e-6_sp	EPS=1.0e-14_sp

C1.2 Numerical Recipes Utilities (nrutil)

The file supplied as nrutil.f90 contains a single module named nrutil, which contains specific implementations for all the Numerical Recipes utility functions described in detail in Chapter 23.

The specific implementations given are something of a compromise between demonstrating parallel techniques (when they can be achieved in Fortran 90) and running efficiently on conventional, serial machines. The parameters at the beginning of the module (names beginning with NPAR_) are typically related to array lengths *below which* the implementations revert to serial operations. On a purely serial machine, these can be set to large values to suppress many parallel constructions.

The length and repetitiveness of the nrutil.f90 file stems in large part from its extensive use of overloading. Indeed, the file would be even longer if we overloaded versions for all the applicable data types that each utility could, in principle, instantiate. The descriptions in Chapter 23 detail both the full set of intended data types and shapes for each routine, and also the types and shapes actually here implemented (which can also be gleaned by examining the file). The intended result of all this overloading is, in essence, to give the utility routines the desirable properties of many of the Fortran 90 intrinsic functions, namely, to be both *generic* (apply to many data types) and *elemental* (apply element-by-element to arbitrary shapes). Fortran 95's provision of user-defined elemental functions will reduce the multiplicity of overloading in some of our routines; unfortunately the necessity to overload for multiple data types will still be present.

Finally, it is worth reemphasizing the following point, already made in Chapter 23: The purpose of the nrutil utilities is to remove from the Numerical Recipes programs just those programming tasks and "idioms" whose efficient implementation is *most* hardware and compiler dependent, so as to allow for specific, efficient implementations on different machines. One should therefore not expect the utmost in efficiency from the general purpose, one-size-fits-all, implementation listed here.

Correspondingly, we would encourage the incorporation of efficient nrutil implementations, and/or comparable capabilities under different names, with as broad as possible a set of overloaded data types, in libraries associated with specific compilers or machines. In support of this goal, we have specifically put this Appendix C1, and the files nrtype.f90 and nrutil.f90, into the public domain.

```
MODULE nrutil
    TABLE OF CONTENTS OF THE NRUTIL MODULE:
        routines that move data:
            array_copy, swap, reallocate
        routines returning a location as an integer value
            ifirstloc, imaxloc, iminloc
        routines for argument checking and error handling:
            assert, assert_eq, nrerror
        routines relating to polynomials and recurrences
            arth, geop, cumsum, cumprod, poly, polyterm,
            zroots_unity
        routines for "outer" operations on vectors
            outerand, outersum, outerdiff, outerprod, outerdiv
        routines for scatter-with-combine
            scatter_add, scatter_max
        routines for skew operations on matrices
            diagadd, diagmult, get_diag, put_diag,
```

unit_matrix, lower_triangle, upper_triangle
miscellaneous routines
vabs

```
USE nrtype
```

Parameters for crossover from serial to parallel algorithms (these are used only within this nrutil module):

```
IMPLICIT NONE
INTEGER(I4B), PARAMETER :: NPAR_ARTH=16,NPAR2_ARTH=8
INTEGER(I4B), PARAMETER :: NPAR_GEOP=4,NPAR2_GEOP=2
INTEGER(I4B), PARAMETER :: NPAR_CUMSUM=16
INTEGER(I4B), PARAMETER :: NPAR_CUMPROD=8
INTEGER(I4B), PARAMETER :: NPAR_POLY=8
INTEGER(I4B), PARAMETER :: NPAR_POLYTERM=8
```

Each NPAR2 must be $\leq$ the corresponding NPAR.

Next, generic interfaces for routines with overloaded versions. Naming conventions for appended codes in the names of overloaded routines are as follows: r=real, d=double precision, i=integer, c=complex, z=double-precision complex, h=character, l=logical. Any of r,d,i,c,z,h,l may be followed by v=vector or m=matrix (v,m suffixes are used only when needed to resolve ambiguities).

Routines that move data:

```
INTERFACE array_copy
    MODULE PROCEDURE array_copy_r, array_copy_d, array_copy_i
END INTERFACE
INTERFACE swap
    MODULE PROCEDURE swap_i,swap_r,swap_rv,swap_c, &
        swap_cv,swap_cm,swap_z,swap_zv,swap_zm, &
        masked_swap_rs,masked_swap_rv,masked_swap_rm
END INTERFACE
INTERFACE reallocate
    MODULE PROCEDURE reallocate_rv,reallocate_rm,&
        reallocate_iv,reallocate_im,reallocate_hv
END INTERFACE
```

Routines returning a location as an integer value (ifirstloc, iminloc are not currently overloaded and so do not have a generic interface here):

```
INTERFACE imaxloc
    MODULE PROCEDURE imaxloc_r,imaxloc_i
END INTERFACE
```

Routines for argument checking and error handling (nrerror is not currently overloaded):

```
INTERFACE assert
    MODULE PROCEDURE assert1,assert2,assert3,assert4,assert_v
END INTERFACE
INTERFACE assert_eq
    MODULE PROCEDURE assert_eq2,assert_eq3,assert_eq4,assert_eqn
END INTERFACE
```

Routines relating to polynomials and recurrences (cumprod, zroots_unity are not currently overloaded):

```
INTERFACE arth
    MODULE PROCEDURE arth_r, arth_d, arth_i
END INTERFACE
INTERFACE geop
    MODULE PROCEDURE geop_r, geop_d, geop_i, geop_c, geop_dv
END INTERFACE
INTERFACE cumsum
    MODULE PROCEDURE cumsum_r,cumsum_i
END INTERFACE
INTERFACE poly
    MODULE PROCEDURE poly_rr,poly_rrv,poly_dd,poly_ddv,&
        poly_rc,poly_cc,poly_msk_rrv,poly_msk_ddv
END INTERFACE
INTERFACE poly_term
    MODULE PROCEDURE poly_term_rr,poly_term_cc
END INTERFACE
```

Routines for "outer" operations on vectors (outerand, outersum, outerdiv are not currently overloaded):

```
INTERFACE outerprod
```

```
    MODULE PROCEDURE outerprod_r,outerprod_d
END INTERFACE
INTERFACE outerdiff
    MODULE PROCEDURE outerdiff_r,outerdiff_d,outerdiff_i
END INTERFACE
```
Routines for scatter-with-combine, scatter_add, scatter_max:
```
INTERFACE scatter_add
    MODULE PROCEDURE scatter_add_r,scatter_add_d
END INTERFACE
INTERFACE scatter_max
    MODULE PROCEDURE scatter_max_r,scatter_max_d
END INTERFACE
```
Routines for skew operations on matrices (unit_matrix, lower_triangle, upper_triangle not currently overloaded):
```
INTERFACE diagadd
    MODULE PROCEDURE diagadd_rv,diagadd_r
END INTERFACE
INTERFACE diagmult
    MODULE PROCEDURE diagmult_rv,diagmult_r
END INTERFACE
INTERFACE get_diag
    MODULE PROCEDURE get_diag_rv, get_diag_dv
END INTERFACE
INTERFACE put_diag
    MODULE PROCEDURE put_diag_rv, put_diag_r
END INTERFACE
```
Other routines (vabs is not currently overloaded):

CONTAINS

Routines that move data:
```
SUBROUTINE array_copy_r(src,dest,n_copied,n_not_copied)
```
Copy array where size of source not known in advance.
```
REAL(SP), DIMENSION(:), INTENT(IN) :: src
REAL(SP), DIMENSION(:), INTENT(OUT) :: dest
INTEGER(I4B), INTENT(OUT) :: n_copied, n_not_copied
n_copied=min(size(src),size(dest))
n_not_copied=size(src)-n_copied
dest(1:n_copied)=src(1:n_copied)
END SUBROUTINE array_copy_r

SUBROUTINE array_copy_d(src,dest,n_copied,n_not_copied)
REAL(DP), DIMENSION(:), INTENT(IN) :: src
REAL(DP), DIMENSION(:), INTENT(OUT) :: dest
INTEGER(I4B), INTENT(OUT) :: n_copied, n_not_copied
n_copied=min(size(src),size(dest))
n_not_copied=size(src)-n_copied
dest(1:n_copied)=src(1:n_copied)
END SUBROUTINE array_copy_d

SUBROUTINE array_copy_i(src,dest,n_copied,n_not_copied)
INTEGER(I4B), DIMENSION(:), INTENT(IN) :: src
INTEGER(I4B), DIMENSION(:), INTENT(OUT) :: dest
INTEGER(I4B), INTENT(OUT) :: n_copied, n_not_copied
n_copied=min(size(src),size(dest))
n_not_copied=size(src)-n_copied
dest(1:n_copied)=src(1:n_copied)
END SUBROUTINE array_copy_i

SUBROUTINE swap_i(a,b)
```
Swap the contents of a and b.
```
INTEGER(I4B), INTENT(INOUT) :: a,b
INTEGER(I4B) :: dum
dum=a
a=b
b=dum
END SUBROUTINE swap_i
```

```
SUBROUTINE swap_r(a,b)
REAL(SP), INTENT(INOUT) :: a,b
REAL(SP) :: dum
dum=a
a=b
b=dum
END SUBROUTINE swap_r

SUBROUTINE swap_rv(a,b)
REAL(SP), DIMENSION(:), INTENT(INOUT) :: a,b
REAL(SP), DIMENSION(SIZE(a)) :: dum
dum=a
a=b
b=dum
END SUBROUTINE swap_rv

SUBROUTINE swap_c(a,b)
COMPLEX(SPC), INTENT(INOUT) :: a,b
COMPLEX(SPC) :: dum
dum=a
a=b
b=dum
END SUBROUTINE swap_c

SUBROUTINE swap_cv(a,b)
COMPLEX(SPC), DIMENSION(:), INTENT(INOUT) :: a,b
COMPLEX(SPC), DIMENSION(SIZE(a)) :: dum
dum=a
a=b
b=dum
END SUBROUTINE swap_cv

SUBROUTINE swap_cm(a,b)
COMPLEX(SPC), DIMENSION(:,:), INTENT(INOUT) :: a,b
COMPLEX(SPC), DIMENSION(size(a,1),size(a,2)) :: dum
dum=a
a=b
b=dum
END SUBROUTINE swap_cm

SUBROUTINE swap_z(a,b)
COMPLEX(DPC), INTENT(INOUT) :: a,b
COMPLEX(DPC) :: dum
dum=a
a=b
b=dum
END SUBROUTINE swap_z

SUBROUTINE swap_zv(a,b)
COMPLEX(DPC), DIMENSION(:), INTENT(INOUT) :: a,b
COMPLEX(DPC), DIMENSION(SIZE(a)) :: dum
dum=a
a=b
b=dum
END SUBROUTINE swap_zv

SUBROUTINE swap_zm(a,b)
COMPLEX(DPC), DIMENSION(:,:), INTENT(INOUT) :: a,b
COMPLEX(DPC), DIMENSION(size(a,1),size(a,2)) :: dum
dum=a
a=b
b=dum
END SUBROUTINE swap_zm

SUBROUTINE masked_swap_rs(a,b,mask)
REAL(SP), INTENT(INOUT) :: a,b
LOGICAL(LGT), INTENT(IN) :: mask
REAL(SP) :: swp
```

```
if (mask) then
    swp=a
    a=b
    b=swp
end if
END SUBROUTINE masked_swap_rs

SUBROUTINE masked_swap_rv(a,b,mask)
REAL(SP), DIMENSION(:), INTENT(INOUT) :: a,b
LOGICAL(LGT), DIMENSION(:), INTENT(IN) :: mask
REAL(SP), DIMENSION(size(a)) :: swp
where (mask)
    swp=a
    a=b
    b=swp
end where
END SUBROUTINE masked_swap_rv

SUBROUTINE masked_swap_rm(a,b,mask)
REAL(SP), DIMENSION(:,:), INTENT(INOUT) :: a,b
LOGICAL(LGT), DIMENSION(:,:), INTENT(IN) :: mask
REAL(SP), DIMENSION(size(a,1),size(a,2)) :: swp
where (mask)
    swp=a
    a=b
    b=swp
end where
END SUBROUTINE masked_swap_rm

FUNCTION reallocate_rv(p,n)
    Reallocate a pointer to a new size, preserving its previous contents.
REAL(SP), DIMENSION(:), POINTER :: p, reallocate_rv
INTEGER(I4B), INTENT(IN) :: n
INTEGER(I4B) :: nold,ierr
allocate(reallocate_rv(n),stat=ierr)
if (ierr /= 0) call &
    nrerror('reallocate_rv: problem in attempt to allocate memory')
if (.not. associated(p)) RETURN
nold=size(p)
reallocate_rv(1:min(nold,n))=p(1:min(nold,n))
deallocate(p)
END FUNCTION reallocate_rv

FUNCTION reallocate_iv(p,n)
INTEGER(I4B), DIMENSION(:), POINTER :: p, reallocate_iv
INTEGER(I4B), INTENT(IN) :: n
INTEGER(I4B) :: nold,ierr
allocate(reallocate_iv(n),stat=ierr)
if (ierr /= 0) call &
    nrerror('reallocate_iv: problem in attempt to allocate memory')
if (.not. associated(p)) RETURN
nold=size(p)
reallocate_iv(1:min(nold,n))=p(1:min(nold,n))
deallocate(p)
END FUNCTION reallocate_iv

FUNCTION reallocate_hv(p,n)
CHARACTER(1), DIMENSION(:), POINTER :: p, reallocate_hv
INTEGER(I4B), INTENT(IN) :: n
INTEGER(I4B) :: nold,ierr
allocate(reallocate_hv(n),stat=ierr)
if (ierr /= 0) call &
    nrerror('reallocate_hv: problem in attempt to allocate memory')
if (.not. associated(p)) RETURN
nold=size(p)
reallocate_hv(1:min(nold,n))=p(1:min(nold,n))
```

```
deallocate(p)
END FUNCTION reallocate_hv

FUNCTION reallocate_rm(p,n,m)
REAL(SP), DIMENSION(:,:), POINTER :: p, reallocate_rm
INTEGER(I4B), INTENT(IN) :: n,m
INTEGER(I4B) :: nold,mold,ierr
allocate(reallocate_rm(n,m),stat=ierr)
if (ierr /= 0) call &
    nrerror('reallocate_rm: problem in attempt to allocate memory')
if (.not. associated(p)) RETURN
nold=size(p,1)
mold=size(p,2)
reallocate_rm(1:min(nold,n),1:min(mold,m))=&
    p(1:min(nold,n),1:min(mold,m))
deallocate(p)
END FUNCTION reallocate_rm

FUNCTION reallocate_im(p,n,m)
INTEGER(I4B), DIMENSION(:,:), POINTER :: p, reallocate_im
INTEGER(I4B), INTENT(IN) :: n,m
INTEGER(I4B) :: nold,mold,ierr
allocate(reallocate_im(n,m),stat=ierr)
if (ierr /= 0) call &
    nrerror('reallocate_im: problem in attempt to allocate memory')
if (.not. associated(p)) RETURN
nold=size(p,1)
mold=size(p,2)
reallocate_im(1:min(nold,n),1:min(mold,m))=&
    p(1:min(nold,n),1:min(mold,m))
deallocate(p)
END FUNCTION reallocate_im
```

Routines returning a location as an integer value:

```
FUNCTION ifirstloc(mask)
```

Index of first occurrence of .true. in a logical vector.

```
LOGICAL(LGT), DIMENSION(:), INTENT(IN) :: mask
INTEGER(I4B) :: ifirstloc
INTEGER(I4B), DIMENSION(1) :: loc
loc=maxloc(merge(1,0,mask))
ifirstloc=loc(1)
if (.not. mask(ifirstloc)) ifirstloc=size(mask)+1
END FUNCTION ifirstloc

FUNCTION imaxloc_r(arr)
```

Index of maxloc on an array.

```
REAL(SP), DIMENSION(:), INTENT(IN) :: arr
INTEGER(I4B) :: imaxloc_r
INTEGER(I4B), DIMENSION(1) :: imax
imax=maxloc(arr(:))
imaxloc_r=imax(1)
END FUNCTION imaxloc_r

FUNCTION imaxloc_i(iarr)
INTEGER(I4B), DIMENSION(:), INTENT(IN) :: iarr
INTEGER(I4B), DIMENSION(1) :: imax
INTEGER(I4B) :: imaxloc_i
imax=maxloc(iarr(:))
imaxloc_i=imax(1)
END FUNCTION imaxloc_i

FUNCTION iminloc(arr)
```

Index of minloc on an array.

```
REAL(SP), DIMENSION(:), INTENT(IN) :: arr
INTEGER(I4B), DIMENSION(1) :: imin
INTEGER(I4B) :: iminloc
imin=minloc(arr(:))
```

```
iminloc=imin(1)
END FUNCTION iminloc
```

Routines for argument checking and error handling:

```
SUBROUTINE assert1(n1,string)
```

Report and die if any logical is false (used for arg range checking).

```
CHARACTER(LEN=*), INTENT(IN) :: string
LOGICAL, INTENT(IN) :: n1
if (.not. n1) then
    write (*,*) 'nrerror: an assertion failed with this tag:', &
        string
    STOP 'program terminated by assert1'
end if
END SUBROUTINE assert1

SUBROUTINE assert2(n1,n2,string)
CHARACTER(LEN=*), INTENT(IN) :: string
LOGICAL, INTENT(IN) :: n1,n2
if (.not. (n1 .and. n2)) then
    write (*,*) 'nrerror: an assertion failed with this tag:', &
        string
    STOP 'program terminated by assert2'
end if
END SUBROUTINE assert2

SUBROUTINE assert3(n1,n2,n3,string)
CHARACTER(LEN=*), INTENT(IN) :: string
LOGICAL, INTENT(IN) :: n1,n2,n3
if (.not. (n1 .and. n2 .and. n3)) then
    write (*,*) 'nrerror: an assertion failed with this tag:', &
        string
    STOP 'program terminated by assert3'
end if
END SUBROUTINE assert3

SUBROUTINE assert4(n1,n2,n3,n4,string)
CHARACTER(LEN=*), INTENT(IN) :: string
LOGICAL, INTENT(IN) :: n1,n2,n3,n4
if (.not. (n1 .and. n2 .and. n3 .and. n4)) then
    write (*,*) 'nrerror: an assertion failed with this tag:', &
        string
    STOP 'program terminated by assert4'
end if
END SUBROUTINE assert4

SUBROUTINE assert_v(n,string)
CHARACTER(LEN=*), INTENT(IN) :: string
LOGICAL, DIMENSION(:), INTENT(IN) :: n
if (.not. all(n)) then
    write (*,*) 'nrerror: an assertion failed with this tag:', &
        string
    STOP 'program terminated by assert_v'
end if
END SUBROUTINE assert_v

FUNCTION assert_eq2(n1,n2,string)
```

Report and die if integers not all equal (used for size checking).

```
CHARACTER(LEN=*), INTENT(IN) :: string
INTEGER, INTENT(IN) :: n1,n2
INTEGER :: assert_eq2
if (n1 == n2) then
    assert_eq2=n1
else
    write (*,*) 'nrerror: an assert_eq failed with this tag:', &
        string
    STOP 'program terminated by assert_eq2'
end if
```

```
END FUNCTION assert_eq2

FUNCTION assert_eq3(n1,n2,n3,string)
CHARACTER(LEN=*), INTENT(IN) :: string
INTEGER, INTENT(IN) :: n1,n2,n3
INTEGER :: assert_eq3
if (n1 == n2 .and. n2 == n3) then
    assert_eq3=n1
else
    write (*,*) 'nrerror: an assert_eq failed with this tag:', &
        string
    STOP 'program terminated by assert_eq3'
end if
END FUNCTION assert_eq3

FUNCTION assert_eq4(n1,n2,n3,n4,string)
CHARACTER(LEN=*), INTENT(IN) :: string
INTEGER, INTENT(IN) :: n1,n2,n3,n4
INTEGER :: assert_eq4
if (n1 == n2 .and. n2 == n3 .and. n3 == n4) then
    assert_eq4=n1
else
    write (*,*) 'nrerror: an assert_eq failed with this tag:', &
        string
    STOP 'program terminated by assert_eq4'
end if
END FUNCTION assert_eq4

FUNCTION assert_eqn(nn,string)
CHARACTER(LEN=*), INTENT(IN) :: string
INTEGER, DIMENSION(:), INTENT(IN) :: nn
INTEGER :: assert_eqn
if (all(nn(2:) == nn(1))) then
    assert_eqn=nn(1)
else
    write (*,*) 'nrerror: an assert_eq failed with this tag:', &
        string
    STOP 'program terminated by assert_eqn'
end if
END FUNCTION assert_eqn

SUBROUTINE nrerror(string)
```

Report a message, then die.

```
CHARACTER(LEN=*), INTENT(IN) :: string
write (*,*) 'nrerror: ',string
STOP 'program terminated by nrerror'
END SUBROUTINE nrerror
```

Routines relating to polynomials and recurrences:

```
FUNCTION arth_r(first,increment,n)
```

Array function returning an arithmetic progression.

```
REAL(SP), INTENT(IN) :: first,increment
INTEGER(I4B), INTENT(IN) :: n
REAL(SP), DIMENSION(n) :: arth_r
INTEGER(I4B) :: k,k2
REAL(SP) :: temp
if (n > 0) arth_r(1)=first
if (n <= NPAR_ARTH) then
    do k=2,n
        arth_r(k)=arth_r(k-1)+increment
    end do
else
    do k=2,NPAR2_ARTH
        arth_r(k)=arth_r(k-1)+increment
    end do
    temp=increment*NPAR2_ARTH
    k=NPAR2_ARTH
```

```
    do
        if (k >= n) exit
        k2=k+k
        arth_r(k+1:min(k2,n))=temp+arth_r(1:min(k,n-k))
        temp=temp+temp
        k=k2
    end do
end if
END FUNCTION arth_r

FUNCTION arth_d(first,increment,n)
REAL(DP), INTENT(IN) :: first,increment
INTEGER(I4B), INTENT(IN) :: n
REAL(DP), DIMENSION(n) :: arth_d
INTEGER(I4B) :: k,k2
REAL(DP) :: temp
if (n > 0) arth_d(1)=first
if (n <= NPAR_ARTH) then
    do k=2,n
        arth_d(k)=arth_d(k-1)+increment
    end do
else
    do k=2,NPAR2_ARTH
        arth_d(k)=arth_d(k-1)+increment
    end do
    temp=increment*NPAR2_ARTH
    k=NPAR2_ARTH
    do
        if (k >= n) exit
        k2=k+k
        arth_d(k+1:min(k2,n))=temp+arth_d(1:min(k,n-k))
        temp=temp+temp
        k=k2
    end do
end if
END FUNCTION arth_d

FUNCTION arth_i(first,increment,n)
INTEGER(I4B), INTENT(IN) :: first,increment,n
INTEGER(I4B), DIMENSION(n) :: arth_i
INTEGER(I4B) :: k,k2,temp
if (n > 0) arth_i(1)=first
if (n <= NPAR_ARTH) then
    do k=2,n
        arth_i(k)=arth_i(k-1)+increment
    end do
else
    do k=2,NPAR2_ARTH
        arth_i(k)=arth_i(k-1)+increment
    end do
    temp=increment*NPAR2_ARTH
    k=NPAR2_ARTH
    do
        if (k >= n) exit
        k2=k+k
        arth_i(k+1:min(k2,n))=temp+arth_i(1:min(k,n-k))
        temp=temp+temp
        k=k2
    end do
end if
END FUNCTION arth_i

FUNCTION geop_r(first,factor,n)
```

Array function returning a geometric progression.

```
REAL(SP), INTENT(IN) :: first,factor
```

```
INTEGER(I4B), INTENT(IN) :: n
REAL(SP), DIMENSION(n) :: geop_r
INTEGER(I4B) :: k,k2
REAL(SP) :: temp
if (n > 0) geop_r(1)=first
if (n <= NPAR_GEOP) then
    do k=2,n
        geop_r(k)=geop_r(k-1)*factor
    end do
else
    do k=2,NPAR2_GEOP
        geop_r(k)=geop_r(k-1)*factor
    end do
    temp=factor**NPAR2_GEOP
    k=NPAR2_GEOP
    do
        if (k >= n) exit
        k2=k+k
        geop_r(k+1:min(k2,n))=temp*geop_r(1:min(k,n-k))
        temp=temp*temp
        k=k2
    end do
end if
END FUNCTION geop_r

FUNCTION geop_d(first,factor,n)
REAL(DP), INTENT(IN) :: first,factor
INTEGER(I4B), INTENT(IN) :: n
REAL(DP), DIMENSION(n) :: geop_d
INTEGER(I4B) :: k,k2
REAL(DP) :: temp
if (n > 0) geop_d(1)=first
if (n <= NPAR_GEOP) then
    do k=2,n
        geop_d(k)=geop_d(k-1)*factor
    end do
else
    do k=2,NPAR2_GEOP
        geop_d(k)=geop_d(k-1)*factor
    end do
    temp=factor**NPAR2_GEOP
    k=NPAR2_GEOP
    do
        if (k >= n) exit
        k2=k+k
        geop_d(k+1:min(k2,n))=temp*geop_d(1:min(k,n-k))
        temp=temp*temp
        k=k2
    end do
end if
END FUNCTION geop_d

FUNCTION geop_i(first,factor,n)
INTEGER(I4B), INTENT(IN) :: first,factor,n
INTEGER(I4B), DIMENSION(n) :: geop_i
INTEGER(I4B) :: k,k2,temp
if (n > 0) geop_i(1)=first
if (n <= NPAR_GEOP) then
    do k=2,n
        geop_i(k)=geop_i(k-1)*factor
    end do
else
    do k=2,NPAR2_GEOP
        geop_i(k)=geop_i(k-1)*factor
    end do
```

```
    temp=factor**NPAR2_GEOP
    k=NPAR2_GEOP
    do
        if (k >= n) exit
        k2=k+k
        geop_i(k+1:min(k2,n))=temp*geop_i(1:min(k,n-k))
        temp=temp*temp
        k=k2
    end do
end if
END FUNCTION geop_i

FUNCTION geop_c(first,factor,n)
COMPLEX(SP), INTENT(IN) :: first,factor
INTEGER(I4B), INTENT(IN) :: n
COMPLEX(SP), DIMENSION(n) :: geop_c
INTEGER(I4B) :: k,k2
COMPLEX(SP) :: temp
if (n > 0) geop_c(1)=first
if (n <= NPAR_GEOP) then
    do k=2,n
        geop_c(k)=geop_c(k-1)*factor
    end do
else
    do k=2,NPAR2_GEOP
        geop_c(k)=geop_c(k-1)*factor
    end do
    temp=factor**NPAR2_GEOP
    k=NPAR2_GEOP
    do
        if (k >= n) exit
        k2=k+k
        geop_c(k+1:min(k2,n))=temp*geop_c(1:min(k,n-k))
        temp=temp*temp
        k=k2
    end do
end if
END FUNCTION geop_c

FUNCTION geop_dv(first,factor,n)
REAL(DP), DIMENSION(:), INTENT(IN) :: first,factor
INTEGER(I4B), INTENT(IN) :: n
REAL(DP), DIMENSION(size(first),n) :: geop_dv
INTEGER(I4B) :: k,k2
REAL(DP), DIMENSION(size(first)) :: temp
if (n > 0) geop_dv(:,1)=first(:)
if (n <= NPAR_GEOP) then
    do k=2,n
        geop_dv(:,k)=geop_dv(:,k-1)*factor(:)
    end do
else
    do k=2,NPAR2_GEOP
        geop_dv(:,k)=geop_dv(:,k-1)*factor(:)
    end do
    temp=factor**NPAR2_GEOP
    k=NPAR2_GEOP
    do
        if (k >= n) exit
        k2=k+k
        geop_dv(:,k+1:min(k2,n))=geop_dv(:,1:min(k,n-k))*&
            spread(temp,2,size(geop_dv(:,1:min(k,n-k)),2))
        temp=temp*temp
        k=k2
    end do
end if
```

```
END FUNCTION geop_dv

RECURSIVE FUNCTION cumsum_r(arr,seed) RESULT(ans)
   Cumulative sum on an array, with optional additive seed.
REAL(SP), DIMENSION(:), INTENT(IN) :: arr
REAL(SP), OPTIONAL, INTENT(IN) :: seed
REAL(SP), DIMENSION(size(arr)) :: ans
INTEGER(I4B) :: n,j
REAL(SP) :: sd
n=size(arr)
if (n == 0_i4b) RETURN
sd=0.0_sp
if (present(seed)) sd=seed
ans(1)=arr(1)+sd
if (n < NPAR_CUMSUM) then
    do j=2,n
        ans(j)=ans(j-1)+arr(j)
    end do
else
    ans(2:n:2)=cumsum_r(arr(2:n:2)+arr(1:n-1:2),sd)
    ans(3:n:2)=ans(2:n-1:2)+arr(3:n:2)
end if
END FUNCTION cumsum_r

RECURSIVE FUNCTION cumsum_i(arr,seed) RESULT(ans)
INTEGER(I4B), DIMENSION(:), INTENT(IN) :: arr
INTEGER(I4B), OPTIONAL, INTENT(IN) :: seed
INTEGER(I4B), DIMENSION(size(arr)) :: ans
INTEGER(I4B) :: n,j,sd
n=size(arr)
if (n == 0_i4b) RETURN
sd=0_i4b
if (present(seed)) sd=seed
ans(1)=arr(1)+sd
if (n < NPAR_CUMSUM) then
    do j=2,n
        ans(j)=ans(j-1)+arr(j)
    end do
else
    ans(2:n:2)=cumsum_i(arr(2:n:2)+arr(1:n-1:2),sd)
    ans(3:n:2)=ans(2:n-1:2)+arr(3:n:2)
end if
END FUNCTION cumsum_i

RECURSIVE FUNCTION cumprod(arr,seed) RESULT(ans)
   Cumulative product on an array, with optional multiplicative seed.
REAL(SP), DIMENSION(:), INTENT(IN) :: arr
REAL(SP), OPTIONAL, INTENT(IN) :: seed
REAL(SP), DIMENSION(size(arr)) :: ans
INTEGER(I4B) :: n,j
REAL(SP) :: sd
n=size(arr)
if (n == 0_i4b) RETURN
sd=1.0_sp
if (present(seed)) sd=seed
ans(1)=arr(1)*sd
if (n < NPAR_CUMPROD) then
    do j=2,n
        ans(j)=ans(j-1)*arr(j)
    end do
else
    ans(2:n:2)=cumprod(arr(2:n:2)*arr(1:n-1:2),sd)
    ans(3:n:2)=ans(2:n-1:2)*arr(3:n:2)
end if
END FUNCTION cumprod
```

```
FUNCTION poly_rr(x,coeffs)
   Polynomial evaluation.
REAL(SP), INTENT(IN) :: x
REAL(SP), DIMENSION(:), INTENT(IN) :: coeffs
REAL(SP) :: poly_rr
REAL(SP) :: pow
REAL(SP), DIMENSION(:), ALLOCATABLE :: vec
INTEGER(I4B) :: i,n,nn
n=size(coeffs)
if (n <= 0) then
    poly_rr=0.0_sp
else if (n < NPAR_POLY) then
    poly_rr=coeffs(n)
    do i=n-1,1,-1
        poly_rr=x*poly_rr+coeffs(i)
    end do
else
    allocate(vec(n+1))
    pow=x
    vec(1:n)=coeffs
    do
        vec(n+1)=0.0_sp
        nn=ishft(n+1,-1)
        vec(1:nn)=vec(1:n:2)+pow*vec(2:n+1:2)
        if (nn == 1) exit
        pow=pow*pow
        n=nn
    end do
    poly_rr=vec(1)
    deallocate(vec)
end if
END FUNCTION poly_rr

FUNCTION poly_dd(x,coeffs)
REAL(DP), INTENT(IN) :: x
REAL(DP), DIMENSION(:), INTENT(IN) :: coeffs
REAL(DP) :: poly_dd
REAL(DP) :: pow
REAL(DP), DIMENSION(:), ALLOCATABLE :: vec
INTEGER(I4B) :: i,n,nn
n=size(coeffs)
if (n <= 0) then
    poly_dd=0.0_dp
else if (n < NPAR_POLY) then
    poly_dd=coeffs(n)
    do i=n-1,1,-1
        poly_dd=x*poly_dd+coeffs(i)
    end do
else
    allocate(vec(n+1))
    pow=x
    vec(1:n)=coeffs
    do
        vec(n+1)=0.0_dp
        nn=ishft(n+1,-1)
        vec(1:nn)=vec(1:n:2)+pow*vec(2:n+1:2)
        if (nn == 1) exit
        pow=pow*pow
        n=nn
    end do
    poly_dd=vec(1)
    deallocate(vec)
end if
END FUNCTION poly_dd
```

```
FUNCTION poly_rc(x,coeffs)
COMPLEX(SPC), INTENT(IN) :: x
REAL(SP), DIMENSION(:), INTENT(IN) :: coeffs
COMPLEX(SPC) :: poly_rc
COMPLEX(SPC) :: pow
COMPLEX(SPC), DIMENSION(:), ALLOCATABLE :: vec
INTEGER(I4B) :: i,n,nn
n=size(coeffs)
if (n <= 0) then
    poly_rc=0.0_sp
else if (n < NPAR_POLY) then
    poly_rc=coeffs(n)
    do i=n-1,1,-1
        poly_rc=x*poly_rc+coeffs(i)
    end do
else
    allocate(vec(n+1))
    pow=x
    vec(1:n)=coeffs
    do
        vec(n+1)=0.0_sp
        nn=ishft(n+1,-1)
        vec(1:nn)=vec(1:n:2)+pow*vec(2:n+1:2)
        if (nn == 1) exit
        pow=pow*pow
        n=nn
    end do
    poly_rc=vec(1)
    deallocate(vec)
end if
END FUNCTION poly_rc

FUNCTION poly_cc(x,coeffs)
COMPLEX(SPC), INTENT(IN) :: x
COMPLEX(SPC), DIMENSION(:), INTENT(IN) :: coeffs
COMPLEX(SPC) :: poly_cc
COMPLEX(SPC) :: pow
COMPLEX(SPC), DIMENSION(:), ALLOCATABLE :: vec
INTEGER(I4B) :: i,n,nn
n=size(coeffs)
if (n <= 0) then
    poly_cc=0.0_sp
else if (n < NPAR_POLY) then
    poly_cc=coeffs(n)
    do i=n-1,1,-1
        poly_cc=x*poly_cc+coeffs(i)
    end do
else
    allocate(vec(n+1))
    pow=x
    vec(1:n)=coeffs
    do
        vec(n+1)=0.0_sp
        nn=ishft(n+1,-1)
        vec(1:nn)=vec(1:n:2)+pow*vec(2:n+1:2)
        if (nn == 1) exit
        pow=pow*pow
        n=nn
    end do
    poly_cc=vec(1)
    deallocate(vec)
end if
END FUNCTION poly_cc

FUNCTION poly_rrv(x,coeffs)
```

```
REAL(SP), DIMENSION(:), INTENT(IN) :: coeffs,x
REAL(SP), DIMENSION(size(x)) :: poly_rrv
INTEGER(I4B) :: i,n,m
m=size(coeffs)
n=size(x)
if (m <= 0) then
    poly_rrv=0.0_sp
else if (m < n .or. m < NPAR_POLY) then
    poly_rrv=coeffs(m)
    do i=m-1,1,-1
        poly_rrv=x*poly_rrv+coeffs(i)
    end do
else
    do i=1,n
        poly_rrv(i)=poly_rr(x(i),coeffs)
    end do
end if
END FUNCTION poly_rrv

FUNCTION poly_ddv(x,coeffs)
REAL(DP), DIMENSION(:), INTENT(IN) :: coeffs,x
REAL(DP), DIMENSION(size(x)) :: poly_ddv
INTEGER(I4B) :: i,n,m
m=size(coeffs)
n=size(x)
if (m <= 0) then
    poly_ddv=0.0_dp
else if (m < n .or. m < NPAR_POLY) then
    poly_ddv=coeffs(m)
    do i=m-1,1,-1
        poly_ddv=x*poly_ddv+coeffs(i)
    end do
else
    do i=1,n
        poly_ddv(i)=poly_dd(x(i),coeffs)
    end do
end if
END FUNCTION poly_ddv

FUNCTION poly_msk_rrv(x,coeffs,mask)
REAL(SP), DIMENSION(:), INTENT(IN) :: coeffs,x
LOGICAL(LGT), DIMENSION(:), INTENT(IN) :: mask
REAL(SP), DIMENSION(size(x)) :: poly_msk_rrv
poly_msk_rrv=unpack(poly_rrv(pack(x,mask),coeffs),mask,0.0_sp)
END FUNCTION poly_msk_rrv

FUNCTION poly_msk_ddv(x,coeffs,mask)
REAL(DP), DIMENSION(:), INTENT(IN) :: coeffs,x
LOGICAL(LGT), DIMENSION(:), INTENT(IN) :: mask
REAL(DP), DIMENSION(size(x)) :: poly_msk_ddv
poly_msk_ddv=unpack(poly_ddv(pack(x,mask),coeffs),mask,0.0_dp)
END FUNCTION poly_msk_ddv

RECURSIVE FUNCTION poly_term_rr(a,b) RESULT(u)
```

Tabulate cumulants of a polynomial.

```
REAL(SP), DIMENSION(:), INTENT(IN) :: a
REAL(SP), INTENT(IN) :: b
REAL(SP), DIMENSION(size(a)) :: u
INTEGER(I4B) :: n,j
n=size(a)
if (n <= 0) RETURN
u(1)=a(1)
if (n < NPAR_POLYTERM) then
    do j=2,n
        u(j)=a(j)+b*u(j-1)
    end do
```

```
else
    u(2:n:2)=poly_term_rr(a(2:n:2)+a(1:n-1:2)*b,b*b)
    u(3:n:2)=a(3:n:2)+b*u(2:n-1:2)
end if
END FUNCTION poly_term_rr

RECURSIVE FUNCTION poly_term_cc(a,b) RESULT(u)
COMPLEX(SPC), DIMENSION(:), INTENT(IN) :: a
COMPLEX(SPC), INTENT(IN) :: b
COMPLEX(SPC), DIMENSION(size(a)) :: u
INTEGER(I4B) :: n,j
n=size(a)
if (n <= 0) RETURN
u(1)=a(1)
if (n < NPAR_POLYTERM) then
    do j=2,n
        u(j)=a(j)+b*u(j-1)
    end do
else
    u(2:n:2)=poly_term_cc(a(2:n:2)+a(1:n-1:2)*b,b*b)
    u(3:n:2)=a(3:n:2)+b*u(2:n-1:2)
end if
END FUNCTION poly_term_cc

FUNCTION zroots_unity(n,nn)
    Complex function returning nn powers of the nth root of unity.
INTEGER(I4B), INTENT(IN) :: n,nn
COMPLEX(SPC), DIMENSION(nn) :: zroots_unity
INTEGER(I4B) :: k
REAL(SP) :: theta
zroots_unity(1)=1.0
theta=TWOPI/n
k=1
do
    if (k >= nn) exit
    zroots_unity(k+1)=cmplx(cos(k*theta),sin(k*theta),SPC)
    zroots_unity(k+2:min(2*k,nn))=zroots_unity(k+1)*&
        zroots_unity(2:min(k,nn-k))
    k=2*k
end do
END FUNCTION zroots_unity
```

Routines for "outer" operations on vectors. The order convention is: result(i,j) = first_operand(i) (op) second_operand(j).

```
FUNCTION outerprod_r(a,b)
REAL(SP), DIMENSION(:), INTENT(IN) :: a,b
REAL(SP), DIMENSION(size(a),size(b)) :: outerprod_r
outerprod_r = spread(a,dim=2,ncopies=size(b)) * &
    spread(b,dim=1,ncopies=size(a))
END FUNCTION outerprod_r

FUNCTION outerprod_d(a,b)
REAL(DP), DIMENSION(:), INTENT(IN) :: a,b
REAL(DP), DIMENSION(size(a),size(b)) :: outerprod_d
outerprod_d = spread(a,dim=2,ncopies=size(b)) * &
    spread(b,dim=1,ncopies=size(a))
END FUNCTION outerprod_d

FUNCTION outerdiv(a,b)
REAL(SP), DIMENSION(:), INTENT(IN) :: a,b
REAL(SP), DIMENSION(size(a),size(b)) :: outerdiv
outerdiv = spread(a,dim=2,ncopies=size(b)) / &
    spread(b,dim=1,ncopies=size(a))
END FUNCTION outerdiv

FUNCTION outersum(a,b)
REAL(SP), DIMENSION(:), INTENT(IN) :: a,b
```

```
REAL(SP), DIMENSION(size(a),size(b)) :: outersum
outersum = spread(a,dim=2,ncopies=size(b)) + &
    spread(b,dim=1,ncopies=size(a))
END FUNCTION outersum

FUNCTION outerdiff_r(a,b)
REAL(SP), DIMENSION(:), INTENT(IN) :: a,b
REAL(SP), DIMENSION(size(a),size(b)) :: outerdiff_r
outerdiff_r = spread(a,dim=2,ncopies=size(b)) - &
    spread(b,dim=1,ncopies=size(a))
END FUNCTION outerdiff_r

FUNCTION outerdiff_d(a,b)
REAL(DP), DIMENSION(:), INTENT(IN) :: a,b
REAL(DP), DIMENSION(size(a),size(b)) :: outerdiff_d
outerdiff_d = spread(a,dim=2,ncopies=size(b)) - &
    spread(b,dim=1,ncopies=size(a))
END FUNCTION outerdiff_d

FUNCTION outerdiff_i(a,b)
INTEGER(I4B), DIMENSION(:), INTENT(IN) :: a,b
INTEGER(I4B), DIMENSION(size(a),size(b)) :: outerdiff_i
outerdiff_i = spread(a,dim=2,ncopies=size(b)) - &
    spread(b,dim=1,ncopies=size(a))
END FUNCTION outerdiff_i

FUNCTION outerand(a,b)
LOGICAL(LGT), DIMENSION(:), INTENT(IN) :: a,b
LOGICAL(LGT), DIMENSION(size(a),size(b)) :: outerand
outerand = spread(a,dim=2,ncopies=size(b)) .and. &
    spread(b,dim=1,ncopies=size(a))
END FUNCTION outerand
```

Routines for scatter-with-combine.

```
SUBROUTINE scatter_add_r(dest,source,dest_index)
REAL(SP), DIMENSION(:), INTENT(OUT) :: dest
REAL(SP), DIMENSION(:), INTENT(IN) :: source
INTEGER(I4B), DIMENSION(:), INTENT(IN) :: dest_index
INTEGER(I4B) :: m,n,j,i
n=assert_eq2(size(source),size(dest_index),'scatter_add_r')
m=size(dest)
do j=1,n
    i=dest_index(j)
    if (i > 0 .and. i <= m) dest(i)=dest(i)+source(j)
end do
END SUBROUTINE scatter_add_r
SUBROUTINE scatter_add_d(dest,source,dest_index)
REAL(DP), DIMENSION(:), INTENT(OUT) :: dest
REAL(DP), DIMENSION(:), INTENT(IN) :: source
INTEGER(I4B), DIMENSION(:), INTENT(IN) :: dest_index
INTEGER(I4B) :: m,n,j,i
n=assert_eq2(size(source),size(dest_index),'scatter_add_d')
m=size(dest)
do j=1,n
    i=dest_index(j)
    if (i > 0 .and. i <= m) dest(i)=dest(i)+source(j)
end do
END SUBROUTINE scatter_add_d
SUBROUTINE scatter_max_r(dest,source,dest_index)
REAL(SP), DIMENSION(:), INTENT(OUT) :: dest
REAL(SP), DIMENSION(:), INTENT(IN) :: source
INTEGER(I4B), DIMENSION(:), INTENT(IN) :: dest_index
INTEGER(I4B) :: m,n,j,i
n=assert_eq2(size(source),size(dest_index),'scatter_max_r')
m=size(dest)
do j=1,n
    i=dest_index(j)
```

```
    if (i > 0 .and. i <= m) dest(i)=max(dest(i),source(j))
end do
END SUBROUTINE scatter_max_r
SUBROUTINE scatter_max_d(dest,source,dest_index)
REAL(DP), DIMENSION(:), INTENT(OUT) :: dest
REAL(DP), DIMENSION(:), INTENT(IN) :: source
INTEGER(I4B), DIMENSION(:), INTENT(IN) :: dest_index
INTEGER(I4B) :: m,n,j,i
n=assert_eq2(size(source),size(dest_index),'scatter_max_d')
m=size(dest)
do j=1,n
    i=dest_index(j)
    if (i > 0 .and. i <= m) dest(i)=max(dest(i),source(j))
end do
END SUBROUTINE scatter_max_d
```

Routines for skew operations on matrices:

```
SUBROUTINE diagadd_rv(mat,diag)
```

Adds vector or scalar diag to the diagonal of matrix mat.

```
REAL(SP), DIMENSION(:,:), INTENT(INOUT) :: mat
REAL(SP), DIMENSION(:), INTENT(IN) :: diag
INTEGER(I4B) :: j,n
n = assert_eq2(size(diag),min(size(mat,1),size(mat,2)),'diagadd_rv')
do j=1,n
    mat(j,j)=mat(j,j)+diag(j)
end do
END SUBROUTINE diagadd_rv

SUBROUTINE diagadd_r(mat,diag)
REAL(SP), DIMENSION(:,:), INTENT(INOUT) :: mat
REAL(SP), INTENT(IN) :: diag
INTEGER(I4B) :: j,n
n = min(size(mat,1),size(mat,2))
do j=1,n
    mat(j,j)=mat(j,j)+diag
end do
END SUBROUTINE diagadd_r

SUBROUTINE diagmult_rv(mat,diag)
```

Multiplies vector or scalar diag into the diagonal of matrix mat.

```
REAL(SP), DIMENSION(:,:), INTENT(INOUT) :: mat
REAL(SP), DIMENSION(:), INTENT(IN) :: diag
INTEGER(I4B) :: j,n
n = assert_eq2(size(diag),min(size(mat,1),size(mat,2)),'diagmult_rv')
do j=1,n
    mat(j,j)=mat(j,j)*diag(j)
end do
END SUBROUTINE diagmult_rv

SUBROUTINE diagmult_r(mat,diag)
REAL(SP), DIMENSION(:,:), INTENT(INOUT) :: mat
REAL(SP), INTENT(IN) :: diag
INTEGER(I4B) :: j,n
n = min(size(mat,1),size(mat,2))
do j=1,n
    mat(j,j)=mat(j,j)*diag
end do
END SUBROUTINE diagmult_r

FUNCTION get_diag_rv(mat)
```

Return as a vector the diagonal of matrix mat.

```
REAL(SP), DIMENSION(:,:), INTENT(IN) :: mat
REAL(SP), DIMENSION(size(mat,1)) :: get_diag_rv
INTEGER(I4B) :: j
j=assert_eq2(size(mat,1),size(mat,2),'get_diag_rv')
do j=1,size(mat,1)
    get_diag_rv(j)=mat(j,j)
```

```
end do
END FUNCTION get_diag_rv

FUNCTION get_diag_dv(mat)
REAL(DP), DIMENSION(:,:), INTENT(IN) :: mat
REAL(DP), DIMENSION(size(mat,1)) :: get_diag_dv
INTEGER(I4B) :: j
j=assert_eq2(size(mat,1),size(mat,2),'get_diag_dv')
do j=1,size(mat,1)
    get_diag_dv(j)=mat(j,j)
end do
END FUNCTION get_diag_dv

SUBROUTINE put_diag_rv(diagv,mat)
    Set the diagonal of matrix mat to the values of a vector or scalar.
REAL(SP), DIMENSION(:), INTENT(IN) :: diagv
REAL(SP), DIMENSION(:,:), INTENT(INOUT) :: mat
INTEGER(I4B) :: j,n
n=assert_eq2(size(diagv),min(size(mat,1),size(mat,2)),'put_diag_rv')
do j=1,n
    mat(j,j)=diagv(j)
end do
END SUBROUTINE put_diag_rv

SUBROUTINE put_diag_r(scal,mat)
REAL(SP), INTENT(IN) :: scal
REAL(SP), DIMENSION(:,:), INTENT(INOUT) :: mat
INTEGER(I4B) :: j,n
n = min(size(mat,1),size(mat,2))
do j=1,n
    mat(j,j)=scal
end do
END SUBROUTINE put_diag_r

SUBROUTINE unit_matrix(mat)
    Set the matrix mat to be a unit matrix (if it is square).
REAL(SP), DIMENSION(:,:), INTENT(OUT) :: mat
INTEGER(I4B) :: i,n
n=min(size(mat,1),size(mat,2))
mat(:,:)=0.0_sp
do i=1,n
    mat(i,i)=1.0_sp
end do
END SUBROUTINE unit_matrix

FUNCTION upper_triangle(j,k,extra)
    Return an upper triangular logical mask.
INTEGER(I4B), INTENT(IN) :: j,k
INTEGER(I4B), OPTIONAL, INTENT(IN) :: extra
LOGICAL(LGT), DIMENSION(j,k) :: upper_triangle
INTEGER(I4B) :: n
n=0
if (present(extra)) n=extra
upper_triangle=(outerdiff(arth_i(1,1,j),arth_i(1,1,k)) < n)
END FUNCTION upper_triangle

FUNCTION lower_triangle(j,k,extra)
    Return a lower triangular logical mask.
INTEGER(I4B), INTENT(IN) :: j,k
INTEGER(I4B), OPTIONAL, INTENT(IN) :: extra
LOGICAL(LGT), DIMENSION(j,k) :: lower_triangle
INTEGER(I4B) :: n
n=0
if (present(extra)) n=extra
lower_triangle=(outerdiff(arth_i(1,1,j),arth_i(1,1,k)) > -n)
END FUNCTION lower_triangle
```

Other routines:

```
FUNCTION vabs(v)
    Return the length (ordinary L2 norm) of a vector.
REAL(SP), DIMENSION(:), INTENT(IN) :: v
REAL(SP) :: vabs
vabs=sqrt(dot_product(v,v))
END FUNCTION vabs

END MODULE nrutil
```

C2. Alphabetical Listing of Explicit Interfaces

The file supplied as `nr.f90` contains explicit interfaces for all the Numerical Recipes routines (except those already in the module `nrutil`). The interfaces are in alphabetical order, by the generic interface name, if one exists, or by the specific routine name if there is no generic name.

The file `nr.f90` is normally invoked via a USE statement within a main program or subroutine that references a Numerical Recipes routine. See §21.1 for an example.

```
MODULE nr
INTERFACE
    SUBROUTINE airy(x,ai,bi,aip,bip)
    USE nrtype
    REAL(SP), INTENT(IN) :: x
    REAL(SP), INTENT(OUT) :: ai,bi,aip,bip
    END SUBROUTINE airy
END INTERFACE
INTERFACE
    SUBROUTINE amebsa(p,y,pb,yb,ftol,func,iter,temptr)
    USE nrtype
    INTEGER(I4B), INTENT(INOUT) :: iter
    REAL(SP), INTENT(INOUT) :: yb
    REAL(SP), INTENT(IN) :: ftol,temptr
    REAL(SP), DIMENSION(:), INTENT(INOUT) :: y,pb
    REAL(SP), DIMENSION(:,:), INTENT(INOUT) :: p
    INTERFACE
        FUNCTION func(x)
        USE nrtype
        REAL(SP), DIMENSION(:), INTENT(IN) :: x
        REAL(SP) :: func
        END FUNCTION func
    END INTERFACE
    END SUBROUTINE amebsa
END INTERFACE
INTERFACE
    SUBROUTINE amoeba(p,y,ftol,func,iter)
    USE nrtype
    INTEGER(I4B), INTENT(OUT) :: iter
    REAL(SP), INTENT(IN) :: ftol
    REAL(SP), DIMENSION(:), INTENT(INOUT) :: y
    REAL(SP), DIMENSION(:,:), INTENT(INOUT) :: p
    INTERFACE
        FUNCTION func(x)
        USE nrtype
        REAL(SP), DIMENSION(:), INTENT(IN) :: x
        REAL(SP) :: func
        END FUNCTION func
    END INTERFACE
```

```
        END SUBROUTINE amoeba
    END INTERFACE
    INTERFACE
        SUBROUTINE anneal(x,y,iorder)
        USE nrtype
        INTEGER(I4B), DIMENSION(:), INTENT(INOUT) :: iorder
        REAL(SP), DIMENSION(:), INTENT(IN) :: x,y
        END SUBROUTINE anneal
    END INTERFACE
    INTERFACE
        SUBROUTINE asolve(b,x,itrnsp)
        USE nrtype
        REAL(DP), DIMENSION(:), INTENT(IN) :: b
        REAL(DP), DIMENSION(:), INTENT(OUT) :: x
        INTEGER(I4B), INTENT(IN) :: itrnsp
        END SUBROUTINE asolve
    END INTERFACE
    INTERFACE
        SUBROUTINE atimes(x,r,itrnsp)
        USE nrtype
        REAL(DP), DIMENSION(:), INTENT(IN) :: x
        REAL(DP), DIMENSION(:), INTENT(OUT) :: r
        INTEGER(I4B), INTENT(IN) :: itrnsp
        END SUBROUTINE atimes
    END INTERFACE
    INTERFACE
        SUBROUTINE avevar(data,ave,var)
        USE nrtype
        REAL(SP), DIMENSION(:), INTENT(IN) :: data
        REAL(SP), INTENT(OUT) :: ave,var
        END SUBROUTINE avevar
    END INTERFACE
    INTERFACE
        SUBROUTINE balanc(a)
        USE nrtype
        REAL(SP), DIMENSION(:,:), INTENT(INOUT) :: a
        END SUBROUTINE balanc
    END INTERFACE
    INTERFACE
        SUBROUTINE banbks(a,m1,m2,al,indx,b)
        USE nrtype
        INTEGER(I4B), INTENT(IN) :: m1,m2
        INTEGER(I4B), DIMENSION(:), INTENT(IN) :: indx
        REAL(SP), DIMENSION(:,:), INTENT(IN) :: a,al
        REAL(SP), DIMENSION(:), INTENT(INOUT) :: b
        END SUBROUTINE banbks
    END INTERFACE
    INTERFACE
        SUBROUTINE bandec(a,m1,m2,al,indx,d)
        USE nrtype
        INTEGER(I4B), INTENT(IN) :: m1,m2
        INTEGER(I4B), DIMENSION(:), INTENT(OUT) :: indx
        REAL(SP), INTENT(OUT) :: d
        REAL(SP), DIMENSION(:,:), INTENT(INOUT) :: a
        REAL(SP), DIMENSION(:,:), INTENT(OUT) :: al
        END SUBROUTINE bandec
    END INTERFACE
    INTERFACE
        SUBROUTINE banmul(a,m1,m2,x,b)
        USE nrtype
        INTEGER(I4B), INTENT(IN) :: m1,m2
        REAL(SP), DIMENSION(:), INTENT(IN) :: x
        REAL(SP), DIMENSION(:), INTENT(OUT) :: b
        REAL(SP), DIMENSION(:,:), INTENT(IN) :: a
```

```
        END SUBROUTINE banmul
    END INTERFACE
    INTERFACE
        SUBROUTINE bcucof(y,y1,y2,y12,d1,d2,c)
        USE nrtype
        REAL(SP), INTENT(IN) :: d1,d2
        REAL(SP), DIMENSION(4), INTENT(IN) :: y,y1,y2,y12
        REAL(SP), DIMENSION(4,4), INTENT(OUT) :: c
        END SUBROUTINE bcucof
    END INTERFACE
    INTERFACE
        SUBROUTINE bcuint(y,y1,y2,y12,x1l,x1u,x2l,x2u,x1,x2,ansy,&
            ansy1,ansy2)
        USE nrtype
        REAL(SP), DIMENSION(4), INTENT(IN) :: y,y1,y2,y12
        REAL(SP), INTENT(IN) :: x1l,x1u,x2l,x2u,x1,x2
        REAL(SP), INTENT(OUT) :: ansy,ansy1,ansy2
        END SUBROUTINE bcuint
    END INTERFACE
    INTERFACE beschb
        SUBROUTINE beschb_s(x,gam1,gam2,gampl,gammi)
        USE nrtype
        REAL(DP), INTENT(IN) :: x
        REAL(DP), INTENT(OUT) :: gam1,gam2,gampl,gammi
        END SUBROUTINE beschb_s

        SUBROUTINE beschb_v(x,gam1,gam2,gampl,gammi)
        USE nrtype
        REAL(DP), DIMENSION(:), INTENT(IN) :: x
        REAL(DP), DIMENSION(:), INTENT(OUT) :: gam1,gam2,gampl,gammi
        END SUBROUTINE beschb_v
    END INTERFACE
    INTERFACE bessi
        FUNCTION bessi_s(n,x)
        USE nrtype
        INTEGER(I4B), INTENT(IN) :: n
        REAL(SP), INTENT(IN) :: x
        REAL(SP) :: bessi_s
        END FUNCTION bessi_s

        FUNCTION bessi_v(n,x)
        USE nrtype
        INTEGER(I4B), INTENT(IN) :: n
        REAL(SP), DIMENSION(:), INTENT(IN) :: x
        REAL(SP), DIMENSION(size(x)) :: bessi_v
        END FUNCTION bessi_v
    END INTERFACE
    INTERFACE bessi0
        FUNCTION bessi0_s(x)
        USE nrtype
        REAL(SP), INTENT(IN) :: x
        REAL(SP) :: bessi0_s
        END FUNCTION bessi0_s

        FUNCTION bessi0_v(x)
        USE nrtype
        REAL(SP), DIMENSION(:), INTENT(IN) :: x
        REAL(SP), DIMENSION(size(x)) :: bessi0_v
        END FUNCTION bessi0_v
    END INTERFACE
    INTERFACE bessi1
        FUNCTION bessi1_s(x)
        USE nrtype
        REAL(SP), INTENT(IN) :: x
        REAL(SP) :: bessi1_s
        END FUNCTION bessi1_s
```

```
        FUNCTION bessi1_v(x)
        USE nrtype
        REAL(SP), DIMENSION(:), INTENT(IN) :: x
        REAL(SP), DIMENSION(size(x)) :: bessi1_v
        END FUNCTION bessi1_v
    END INTERFACE
    INTERFACE
        SUBROUTINE bessik(x,xnu,ri,rk,rip,rkp)
        USE nrtype
        REAL(SP), INTENT(IN) :: x,xnu
        REAL(SP), INTENT(OUT) :: ri,rk,rip,rkp
        END SUBROUTINE bessik
    END INTERFACE
    INTERFACE bessj
        FUNCTION bessj_s(n,x)
        USE nrtype
        INTEGER(I4B), INTENT(IN) :: n
        REAL(SP), INTENT(IN) :: x
        REAL(SP) :: bessj_s
        END FUNCTION bessj_s

        FUNCTION bessj_v(n,x)
        USE nrtype
        INTEGER(I4B), INTENT(IN) :: n
        REAL(SP), DIMENSION(:), INTENT(IN) :: x
        REAL(SP), DIMENSION(size(x)) :: bessj_v
        END FUNCTION bessj_v
    END INTERFACE
    INTERFACE bessj0
        FUNCTION bessj0_s(x)
        USE nrtype
        REAL(SP), INTENT(IN) :: x
        REAL(SP) :: bessj0_s
        END FUNCTION bessj0_s

        FUNCTION bessj0_v(x)
        USE nrtype
        REAL(SP), DIMENSION(:), INTENT(IN) :: x
        REAL(SP), DIMENSION(size(x)) :: bessj0_v
        END FUNCTION bessj0_v
    END INTERFACE
    INTERFACE bessj1
        FUNCTION bessj1_s(x)
        USE nrtype
        REAL(SP), INTENT(IN) :: x
        REAL(SP) :: bessj1_s
        END FUNCTION bessj1_s

        FUNCTION bessj1_v(x)
        USE nrtype
        REAL(SP), DIMENSION(:), INTENT(IN) :: x
        REAL(SP), DIMENSION(size(x)) :: bessj1_v
        END FUNCTION bessj1_v
    END INTERFACE
    INTERFACE bessjy
        SUBROUTINE bessjy_s(x,xnu,rj,ry,rjp,ryp)
        USE nrtype
        REAL(SP), INTENT(IN) :: x,xnu
        REAL(SP), INTENT(OUT) :: rj,ry,rjp,ryp
        END SUBROUTINE bessjy_s

        SUBROUTINE bessjy_v(x,xnu,rj,ry,rjp,ryp)
        USE nrtype
        REAL(SP), INTENT(IN) :: xnu
        REAL(SP), DIMENSION(:), INTENT(IN) :: x
        REAL(SP), DIMENSION(:), INTENT(OUT) :: rj,rjp,ry,ryp
```

```
        END SUBROUTINE bessjy_v
    END INTERFACE
    INTERFACE bessk
        FUNCTION bessk_s(n,x)
        USE nrtype
        INTEGER(I4B), INTENT(IN) :: n
        REAL(SP), INTENT(IN) :: x
        REAL(SP) :: bessk_s
        END FUNCTION bessk_s

        FUNCTION bessk_v(n,x)
        USE nrtype
        INTEGER(I4B), INTENT(IN) :: n
        REAL(SP), DIMENSION(:), INTENT(IN) :: x
        REAL(SP), DIMENSION(size(x)) :: bessk_v
        END FUNCTION bessk_v
    END INTERFACE
    INTERFACE bessk0
        FUNCTION bessk0_s(x)
        USE nrtype
        REAL(SP), INTENT(IN) :: x
        REAL(SP) :: bessk0_s
        END FUNCTION bessk0_s

        FUNCTION bessk0_v(x)
        USE nrtype
        REAL(SP), DIMENSION(:), INTENT(IN) :: x
        REAL(SP), DIMENSION(size(x)) :: bessk0_v
        END FUNCTION bessk0_v
    END INTERFACE
    INTERFACE bessk1
        FUNCTION bessk1_s(x)
        USE nrtype
        REAL(SP), INTENT(IN) :: x
        REAL(SP) :: bessk1_s
        END FUNCTION bessk1_s

        FUNCTION bessk1_v(x)
        USE nrtype
        REAL(SP), DIMENSION(:), INTENT(IN) :: x
        REAL(SP), DIMENSION(size(x)) :: bessk1_v
        END FUNCTION bessk1_v
    END INTERFACE
    INTERFACE bessy
        FUNCTION bessy_s(n,x)
        USE nrtype
        INTEGER(I4B), INTENT(IN) :: n
        REAL(SP), INTENT(IN) :: x
        REAL(SP) :: bessy_s
        END FUNCTION bessy_s

        FUNCTION bessy_v(n,x)
        USE nrtype
        INTEGER(I4B), INTENT(IN) :: n
        REAL(SP), DIMENSION(:), INTENT(IN) :: x
        REAL(SP), DIMENSION(size(x)) :: bessy_v
        END FUNCTION bessy_v
    END INTERFACE
    INTERFACE bessy0
        FUNCTION bessy0_s(x)
        USE nrtype
        REAL(SP), INTENT(IN) :: x
        REAL(SP) :: bessy0_s
        END FUNCTION bessy0_s

        FUNCTION bessy0_v(x)
        USE nrtype
```

```
      REAL(SP), DIMENSION(:), INTENT(IN) :: x
      REAL(SP), DIMENSION(size(x)) :: bessy0_v
      END FUNCTION bessy0_v
   END INTERFACE
   INTERFACE bessy1
      FUNCTION bessy1_s(x)
      USE nrtype
      REAL(SP), INTENT(IN) :: x
      REAL(SP) :: bessy1_s
      END FUNCTION bessy1_s

      FUNCTION bessy1_v(x)
      USE nrtype
      REAL(SP), DIMENSION(:), INTENT(IN) :: x
      REAL(SP), DIMENSION(size(x)) :: bessy1_v
      END FUNCTION bessy1_v
   END INTERFACE
   INTERFACE beta
      FUNCTION beta_s(z,w)
      USE nrtype
      REAL(SP), INTENT(IN) :: z,w
      REAL(SP) :: beta_s
      END FUNCTION beta_s

      FUNCTION beta_v(z,w)
      USE nrtype
      REAL(SP), DIMENSION(:), INTENT(IN) :: z,w
      REAL(SP), DIMENSION(size(z)) :: beta_v
      END FUNCTION beta_v
   END INTERFACE
   INTERFACE betacf
      FUNCTION betacf_s(a,b,x)
      USE nrtype
      REAL(SP), INTENT(IN) :: a,b,x
      REAL(SP) :: betacf_s
      END FUNCTION betacf_s

      FUNCTION betacf_v(a,b,x)
      USE nrtype
      REAL(SP), DIMENSION(:), INTENT(IN) :: a,b,x
      REAL(SP), DIMENSION(size(x)) :: betacf_v
      END FUNCTION betacf_v
   END INTERFACE
   INTERFACE betai
      FUNCTION betai_s(a,b,x)
      USE nrtype
      REAL(SP), INTENT(IN) :: a,b,x
      REAL(SP) :: betai_s
      END FUNCTION betai_s

      FUNCTION betai_v(a,b,x)
      USE nrtype
      REAL(SP), DIMENSION(:), INTENT(IN) :: a,b,x
      REAL(SP), DIMENSION(size(a)) :: betai_v
      END FUNCTION betai_v
   END INTERFACE
   INTERFACE bico
      FUNCTION bico_s(n,k)
      USE nrtype
      INTEGER(I4B), INTENT(IN) :: n,k
      REAL(SP) :: bico_s
      END FUNCTION bico_s

      FUNCTION bico_v(n,k)
      USE nrtype
      INTEGER(I4B), DIMENSION(:), INTENT(IN) :: n,k
      REAL(SP), DIMENSION(size(n)) :: bico_v
```

```
        END FUNCTION bico_v
    END INTERFACE
    INTERFACE
        FUNCTION bnldev(pp,n)
        USE nrtype
        REAL(SP), INTENT(IN) :: pp
        INTEGER(I4B), INTENT(IN) :: n
        REAL(SP) :: bnldev
        END FUNCTION bnldev
    END INTERFACE
    INTERFACE
        FUNCTION brent(ax,bx,cx,func,tol,xmin)
        USE nrtype
        REAL(SP), INTENT(IN) :: ax,bx,cx,tol
        REAL(SP), INTENT(OUT) :: xmin
        REAL(SP) :: brent
        INTERFACE
            FUNCTION func(x)
            USE nrtype
            REAL(SP), INTENT(IN) :: x
            REAL(SP) :: func
            END FUNCTION func
        END INTERFACE
        END FUNCTION brent
    END INTERFACE
    INTERFACE
        SUBROUTINE broydn(x,check)
        USE nrtype
        REAL(SP), DIMENSION(:), INTENT(INOUT) :: x
        LOGICAL(LGT), INTENT(OUT) :: check
        END SUBROUTINE broydn
    END INTERFACE
    INTERFACE
        SUBROUTINE bsstep(y,dydx,x,htry,eps,yscal,hdid,hnext,derivs)
        USE nrtype
        REAL(SP), DIMENSION(:), INTENT(INOUT) :: y
        REAL(SP), DIMENSION(:), INTENT(IN) :: dydx,yscal
        REAL(SP), INTENT(INOUT) :: x
        REAL(SP), INTENT(IN) :: htry,eps
        REAL(SP), INTENT(OUT) :: hdid,hnext
        INTERFACE
            SUBROUTINE derivs(x,y,dydx)
            USE nrtype
            REAL(SP), INTENT(IN) :: x
            REAL(SP), DIMENSION(:), INTENT(IN) :: y
            REAL(SP), DIMENSION(:), INTENT(OUT) :: dydx
            END SUBROUTINE derivs
        END INTERFACE
        END SUBROUTINE bsstep
    END INTERFACE
    INTERFACE
        SUBROUTINE caldat(julian,mm,id,iyyy)
        USE nrtype
        INTEGER(I4B), INTENT(IN) :: julian
        INTEGER(I4B), INTENT(OUT) :: mm,id,iyyy
        END SUBROUTINE caldat
    END INTERFACE
    INTERFACE
        FUNCTION chder(a,b,c)
        USE nrtype
        REAL(SP), INTENT(IN) :: a,b
        REAL(SP), DIMENSION(:), INTENT(IN) :: c
        REAL(SP), DIMENSION(size(c)) :: chder
        END FUNCTION chder
```

```
END INTERFACE
INTERFACE chebev
    FUNCTION chebev_s(a,b,c,x)
    USE nrtype
    REAL(SP), INTENT(IN) :: a,b,x
    REAL(SP), DIMENSION(:), INTENT(IN) :: c
    REAL(SP) :: chebev_s
    END FUNCTION chebev_s

    FUNCTION chebev_v(a,b,c,x)
    USE nrtype
    REAL(SP), INTENT(IN) :: a,b
    REAL(SP), DIMENSION(:), INTENT(IN) :: c,x
    REAL(SP), DIMENSION(size(x)) :: chebev_v
    END FUNCTION chebev_v
END INTERFACE
INTERFACE
    FUNCTION chebft(a,b,n,func)
    USE nrtype
    REAL(SP), INTENT(IN) :: a,b
    INTEGER(I4B), INTENT(IN) :: n
    REAL(SP), DIMENSION(n) :: chebft
    INTERFACE
        FUNCTION func(x)
        USE nrtype
        REAL(SP), DIMENSION(:), INTENT(IN) :: x
        REAL(SP), DIMENSION(size(x)) :: func
        END FUNCTION func
    END INTERFACE
    END FUNCTION chebft
END INTERFACE
INTERFACE
    FUNCTION chebpc(c)
    USE nrtype
    REAL(SP), DIMENSION(:), INTENT(IN) :: c
    REAL(SP), DIMENSION(size(c)) :: chebpc
    END FUNCTION chebpc
END INTERFACE
INTERFACE
    FUNCTION chint(a,b,c)
    USE nrtype
    REAL(SP), INTENT(IN) :: a,b
    REAL(SP), DIMENSION(:), INTENT(IN) :: c
    REAL(SP), DIMENSION(size(c)) :: chint
    END FUNCTION chint
END INTERFACE
INTERFACE
    SUBROUTINE choldc(a,p)
    USE nrtype
    REAL(SP), DIMENSION(:,:), INTENT(INOUT) :: a
    REAL(SP), DIMENSION(:), INTENT(OUT) :: p
    END SUBROUTINE choldc
END INTERFACE
INTERFACE
    SUBROUTINE cholsl(a,p,b,x)
    USE nrtype
    REAL(SP), DIMENSION(:,:), INTENT(IN) :: a
    REAL(SP), DIMENSION(:), INTENT(IN) :: p,b
    REAL(SP), DIMENSION(:), INTENT(INOUT) :: x
    END SUBROUTINE cholsl
END INTERFACE
INTERFACE
    SUBROUTINE chsone(bins,ebins,knstrn,df,chsq,prob)
    USE nrtype
```

```
		INTEGER(I4B), INTENT(IN) :: knstrn
		REAL(SP), INTENT(OUT) :: df,chsq,prob
		REAL(SP), DIMENSION(:), INTENT(IN) :: bins,ebins
		END SUBROUTINE chsone
	END INTERFACE
	INTERFACE
		SUBROUTINE chstwo(bins1,bins2,knstrn,df,chsq,prob)
		USE nrtype
		INTEGER(I4B), INTENT(IN) :: knstrn
		REAL(SP), INTENT(OUT) :: df,chsq,prob
		REAL(SP), DIMENSION(:), INTENT(IN) :: bins1,bins2
		END SUBROUTINE chstwo
	END INTERFACE
	INTERFACE
		SUBROUTINE cisi(x,ci,si)
		USE nrtype
		REAL(SP), INTENT(IN) :: x
		REAL(SP), INTENT(OUT) :: ci,si
		END SUBROUTINE cisi
	END INTERFACE
	INTERFACE
		SUBROUTINE cntab1(nn,chisq,df,prob,cramrv,ccc)
		USE nrtype
		INTEGER(I4B), DIMENSION(:,:), INTENT(IN) :: nn
		REAL(SP), INTENT(OUT) :: chisq,df,prob,cramrv,ccc
		END SUBROUTINE cntab1
	END INTERFACE
	INTERFACE
		SUBROUTINE cntab2(nn,h,hx,hy,hygx,hxgy,uygx,uxgy,uxy)
		USE nrtype
		INTEGER(I4B), DIMENSION(:,:), INTENT(IN) :: nn
		REAL(SP), INTENT(OUT) :: h,hx,hy,hygx,hxgy,uygx,uxgy,uxy
		END SUBROUTINE cntab2
	END INTERFACE
	INTERFACE
		FUNCTION convlv(data,respns,isign)
		USE nrtype
		REAL(SP), DIMENSION(:), INTENT(IN) :: data
		REAL(SP), DIMENSION(:), INTENT(IN) :: respns
		INTEGER(I4B), INTENT(IN) :: isign
		REAL(SP), DIMENSION(size(data)) :: convlv
		END FUNCTION convlv
	END INTERFACE
	INTERFACE
		FUNCTION correl(data1,data2)
		USE nrtype
		REAL(SP), DIMENSION(:), INTENT(IN) :: data1,data2
		REAL(SP), DIMENSION(size(data1)) :: correl
		END FUNCTION correl
	END INTERFACE
	INTERFACE
		SUBROUTINE cosft1(y)
		USE nrtype
		REAL(SP), DIMENSION(:), INTENT(INOUT) :: y
		END SUBROUTINE cosft1
	END INTERFACE
	INTERFACE
		SUBROUTINE cosft2(y,isign)
		USE nrtype
		REAL(SP), DIMENSION(:), INTENT(INOUT) :: y
		INTEGER(I4B), INTENT(IN) :: isign
		END SUBROUTINE cosft2
	END INTERFACE
	INTERFACE
```

```
    SUBROUTINE covsrt(covar,maska)
    USE nrtype
    REAL(SP), DIMENSION(:,:), INTENT(INOUT) :: covar
    LOGICAL(LGT), DIMENSION(:), INTENT(IN) :: maska
    END SUBROUTINE covsrt
END INTERFACE
INTERFACE
    SUBROUTINE cyclic(a,b,c,alpha,beta,r,x)
    USE nrtype
    REAL(SP), DIMENSION(:), INTENT(IN):: a,b,c,r
    REAL(SP), INTENT(IN) :: alpha,beta
    REAL(SP), DIMENSION(:), INTENT(OUT):: x
    END SUBROUTINE cyclic
END INTERFACE
INTERFACE
    SUBROUTINE daub4(a,isign)
    USE nrtype
    REAL(SP), DIMENSION(:), INTENT(INOUT) :: a
    INTEGER(I4B), INTENT(IN) :: isign
    END SUBROUTINE daub4
END INTERFACE
INTERFACE dawson
    FUNCTION dawson_s(x)
    USE nrtype
    REAL(SP), INTENT(IN) :: x
    REAL(SP) :: dawson_s
    END FUNCTION dawson_s

    FUNCTION dawson_v(x)
    USE nrtype
    REAL(SP), DIMENSION(:), INTENT(IN) :: x
    REAL(SP), DIMENSION(size(x)) :: dawson_v
    END FUNCTION dawson_v
END INTERFACE
INTERFACE
    FUNCTION dbrent(ax,bx,cx,func,dbrent_dfunc,tol,xmin)
    USE nrtype
    REAL(SP), INTENT(IN) :: ax,bx,cx,tol
    REAL(SP), INTENT(OUT) :: xmin
    REAL(SP) :: dbrent
    INTERFACE
        FUNCTION func(x)
        USE nrtype
        REAL(SP), INTENT(IN) :: x
        REAL(SP) :: func
        END FUNCTION func

        FUNCTION dbrent_dfunc(x)
        USE nrtype
        REAL(SP), INTENT(IN) :: x
        REAL(SP) :: dbrent_dfunc
        END FUNCTION dbrent_dfunc
    END INTERFACE
    END FUNCTION dbrent
END INTERFACE
INTERFACE
    SUBROUTINE ddpoly(c,x,pd)
    USE nrtype
    REAL(SP), INTENT(IN) :: x
    REAL(SP), DIMENSION(:), INTENT(IN) :: c
    REAL(SP), DIMENSION(:), INTENT(OUT) :: pd
    END SUBROUTINE ddpoly
END INTERFACE
INTERFACE
    FUNCTION decchk(string,ch)
```

```
    USE nrtype
    CHARACTER(1), DIMENSION(:), INTENT(IN) :: string
    CHARACTER(1), INTENT(OUT) :: ch
    LOGICAL(LGT) :: decchk
    END FUNCTION decchk
END INTERFACE
INTERFACE
    SUBROUTINE dfpmin(p,gtol,iter,fret,func,dfunc)
    USE nrtype
    INTEGER(I4B), INTENT(OUT) :: iter
    REAL(SP), INTENT(IN) :: gtol
    REAL(SP), INTENT(OUT) :: fret
    REAL(SP), DIMENSION(:), INTENT(INOUT) :: p
    INTERFACE
        FUNCTION func(p)
        USE nrtype
        REAL(SP), DIMENSION(:), INTENT(IN) :: p
        REAL(SP) :: func
        END FUNCTION func

        FUNCTION dfunc(p)
        USE nrtype
        REAL(SP), DIMENSION(:), INTENT(IN) :: p
        REAL(SP), DIMENSION(size(p)) :: dfunc
        END FUNCTION dfunc
    END INTERFACE
    END SUBROUTINE dfpmin
END INTERFACE
INTERFACE
    FUNCTION dfridr(func,x,h,err)
    USE nrtype
    REAL(SP), INTENT(IN) :: x,h
    REAL(SP), INTENT(OUT) :: err
    REAL(SP) :: dfridr
    INTERFACE
        FUNCTION func(x)
        USE nrtype
        REAL(SP), INTENT(IN) :: x
        REAL(SP) :: func
        END FUNCTION func
    END INTERFACE
    END FUNCTION dfridr
END INTERFACE
INTERFACE
    SUBROUTINE dftcor(w,delta,a,b,endpts,corre,corim,corfac)
    USE nrtype
    REAL(SP), INTENT(IN) :: w,delta,a,b
    REAL(SP), INTENT(OUT) :: corre,corim,corfac
    REAL(SP), DIMENSION(:), INTENT(IN) :: endpts
    END SUBROUTINE dftcor
END INTERFACE
INTERFACE
    SUBROUTINE dftint(func,a,b,w,cosint,sinint)
    USE nrtype
    REAL(SP), INTENT(IN) :: a,b,w
    REAL(SP), INTENT(OUT) :: cosint,sinint
    INTERFACE
        FUNCTION func(x)
        USE nrtype
        REAL(SP), DIMENSION(:), INTENT(IN) :: x
        REAL(SP), DIMENSION(size(x)) :: func
        END FUNCTION func
    END INTERFACE
    END SUBROUTINE dftint
```

```
END INTERFACE
INTERFACE
    SUBROUTINE difeq(k,k1,k2,jsf,is1,isf,indexv,s,y)
    USE nrtype
    INTEGER(I4B), INTENT(IN) :: is1,isf,jsf,k,k1,k2
    INTEGER(I4B), DIMENSION(:), INTENT(IN) :: indexv
    REAL(SP), DIMENSION(:,:), INTENT(OUT) :: s
    REAL(SP), DIMENSION(:,:), INTENT(IN) :: y
    END SUBROUTINE difeq
END INTERFACE
INTERFACE
    FUNCTION eclass(lista,listb,n)
    USE nrtype
    INTEGER(I4B), DIMENSION(:), INTENT(IN) :: lista,listb
    INTEGER(I4B), INTENT(IN) :: n
    INTEGER(I4B), DIMENSION(n) :: eclass
    END FUNCTION eclass
END INTERFACE
INTERFACE
    FUNCTION eclazz(equiv,n)
    USE nrtype
    INTERFACE
        FUNCTION equiv(i,j)
        USE nrtype
        LOGICAL(LGT) :: equiv
        INTEGER(I4B), INTENT(IN) :: i,j
        END FUNCTION equiv
    END INTERFACE
    INTEGER(I4B), INTENT(IN) :: n
    INTEGER(I4B), DIMENSION(n) :: eclazz
    END FUNCTION eclazz
END INTERFACE
INTERFACE
    FUNCTION ei(x)
    USE nrtype
    REAL(SP), INTENT(IN) :: x
    REAL(SP) :: ei
    END FUNCTION ei
END INTERFACE
INTERFACE
    SUBROUTINE eigsrt(d,v)
    USE nrtype
    REAL(SP), DIMENSION(:), INTENT(INOUT) :: d
    REAL(SP), DIMENSION(:,:), INTENT(INOUT) :: v
    END SUBROUTINE eigsrt
END INTERFACE
INTERFACE elle
    FUNCTION elle_s(phi,ak)
    USE nrtype
    REAL(SP), INTENT(IN) :: phi,ak
    REAL(SP) :: elle_s
    END FUNCTION elle_s

    FUNCTION elle_v(phi,ak)
    USE nrtype
    REAL(SP), DIMENSION(:), INTENT(IN) :: phi,ak
    REAL(SP), DIMENSION(size(phi)) :: elle_v
    END FUNCTION elle_v
END INTERFACE
INTERFACE ellf
    FUNCTION ellf_s(phi,ak)
    USE nrtype
    REAL(SP), INTENT(IN) :: phi,ak
    REAL(SP) :: ellf_s
```

```
        END FUNCTION ellf_s

        FUNCTION ellf_v(phi,ak)
        USE nrtype
        REAL(SP), DIMENSION(:), INTENT(IN) :: phi,ak
        REAL(SP), DIMENSION(size(phi)) :: ellf_v
        END FUNCTION ellf_v
    END INTERFACE
    INTERFACE ellpi
        FUNCTION ellpi_s(phi,en,ak)
        USE nrtype
        REAL(SP), INTENT(IN) :: phi,en,ak
        REAL(SP) :: ellpi_s
        END FUNCTION ellpi_s

        FUNCTION ellpi_v(phi,en,ak)
        USE nrtype
        REAL(SP), DIMENSION(:), INTENT(IN) :: phi,en,ak
        REAL(SP), DIMENSION(size(phi)) :: ellpi_v
        END FUNCTION ellpi_v
    END INTERFACE
    INTERFACE
        SUBROUTINE elmhes(a)
        USE nrtype
        REAL(SP), DIMENSION(:,:), INTENT(INOUT) :: a
        END SUBROUTINE elmhes
    END INTERFACE
    INTERFACE erf
        FUNCTION erf_s(x)
        USE nrtype
        REAL(SP), INTENT(IN) :: x
        REAL(SP) :: erf_s
        END FUNCTION erf_s

        FUNCTION erf_v(x)
        USE nrtype
        REAL(SP), DIMENSION(:), INTENT(IN) :: x
        REAL(SP), DIMENSION(size(x)) :: erf_v
        END FUNCTION erf_v
    END INTERFACE
    INTERFACE erfc
        FUNCTION erfc_s(x)
        USE nrtype
        REAL(SP), INTENT(IN) :: x
        REAL(SP) :: erfc_s
        END FUNCTION erfc_s

        FUNCTION erfc_v(x)
        USE nrtype
        REAL(SP), DIMENSION(:), INTENT(IN) :: x
        REAL(SP), DIMENSION(size(x)) :: erfc_v
        END FUNCTION erfc_v
    END INTERFACE
    INTERFACE erfcc
        FUNCTION erfcc_s(x)
        USE nrtype
        REAL(SP), INTENT(IN) :: x
        REAL(SP) :: erfcc_s
        END FUNCTION erfcc_s

        FUNCTION erfcc_v(x)
        USE nrtype
        REAL(SP), DIMENSION(:), INTENT(IN) :: x
        REAL(SP), DIMENSION(size(x)) :: erfcc_v
        END FUNCTION erfcc_v
    END INTERFACE
    INTERFACE
```

```
        SUBROUTINE eulsum(sum,term,jterm)
        USE nrtype
        REAL(SP), INTENT(INOUT) :: sum
        REAL(SP), INTENT(IN) :: term
        INTEGER(I4B), INTENT(IN) :: jterm
        END SUBROUTINE eulsum
    END INTERFACE
    INTERFACE
        FUNCTION evlmem(fdt,d,xms)
        USE nrtype
        REAL(SP), INTENT(IN) :: fdt,xms
        REAL(SP), DIMENSION(:), INTENT(IN) :: d
        REAL(SP) :: evlmem
        END FUNCTION evlmem
    END INTERFACE
    INTERFACE expdev
        SUBROUTINE expdev_s(harvest)
        USE nrtype
        REAL(SP), INTENT(OUT) :: harvest
        END SUBROUTINE expdev_s

        SUBROUTINE expdev_v(harvest)
        USE nrtype
        REAL(SP), DIMENSION(:), INTENT(OUT) :: harvest
        END SUBROUTINE expdev_v
    END INTERFACE
    INTERFACE
        FUNCTION expint(n,x)
        USE nrtype
        INTEGER(I4B), INTENT(IN) :: n
        REAL(SP), INTENT(IN) :: x
        REAL(SP) :: expint
        END FUNCTION expint
    END INTERFACE
    INTERFACE factln
        FUNCTION factln_s(n)
        USE nrtype
        INTEGER(I4B), INTENT(IN) :: n
        REAL(SP) :: factln_s
        END FUNCTION factln_s

        FUNCTION factln_v(n)
        USE nrtype
        INTEGER(I4B), DIMENSION(:), INTENT(IN) :: n
        REAL(SP), DIMENSION(size(n)) :: factln_v
        END FUNCTION factln_v
    END INTERFACE
    INTERFACE factrl
        FUNCTION factrl_s(n)
        USE nrtype
        INTEGER(I4B), INTENT(IN) :: n
        REAL(SP) :: factrl_s
        END FUNCTION factrl_s

        FUNCTION factrl_v(n)
        USE nrtype
        INTEGER(I4B), DIMENSION(:), INTENT(IN) :: n
        REAL(SP), DIMENSION(size(n)) :: factrl_v
        END FUNCTION factrl_v
    END INTERFACE
    INTERFACE
        SUBROUTINE fasper(x,y,ofac,hifac,px,py,jmax,prob)
        USE nrtype
        REAL(SP), DIMENSION(:), INTENT(IN) :: x,y
        REAL(SP), INTENT(IN) :: ofac,hifac
        INTEGER(I4B), INTENT(OUT) :: jmax
```

```
        REAL(SP), INTENT(OUT) :: prob
        REAL(SP), DIMENSION(:), POINTER :: px,py
        END SUBROUTINE fasper
    END INTERFACE
    INTERFACE
        SUBROUTINE fdjac(x,fvec,df)
        USE nrtype
        REAL(SP), DIMENSION(:), INTENT(IN) :: fvec
        REAL(SP), DIMENSION(:), INTENT(INOUT) :: x
        REAL(SP), DIMENSION(:,:), INTENT(OUT) :: df
        END SUBROUTINE fdjac
    END INTERFACE
    INTERFACE
        SUBROUTINE fgauss(x,a,y,dyda)
        USE nrtype
        REAL(SP), DIMENSION(:), INTENT(IN) :: x,a
        REAL(SP), DIMENSION(:), INTENT(OUT) :: y
        REAL(SP), DIMENSION(:,:), INTENT(OUT) :: dyda
        END SUBROUTINE fgauss
    END INTERFACE
    INTERFACE
        SUBROUTINE fit(x,y,a,b,siga,sigb,chi2,q,sig)
        USE nrtype
        REAL(SP), DIMENSION(:), INTENT(IN) :: x,y
        REAL(SP), INTENT(OUT) :: a,b,siga,sigb,chi2,q
        REAL(SP), DIMENSION(:), OPTIONAL, INTENT(IN) :: sig
        END SUBROUTINE fit
    END INTERFACE
    INTERFACE
        SUBROUTINE fitexy(x,y,sigx,sigy,a,b,siga,sigb,chi2,q)
        USE nrtype
        REAL(SP), DIMENSION(:), INTENT(IN) :: x,y,sigx,sigy
        REAL(SP), INTENT(OUT) :: a,b,siga,sigb,chi2,q
        END SUBROUTINE fitexy
    END INTERFACE
    INTERFACE
        SUBROUTINE fixrts(d)
        USE nrtype
        REAL(SP), DIMENSION(:), INTENT(INOUT) :: d
        END SUBROUTINE fixrts
    END INTERFACE
    INTERFACE
        FUNCTION fleg(x,n)
        USE nrtype
        REAL(SP), INTENT(IN) :: x
        INTEGER(I4B), INTENT(IN) :: n
        REAL(SP), DIMENSION(n) :: fleg
        END FUNCTION fleg
    END INTERFACE
    INTERFACE
        SUBROUTINE flmoon(n,nph,jd,frac)
        USE nrtype
        INTEGER(I4B), INTENT(IN) :: n,nph
        INTEGER(I4B), INTENT(OUT) :: jd
        REAL(SP), INTENT(OUT) :: frac
        END SUBROUTINE flmoon
    END INTERFACE
    INTERFACE four1
        SUBROUTINE four1_dp(data,isign)
        USE nrtype
        COMPLEX(DPC), DIMENSION(:), INTENT(INOUT) :: data
        INTEGER(I4B), INTENT(IN) :: isign
        END SUBROUTINE four1_dp
```

```
        SUBROUTINE four1_sp(data,isign)
        USE nrtype
        COMPLEX(SPC), DIMENSION(:), INTENT(INOUT) :: data
        INTEGER(I4B), INTENT(IN) :: isign
        END SUBROUTINE four1_sp
    END INTERFACE
    INTERFACE
        SUBROUTINE four1_alt(data,isign)
        USE nrtype
        COMPLEX(SPC), DIMENSION(:), INTENT(INOUT) :: data
        INTEGER(I4B), INTENT(IN) :: isign
        END SUBROUTINE four1_alt
    END INTERFACE
    INTERFACE
        SUBROUTINE four1_gather(data,isign)
        USE nrtype
        COMPLEX(SPC), DIMENSION(:), INTENT(INOUT) :: data
        INTEGER(I4B), INTENT(IN) :: isign
        END SUBROUTINE four1_gather
    END INTERFACE
    INTERFACE
        SUBROUTINE four2(data,isign)
        USE nrtype
        COMPLEX(SPC), DIMENSION(:,:), INTENT(INOUT) :: data
        INTEGER(I4B),INTENT(IN) :: isign
        END SUBROUTINE four2
    END INTERFACE
    INTERFACE
        SUBROUTINE four2_alt(data,isign)
        USE nrtype
        COMPLEX(SPC), DIMENSION(:,:), INTENT(INOUT) :: data
        INTEGER(I4B), INTENT(IN) :: isign
        END SUBROUTINE four2_alt
    END INTERFACE
    INTERFACE
        SUBROUTINE four3(data,isign)
        USE nrtype
        COMPLEX(SPC), DIMENSION(:,:,:), INTENT(INOUT) :: data
        INTEGER(I4B),INTENT(IN) :: isign
        END SUBROUTINE four3
    END INTERFACE
    INTERFACE
        SUBROUTINE four3_alt(data,isign)
        USE nrtype
        COMPLEX(SPC), DIMENSION(:,:,:), INTENT(INOUT) :: data
        INTEGER(I4B), INTENT(IN) :: isign
        END SUBROUTINE four3_alt
    END INTERFACE
    INTERFACE
        SUBROUTINE fourcol(data,isign)
        USE nrtype
        COMPLEX(SPC), DIMENSION(:,:), INTENT(INOUT) :: data
        INTEGER(I4B), INTENT(IN) :: isign
        END SUBROUTINE fourcol
    END INTERFACE
    INTERFACE
        SUBROUTINE fourcol_3d(data,isign)
        USE nrtype
        COMPLEX(SPC), DIMENSION(:,:,:), INTENT(INOUT) :: data
        INTEGER(I4B), INTENT(IN) :: isign
        END SUBROUTINE fourcol_3d
    END INTERFACE
    INTERFACE
        SUBROUTINE fourn_gather(data,nn,isign)
```

```
        USE nrtype
        COMPLEX(SPC), DIMENSION(:), INTENT(INOUT) :: data
        INTEGER(I4B), DIMENSION(:), INTENT(IN) :: nn
        INTEGER(I4B), INTENT(IN) :: isign
        END SUBROUTINE fourn_gather
    END INTERFACE
    INTERFACE fourrow
        SUBROUTINE fourrow_dp(data,isign)
        USE nrtype
        COMPLEX(DPC), DIMENSION(:,:), INTENT(INOUT) :: data
        INTEGER(I4B), INTENT(IN) :: isign
        END SUBROUTINE fourrow_dp

        SUBROUTINE fourrow_sp(data,isign)
        USE nrtype
        COMPLEX(SPC), DIMENSION(:,:), INTENT(INOUT) :: data
        INTEGER(I4B), INTENT(IN) :: isign
        END SUBROUTINE fourrow_sp
    END INTERFACE
    INTERFACE
        SUBROUTINE fourrow_3d(data,isign)
        USE nrtype
        COMPLEX(SPC), DIMENSION(:,:,:), INTENT(INOUT) :: data
        INTEGER(I4B), INTENT(IN) :: isign
        END SUBROUTINE fourrow_3d
    END INTERFACE
    INTERFACE
        FUNCTION fpoly(x,n)
        USE nrtype
        REAL(SP), INTENT(IN) :: x
        INTEGER(I4B), INTENT(IN) :: n
        REAL(SP), DIMENSION(n) :: fpoly
        END FUNCTION fpoly
    END INTERFACE
    INTERFACE
        SUBROUTINE fred2(a,b,t,f,w,g,ak)
        USE nrtype
        REAL(SP), INTENT(IN) :: a,b
        REAL(SP), DIMENSION(:), INTENT(OUT) :: t,f,w
        INTERFACE
            FUNCTION g(t)
            USE nrtype
            REAL(SP), DIMENSION(:), INTENT(IN) :: t
            REAL(SP), DIMENSION(size(t)) :: g
            END FUNCTION g

            FUNCTION ak(t,s)
            USE nrtype
            REAL(SP), DIMENSION(:), INTENT(IN) :: t,s
            REAL(SP), DIMENSION(size(t),size(s)) :: ak
            END FUNCTION ak
        END INTERFACE
        END SUBROUTINE fred2
    END INTERFACE
    INTERFACE
        FUNCTION fredin(x,a,b,t,f,w,g,ak)
        USE nrtype
        REAL(SP), INTENT(IN) :: a,b
        REAL(SP), DIMENSION(:), INTENT(IN) :: x,t,f,w
        REAL(SP), DIMENSION(size(x)) :: fredin
        INTERFACE
            FUNCTION g(t)
            USE nrtype
            REAL(SP), DIMENSION(:), INTENT(IN) :: t
            REAL(SP), DIMENSION(size(t)) :: g
```

```
            END FUNCTION g

            FUNCTION ak(t,s)
            USE nrtype
            REAL(SP), DIMENSION(:), INTENT(IN) :: t,s
            REAL(SP), DIMENSION(size(t),size(s)) :: ak
            END FUNCTION ak
        END INTERFACE
        END FUNCTION fredin
    END INTERFACE
    INTERFACE
        SUBROUTINE frenel(x,s,c)
        USE nrtype
        REAL(SP), INTENT(IN) :: x
        REAL(SP), INTENT(OUT) :: s,c
        END SUBROUTINE frenel
    END INTERFACE
    INTERFACE
        SUBROUTINE frprmn(p,ftol,iter,fret)
        USE nrtype
        INTEGER(I4B), INTENT(OUT) :: iter
        REAL(SP), INTENT(IN) :: ftol
        REAL(SP), INTENT(OUT) :: fret
        REAL(SP), DIMENSION(:), INTENT(INOUT) :: p
        END SUBROUTINE frprmn
    END INTERFACE
    INTERFACE
        SUBROUTINE ftest(data1,data2,f,prob)
        USE nrtype
        REAL(SP), INTENT(OUT) :: f,prob
        REAL(SP), DIMENSION(:), INTENT(IN) :: data1,data2
        END SUBROUTINE ftest
    END INTERFACE
    INTERFACE
        FUNCTION gamdev(ia)
        USE nrtype
        INTEGER(I4B), INTENT(IN) :: ia
        REAL(SP) :: gamdev
        END FUNCTION gamdev
    END INTERFACE
    INTERFACE gammln
        FUNCTION gammln_s(xx)
        USE nrtype
        REAL(SP), INTENT(IN) :: xx
        REAL(SP) :: gammln_s
        END FUNCTION gammln_s

        FUNCTION gammln_v(xx)
        USE nrtype
        REAL(SP), DIMENSION(:), INTENT(IN) :: xx
        REAL(SP), DIMENSION(size(xx)) :: gammln_v
        END FUNCTION gammln_v
    END INTERFACE
    INTERFACE gammp
        FUNCTION gammp_s(a,x)
        USE nrtype
        REAL(SP), INTENT(IN) :: a,x
        REAL(SP) :: gammp_s
        END FUNCTION gammp_s

        FUNCTION gammp_v(a,x)
        USE nrtype
        REAL(SP), DIMENSION(:), INTENT(IN) :: a,x
        REAL(SP), DIMENSION(size(a)) :: gammp_v
        END FUNCTION gammp_v
    END INTERFACE
```

```
INTERFACE gammq
    FUNCTION gammq_s(a,x)
    USE nrtype
    REAL(SP), INTENT(IN) :: a,x
    REAL(SP) :: gammq_s
    END FUNCTION gammq_s

    FUNCTION gammq_v(a,x)
    USE nrtype
    REAL(SP), DIMENSION(:), INTENT(IN) :: a,x
    REAL(SP), DIMENSION(size(a)) :: gammq_v
    END FUNCTION gammq_v
END INTERFACE
INTERFACE gasdev
    SUBROUTINE gasdev_s(harvest)
    USE nrtype
    REAL(SP), INTENT(OUT) :: harvest
    END SUBROUTINE gasdev_s

    SUBROUTINE gasdev_v(harvest)
    USE nrtype
    REAL(SP), DIMENSION(:), INTENT(OUT) :: harvest
    END SUBROUTINE gasdev_v
END INTERFACE
INTERFACE
    SUBROUTINE gaucof(a,b,amu0,x,w)
    USE nrtype
    REAL(SP), INTENT(IN) :: amu0
    REAL(SP), DIMENSION(:), INTENT(INOUT) :: a,b
    REAL(SP), DIMENSION(:), INTENT(OUT) :: x,w
    END SUBROUTINE gaucof
END INTERFACE
INTERFACE
    SUBROUTINE gauher(x,w)
    USE nrtype
    REAL(SP), DIMENSION(:), INTENT(OUT) :: x,w
    END SUBROUTINE gauher
END INTERFACE
INTERFACE
    SUBROUTINE gaujac(x,w,alf,bet)
    USE nrtype
    REAL(SP), INTENT(IN) :: alf,bet
    REAL(SP), DIMENSION(:), INTENT(OUT) :: x,w
    END SUBROUTINE gaujac
END INTERFACE
INTERFACE
    SUBROUTINE gaulag(x,w,alf)
    USE nrtype
    REAL(SP), INTENT(IN) :: alf
    REAL(SP), DIMENSION(:), INTENT(OUT) :: x,w
    END SUBROUTINE gaulag
END INTERFACE
INTERFACE
    SUBROUTINE gauleg(x1,x2,x,w)
    USE nrtype
    REAL(SP), INTENT(IN) :: x1,x2
    REAL(SP), DIMENSION(:), INTENT(OUT) :: x,w
    END SUBROUTINE gauleg
END INTERFACE
INTERFACE
    SUBROUTINE gaussj(a,b)
    USE nrtype
    REAL(SP), DIMENSION(:,:), INTENT(INOUT) :: a,b
    END SUBROUTINE gaussj
END INTERFACE
```

```
INTERFACE gcf
    FUNCTION gcf_s(a,x,gln)
    USE nrtype
    REAL(SP), INTENT(IN) :: a,x
    REAL(SP), OPTIONAL, INTENT(OUT) :: gln
    REAL(SP) :: gcf_s
    END FUNCTION gcf_s

    FUNCTION gcf_v(a,x,gln)
    USE nrtype
    REAL(SP), DIMENSION(:), INTENT(IN) :: a,x
    REAL(SP), DIMENSION(:), OPTIONAL, INTENT(OUT) :: gln
    REAL(SP), DIMENSION(size(a)) :: gcf_v
    END FUNCTION gcf_v
END INTERFACE
INTERFACE
    FUNCTION golden(ax,bx,cx,func,tol,xmin)
    USE nrtype
    REAL(SP), INTENT(IN) :: ax,bx,cx,tol
    REAL(SP), INTENT(OUT) :: xmin
    REAL(SP) :: golden
    INTERFACE
        FUNCTION func(x)
        USE nrtype
        REAL(SP), INTENT(IN) :: x
        REAL(SP) :: func
        END FUNCTION func
    END INTERFACE
    END FUNCTION golden
END INTERFACE
INTERFACE gser
    FUNCTION gser_s(a,x,gln)
    USE nrtype
    REAL(SP), INTENT(IN) :: a,x
    REAL(SP), OPTIONAL, INTENT(OUT) :: gln
    REAL(SP) :: gser_s
    END FUNCTION gser_s

    FUNCTION gser_v(a,x,gln)
    USE nrtype
    REAL(SP), DIMENSION(:), INTENT(IN) :: a,x
    REAL(SP), DIMENSION(:), OPTIONAL, INTENT(OUT) :: gln
    REAL(SP), DIMENSION(size(a)) :: gser_v
    END FUNCTION gser_v
END INTERFACE
INTERFACE
    SUBROUTINE hqr(a,wr,wi)
    USE nrtype
    REAL(SP), DIMENSION(:), INTENT(OUT) :: wr,wi
    REAL(SP), DIMENSION(:,:), INTENT(INOUT) :: a
    END SUBROUTINE hqr
END INTERFACE
INTERFACE
    SUBROUTINE hunt(xx,x,jlo)
    USE nrtype
    INTEGER(I4B), INTENT(INOUT) :: jlo
    REAL(SP), INTENT(IN) :: x
    REAL(SP), DIMENSION(:), INTENT(IN) :: xx
    END SUBROUTINE hunt
END INTERFACE
INTERFACE
    SUBROUTINE hypdrv(s,ry,rdyds)
    USE nrtype
    REAL(SP), INTENT(IN) :: s
    REAL(SP), DIMENSION(:), INTENT(IN) :: ry
```

```
    REAL(SP), DIMENSION(:), INTENT(OUT) :: rdyds
    END SUBROUTINE hypdrv
END INTERFACE
INTERFACE
    FUNCTION hypgeo(a,b,c,z)
    USE nrtype
    COMPLEX(SPC), INTENT(IN) :: a,b,c,z
    COMPLEX(SPC) :: hypgeo
    END FUNCTION hypgeo
END INTERFACE
INTERFACE
    SUBROUTINE hypser(a,b,c,z,series,deriv)
    USE nrtype
    COMPLEX(SPC), INTENT(IN) :: a,b,c,z
    COMPLEX(SPC), INTENT(OUT) :: series,deriv
    END SUBROUTINE hypser
END INTERFACE
INTERFACE
    FUNCTION icrc(crc,buf,jinit,jrev)
    USE nrtype
    CHARACTER(1), DIMENSION(:), INTENT(IN) :: buf
    INTEGER(I2B), INTENT(IN) :: crc,jinit
    INTEGER(I4B), INTENT(IN) :: jrev
    INTEGER(I2B) :: icrc
    END FUNCTION icrc
END INTERFACE
INTERFACE
    FUNCTION igray(n,is)
    USE nrtype
    INTEGER(I4B), INTENT(IN) :: n,is
    INTEGER(I4B) :: igray
    END FUNCTION igray
END INTERFACE
INTERFACE
    RECURSIVE SUBROUTINE index_bypack(arr,index,partial)
    USE nrtype
    REAL(SP), DIMENSION(:), INTENT(IN) :: arr
    INTEGER(I4B), DIMENSION(:), INTENT(INOUT) :: index
    INTEGER, OPTIONAL, INTENT(IN) :: partial
    END SUBROUTINE index_bypack
END INTERFACE
INTERFACE indexx
    SUBROUTINE indexx_sp(arr,index)
    USE nrtype
    REAL(SP), DIMENSION(:), INTENT(IN) :: arr
    INTEGER(I4B), DIMENSION(:), INTENT(OUT) :: index
    END SUBROUTINE indexx_sp
    SUBROUTINE indexx_i4b(iarr,index)
    USE nrtype
    INTEGER(I4B), DIMENSION(:), INTENT(IN) :: iarr
    INTEGER(I4B), DIMENSION(:), INTENT(OUT) :: index
    END SUBROUTINE indexx_i4b
END INTERFACE
INTERFACE
    FUNCTION interp(uc)
    USE nrtype
    REAL(DP), DIMENSION(:,:), INTENT(IN) :: uc
    REAL(DP), DIMENSION(2*size(uc,1)-1,2*size(uc,1)-1) :: interp
    END FUNCTION interp
END INTERFACE
INTERFACE
    FUNCTION rank(indx)
    USE nrtype
    INTEGER(I4B), DIMENSION(:), INTENT(IN) :: indx
```

```
        INTEGER(I4B), DIMENSION(size(indx)) :: rank
        END FUNCTION rank
    END INTERFACE
    INTERFACE
        FUNCTION irbit1(iseed)
        USE nrtype
        INTEGER(I4B), INTENT(INOUT) :: iseed
        INTEGER(I4B) :: irbit1
        END FUNCTION irbit1
    END INTERFACE
    INTERFACE
        FUNCTION irbit2(iseed)
        USE nrtype
        INTEGER(I4B), INTENT(INOUT) :: iseed
        INTEGER(I4B) :: irbit2
        END FUNCTION irbit2
    END INTERFACE
    INTERFACE
        SUBROUTINE jacobi(a,d,v,nrot)
        USE nrtype
        INTEGER(I4B), INTENT(OUT) :: nrot
        REAL(SP), DIMENSION(:), INTENT(OUT) :: d
        REAL(SP), DIMENSION(:,:), INTENT(INOUT) :: a
        REAL(SP), DIMENSION(:,:), INTENT(OUT) :: v
        END SUBROUTINE jacobi
    END INTERFACE
    INTERFACE
        SUBROUTINE jacobn(x,y,dfdx,dfdy)
        USE nrtype
        REAL(SP), INTENT(IN) :: x
        REAL(SP), DIMENSION(:), INTENT(IN) :: y
        REAL(SP), DIMENSION(:), INTENT(OUT) :: dfdx
        REAL(SP), DIMENSION(:,:), INTENT(OUT) :: dfdy
        END SUBROUTINE jacobn
    END INTERFACE
    INTERFACE
        FUNCTION julday(mm,id,iyyy)
        USE nrtype
        INTEGER(I4B), INTENT(IN) :: mm,id,iyyy
        INTEGER(I4B) :: julday
        END FUNCTION julday
    END INTERFACE
    INTERFACE
        SUBROUTINE kendl1(data1,data2,tau,z,prob)
        USE nrtype
        REAL(SP), INTENT(OUT) :: tau,z,prob
        REAL(SP), DIMENSION(:), INTENT(IN) :: data1,data2
        END SUBROUTINE kendl1
    END INTERFACE
    INTERFACE
        SUBROUTINE kendl2(tab,tau,z,prob)
        USE nrtype
        REAL(SP), DIMENSION(:,:), INTENT(IN) :: tab
        REAL(SP), INTENT(OUT) :: tau,z,prob
        END SUBROUTINE kendl2
    END INTERFACE
    INTERFACE
        FUNCTION kermom(y,m)
        USE nrtype
        REAL(DP), INTENT(IN) :: y
        INTEGER(I4B), INTENT(IN) :: m
        REAL(DP), DIMENSION(m) :: kermom
        END FUNCTION kermom
    END INTERFACE
```

```
INTERFACE
    SUBROUTINE ks2d1s(x1,y1,quadvl,d1,prob)
    USE nrtype
    REAL(SP), DIMENSION(:), INTENT(IN) :: x1,y1
    REAL(SP), INTENT(OUT) :: d1,prob
    INTERFACE
        SUBROUTINE quadvl(x,y,fa,fb,fc,fd)
        USE nrtype
        REAL(SP), INTENT(IN) :: x,y
        REAL(SP), INTENT(OUT) :: fa,fb,fc,fd
        END SUBROUTINE quadvl
    END INTERFACE
    END SUBROUTINE ks2d1s
END INTERFACE
INTERFACE
    SUBROUTINE ks2d2s(x1,y1,x2,y2,d,prob)
    USE nrtype
    REAL(SP), DIMENSION(:), INTENT(IN) :: x1,y1,x2,y2
    REAL(SP), INTENT(OUT) :: d,prob
    END SUBROUTINE ks2d2s
END INTERFACE
INTERFACE
    SUBROUTINE ksone(data,func,d,prob)
    USE nrtype
    REAL(SP), INTENT(OUT) :: d,prob
    REAL(SP), DIMENSION(:), INTENT(INOUT) :: data
    INTERFACE
        FUNCTION func(x)
        USE nrtype
        REAL(SP), DIMENSION(:), INTENT(IN) :: x
        REAL(SP), DIMENSION(size(x)) :: func
        END FUNCTION func
    END INTERFACE
    END SUBROUTINE ksone
END INTERFACE
INTERFACE
    SUBROUTINE kstwo(data1,data2,d,prob)
    USE nrtype
    REAL(SP), INTENT(OUT) :: d,prob
    REAL(SP), DIMENSION(:), INTENT(IN) :: data1,data2
    END SUBROUTINE kstwo
END INTERFACE
INTERFACE
    SUBROUTINE laguer(a,x,its)
    USE nrtype
    INTEGER(I4B), INTENT(OUT) :: its
    COMPLEX(SPC), INTENT(INOUT) :: x
    COMPLEX(SPC), DIMENSION(:), INTENT(IN) :: a
    END SUBROUTINE laguer
END INTERFACE
INTERFACE
    SUBROUTINE lfit(x,y,sig,a,maska,covar,chisq,funcs)
    USE nrtype
    REAL(SP), DIMENSION(:), INTENT(IN) :: x,y,sig
    REAL(SP), DIMENSION(:), INTENT(INOUT) :: a
    LOGICAL(LGT), DIMENSION(:), INTENT(IN) :: maska
    REAL(SP), DIMENSION(:,:), INTENT(INOUT) :: covar
    REAL(SP), INTENT(OUT) :: chisq
    INTERFACE
        SUBROUTINE funcs(x,arr)
        USE nrtype
        REAL(SP),INTENT(IN) :: x
        REAL(SP), DIMENSION(:), INTENT(OUT) :: arr
        END SUBROUTINE funcs
```

```
        END INTERFACE
        END SUBROUTINE lfit
    END INTERFACE
    INTERFACE
        SUBROUTINE linbcg(b,x,itol,tol,itmax,iter,err)
        USE nrtype
        REAL(DP), DIMENSION(:), INTENT(IN) :: b
        REAL(DP), DIMENSION(:), INTENT(INOUT) :: x
        INTEGER(I4B), INTENT(IN) :: itol,itmax
        REAL(DP), INTENT(IN) :: tol
        INTEGER(I4B), INTENT(OUT) :: iter
        REAL(DP), INTENT(OUT) :: err
        END SUBROUTINE linbcg
    END INTERFACE
    INTERFACE
        SUBROUTINE linmin(p,xi,fret)
        USE nrtype
        REAL(SP), INTENT(OUT) :: fret
        REAL(SP), DIMENSION(:), TARGET, INTENT(INOUT) :: p,xi
        END SUBROUTINE linmin
    END INTERFACE
    INTERFACE
        SUBROUTINE lnsrch(xold,fold,g,p,x,f,stpmax,check,func)
        USE nrtype
        REAL(SP), DIMENSION(:), INTENT(IN) :: xold,g
        REAL(SP), DIMENSION(:), INTENT(INOUT) :: p
        REAL(SP), INTENT(IN) :: fold,stpmax
        REAL(SP), DIMENSION(:), INTENT(OUT) :: x
        REAL(SP), INTENT(OUT) :: f
        LOGICAL(LGT), INTENT(OUT) :: check
        INTERFACE
            FUNCTION func(x)
            USE nrtype
            REAL(SP) :: func
            REAL(SP), DIMENSION(:), INTENT(IN) :: x
            END FUNCTION func
        END INTERFACE
        END SUBROUTINE lnsrch
    END INTERFACE
    INTERFACE
        FUNCTION locate(xx,x)
        USE nrtype
        REAL(SP), DIMENSION(:), INTENT(IN) :: xx
        REAL(SP), INTENT(IN) :: x
        INTEGER(I4B) :: locate
        END FUNCTION locate
    END INTERFACE
    INTERFACE
        FUNCTION lop(u)
        USE nrtype
        REAL(DP), DIMENSION(:,:), INTENT(IN) :: u
        REAL(DP), DIMENSION(size(u,1),size(u,1)) :: lop
        END FUNCTION lop
    END INTERFACE
    INTERFACE
        SUBROUTINE lubksb(a,indx,b)
        USE nrtype
        REAL(SP), DIMENSION(:,:), INTENT(IN) :: a
        INTEGER(I4B), DIMENSION(:), INTENT(IN) :: indx
        REAL(SP), DIMENSION(:), INTENT(INOUT) :: b
        END SUBROUTINE lubksb
    END INTERFACE
    INTERFACE
        SUBROUTINE ludcmp(a,indx,d)
```

```
		USE nrtype
		REAL(SP), DIMENSION(:,:), INTENT(INOUT) :: a
		INTEGER(I4B), DIMENSION(:), INTENT(OUT) :: indx
		REAL(SP), INTENT(OUT) :: d
		END SUBROUTINE ludcmp
	END INTERFACE
	INTERFACE
		SUBROUTINE machar(ibeta,it,irnd,ngrd,machep,negep,iexp,minexp,&
			maxexp,eps,epsneg,xmin,xmax)
		USE nrtype
		INTEGER(I4B), INTENT(OUT) :: ibeta,iexp,irnd,it,machep,maxexp,&
			minexp,negep,ngrd
		REAL(SP), INTENT(OUT) :: eps,epsneg,xmax,xmin
		END SUBROUTINE machar
	END INTERFACE
	INTERFACE
		SUBROUTINE medfit(x,y,a,b,abdev)
		USE nrtype
		REAL(SP), DIMENSION(:), INTENT(IN) :: x,y
		REAL(SP), INTENT(OUT) :: a,b,abdev
		END SUBROUTINE medfit
	END INTERFACE
	INTERFACE
		SUBROUTINE memcof(data,xms,d)
		USE nrtype
		REAL(SP), INTENT(OUT) :: xms
		REAL(SP), DIMENSION(:), INTENT(IN) :: data
		REAL(SP), DIMENSION(:), INTENT(OUT) :: d
		END SUBROUTINE memcof
	END INTERFACE
	INTERFACE
		SUBROUTINE mgfas(u,maxcyc)
		USE nrtype
		REAL(DP), DIMENSION(:,:), INTENT(INOUT) :: u
		INTEGER(I4B), INTENT(IN) :: maxcyc
		END SUBROUTINE mgfas
	END INTERFACE
	INTERFACE
		SUBROUTINE mglin(u,ncycle)
		USE nrtype
		REAL(DP), DIMENSION(:,:), INTENT(INOUT) :: u
		INTEGER(I4B), INTENT(IN) :: ncycle
		END SUBROUTINE mglin
	END INTERFACE
	INTERFACE
		SUBROUTINE midexp(funk,aa,bb,s,n)
		USE nrtype
		REAL(SP), INTENT(IN) :: aa,bb
		REAL(SP), INTENT(INOUT) :: s
		INTEGER(I4B), INTENT(IN) :: n
		INTERFACE
			FUNCTION funk(x)
			USE nrtype
			REAL(SP), DIMENSION(:), INTENT(IN) :: x
			REAL(SP), DIMENSION(size(x)) :: funk
			END FUNCTION funk
		END INTERFACE
		END SUBROUTINE midexp
	END INTERFACE
	INTERFACE
		SUBROUTINE midinf(funk,aa,bb,s,n)
		USE nrtype
		REAL(SP), INTENT(IN) :: aa,bb
		REAL(SP), INTENT(INOUT) :: s
```

```
        INTEGER(I4B), INTENT(IN) :: n
        INTERFACE
            FUNCTION funk(x)
            USE nrtype
            REAL(SP), DIMENSION(:), INTENT(IN) :: x
            REAL(SP), DIMENSION(size(x)) :: funk
            END FUNCTION funk
        END INTERFACE
        END SUBROUTINE midinf
    END INTERFACE
    INTERFACE
        SUBROUTINE midpnt(func,a,b,s,n)
        USE nrtype
        REAL(SP), INTENT(IN) :: a,b
        REAL(SP), INTENT(INOUT) :: s
        INTEGER(I4B), INTENT(IN) :: n
        INTERFACE
            FUNCTION func(x)
            USE nrtype
            REAL(SP), DIMENSION(:), INTENT(IN) :: x
            REAL(SP), DIMENSION(size(x)) :: func
            END FUNCTION func
        END INTERFACE
        END SUBROUTINE midpnt
    END INTERFACE
    INTERFACE
        SUBROUTINE midsql(funk,aa,bb,s,n)
        USE nrtype
        REAL(SP), INTENT(IN) :: aa,bb
        REAL(SP), INTENT(INOUT) :: s
        INTEGER(I4B), INTENT(IN) :: n
        INTERFACE
            FUNCTION funk(x)
            USE nrtype
            REAL(SP), DIMENSION(:), INTENT(IN) :: x
            REAL(SP), DIMENSION(size(x)) :: funk
            END FUNCTION funk
        END INTERFACE
        END SUBROUTINE midsql
    END INTERFACE
    INTERFACE
        SUBROUTINE midsqu(funk,aa,bb,s,n)
        USE nrtype
        REAL(SP), INTENT(IN) :: aa,bb
        REAL(SP), INTENT(INOUT) :: s
        INTEGER(I4B), INTENT(IN) :: n
        INTERFACE
            FUNCTION funk(x)
            USE nrtype
            REAL(SP), DIMENSION(:), INTENT(IN) :: x
            REAL(SP), DIMENSION(size(x)) :: funk
            END FUNCTION funk
        END INTERFACE
        END SUBROUTINE midsqu
    END INTERFACE
    INTERFACE
        RECURSIVE SUBROUTINE miser(func,regn,ndim,npts,dith,ave,var)
        USE nrtype
        INTERFACE
            FUNCTION func(x)
            USE nrtype
            REAL(SP) :: func
            REAL(SP), DIMENSION(:), INTENT(IN) :: x
            END FUNCTION func
```

```
		END INTERFACE
		REAL(SP), DIMENSION(:), INTENT(IN) :: regn
		INTEGER(I4B), INTENT(IN) :: ndim,npts
		REAL(SP), INTENT(IN) :: dith
		REAL(SP), INTENT(OUT) :: ave,var
		END SUBROUTINE miser
	END INTERFACE
	INTERFACE
		SUBROUTINE mmid(y,dydx,xs,htot,nstep,yout,derivs)
		USE nrtype
		INTEGER(I4B), INTENT(IN) :: nstep
		REAL(SP), INTENT(IN) :: xs,htot
		REAL(SP), DIMENSION(:), INTENT(IN) :: y,dydx
		REAL(SP), DIMENSION(:), INTENT(OUT) :: yout
		INTERFACE
			SUBROUTINE derivs(x,y,dydx)
			USE nrtype
			REAL(SP), INTENT(IN) :: x
			REAL(SP), DIMENSION(:), INTENT(IN) :: y
			REAL(SP), DIMENSION(:), INTENT(OUT) :: dydx
			END SUBROUTINE derivs
		END INTERFACE
		END SUBROUTINE mmid
	END INTERFACE
	INTERFACE
		SUBROUTINE mnbrak(ax,bx,cx,fa,fb,fc,func)
		USE nrtype
		REAL(SP), INTENT(INOUT) :: ax,bx
		REAL(SP), INTENT(OUT) :: cx,fa,fb,fc
		INTERFACE
			FUNCTION func(x)
			USE nrtype
			REAL(SP), INTENT(IN) :: x
			REAL(SP) :: func
			END FUNCTION func
		END INTERFACE
		END SUBROUTINE mnbrak
	END INTERFACE
	INTERFACE
		SUBROUTINE mnewt(ntrial,x,tolx,tolf,usrfun)
		USE nrtype
		INTEGER(I4B), INTENT(IN) :: ntrial
		REAL(SP), INTENT(IN) :: tolx,tolf
		REAL(SP), DIMENSION(:), INTENT(INOUT) :: x
		INTERFACE
			SUBROUTINE usrfun(x,fvec,fjac)
			USE nrtype
			REAL(SP), DIMENSION(:), INTENT(IN) :: x
			REAL(SP), DIMENSION(:), INTENT(OUT) :: fvec
			REAL(SP), DIMENSION(:,:), INTENT(OUT) :: fjac
			END SUBROUTINE usrfun
		END INTERFACE
		END SUBROUTINE mnewt
	END INTERFACE
	INTERFACE
		SUBROUTINE moment(data,ave,adev,sdev,var,skew,curt)
		USE nrtype
		REAL(SP), INTENT(OUT) :: ave,adev,sdev,var,skew,curt
		REAL(SP), DIMENSION(:), INTENT(IN) :: data
		END SUBROUTINE moment
	END INTERFACE
	INTERFACE
		SUBROUTINE mp2dfr(a,s,n,m)
		USE nrtype
```

```
    INTEGER(I4B), INTENT(IN) :: n
    INTEGER(I4B), INTENT(OUT) :: m
    CHARACTER(1), DIMENSION(:), INTENT(INOUT) :: a
    CHARACTER(1), DIMENSION(:), INTENT(OUT) :: s
    END SUBROUTINE mp2dfr
END INTERFACE
INTERFACE
    SUBROUTINE mpdiv(q,r,u,v,n,m)
    USE nrtype
    CHARACTER(1), DIMENSION(:), INTENT(OUT) :: q,r
    CHARACTER(1), DIMENSION(:), INTENT(IN) :: u,v
    INTEGER(I4B), INTENT(IN) :: n,m
    END SUBROUTINE mpdiv
END INTERFACE
INTERFACE
    SUBROUTINE mpinv(u,v,n,m)
    USE nrtype
    CHARACTER(1), DIMENSION(:), INTENT(OUT) :: u
    CHARACTER(1), DIMENSION(:), INTENT(IN) :: v
    INTEGER(I4B), INTENT(IN) :: n,m
    END SUBROUTINE mpinv
END INTERFACE
INTERFACE
    SUBROUTINE mpmul(w,u,v,n,m)
    USE nrtype
    CHARACTER(1), DIMENSION(:), INTENT(IN) :: u,v
    CHARACTER(1), DIMENSION(:), INTENT(OUT) :: w
    INTEGER(I4B), INTENT(IN) :: n,m
    END SUBROUTINE mpmul
END INTERFACE
INTERFACE
    SUBROUTINE mppi(n)
    USE nrtype
    INTEGER(I4B), INTENT(IN) :: n
    END SUBROUTINE mppi
END INTERFACE
INTERFACE
    SUBROUTINE mprove(a,alud,indx,b,x)
    USE nrtype
    REAL(SP), DIMENSION(:,:), INTENT(IN) :: a,alud
    INTEGER(I4B), DIMENSION(:), INTENT(IN) :: indx
    REAL(SP), DIMENSION(:), INTENT(IN) :: b
    REAL(SP), DIMENSION(:), INTENT(INOUT) :: x
    END SUBROUTINE mprove
END INTERFACE
INTERFACE
    SUBROUTINE mpsqrt(w,u,v,n,m)
    USE nrtype
    CHARACTER(1), DIMENSION(:), INTENT(OUT) :: w,u
    CHARACTER(1), DIMENSION(:), INTENT(IN) :: v
    INTEGER(I4B), INTENT(IN) :: n,m
    END SUBROUTINE mpsqrt
END INTERFACE
INTERFACE
    SUBROUTINE mrqcof(x,y,sig,a,maska,alpha,beta,chisq,funcs)
    USE nrtype
    REAL(SP), DIMENSION(:), INTENT(IN) :: x,y,a,sig
    REAL(SP), DIMENSION(:), INTENT(OUT) :: beta
    REAL(SP), DIMENSION(:,:), INTENT(OUT) :: alpha
    REAL(SP), INTENT(OUT) :: chisq
    LOGICAL(LGT), DIMENSION(:), INTENT(IN) :: maska
    INTERFACE
        SUBROUTINE funcs(x,a,yfit,dyda)
        USE nrtype
```

```
            REAL(SP), DIMENSION(:), INTENT(IN) :: x,a
            REAL(SP), DIMENSION(:), INTENT(OUT) :: yfit
            REAL(SP), DIMENSION(:,:), INTENT(OUT) :: dyda
            END SUBROUTINE funcs
        END INTERFACE
        END SUBROUTINE mrqcof
    END INTERFACE
    INTERFACE
        SUBROUTINE mrqmin(x,y,sig,a,maska,covar,alpha,chisq,funcs,alamda)
        USE nrtype
        REAL(SP), DIMENSION(:), INTENT(IN) :: x,y,sig
        REAL(SP), DIMENSION(:), INTENT(INOUT) :: a
        REAL(SP), DIMENSION(:,:), INTENT(OUT) :: covar,alpha
        REAL(SP), INTENT(OUT) :: chisq
        REAL(SP), INTENT(INOUT) :: alamda
        LOGICAL(LGT), DIMENSION(:), INTENT(IN) :: maska
        INTERFACE
            SUBROUTINE funcs(x,a,yfit,dyda)
            USE nrtype
            REAL(SP), DIMENSION(:), INTENT(IN) :: x,a
            REAL(SP), DIMENSION(:), INTENT(OUT) :: yfit
            REAL(SP), DIMENSION(:,:), INTENT(OUT) :: dyda
            END SUBROUTINE funcs
        END INTERFACE
        END SUBROUTINE mrqmin
    END INTERFACE
    INTERFACE
        SUBROUTINE newt(x,check)
        USE nrtype
        REAL(SP), DIMENSION(:), INTENT(INOUT) :: x
        LOGICAL(LGT), INTENT(OUT) :: check
        END SUBROUTINE newt
    END INTERFACE
    INTERFACE
        SUBROUTINE odeint(ystart,x1,x2,eps,h1,hmin,derivs,rkqs)
        USE nrtype
        REAL(SP), DIMENSION(:), INTENT(INOUT) :: ystart
        REAL(SP), INTENT(IN) :: x1,x2,eps,h1,hmin
        INTERFACE
            SUBROUTINE derivs(x,y,dydx)
            USE nrtype
            REAL(SP), INTENT(IN) :: x
            REAL(SP), DIMENSION(:), INTENT(IN) :: y
            REAL(SP), DIMENSION(:), INTENT(OUT) :: dydx
            END SUBROUTINE derivs

            SUBROUTINE rkqs(y,dydx,x,htry,eps,yscal,hdid,hnext,derivs)
            USE nrtype
            REAL(SP), DIMENSION(:), INTENT(INOUT) :: y
            REAL(SP), DIMENSION(:), INTENT(IN) :: dydx,yscal
            REAL(SP), INTENT(INOUT) :: x
            REAL(SP), INTENT(IN) :: htry,eps
            REAL(SP), INTENT(OUT) :: hdid,hnext
                INTERFACE
                SUBROUTINE derivs(x,y,dydx)
                    USE nrtype
                    REAL(SP), INTENT(IN) :: x
                    REAL(SP), DIMENSION(:), INTENT(IN) :: y
                    REAL(SP), DIMENSION(:), INTENT(OUT) :: dydx
                    END SUBROUTINE derivs
                END INTERFACE
            END SUBROUTINE rkqs
        END INTERFACE
        END SUBROUTINE odeint
```

```
END INTERFACE
INTERFACE
    SUBROUTINE orthog(anu,alpha,beta,a,b)
    USE nrtype
    REAL(SP), DIMENSION(:), INTENT(IN) :: anu,alpha,beta
    REAL(SP), DIMENSION(:), INTENT(OUT) :: a,b
    END SUBROUTINE orthog
END INTERFACE
INTERFACE
    SUBROUTINE pade(cof,resid)
    USE nrtype
    REAL(DP), DIMENSION(:), INTENT(INOUT) :: cof
    REAL(SP), INTENT(OUT) :: resid
    END SUBROUTINE pade
END INTERFACE
INTERFACE
    FUNCTION pccheb(d)
    USE nrtype
    REAL(SP), DIMENSION(:), INTENT(IN) :: d
    REAL(SP), DIMENSION(size(d)) :: pccheb
    END FUNCTION pccheb
END INTERFACE
INTERFACE
    SUBROUTINE pcshft(a,b,d)
    USE nrtype
    REAL(SP), INTENT(IN) :: a,b
    REAL(SP), DIMENSION(:), INTENT(INOUT) :: d
    END SUBROUTINE pcshft
END INTERFACE
INTERFACE
    SUBROUTINE pearsn(x,y,r,prob,z)
    USE nrtype
    REAL(SP), INTENT(OUT) :: r,prob,z
    REAL(SP), DIMENSION(:), INTENT(IN) :: x,y
    END SUBROUTINE pearsn
END INTERFACE
INTERFACE
    SUBROUTINE period(x,y,ofac,hifac,px,py,jmax,prob)
    USE nrtype
    INTEGER(I4B), INTENT(OUT) :: jmax
    REAL(SP), INTENT(IN) :: ofac,hifac
    REAL(SP), INTENT(OUT) :: prob
    REAL(SP), DIMENSION(:), INTENT(IN) :: x,y
    REAL(SP), DIMENSION(:), POINTER :: px,py
    END SUBROUTINE period
END INTERFACE
INTERFACE plgndr
    FUNCTION plgndr_s(l,m,x)
    USE nrtype
    INTEGER(I4B), INTENT(IN) :: l,m
    REAL(SP), INTENT(IN) :: x
    REAL(SP) :: plgndr_s
    END FUNCTION plgndr_s

    FUNCTION plgndr_v(l,m,x)
    USE nrtype
    INTEGER(I4B), INTENT(IN) :: l,m
    REAL(SP), DIMENSION(:), INTENT(IN) :: x
    REAL(SP), DIMENSION(size(x)) :: plgndr_v
    END FUNCTION plgndr_v
END INTERFACE
INTERFACE
    FUNCTION poidev(xm)
    USE nrtype
```

```
    REAL(SP), INTENT(IN) :: xm
    REAL(SP) :: poidev
    END FUNCTION poidev
END INTERFACE
INTERFACE
    FUNCTION polcoe(x,y)
    USE nrtype
    REAL(SP), DIMENSION(:), INTENT(IN) :: x,y
    REAL(SP), DIMENSION(size(x)) :: polcoe
    END FUNCTION polcoe
END INTERFACE
INTERFACE
    FUNCTION polcof(xa,ya)
    USE nrtype
    REAL(SP), DIMENSION(:), INTENT(IN) :: xa,ya
    REAL(SP), DIMENSION(size(xa)) :: polcof
    END FUNCTION polcof
END INTERFACE
INTERFACE
    SUBROUTINE poldiv(u,v,q,r)
    USE nrtype
    REAL(SP), DIMENSION(:), INTENT(IN) :: u,v
    REAL(SP), DIMENSION(:), INTENT(OUT) :: q,r
    END SUBROUTINE poldiv
END INTERFACE
INTERFACE
    SUBROUTINE polin2(x1a,x2a,ya,x1,x2,y,dy)
    USE nrtype
    REAL(SP), DIMENSION(:), INTENT(IN) :: x1a,x2a
    REAL(SP), DIMENSION(:,:), INTENT(IN) :: ya
    REAL(SP), INTENT(IN) :: x1,x2
    REAL(SP), INTENT(OUT) :: y,dy
    END SUBROUTINE polin2
END INTERFACE
INTERFACE
    SUBROUTINE polint(xa,ya,x,y,dy)
    USE nrtype
    REAL(SP), DIMENSION(:), INTENT(IN) :: xa,ya
    REAL(SP), INTENT(IN) :: x
    REAL(SP), INTENT(OUT) :: y,dy
    END SUBROUTINE polint
END INTERFACE
INTERFACE
    SUBROUTINE powell(p,xi,ftol,iter,fret)
    USE nrtype
    REAL(SP), DIMENSION(:), INTENT(INOUT) :: p
    REAL(SP), DIMENSION(:,:), INTENT(INOUT) :: xi
    INTEGER(I4B), INTENT(OUT) :: iter
    REAL(SP), INTENT(IN) :: ftol
    REAL(SP), INTENT(OUT) :: fret
    END SUBROUTINE powell
END INTERFACE
INTERFACE
    FUNCTION predic(data,d,nfut)
    USE nrtype
    REAL(SP), DIMENSION(:), INTENT(IN) :: data,d
    INTEGER(I4B), INTENT(IN) :: nfut
    REAL(SP), DIMENSION(nfut) :: predic
    END FUNCTION predic
END INTERFACE
INTERFACE
    FUNCTION probks(alam)
    USE nrtype
    REAL(SP), INTENT(IN) :: alam
```

```
        REAL(SP) :: probks
        END FUNCTION probks
    END INTERFACE
    INTERFACE psdes
        SUBROUTINE psdes_s(lword,rword)
        USE nrtype
        INTEGER(I4B), INTENT(INOUT) :: lword,rword
        END SUBROUTINE psdes_s

        SUBROUTINE psdes_v(lword,rword)
        USE nrtype
        INTEGER(I4B), DIMENSION(:), INTENT(INOUT) :: lword,rword
        END SUBROUTINE psdes_v
    END INTERFACE
    INTERFACE
        SUBROUTINE pwt(a,isign)
        USE nrtype
        REAL(SP), DIMENSION(:), INTENT(INOUT) :: a
        INTEGER(I4B), INTENT(IN) :: isign
        END SUBROUTINE pwt
    END INTERFACE
    INTERFACE
        SUBROUTINE pwtset(n)
        USE nrtype
        INTEGER(I4B), INTENT(IN) :: n
        END SUBROUTINE pwtset
    END INTERFACE
    INTERFACE pythag
        FUNCTION pythag_dp(a,b)
        USE nrtype
        REAL(DP), INTENT(IN) :: a,b
        REAL(DP) :: pythag_dp
        END FUNCTION pythag_dp

        FUNCTION pythag_sp(a,b)
        USE nrtype
        REAL(SP), INTENT(IN) :: a,b
        REAL(SP) :: pythag_sp
        END FUNCTION pythag_sp
    END INTERFACE
    INTERFACE
        SUBROUTINE pzextr(iest,xest,yest,yz,dy)
        USE nrtype
        INTEGER(I4B), INTENT(IN) :: iest
        REAL(SP), INTENT(IN) :: xest
        REAL(SP), DIMENSION(:), INTENT(IN) :: yest
        REAL(SP), DIMENSION(:), INTENT(OUT) :: yz,dy
        END SUBROUTINE pzextr
    END INTERFACE
    INTERFACE
        SUBROUTINE qrdcmp(a,c,d,sing)
        USE nrtype
        REAL(SP), DIMENSION(:,:), INTENT(INOUT) :: a
        REAL(SP), DIMENSION(:), INTENT(OUT) :: c,d
        LOGICAL(LGT), INTENT(OUT) :: sing
        END SUBROUTINE qrdcmp
    END INTERFACE
    INTERFACE
        FUNCTION qromb(func,a,b)
        USE nrtype
        REAL(SP), INTENT(IN) :: a,b
        REAL(SP) :: qromb
        INTERFACE
            FUNCTION func(x)
            USE nrtype
```

```
        REAL(SP), DIMENSION(:), INTENT(IN) :: x
        REAL(SP), DIMENSION(size(x)) :: func
        END FUNCTION func
    END INTERFACE
    END FUNCTION qromb
END INTERFACE
INTERFACE
    FUNCTION qromo(func,a,b,choose)
    USE nrtype
    REAL(SP), INTENT(IN) :: a,b
    REAL(SP) :: qromo
    INTERFACE
        FUNCTION func(x)
        USE nrtype
        REAL(SP), DIMENSION(:), INTENT(IN) :: x
        REAL(SP), DIMENSION(size(x)) :: func
        END FUNCTION func
    END INTERFACE
    INTERFACE
        SUBROUTINE choose(funk,aa,bb,s,n)
        USE nrtype
        REAL(SP), INTENT(IN) :: aa,bb
        REAL(SP), INTENT(INOUT) :: s
        INTEGER(I4B), INTENT(IN) :: n
        INTERFACE
            FUNCTION funk(x)
            USE nrtype
            REAL(SP), DIMENSION(:), INTENT(IN) :: x
            REAL(SP), DIMENSION(size(x)) :: funk
            END FUNCTION funk
        END INTERFACE
        END SUBROUTINE choose
    END INTERFACE
    END FUNCTION qromo
END INTERFACE
INTERFACE
    SUBROUTINE qroot(p,b,c,eps)
    USE nrtype
    REAL(SP), DIMENSION(:), INTENT(IN) :: p
    REAL(SP), INTENT(INOUT) :: b,c
    REAL(SP), INTENT(IN) :: eps
    END SUBROUTINE qroot
END INTERFACE
INTERFACE
    SUBROUTINE qrsolv(a,c,d,b)
    USE nrtype
    REAL(SP), DIMENSION(:,:), INTENT(IN) :: a
    REAL(SP), DIMENSION(:), INTENT(IN) :: c,d
    REAL(SP), DIMENSION(:), INTENT(INOUT) :: b
    END SUBROUTINE qrsolv
END INTERFACE
INTERFACE
    SUBROUTINE qrupdt(r,qt,u,v)
    USE nrtype
    REAL(SP), DIMENSION(:,:), INTENT(INOUT) :: r,qt
    REAL(SP), DIMENSION(:), INTENT(INOUT) :: u
    REAL(SP), DIMENSION(:), INTENT(IN) :: v
    END SUBROUTINE qrupdt
END INTERFACE
INTERFACE
    FUNCTION qsimp(func,a,b)
    USE nrtype
    REAL(SP), INTENT(IN) :: a,b
    REAL(SP) :: qsimp
```

```
        INTERFACE
            FUNCTION func(x)
            USE nrtype
            REAL(SP), DIMENSION(:), INTENT(IN) :: x
            REAL(SP), DIMENSION(size(x)) :: func
            END FUNCTION func
        END INTERFACE
        END FUNCTION qsimp
    END INTERFACE
    INTERFACE
        FUNCTION qtrap(func,a,b)
        USE nrtype
        REAL(SP), INTENT(IN) :: a,b
        REAL(SP) :: qtrap
        INTERFACE
            FUNCTION func(x)
            USE nrtype
            REAL(SP), DIMENSION(:), INTENT(IN) :: x
            REAL(SP), DIMENSION(size(x)) :: func
            END FUNCTION func
        END INTERFACE
        END FUNCTION qtrap
    END INTERFACE
    INTERFACE
        SUBROUTINE quadct(x,y,xx,yy,fa,fb,fc,fd)
        USE nrtype
        REAL(SP), INTENT(IN) :: x,y
        REAL(SP), DIMENSION(:), INTENT(IN) :: xx,yy
        REAL(SP), INTENT(OUT) :: fa,fb,fc,fd
        END SUBROUTINE quadct
    END INTERFACE
    INTERFACE
        SUBROUTINE quadmx(a)
        USE nrtype
        REAL(SP), DIMENSION(:,:), INTENT(OUT) :: a
        END SUBROUTINE quadmx
    END INTERFACE
    INTERFACE
        SUBROUTINE quadvl(x,y,fa,fb,fc,fd)
        USE nrtype
        REAL(SP), INTENT(IN) :: x,y
        REAL(SP), INTENT(OUT) :: fa,fb,fc,fd
        END SUBROUTINE quadvl
    END INTERFACE
    INTERFACE
        FUNCTION ran(idum)
        INTEGER(selected_int_kind(9)), INTENT(INOUT) :: idum
        REAL :: ran
        END FUNCTION ran
    END INTERFACE
    INTERFACE ran0
        SUBROUTINE ran0_s(harvest)
        USE nrtype
        REAL(SP), INTENT(OUT) :: harvest
        END SUBROUTINE ran0_s

        SUBROUTINE ran0_v(harvest)
        USE nrtype
        REAL(SP), DIMENSION(:), INTENT(OUT) :: harvest
        END SUBROUTINE ran0_v
    END INTERFACE
    INTERFACE ran1
        SUBROUTINE ran1_s(harvest)
        USE nrtype
```

```
        REAL(SP), INTENT(OUT) :: harvest
        END SUBROUTINE ran1_s

        SUBROUTINE ran1_v(harvest)
        USE nrtype
        REAL(SP), DIMENSION(:), INTENT(OUT) :: harvest
        END SUBROUTINE ran1_v
    END INTERFACE
    INTERFACE ran2
        SUBROUTINE ran2_s(harvest)
        USE nrtype
        REAL(SP), INTENT(OUT) :: harvest
        END SUBROUTINE ran2_s

        SUBROUTINE ran2_v(harvest)
        USE nrtype
        REAL(SP), DIMENSION(:), INTENT(OUT) :: harvest
        END SUBROUTINE ran2_v
    END INTERFACE
    INTERFACE ran3
        SUBROUTINE ran3_s(harvest)
        USE nrtype
        REAL(SP), INTENT(OUT) :: harvest
        END SUBROUTINE ran3_s

        SUBROUTINE ran3_v(harvest)
        USE nrtype
        REAL(SP), DIMENSION(:), INTENT(OUT) :: harvest
        END SUBROUTINE ran3_v
    END INTERFACE
    INTERFACE
        SUBROUTINE ratint(xa,ya,x,y,dy)
        USE nrtype
        REAL(SP), DIMENSION(:), INTENT(IN) :: xa,ya
        REAL(SP), INTENT(IN) :: x
        REAL(SP), INTENT(OUT) :: y,dy
        END SUBROUTINE ratint
    END INTERFACE
    INTERFACE
        SUBROUTINE ratlsq(func,a,b,mm,kk,cof,dev)
        USE nrtype
        REAL(DP), INTENT(IN) :: a,b
        INTEGER(I4B), INTENT(IN) :: mm,kk
        REAL(DP), DIMENSION(:), INTENT(OUT) :: cof
        REAL(DP), INTENT(OUT) :: dev
        INTERFACE
            FUNCTION func(x)
            USE nrtype
            REAL(DP), DIMENSION(:), INTENT(IN) :: x
            REAL(DP), DIMENSION(size(x)) :: func
            END FUNCTION func
        END INTERFACE
        END SUBROUTINE ratlsq
    END INTERFACE
    INTERFACE ratval
        FUNCTION ratval_s(x,cof,mm,kk)
        USE nrtype
        REAL(DP), INTENT(IN) :: x
        INTEGER(I4B), INTENT(IN) :: mm,kk
        REAL(DP), DIMENSION(mm+kk+1), INTENT(IN) :: cof
        REAL(DP) :: ratval_s
        END FUNCTION ratval_s

        FUNCTION ratval_v(x,cof,mm,kk)
        USE nrtype
        REAL(DP), DIMENSION(:), INTENT(IN) :: x
```

```
        INTEGER(I4B), INTENT(IN) :: mm,kk
        REAL(DP), DIMENSION(mm+kk+1), INTENT(IN) :: cof
        REAL(DP), DIMENSION(size(x)) :: ratval_v
        END FUNCTION ratval_v
    END INTERFACE
    INTERFACE rc
        FUNCTION rc_s(x,y)
        USE nrtype
        REAL(SP), INTENT(IN) :: x,y
        REAL(SP) :: rc_s
        END FUNCTION rc_s

        FUNCTION rc_v(x,y)
        USE nrtype
        REAL(SP), DIMENSION(:), INTENT(IN) :: x,y
        REAL(SP), DIMENSION(size(x)) :: rc_v
        END FUNCTION rc_v
    END INTERFACE
    INTERFACE rd
        FUNCTION rd_s(x,y,z)
        USE nrtype
        REAL(SP), INTENT(IN) :: x,y,z
        REAL(SP) :: rd_s
        END FUNCTION rd_s

        FUNCTION rd_v(x,y,z)
        USE nrtype
        REAL(SP), DIMENSION(:), INTENT(IN) :: x,y,z
        REAL(SP), DIMENSION(size(x)) :: rd_v
        END FUNCTION rd_v
    END INTERFACE
    INTERFACE realft
        SUBROUTINE realft_dp(data,isign,zdata)
        USE nrtype
        REAL(DP), DIMENSION(:), INTENT(INOUT) :: data
        INTEGER(I4B), INTENT(IN) :: isign
        COMPLEX(DPC), DIMENSION(:), OPTIONAL, TARGET :: zdata
        END SUBROUTINE realft_dp

        SUBROUTINE realft_sp(data,isign,zdata)
        USE nrtype
        REAL(SP), DIMENSION(:), INTENT(INOUT) :: data
        INTEGER(I4B), INTENT(IN) :: isign
        COMPLEX(SPC), DIMENSION(:), OPTIONAL, TARGET :: zdata
        END SUBROUTINE realft_sp
    END INTERFACE
    INTERFACE
        RECURSIVE FUNCTION recur1(a,b) RESULT(u)
        USE nrtype
        REAL(SP), DIMENSION(:), INTENT(IN) :: a,b
        REAL(SP), DIMENSION(size(a)) :: u
        END FUNCTION recur1
    END INTERFACE
    INTERFACE
        FUNCTION recur2(a,b,c)
        USE nrtype
        REAL(SP), DIMENSION(:), INTENT(IN) :: a,b,c
        REAL(SP), DIMENSION(size(a)) :: recur2
        END FUNCTION recur2
    END INTERFACE
    INTERFACE
        SUBROUTINE relax(u,rhs)
        USE nrtype
        REAL(DP), DIMENSION(:,:), INTENT(INOUT) :: u
        REAL(DP), DIMENSION(:,:), INTENT(IN) :: rhs
        END SUBROUTINE relax
```

```
END INTERFACE
INTERFACE
    SUBROUTINE relax2(u,rhs)
    USE nrtype
    REAL(DP), DIMENSION(:,:), INTENT(INOUT) :: u
    REAL(DP), DIMENSION(:,:), INTENT(IN) :: rhs
    END SUBROUTINE relax2
END INTERFACE
INTERFACE
FUNCTION resid(u,rhs)
    USE nrtype
    REAL(DP), DIMENSION(:,:), INTENT(IN) :: u,rhs
    REAL(DP), DIMENSION(size(u,1),size(u,1)) :: resid
    END FUNCTION resid
END INTERFACE
INTERFACE rf
    FUNCTION rf_s(x,y,z)
    USE nrtype
    REAL(SP), INTENT(IN) :: x,y,z
    REAL(SP) :: rf_s
    END FUNCTION rf_s

    FUNCTION rf_v(x,y,z)
    USE nrtype
    REAL(SP), DIMENSION(:), INTENT(IN) :: x,y,z
    REAL(SP), DIMENSION(size(x)) :: rf_v
    END FUNCTION rf_v
END INTERFACE
INTERFACE rj
    FUNCTION rj_s(x,y,z,p)
    USE nrtype
    REAL(SP), INTENT(IN) :: x,y,z,p
    REAL(SP) :: rj_s
    END FUNCTION rj_s

    FUNCTION rj_v(x,y,z,p)
    USE nrtype
    REAL(SP), DIMENSION(:), INTENT(IN) :: x,y,z,p
    REAL(SP), DIMENSION(size(x)) :: rj_v
    END FUNCTION rj_v
END INTERFACE
INTERFACE
    SUBROUTINE rk4(y,dydx,x,h,yout,derivs)
    USE nrtype
    REAL(SP), DIMENSION(:), INTENT(IN) :: y,dydx
    REAL(SP), INTENT(IN) :: x,h
    REAL(SP), DIMENSION(:), INTENT(OUT) :: yout
    INTERFACE
        SUBROUTINE derivs(x,y,dydx)
        USE nrtype
        REAL(SP), INTENT(IN) :: x
        REAL(SP), DIMENSION(:), INTENT(IN) :: y
        REAL(SP), DIMENSION(:), INTENT(OUT) :: dydx
        END SUBROUTINE derivs
    END INTERFACE
    END SUBROUTINE rk4
END INTERFACE
INTERFACE
    SUBROUTINE rkck(y,dydx,x,h,yout,yerr,derivs)
    USE nrtype
    REAL(SP), DIMENSION(:), INTENT(IN) :: y,dydx
    REAL(SP), INTENT(IN) :: x,h
    REAL(SP), DIMENSION(:), INTENT(OUT) :: yout,yerr
    INTERFACE
        SUBROUTINE derivs(x,y,dydx)
```

```
        USE nrtype
        REAL(SP), INTENT(IN) :: x
        REAL(SP), DIMENSION(:), INTENT(IN) :: y
        REAL(SP), DIMENSION(:), INTENT(OUT) :: dydx
        END SUBROUTINE derivs
    END INTERFACE
    END SUBROUTINE rkck
END INTERFACE
INTERFACE
    SUBROUTINE rkdumb(vstart,x1,x2,nstep,derivs)
    USE nrtype
    REAL(SP), DIMENSION(:), INTENT(IN) :: vstart
    REAL(SP), INTENT(IN) :: x1,x2
    INTEGER(I4B), INTENT(IN) :: nstep
    INTERFACE
        SUBROUTINE derivs(x,y,dydx)
        USE nrtype
        REAL(SP), INTENT(IN) :: x
        REAL(SP), DIMENSION(:), INTENT(IN) :: y
        REAL(SP), DIMENSION(:), INTENT(OUT) :: dydx
        END SUBROUTINE derivs
    END INTERFACE
    END SUBROUTINE rkdumb
END INTERFACE
INTERFACE
    SUBROUTINE rkqs(y,dydx,x,htry,eps,yscal,hdid,hnext,derivs)
    USE nrtype
    REAL(SP), DIMENSION(:), INTENT(INOUT) :: y
    REAL(SP), DIMENSION(:), INTENT(IN) :: dydx,yscal
    REAL(SP), INTENT(INOUT) :: x
    REAL(SP), INTENT(IN) :: htry,eps
    REAL(SP), INTENT(OUT) :: hdid,hnext
    INTERFACE
        SUBROUTINE derivs(x,y,dydx)
        USE nrtype
        REAL(SP), INTENT(IN) :: x
        REAL(SP), DIMENSION(:), INTENT(IN) :: y
        REAL(SP), DIMENSION(:), INTENT(OUT) :: dydx
        END SUBROUTINE derivs
    END INTERFACE
    END SUBROUTINE rkqs
END INTERFACE
INTERFACE
    SUBROUTINE rlft2(data,spec,speq,isign)
    USE nrtype
    REAL(SP), DIMENSION(:,:), INTENT(INOUT) :: data
    COMPLEX(SPC), DIMENSION(:,:), INTENT(OUT) :: spec
    COMPLEX(SPC), DIMENSION(:), INTENT(OUT) :: speq
    INTEGER(I4B), INTENT(IN) :: isign
    END SUBROUTINE rlft2
END INTERFACE
INTERFACE
    SUBROUTINE rlft3(data,spec,speq,isign)
    USE nrtype
    REAL(SP), DIMENSION(:,:,:), INTENT(INOUT) :: data
    COMPLEX(SPC), DIMENSION(:,:,:), INTENT(OUT) :: spec
    COMPLEX(SPC), DIMENSION(:,:), INTENT(OUT) :: speq
    INTEGER(I4B), INTENT(IN) :: isign
    END SUBROUTINE rlft3
END INTERFACE
INTERFACE
    SUBROUTINE rotate(r,qt,i,a,b)
    USE nrtype
    REAL(SP), DIMENSION(:,:), TARGET, INTENT(INOUT) :: r,qt
```

```
        INTEGER(I4B), INTENT(IN) :: i
        REAL(SP), INTENT(IN) :: a,b
        END SUBROUTINE rotate
    END INTERFACE
    INTERFACE
        SUBROUTINE rsolv(a,d,b)
        USE nrtype
        REAL(SP), DIMENSION(:,:), INTENT(IN) :: a
        REAL(SP), DIMENSION(:), INTENT(IN) :: d
        REAL(SP), DIMENSION(:), INTENT(INOUT) :: b
        END SUBROUTINE rsolv
    END INTERFACE
    INTERFACE
        FUNCTION rstrct(uf)
        USE nrtype
        REAL(DP), DIMENSION(:,:), INTENT(IN) :: uf
        REAL(DP), DIMENSION((size(uf,1)+1)/2,(size(uf,1)+1)/2) :: rstrct
        END FUNCTION rstrct
    END INTERFACE
    INTERFACE
        FUNCTION rtbis(func,x1,x2,xacc)
        USE nrtype
        REAL(SP), INTENT(IN) :: x1,x2,xacc
        REAL(SP) :: rtbis
        INTERFACE
            FUNCTION func(x)
            USE nrtype
            REAL(SP), INTENT(IN) :: x
            REAL(SP) :: func
            END FUNCTION func
        END INTERFACE
        END FUNCTION rtbis
    END INTERFACE
    INTERFACE
        FUNCTION rtflsp(func,x1,x2,xacc)
        USE nrtype
        REAL(SP), INTENT(IN) :: x1,x2,xacc
        REAL(SP) :: rtflsp
        INTERFACE
            FUNCTION func(x)
            USE nrtype
            REAL(SP), INTENT(IN) :: x
            REAL(SP) :: func
            END FUNCTION func
        END INTERFACE
        END FUNCTION rtflsp
    END INTERFACE
    INTERFACE
        FUNCTION rtnewt(funcd,x1,x2,xacc)
        USE nrtype
        REAL(SP), INTENT(IN) :: x1,x2,xacc
        REAL(SP) :: rtnewt
        INTERFACE
            SUBROUTINE funcd(x,fval,fderiv)
            USE nrtype
            REAL(SP), INTENT(IN) :: x
            REAL(SP), INTENT(OUT) :: fval,fderiv
            END SUBROUTINE funcd
        END INTERFACE
        END FUNCTION rtnewt
    END INTERFACE
    INTERFACE
        FUNCTION rtsafe(funcd,x1,x2,xacc)
        USE nrtype
```

```
        REAL(SP), INTENT(IN) :: x1,x2,xacc
        REAL(SP) :: rtsafe
        INTERFACE
            SUBROUTINE funcd(x,fval,fderiv)
            USE nrtype
            REAL(SP), INTENT(IN) :: x
            REAL(SP), INTENT(OUT) :: fval,fderiv
            END SUBROUTINE funcd
        END INTERFACE
        END FUNCTION rtsafe
    END INTERFACE
    INTERFACE
        FUNCTION rtsec(func,x1,x2,xacc)
        USE nrtype
        REAL(SP), INTENT(IN) :: x1,x2,xacc
        REAL(SP) :: rtsec
        INTERFACE
            FUNCTION func(x)
            USE nrtype
            REAL(SP), INTENT(IN) :: x
            REAL(SP) :: func
            END FUNCTION func
        END INTERFACE
        END FUNCTION rtsec
    END INTERFACE
    INTERFACE
        SUBROUTINE rzextr(iest,xest,yest,yz,dy)
        USE nrtype
        INTEGER(I4B), INTENT(IN) :: iest
        REAL(SP), INTENT(IN) :: xest
        REAL(SP), DIMENSION(:), INTENT(IN) :: yest
        REAL(SP), DIMENSION(:), INTENT(OUT) :: yz,dy
        END SUBROUTINE rzextr
    END INTERFACE
    INTERFACE
        FUNCTION savgol(nl,nrr,ld,m)
        USE nrtype
        INTEGER(I4B), INTENT(IN) :: nl,nrr,ld,m
        REAL(SP), DIMENSION(nl+nrr+1) :: savgol
        END FUNCTION savgol
    END INTERFACE
    INTERFACE
        SUBROUTINE scrsho(func)
        USE nrtype
        INTERFACE
            FUNCTION func(x)
            USE nrtype
            REAL(SP), INTENT(IN) :: x
            REAL(SP) :: func
            END FUNCTION func
        END INTERFACE
        END SUBROUTINE scrsho
    END INTERFACE
    INTERFACE
        FUNCTION select(k,arr)
        USE nrtype
        INTEGER(I4B), INTENT(IN) :: k
        REAL(SP), DIMENSION(:), INTENT(INOUT) :: arr
        REAL(SP) :: select
        END FUNCTION select
    END INTERFACE
    INTERFACE
        FUNCTION select_bypack(k,arr)
        USE nrtype
```

```
      INTEGER(I4B), INTENT(IN) :: k
      REAL(SP), DIMENSION(:), INTENT(INOUT) :: arr
      REAL(SP) :: select_bypack
      END FUNCTION select_bypack
   END INTERFACE
   INTERFACE
      SUBROUTINE select_heap(arr,heap)
      USE nrtype
      REAL(SP), DIMENSION(:), INTENT(IN) :: arr
      REAL(SP), DIMENSION(:), INTENT(OUT) :: heap
      END SUBROUTINE select_heap
   END INTERFACE
   INTERFACE
      FUNCTION select_inplace(k,arr)
      USE nrtype
      INTEGER(I4B), INTENT(IN) :: k
      REAL(SP), DIMENSION(:), INTENT(IN) :: arr
      REAL(SP) :: select_inplace
      END FUNCTION select_inplace
   END INTERFACE
   INTERFACE
      SUBROUTINE simplx(a,m1,m2,m3,icase,izrov,iposv)
      USE nrtype
      REAL(SP), DIMENSION(:,:), INTENT(INOUT) :: a
      INTEGER(I4B), INTENT(IN) :: m1,m2,m3
      INTEGER(I4B), INTENT(OUT) :: icase
      INTEGER(I4B), DIMENSION(:), INTENT(OUT) :: izrov,iposv
      END SUBROUTINE simplx
   END INTERFACE
   INTERFACE
      SUBROUTINE simpr(y,dydx,dfdx,dfdy,xs,htot,nstep,yout,derivs)
      USE nrtype
      REAL(SP), INTENT(IN) :: xs,htot
      REAL(SP), DIMENSION(:), INTENT(IN) :: y,dydx,dfdx
      REAL(SP), DIMENSION(:,:), INTENT(IN) :: dfdy
      INTEGER(I4B), INTENT(IN) :: nstep
      REAL(SP), DIMENSION(:), INTENT(OUT) :: yout
      INTERFACE
         SUBROUTINE derivs(x,y,dydx)
         USE nrtype
         REAL(SP), INTENT(IN) :: x
         REAL(SP), DIMENSION(:), INTENT(IN) :: y
         REAL(SP), DIMENSION(:), INTENT(OUT) :: dydx
         END SUBROUTINE derivs
      END INTERFACE
      END SUBROUTINE simpr
   END INTERFACE
   INTERFACE
      SUBROUTINE sinft(y)
      USE nrtype
      REAL(SP), DIMENSION(:), INTENT(INOUT) :: y
      END SUBROUTINE sinft
   END INTERFACE
   INTERFACE
      SUBROUTINE slvsm2(u,rhs)
      USE nrtype
      REAL(DP), DIMENSION(3,3), INTENT(OUT) :: u
      REAL(DP), DIMENSION(3,3), INTENT(IN) :: rhs
      END SUBROUTINE slvsm2
   END INTERFACE
   INTERFACE
      SUBROUTINE slvsml(u,rhs)
      USE nrtype
      REAL(DP), DIMENSION(3,3), INTENT(OUT) :: u
```

```
        REAL(DP), DIMENSION(3,3), INTENT(IN) :: rhs
        END SUBROUTINE slvsml
    END INTERFACE
    INTERFACE
        SUBROUTINE sncndn(uu,emmc,sn,cn,dn)
        USE nrtype
        REAL(SP), INTENT(IN) :: uu,emmc
        REAL(SP), INTENT(OUT) :: sn,cn,dn
        END SUBROUTINE sncndn
    END INTERFACE
    INTERFACE
        FUNCTION snrm(sx,itol)
        USE nrtype
        REAL(DP), DIMENSION(:), INTENT(IN) :: sx
        INTEGER(I4B), INTENT(IN) :: itol
        REAL(DP) :: snrm
        END FUNCTION snrm
    END INTERFACE
    INTERFACE
        SUBROUTINE sobseq(x,init)
        USE nrtype
        REAL(SP), DIMENSION(:), INTENT(OUT) :: x
        INTEGER(I4B), OPTIONAL, INTENT(IN) :: init
        END SUBROUTINE sobseq
    END INTERFACE
    INTERFACE
        SUBROUTINE solvde(itmax,conv,slowc,scalv,indexv,nb,y)
        USE nrtype
        INTEGER(I4B), INTENT(IN) :: itmax,nb
        REAL(SP), INTENT(IN) :: conv,slowc
        REAL(SP), DIMENSION(:), INTENT(IN) :: scalv
        INTEGER(I4B), DIMENSION(:), INTENT(IN) :: indexv
        REAL(SP), DIMENSION(:,:), INTENT(INOUT) :: y
        END SUBROUTINE solvde
    END INTERFACE
    INTERFACE
        SUBROUTINE sor(a,b,c,d,e,f,u,rjac)
        USE nrtype
        REAL(DP), DIMENSION(:,:), INTENT(IN) :: a,b,c,d,e,f
        REAL(DP), DIMENSION(:,:), INTENT(INOUT) :: u
        REAL(DP), INTENT(IN) :: rjac
        END SUBROUTINE sor
    END INTERFACE
    INTERFACE
        SUBROUTINE sort(arr)
        USE nrtype
        REAL(SP), DIMENSION(:), INTENT(INOUT) :: arr
        END SUBROUTINE sort
    END INTERFACE
    INTERFACE
        SUBROUTINE sort2(arr,slave)
        USE nrtype
        REAL(SP), DIMENSION(:), INTENT(INOUT) :: arr,slave
        END SUBROUTINE sort2
    END INTERFACE
    INTERFACE
        SUBROUTINE sort3(arr,slave1,slave2)
        USE nrtype
        REAL(SP), DIMENSION(:), INTENT(INOUT) :: arr,slave1,slave2
        END SUBROUTINE sort3
    END INTERFACE
    INTERFACE
        SUBROUTINE sort_bypack(arr)
        USE nrtype
```

```
        REAL(SP), DIMENSION(:), INTENT(INOUT) :: arr
        END SUBROUTINE sort_bypack
    END INTERFACE
    INTERFACE
        SUBROUTINE sort_byreshape(arr)
        USE nrtype
        REAL(SP), DIMENSION(:), INTENT(INOUT) :: arr
        END SUBROUTINE sort_byreshape
    END INTERFACE
    INTERFACE
        SUBROUTINE sort_heap(arr)
        USE nrtype
        REAL(SP), DIMENSION(:), INTENT(INOUT) :: arr
        END SUBROUTINE sort_heap
    END INTERFACE
    INTERFACE
        SUBROUTINE sort_pick(arr)
        USE nrtype
        REAL(SP), DIMENSION(:), INTENT(INOUT) :: arr
        END SUBROUTINE sort_pick
    END INTERFACE
    INTERFACE
        SUBROUTINE sort_radix(arr)
        USE nrtype
        REAL(SP), DIMENSION(:), INTENT(INOUT) :: arr
        END SUBROUTINE sort_radix
    END INTERFACE
    INTERFACE
        SUBROUTINE sort_shell(arr)
        USE nrtype
        REAL(SP), DIMENSION(:), INTENT(INOUT) :: arr
        END SUBROUTINE sort_shell
    END INTERFACE
    INTERFACE
        SUBROUTINE spctrm(p,k,ovrlap,unit,n_window)
        USE nrtype
        REAL(SP), DIMENSION(:), INTENT(OUT) :: p
        INTEGER(I4B), INTENT(IN) :: k
        LOGICAL(LGT), INTENT(IN) :: ovrlap
        INTEGER(I4B), OPTIONAL, INTENT(IN) :: n_window,unit
        END SUBROUTINE spctrm
    END INTERFACE
    INTERFACE
        SUBROUTINE spear(data1,data2,d,zd,probd,rs,probrs)
        USE nrtype
        REAL(SP), DIMENSION(:), INTENT(IN) :: data1,data2
        REAL(SP), INTENT(OUT) :: d,zd,probd,rs,probrs
        END SUBROUTINE spear
    END INTERFACE
    INTERFACE sphbes
        SUBROUTINE sphbes_s(n,x,sj,sy,sjp,syp)
        USE nrtype
        INTEGER(I4B), INTENT(IN) :: n
        REAL(SP), INTENT(IN) :: x
        REAL(SP), INTENT(OUT) :: sj,sy,sjp,syp
        END SUBROUTINE sphbes_s

        SUBROUTINE sphbes_v(n,x,sj,sy,sjp,syp)
        USE nrtype
        INTEGER(I4B), INTENT(IN) :: n
        REAL(SP), DIMENSION(:), INTENT(IN) :: x
        REAL(SP), DIMENSION(:), INTENT(OUT) :: sj,sy,sjp,syp
        END SUBROUTINE sphbes_v
    END INTERFACE
```

```
INTERFACE
    SUBROUTINE splie2(x1a,x2a,ya,y2a)
    USE nrtype
    REAL(SP), DIMENSION(:), INTENT(IN) :: x1a,x2a
    REAL(SP), DIMENSION(:,:), INTENT(IN) :: ya
    REAL(SP), DIMENSION(:,:), INTENT(OUT) :: y2a
    END SUBROUTINE splie2
END INTERFACE
INTERFACE
    FUNCTION splin2(x1a,x2a,ya,y2a,x1,x2)
    USE nrtype
    REAL(SP), DIMENSION(:), INTENT(IN) :: x1a,x2a
    REAL(SP), DIMENSION(:,:), INTENT(IN) :: ya,y2a
    REAL(SP), INTENT(IN) :: x1,x2
    REAL(SP) :: splin2
    END FUNCTION splin2
END INTERFACE
INTERFACE
    SUBROUTINE spline(x,y,yp1,ypn,y2)
    USE nrtype
    REAL(SP), DIMENSION(:), INTENT(IN) :: x,y
    REAL(SP), INTENT(IN) :: yp1,ypn
    REAL(SP), DIMENSION(:), INTENT(OUT) :: y2
    END SUBROUTINE spline
END INTERFACE
INTERFACE
    FUNCTION splint(xa,ya,y2a,x)
    USE nrtype
    REAL(SP), DIMENSION(:), INTENT(IN) :: xa,ya,y2a
    REAL(SP), INTENT(IN) :: x
    REAL(SP) :: splint
    END FUNCTION splint
END INTERFACE
INTERFACE sprsax
    SUBROUTINE sprsax_dp(sa,x,b)
    USE nrtype
    TYPE(sprs2_dp), INTENT(IN) :: sa
    REAL(DP), DIMENSION (:), INTENT(IN) :: x
    REAL(DP), DIMENSION (:), INTENT(OUT) :: b
    END SUBROUTINE sprsax_dp

    SUBROUTINE sprsax_sp(sa,x,b)
    USE nrtype
    TYPE(sprs2_sp), INTENT(IN) :: sa
    REAL(SP), DIMENSION (:), INTENT(IN) :: x
    REAL(SP), DIMENSION (:), INTENT(OUT) :: b
    END SUBROUTINE sprsax_sp
END INTERFACE
INTERFACE sprsdiag
    SUBROUTINE sprsdiag_dp(sa,b)
    USE nrtype
    TYPE(sprs2_dp), INTENT(IN) :: sa
    REAL(DP), DIMENSION(:), INTENT(OUT) :: b
    END SUBROUTINE sprsdiag_dp

    SUBROUTINE sprsdiag_sp(sa,b)
    USE nrtype
    TYPE(sprs2_sp), INTENT(IN) :: sa
    REAL(SP), DIMENSION(:), INTENT(OUT) :: b
    END SUBROUTINE sprsdiag_sp
END INTERFACE
INTERFACE sprsin
    SUBROUTINE sprsin_sp(a,thresh,sa)
    USE nrtype
    REAL(SP), DIMENSION(:,:), INTENT(IN) :: a
```

```
    REAL(SP), INTENT(IN) :: thresh
    TYPE(sprs2_sp), INTENT(OUT) :: sa
    END SUBROUTINE sprsin_sp

    SUBROUTINE sprsin_dp(a,thresh,sa)
    USE nrtype
    REAL(DP), DIMENSION(:,:), INTENT(IN) :: a
    REAL(DP), INTENT(IN) :: thresh
    TYPE(sprs2_dp), INTENT(OUT) :: sa
    END SUBROUTINE sprsin_dp
END INTERFACE
INTERFACE
    SUBROUTINE sprstp(sa)
    USE nrtype
    TYPE(sprs2_sp), INTENT(INOUT) :: sa
    END SUBROUTINE sprstp
END INTERFACE
INTERFACE sprstx
    SUBROUTINE sprstx_dp(sa,x,b)
    USE nrtype
    TYPE(sprs2_dp), INTENT(IN) :: sa
    REAL(DP), DIMENSION (:), INTENT(IN) :: x
    REAL(DP), DIMENSION (:), INTENT(OUT) :: b
    END SUBROUTINE sprstx_dp

    SUBROUTINE sprstx_sp(sa,x,b)
    USE nrtype
    TYPE(sprs2_sp), INTENT(IN) :: sa
    REAL(SP), DIMENSION (:), INTENT(IN) :: x
    REAL(SP), DIMENSION (:), INTENT(OUT) :: b
    END SUBROUTINE sprstx_sp
END INTERFACE
INTERFACE
    SUBROUTINE stifbs(y,dydx,x,htry,eps,yscal,hdid,hnext,derivs)
    USE nrtype
    REAL(SP), DIMENSION(:), INTENT(INOUT) :: y
    REAL(SP), DIMENSION(:), INTENT(IN) :: dydx,yscal
    REAL(SP), INTENT(IN) :: htry,eps
    REAL(SP), INTENT(INOUT) :: x
    REAL(SP), INTENT(OUT) :: hdid,hnext
    INTERFACE
        SUBROUTINE derivs(x,y,dydx)
        USE nrtype
        REAL(SP), INTENT(IN) :: x
        REAL(SP), DIMENSION(:), INTENT(IN) :: y
        REAL(SP), DIMENSION(:), INTENT(OUT) :: dydx
        END SUBROUTINE derivs
    END INTERFACE
    END SUBROUTINE stifbs
END INTERFACE
INTERFACE
    SUBROUTINE stiff(y,dydx,x,htry,eps,yscal,hdid,hnext,derivs)
    USE nrtype
    REAL(SP), DIMENSION(:), INTENT(INOUT) :: y
    REAL(SP), DIMENSION(:), INTENT(IN) :: dydx,yscal
    REAL(SP), INTENT(INOUT) :: x
    REAL(SP), INTENT(IN) :: htry,eps
    REAL(SP), INTENT(OUT) :: hdid,hnext
    INTERFACE
        SUBROUTINE derivs(x,y,dydx)
        USE nrtype
        REAL(SP), INTENT(IN) :: x
        REAL(SP), DIMENSION(:), INTENT(IN) :: y
        REAL(SP), DIMENSION(:), INTENT(OUT) :: dydx
        END SUBROUTINE derivs
```

```
        END INTERFACE
        END SUBROUTINE stiff
    END INTERFACE
    INTERFACE
        SUBROUTINE stoerm(y,d2y,xs,htot,nstep,yout,derivs)
        USE nrtype
        REAL(SP), DIMENSION(:), INTENT(IN) :: y,d2y
        REAL(SP), INTENT(IN) :: xs,htot
        INTEGER(I4B), INTENT(IN) :: nstep
        REAL(SP), DIMENSION(:), INTENT(OUT) :: yout
        INTERFACE
            SUBROUTINE derivs(x,y,dydx)
            USE nrtype
            REAL(SP), INTENT(IN) :: x
            REAL(SP), DIMENSION(:), INTENT(IN) :: y
            REAL(SP), DIMENSION(:), INTENT(OUT) :: dydx
            END SUBROUTINE derivs
        END INTERFACE
        END SUBROUTINE stoerm
    END INTERFACE
    INTERFACE svbksb
        SUBROUTINE svbksb_dp(u,w,v,b,x)
        USE nrtype
        REAL(DP), DIMENSION(:,:), INTENT(IN) :: u,v
        REAL(DP), DIMENSION(:), INTENT(IN) :: w,b
        REAL(DP), DIMENSION(:), INTENT(OUT) :: x
        END SUBROUTINE svbksb_dp

        SUBROUTINE svbksb_sp(u,w,v,b,x)
        USE nrtype
        REAL(SP), DIMENSION(:,:), INTENT(IN) :: u,v
        REAL(SP), DIMENSION(:), INTENT(IN) :: w,b
        REAL(SP), DIMENSION(:), INTENT(OUT) :: x
        END SUBROUTINE svbksb_sp
    END INTERFACE
    INTERFACE svdcmp
        SUBROUTINE svdcmp_dp(a,w,v)
        USE nrtype
        REAL(DP), DIMENSION(:,:), INTENT(INOUT) :: a
        REAL(DP), DIMENSION(:), INTENT(OUT) :: w
        REAL(DP), DIMENSION(:,:), INTENT(OUT) :: v
        END SUBROUTINE svdcmp_dp

        SUBROUTINE svdcmp_sp(a,w,v)
        USE nrtype
        REAL(SP), DIMENSION(:,:), INTENT(INOUT) :: a
        REAL(SP), DIMENSION(:), INTENT(OUT) :: w
        REAL(SP), DIMENSION(:,:), INTENT(OUT) :: v
        END SUBROUTINE svdcmp_sp
    END INTERFACE
    INTERFACE
        SUBROUTINE svdfit(x,y,sig,a,v,w,chisq,funcs)
        USE nrtype
        REAL(SP), DIMENSION(:), INTENT(IN) :: x,y,sig
        REAL(SP), DIMENSION(:), INTENT(OUT) :: a,w
        REAL(SP), DIMENSION(:,:), INTENT(OUT) :: v
        REAL(SP), INTENT(OUT) :: chisq
        INTERFACE
            FUNCTION funcs(x,n)
            USE nrtype
            REAL(SP), INTENT(IN) :: x
            INTEGER(I4B), INTENT(IN) :: n
            REAL(SP), DIMENSION(n) :: funcs
            END FUNCTION funcs
        END INTERFACE
```

```
        END SUBROUTINE svdfit
    END INTERFACE
    INTERFACE
        SUBROUTINE svdvar(v,w,cvm)
        USE nrtype
        REAL(SP), DIMENSION(:,:), INTENT(IN) :: v
        REAL(SP), DIMENSION(:), INTENT(IN) :: w
        REAL(SP), DIMENSION(:,:), INTENT(OUT) :: cvm
        END SUBROUTINE svdvar
    END INTERFACE
    INTERFACE
        FUNCTION toeplz(r,y)
        USE nrtype
        REAL(SP), DIMENSION(:), INTENT(IN) :: r,y
        REAL(SP), DIMENSION(size(y)) :: toeplz
        END FUNCTION toeplz
    END INTERFACE
    INTERFACE
        SUBROUTINE tptest(data1,data2,t,prob)
        USE nrtype
        REAL(SP), DIMENSION(:), INTENT(IN) :: data1,data2
        REAL(SP), INTENT(OUT) :: t,prob
        END SUBROUTINE tptest
    END INTERFACE
    INTERFACE
        SUBROUTINE tqli(d,e,z)
        USE nrtype
        REAL(SP), DIMENSION(:), INTENT(INOUT) :: d,e
        REAL(SP), DIMENSION(:,:), OPTIONAL, INTENT(INOUT) :: z
        END SUBROUTINE tqli
    END INTERFACE
    INTERFACE
        SUBROUTINE trapzd(func,a,b,s,n)
        USE nrtype
        REAL(SP), INTENT(IN) :: a,b
        REAL(SP), INTENT(INOUT) :: s
        INTEGER(I4B), INTENT(IN) :: n
        INTERFACE
            FUNCTION func(x)
            USE nrtype
            REAL(SP), DIMENSION(:), INTENT(IN) :: x
            REAL(SP), DIMENSION(size(x)) :: func
            END FUNCTION func
        END INTERFACE
        END SUBROUTINE trapzd
    END INTERFACE
    INTERFACE
        SUBROUTINE tred2(a,d,e,novectors)
        USE nrtype
        REAL(SP), DIMENSION(:,:), INTENT(INOUT) :: a
        REAL(SP), DIMENSION(:), INTENT(OUT) :: d,e
        LOGICAL(LGT), OPTIONAL, INTENT(IN) :: novectors
        END SUBROUTINE tred2
    END INTERFACE
!   On a purely serial machine, for greater efficiency, remove
!   the generic name tridag from the following interface,
!   and put it on the next one after that.
    INTERFACE tridag
        RECURSIVE SUBROUTINE tridag_par(a,b,c,r,u)
        USE nrtype
        REAL(SP), DIMENSION(:), INTENT(IN) :: a,b,c,r
        REAL(SP), DIMENSION(:), INTENT(OUT) :: u
        END SUBROUTINE tridag_par
    END INTERFACE
```

```
INTERFACE
    SUBROUTINE tridag_ser(a,b,c,r,u)
    USE nrtype
    REAL(SP), DIMENSION(:), INTENT(IN) :: a,b,c,r
    REAL(SP), DIMENSION(:), INTENT(OUT) :: u
    END SUBROUTINE tridag_ser
END INTERFACE
INTERFACE
    SUBROUTINE ttest(data1,data2,t,prob)
    USE nrtype
    REAL(SP), DIMENSION(:), INTENT(IN) :: data1,data2
    REAL(SP), INTENT(OUT) :: t,prob
    END SUBROUTINE ttest
END INTERFACE
INTERFACE
    SUBROUTINE tutest(data1,data2,t,prob)
    USE nrtype
    REAL(SP), DIMENSION(:), INTENT(IN) :: data1,data2
    REAL(SP), INTENT(OUT) :: t,prob
    END SUBROUTINE tutest
END INTERFACE
INTERFACE
    SUBROUTINE twofft(data1,data2,fft1,fft2)
    USE nrtype
    REAL(SP), DIMENSION(:), INTENT(IN) :: data1,data2
    COMPLEX(SPC), DIMENSION(:), INTENT(OUT) :: fft1,fft2
    END SUBROUTINE twofft
END INTERFACE
INTERFACE
    FUNCTION vander(x,q)
    USE nrtype
    REAL(DP), DIMENSION(:), INTENT(IN) :: x,q
    REAL(DP), DIMENSION(size(x)) :: vander
    END FUNCTION vander
END INTERFACE
INTERFACE
    SUBROUTINE vegas(region,func,init,ncall,itmx,nprn,tgral,sd,chi2a)
    USE nrtype
    REAL(SP), DIMENSION(:), INTENT(IN) :: region
    INTEGER(I4B), INTENT(IN) :: init,ncall,itmx,nprn
    REAL(SP), INTENT(OUT) :: tgral,sd,chi2a
    INTERFACE
        FUNCTION func(pt,wgt)
        USE nrtype
        REAL(SP), DIMENSION(:), INTENT(IN) :: pt
        REAL(SP), INTENT(IN) :: wgt
        REAL(SP) :: func
        END FUNCTION func
    END INTERFACE
    END SUBROUTINE vegas
END INTERFACE
INTERFACE
    SUBROUTINE voltra(t0,h,t,f,g,ak)
    USE nrtype
    REAL(SP), INTENT(IN) :: t0,h
    REAL(SP), DIMENSION(:), INTENT(OUT) :: t
    REAL(SP), DIMENSION(:,:), INTENT(OUT) :: f
    INTERFACE
        FUNCTION g(t)
        USE nrtype
        REAL(SP), INTENT(IN) :: t
        REAL(SP), DIMENSION(:), POINTER :: g
        END FUNCTION g
```

```
            FUNCTION ak(t,s)
            USE nrtype
            REAL(SP), INTENT(IN) :: t,s
            REAL(SP), DIMENSION(:,:), POINTER :: ak
            END FUNCTION ak
        END INTERFACE
        END SUBROUTINE voltra
    END INTERFACE
    INTERFACE
        SUBROUTINE wt1(a,isign,wtstep)
        USE nrtype
        REAL(SP), DIMENSION(:), INTENT(INOUT) :: a
        INTEGER(I4B), INTENT(IN) :: isign
        INTERFACE
            SUBROUTINE wtstep(a,isign)
            USE nrtype
            REAL(SP), DIMENSION(:), INTENT(INOUT) :: a
            INTEGER(I4B), INTENT(IN) :: isign
            END SUBROUTINE wtstep
        END INTERFACE
        END SUBROUTINE wt1
    END INTERFACE
    INTERFACE
        SUBROUTINE wtn(a,nn,isign,wtstep)
        USE nrtype
        REAL(SP), DIMENSION(:), INTENT(INOUT) :: a
        INTEGER(I4B), DIMENSION(:), INTENT(IN) :: nn
        INTEGER(I4B), INTENT(IN) :: isign
        INTERFACE
            SUBROUTINE wtstep(a,isign)
            USE nrtype
            REAL(SP), DIMENSION(:), INTENT(INOUT) :: a
            INTEGER(I4B), INTENT(IN) :: isign
            END SUBROUTINE wtstep
        END INTERFACE
        END SUBROUTINE wtn
    END INTERFACE
    INTERFACE
        FUNCTION wwghts(n,h,kermom)
        USE nrtype
        INTEGER(I4B), INTENT(IN) :: n
        REAL(SP), INTENT(IN) :: h
        REAL(SP), DIMENSION(n) :: wwghts
        INTERFACE
            FUNCTION kermom(y,m)
            USE nrtype
            REAL(DP), INTENT(IN) :: y
            INTEGER(I4B), INTENT(IN) :: m
            REAL(DP), DIMENSION(m) :: kermom
            END FUNCTION kermom
        END INTERFACE
        END FUNCTION wwghts
    END INTERFACE
    INTERFACE
        SUBROUTINE zbrac(func,x1,x2,succes)
        USE nrtype
        REAL(SP), INTENT(INOUT) :: x1,x2
        LOGICAL(LGT), INTENT(OUT) :: succes
        INTERFACE
            FUNCTION func(x)
            USE nrtype
            REAL(SP), INTENT(IN) :: x
            REAL(SP) :: func
            END FUNCTION func
```

```
        END INTERFACE
        END SUBROUTINE zbrac
    END INTERFACE
    INTERFACE
        SUBROUTINE zbrak(func,x1,x2,n,xb1,xb2,nb)
        USE nrtype
        INTEGER(I4B), INTENT(IN) :: n
        INTEGER(I4B), INTENT(OUT) :: nb
        REAL(SP), INTENT(IN) :: x1,x2
        REAL(SP), DIMENSION(:), POINTER :: xb1,xb2
        INTERFACE
            FUNCTION func(x)
            USE nrtype
            REAL(SP), INTENT(IN) :: x
            REAL(SP) :: func
            END FUNCTION func
        END INTERFACE
        END SUBROUTINE zbrak
    END INTERFACE
    INTERFACE
        FUNCTION zbrent(func,x1,x2,tol)
        USE nrtype
        REAL(SP), INTENT(IN) :: x1,x2,tol
        REAL(SP) :: zbrent
        INTERFACE
            FUNCTION func(x)
            USE nrtype
            REAL(SP), INTENT(IN) :: x
            REAL(SP) :: func
            END FUNCTION func
        END INTERFACE
        END FUNCTION zbrent
    END INTERFACE
    INTERFACE
        SUBROUTINE zrhqr(a,rtr,rti)
        USE nrtype
        REAL(SP), DIMENSION(:), INTENT(IN) :: a
        REAL(SP), DIMENSION(:), INTENT(OUT) :: rtr,rti
        END SUBROUTINE zrhqr
    END INTERFACE
    INTERFACE
        FUNCTION zriddr(func,x1,x2,xacc)
        USE nrtype
        REAL(SP), INTENT(IN) :: x1,x2,xacc
        REAL(SP) :: zriddr
        INTERFACE
            FUNCTION func(x)
            USE nrtype
            REAL(SP), INTENT(IN) :: x
            REAL(SP) :: func
            END FUNCTION func
        END INTERFACE
        END FUNCTION zriddr
    END INTERFACE
    INTERFACE
        SUBROUTINE zroots(a,roots,polish)
        USE nrtype
        COMPLEX(SPC), DIMENSION(:), INTENT(IN) :: a
        COMPLEX(SPC), DIMENSION(:), INTENT(OUT) :: roots
        LOGICAL(LGT), INTENT(IN) :: polish
        END SUBROUTINE zroots
    END INTERFACE
    END MODULE nr
```

C3. Index of Programs and Dependencies

The following table lists, in alphabetical order, all the routines in Volume 2 of *Numerical Recipes*. When a routine requires subsidiary routines, either from this book or else user-supplied, the full dependency tree is shown: A routine calls directly all routines to which it is connected by a solid line in the column immediately to its right; it calls indirectly the connected routines in all columns to its right. Typographical conventions: Routines from this book are in typewriter font (e.g., `eulsum`, *`gammln`*). The smaller, slanted font is used for the second and subsequent occurrences of a routine in a single dependency tree. (When you are getting routines from the *Numerical Recipes* machine-readable media or hypertext archives, you need specify names only in the larger, upright font.) User-supplied routines are indicated by the use of text font and square brackets, e.g., [funcv]. Consult the text for individual specifications of these routines. The right-hand side of the table lists chapter and page numbers for each program.

General Index to Volumes 1 and 2

In this index, page numbers 1 through 934 refer to Volume 1, *Numerical Recipes in Fortran 77*, while page numbers 935 through 1446 refer to Volume 2, *Numerical Recipes in Fortran 90*. Front matter in Volume 1 is indicated by page numbers in the range 1/i through 1/xxxi, while front matter in Volume 2 is indicated 2/i through 2/xx.

R

Following is a brief explanation of each *Numerical Recipes* product, plus two order forms (one for North American residents, one for all other) that may be used to order these items directly from the publisher if you cannot obtain them from your local bookstore.

Numerical Recipes in Fortran 77, Second Edition and *Numerical Recipes in C, Second Edition* are the main text and reference component of the Numerical Recipes package. Each book contains over 300 programs, in the language of the reader's choice, and constitutes a complete subroutine library for scientific computation. Both versions contain equivalent tutorial, mathematical, and practical discussions.

The second volume of the Fortran Numerical Recipes series, *Numerical Recipes in Fortran 90,* contains a detailed introduction to the Fortran 90 language and to the basic concepts of parallel programming, plus source code for all routines from the second edition of Numerical Recipes. This volume does not repeat any of the discussion of what individual programs actually do, the mathematical methods they utilize, or how to use them.

Two example books contain Fortran 77 or C source programs respectively, that exercise and demonstrate all of the *Numerical Recipes, Second Edition* programs. Each example program contains comments and is prefaced by a short description. Input and output data are supplied in many cases. The example books are designed to help readers incorporate procedures and subroutines and conduct simple validation tests.

The programs contained in both the second edition main books and the example books are available in several machine-readable formats that will save users from hours of tedious keyboarding. The *Numerical Recipes Code CDROM* contains, in a single omnibus edition, all the Numerical Recipes code in Fortran 90, Fortran 77, C, Pascal, BASIC, and Lisp. Many extras include the complete SLATEC, NUMAL, and other major source code program collections. It will run on the Windows, DOS, OS/2, or Macintosh operating systems and comes with a license to use all the copyrighted Numerical Recipes code on a single IBM PC compatible or Apple Macintosh computer. An ISO-9660 CDROM version for UNIX, with a license for single-screen UNIX use, is also available.

Diskettes for the Apple Macintosh are available in either C or Fortran 77. 3 1/2" diskettes for IBM PC compatible machines running Windows are available in C and in combined Fortran 90/Fortran 77 editions.

Some selected first edition products are also still available:

Numerical Recipes in Pascal contains the original 200 *Numerical Recipes* routines translated into Pascal along with the tutorial text.

Numerical Recipes Routines and Examples in BASIC contains all the routines from the original Numerical Recipes plus the exercise programs from the example book, all translated into BASIC, along with the text from the example book.

Instructions

To obtain the books or the latest version of the disks, please order from your bookstore or complete the information on the order form in this book and mail it to Cambridge University Press in Port Chester, New York or Cambridge, England. Please note that there is a separate order form for each location. All orders must be prepaid.

NB: Technical questions, corrections, and requests for information on mainframe and workstation licenses should be directed to Numerical Recipes Software, P.O. Box 243, Cambridge, MA 02238, USA. Please do not write the publisher.

There are no cash refunds for diskettes. Only diskettes with manufacturing defects may be returned to the publisher for replacement.

ORDER FORM (Outside North America)

Order from your bookstore or complete this order form and fax to 223 325959 or mail to: Customer Services Department, Cambridge University Press, Edinburgh Building, Shaftesbury Road, Cambridge CB2 2RU, U.K.

____ 43064-X Numerical Recipes in Fortran 77: The Art of Scientific Computing, Second Edition £37.50

____ 57439-0 Numerical Recipes in Fortran 90: The Art of *Parallel* Scientific Computing (for use with 43064-X, above) £29.95

____ 57440-4 Fortran 90 and Fortran 77 Diskette (3.5" for Windows 3.1, 95, and NT) £24.95

____ 43721-0 Fortran 77 Example Book, Second Edition £19.95

____ 43108-5 Numerical Recipes in C: The Art of Scientific Computing, Second Edition £37.50

____ 43724-5 C Diskette (3.5" for Windows 3.1, 95, or NT), Second Edition £24.95

____ 43720-2 C Example Book, Second Edition £19.95

Numerical Recipes Code CDROM ISO 9660 (includes all computer language versions)

____ 57608-3 With Windows, DOS, or Macintosh Single Screen License £59.95

____ 57607-5 With UNIX Single Screen License £99.95

The following first edition books are still available:

____ 37516-9 Numerical Recipes in Pascal: The Art of Scientific Computing £37.50

____ 40689-7 Numerical Recipes Routines and Examples in BASIC £22.95

Total your order

£ ________ Shipping (£2.50 per order for delivery by surface mail. For delivery by airmail, add an additional £2.50 per item)

£ ________ VAT (as applicable)

£ ________ Net Total

Name ______________________________(Block capitals please)

Address______________________________

____

____

Please accept my payment by cheque or money order in pounds sterling:
I enclose (circle one) a Cheque (made payable to Cambridge University Press)/UK Postal Order/International Money Order/Bank Draft/Post Office Giro

Please accept my payment by credit card:
Charge my (circle one) VISA/Barclaycard/MasterCard/Access/Eurocard/American Express/any other other credit card bearing the Interbank symbol (please specify):

Card No. ______________________________Expiry Date __________

Signature ______________________________

Address as registered with your credit card company

____

____

ORDER FORM (United States, Canada, and Mexico)
Order from your bookstore or complete this order form and fax it to (914-937-4712) or mail it to:
Cambridge University Press, Order Department, 110 Midland Avenue, Port Chester, New York 10573
Call our toll-free order number for current prices: 1-800-431-1580

____ 43064-X Numerical Recipes in Fortran 77: The Art of Scientific Computing, Second Edition
____ 57439-0 Numerical Recipes in Fortran 90: The Art of *Parallel* Scientific Computing (for use with 43064-X, above)
____ 57440-4 Fortran 90 and Fortran 77 Diskette (3.5" for Windows 3.1, 95, or NT)
____ 43721-0 Fortran 77 Example Book, Second Edition

____ 43108-5 Numerical Recipes in C: The Art of Scientific Computing, Second Edition
____ 43724-5 C Diskette (3.5" for Windows 3.1, 95, or NT), Second Edition
____ 43720-2 C Example Book, Second Edition

Numerical Recipes Code CDROM ISO 9660 (includes all computer language versions)
____ 57608-3 With Windows, DOS, or Macintosh Single Screen License
____ 57607-5 With UNIX Single Screen License

The following first edition books are still available:
____ 37516-9 Numerical Recipes in Pascal: The Art of Scientific Computing
____ 40689-7 Numerical Recipes Routines and Examples in BASIC

_____ Please indicate the total number of items ordered.

$ ______ Shipping ($4.00 for the first item; $1.00 for each additional item. Most orders will be delivered by UPS.)
$ ______ Tax (Residents of New York and California, please add appropriate sales tax for your place of residence. Canadian residents please add 7% GST.)
$ ______ Net Total

Please indicate method of payment: check _____, Mastercard _____, or Visa _____

Name ____

Address____

____

____

Charge Card No. ______________________________ Exp. Date______

Signature _______________________________________

Lindsey Wilson College
30285210